T0263894

A History of Immunology

For Fran, Alison, Mark, and Judy

Cover image: Paul Ehrlich's side-chain theory of antibody formation. Note the use of geometric shapes to indicate different specificities. Antigen selects for the production and release of the appropriate membrane receptors.

A History of Immunology

Second Edition

Arthur M. Silverstein

Institute of the History of Medicine
The Johns Hopkins University
School of Medicine
Baltimore, Maryland, USA

AMSTERDAM · BOSTON · HEIDELBERG · LONDON · NEW YORK · OXFORD
PARIS · SAN DIEGO · SAN FRANCISCO · SINGAPORE · SYDNEY · TOKYO
Academic Press is an imprint of Elsevier

Academic Press is an imprint of Elsevier
32 Jamestown Road, London NW1 7BY, UK
30 Corporate Drive, Suite 400, Burlington, MA 01803, USA
525 B Street, Suite 1900, San Diego, CA 92101-4495, USA
360 Park Avenue South, New York, NY 10010-170, USA

First edition 1989
Second edition 2009

Copyright © 1989, 2009, Elsevier Inc. All rights reserved

No part of this publication may be reproduced, stored in a retrieval system
or transmitted in any form or by any means electronic, mechanical, photocopying,
recording or otherwise without the prior written permission of the publisher

Permissions may be sought directly from Elsevier's Science & Technology Rights
Department in Oxford, UK: phone (+44) (0) 1865 843830; fax (+44) (0) 1865 853333;
email: permissions@elsevier.com. Alternatively visit the Science and Technology Books
website at www.elsevierdirect.com/rights for further information

Notice
No responsibility is assumed by the publisher for any injury and/or damage to persons or
property as a matter of products liability, negligence or otherwise, or from any use or
operation of any methods, products, instructions or ideas contained in the material
herein. Because of rapid advances in the medical sciences, in particular, independent
verification of diagnoses and drug dosages should be made

Library of Congress Cataloging-in-Publication Data
A catalog record for this book is available from the Library of Congress

British Library Cataloguing in Publication Data
A catalogue record for this book is available from the British Library

ISBN: 978-0-12-370586-0

For information on all Academic Press publications
visit our web site at www.elsevierdirect.com

Typeset by TNQ Books and Journals, Chennai, India

Printed and bound in the United States of America

09 10 11 12 13   10 9 8 7 6 5 4 3 2 1

Working together to grow
libraries in developing countries
www.elsevier.com | www.bookaid.org | www.sabre.org

ELSEVIER   BOOK AID International   Sabre Foundation

...I fancy you as coming to the acquisition of the myriad facts of medicine with little to tell you of the intellectual forces and historical sequences by which these facts have emerged...

Christian A. Herter

Imagination and Idealism in Medicine (*J. Am. Med. Assoc.* 54:423, 1910)

Contents

List of plates   ix
Foreword: On history and historians   xi
Preface to the second edition   xvii
Preface to the first edition   xix

Part One: Intellectual History   1

1 Theories of acquired immunity   3

2 Cellular vs humoral immunity   25

3 Theories of antibody formation   43

4 The generation of antibody diversity: the germline/somatic
  mutation debate   69

5 The clonal selection theory challenged: the "immunological self" 85

6 The concept of immunologic specificity   97

7 Immunologic specificity: solutions   129

8 *Horror autotoxicus*: the concept of autoimmunity   153

9 Allergy and immunopathology: the "price" of immunity   177

10 Anti-antibodies and anti-idiotype immunoregulation 1899–1904   211

11 Transplantation and immunogenetics   231

Part Two: Social History   259

12 The uses of antibody: magic bullets and magic markers   261

13 The royal experiment: 1721–1722   291

14 The languages of immunological dispute   305

15 The search for cell-bound antibodies: on the influence of dogma   325

16 "Natural" antibodies and "virgin" lymphocytes: the importance
     of context   337

17 The dynamics of conceptual change in immunology   347

18 Immunology in transition 1951–1972: the role of international
     meetings and discipline leaders   367

19 The emergence of subdisciplines   395

20 Immune hemolysis: on the value of experimental systems   421

21 Darwinism and immunology: from Metchnikoff to Burnet   431

22 The end of immunology?   441

Appendix A: The calendar of immunologic progress   455
    Appendix A1: Epochs in immunology   455
    Appendix A2: Seminal discoveries   458
    Appendix A3: Important books in immunology, 1892–1968   460
Appendix B: Nobel Prize highlights in immunology   465
Appendix C: Biographical dictionary   475
Author index   503
Subject index   517

List of plates

Plate 1 Louis Pasteur (1822–1895). Pasteur was honoured at the Sorbonne at the Jubilee celebration of his seventieth birthday in 1892   16

Plate 2 Louis Pasteur. Pasteur was so popular, *Vanity Fair* published this caricature with the legend "Hydropobia"   17

Plate 3 Robert Koch (1843–1910)   26

Plate 4 Ilya (Elie) Metchnikoff (1845–1916)   28

Plate 5 Emil von Behring (1854–1917)   36

Plate 6 Paul Ehrlich (1854–1915)   103

Plate 7 Jules Bordet (1870 – 1961)   109

Plate 8 Karl Landsteiner (1868–1943)   114

Plate 9 Clemens von Pirquet (1874–1929)   184

Plate 10 Niels Jerne (1911–1994)   214

Plate 11 Macfarlane Burnet (1899–1985) and Peter Medawar (1915–1987)   231

Plate 12 Michael Heidelberger (1888–1991)   274

Foreword: On history and historians

History is not the study of origins; rather it is the analysis of all the mediations by which the past was turned into our present.

H. Butterfield

The working scientist who entertains the notion of writing a history of his discipline must do so with diffidence and no little trepidation. While he may know more of the facts and scientific interrelationships within his specialty than does the professional historian, nothing in his training or experience has prepared him to deal in the special currencies so familiar to the historian in general, and to the historian of science in particular. If he is to write more than a mere encyclopedia of names, dates, places, and facts – an unappealing venture – then he must deal with such unfamiliar concepts as the sociology and epistemology of science, cultural relativism, etc. Such recondite ideas rarely enter into the formal training of the biomedical scientist, and never into his scientific practice. Indeed, if he considers such concepts at all, it is probably with suspicion and perhaps disdain, relegating them to that special limbo which he maintains for the "impure" social sciences, firm in the conviction that his is a dependably precise "pure" science.

But this is not the most serious challenge to the practicing scientist-turned-historian. Assuming that he has overcome the typical scientist's feeling that Santayana's maxim "Those who cannot remember the past are condemned to repeat it" applies only to politicians, diplomats, and economists, he has a yet more difficult preparatory task before him. This involves nothing less than a re-examination *and perhaps rejection* of some of his most cherished beliefs – beliefs rarely stated explicitly, but so implicit in all of the scientist's training and education and so permeating his environment as to have become almost the unwritten rules of the game.

The first of the beliefs to be re-examined is that of *the continuity of scientific development*. By this I mean that most mature scientists, and all students and members of the novitiate, tend to suppose that all that has gone before in a field was somehow aimed logically at providing the base for current work in that field. Thus, there is a general view that the history of a discipline involves an almost inexorable progression of facts and theories leading *in a straight and unbroken line* to our own present view of the workings of nature. (Historians refer to this as "Whig history,"[1] and condemn its practice.) Put in other terms, the scientist is tempted to regard the development of his science in much the same way that most of us seem to regard the origin of species – as a sort of melioristic evolution, following a preordained path toward the acme of perfection and logical unity: in the one case man, and in the other our present science.

But this is not really surprising, when we consider how most science is practiced and reported, and especially how scientists are trained. In the first instance, the scientist chooses a problem to work on that could scarcely be justified as other than the next logical step in the progress of his discipline – i.e., the next obvious question to be asked and problem to be solved. Then, having successfully seen the research to its conclusion, he submits the work to the scientific literature (the unsuccessful excursions generally going unreported). Now, for a variety of reasons, including ego, space limitations, and the implicit cultural view of how science *ought* to function, our author prepares his manuscript so that not only is the work presented as internally logical and the result of an ordered sequence from start to finish, but the background introduction and its supporting references from past literature are also carefully chosen to demonstrate that this work was eminently justified in its choice, and in fact was the next obvious step forward in a well-ordered history. Each communication in the scientific literature thus contributes modestly and subtly, *but cumulatively*, to a revision of the reader's understanding of the history of his discipline.[2]

There is, however, a far greater force in science which operates to impose an order and continuity on its history, manifested not only by an influence on the types of problems deemed worthy of pursuit, but more importantly in the way in which young scientists are educated. There is in any scientific discipline, and there ought to be, a priesthood of the elite. These are the guardians of the scientific temple in which resides the current set of received wisdoms. These are the trend-setters and the arbiters of contemporary scientific values. They are also, not coincidently, the principal writers of textbooks and the most sought-after lecturers, as well as the principal researchers in whose laboratories young people serve their scientific apprenticeships. They are, in brief, the strongest and most vocal adherents of what Thomas Kuhn, in his provocative book *The Structure of Scientific Revolution*,[3] has called "the current paradigm." In Kuhn's usage, a paradigm in any field is the current model system and the accepted body of theories, rules, and technics that guide the thinking and determine the problems within that field. Kuhn points out that when a change in paradigm occurs within a discipline (he insists that this is inevitably the result of an abrupt revolution), the textbooks must be rewritten to reflect the new wisdom. This invariably involves a revision in the interpretation of what went before, so that the new paradigm can be shown to be fully justified as a step forward in scientific progress, and worthy in all respects to command the attention of the current community of scholars. Since the object of a text is pedagogy, the facts many and the concepts complex, what went before must necessarily be winnowed, abstracted, and digested, in order to provide the student with what is required to follow in the illustrious footsteps of the current priests. Therefore, the modest history that is included in most texts, and the routine appeals to the idols and heroes of earlier times, are more often than not subconsciously slanted to help justify the current paradigm and its proponents; they serve to reinforce the impression of a uniform continuity of scientific development. Assuming that one is a reputable member of a current scientific community, and thus a subscriber to the current paradigm, the scientist-turned-historian must be especially on guard not to contribute also to a revisionist history of the field. One might then be rightly accused of *presentism*,[4] the interpretation of yesterday's events in today's more modern terms and context.

The second of the beliefs that require re-examination – one also nurtured by our traditional system of scientific pedagogy – is that of *the logic of scientific development*. We have already seen that the investigator justifies the choice of a research problem (not only to scientific peers but also to the sources of financial support) by demonstrating its logic within the context of the accepted paradigm. This is, of course, eminently reasonable, since a paradigm lacking in inner logic (i.e., unable to define the nature of the problems to be asked within its context or to assimilate the results obtained) would scarcely merit support. But the existence of a logical order of development during the limited lifetime of a paradigm is often extended to imply an overall logical development of the entire scientific discipline. Moreover, the concept examined above of a smoothly continuous maturation of a science implies also that its progression has been logical – the step-by-step movement of fact and theory from A to B to C, as the Secrets of Nature are unfolded and Ultimate Truth is approached. Indeed, to accuse science of illogic in its development would, to many, imply the absence of a coherent unity underlying the object of science's quest – the description and understanding of the physical world.

And yet, there is so much that is discontinuous and illogical in the development of any science. On the level of the individual research activity, much attention is paid to the beauty and strength of that eminently logical process, the Inductive Scientific Method. The working scientist, however, who thinks about the course of his own research must wonder sometimes whether the description is apt. One of the few *biologists* who reflected aloud on this problem was Sir Peter Medawar, in his Jayne Lectures before the American Philosophical Society. Following the lead of philosopher Karl Popper,[5] Medawar[6] challenges the popular notion:

> ...*Deductivism in mathematical literature and inductivism in scientific papers are simply the postures we choose to be seen in when the curtain goes up and the public sees us. The theatrical illusion is shattered if we ask what goes on behind the scenes. In real life discovery and justification are almost always different processes... [and later] Methodologists who have no personal experience of scientific research have been gravely handicapped by their failure to realize that nearly all scientific research leads nowhere – or if it does lead somewhere, then not in the direction it started off with. In retrospect, we tend to forget the errors, so that "The Scientific Method" appears very much more powerful than it really is, particularly when it is presented to the public in the terminology of breakthroughs, and to fellow scientists with the studied hypocrisy expected of*

a contribution to a learned journal. I reckon that for all the use it has been to science about four-fifths of my time has been wasted, and I believe this to be the common lot of people who are not merely playing follow-my-leader in research. [And finally]…science in its forward motion is not logically propelled. …The process by which we come to formulate a hypothesis is not illogical, but non-logical, i.e., outside logic. But once we have formed an opinion, we can expose it to criticism, usually by experimentation; this episode lies within and makes use of logic…

Even this last concession to the logic and continuity of the scientific method may overstate the case somewhat. But in any event, it certainly must be restricted in its application to the micro-environment of the normative science of a given time – that is, to a working hypothesis developed within the context of the accepted beliefs (paradigm) of the day. Within the macro-environment of a scientific discipline in transition, these rules often fail. Not only may bold new formulations be insusceptible of formal "proof" by logical application of the scientific method, but the bases for their acceptance or rejection by individual members of the community are generally anything but logical: witness, in chemistry, the transition from the phlogiston theory to Lavoisier's oxygen theory (Priestley went to his grave denying that oxygen was a separate entity); in optics, the transition from corpuscular to wave theory to an ineffable something in between; or in bacteriology, the century-long dispute between believers in spontaneous generation and those who claimed *omnis organismus ex organismo* (Pasteur carried the day less for the compelling logic of his experiments – most had been done before him – than by his reputation and forceful disputation). In the field of dynamics, also, it is difficult to subscribe to the idea that Newtonian theories represented a smoothly continuous development over Aristotelian dynamics, or that Einstein's theories emerged smoothly and logically from Newtonian requirements. Again, in immunology, the transitions represented by Pasteur in 1880, by the conflict between theories of cellular and humoral immunity in the 1890s, and by Burnet and the onset of the immunobiological revolution in the 1960s were hardly smooth evolutions, and perhaps not even logical progressions.

Many of the great advances in the sciences, whether arising from a new theoretical concept or from a discovery which redirects a discipline, are in fact quantum leaps – daring formulations or unexpected findings hardly anticipated or predictable within the context of the rules and traditions of the day. Kuhn makes the interesting suggestion that it is only when the normal state of affairs in a science becomes unsettled, when the accepted paradigm no longer provides satisfying explanations for new anomalies which perplex its theories, when, in fact, the paradigm may no longer even suggest the proper questions to be asked, that a crisis stage is reached, and the old paradigm is likely to be replaced – abruptly and discontinuously – by a new one. And often, the critical discovery or novel formulation is made by someone not committed to the old paradigm and to the old approaches and mind-set that it enforced – by the uncommitted young, or the unconfined outsider from another discipline. At such times, members of the "old guard" seem to view their science through lenses ground during the previous era. One is reminded of the hero in Voltaire's *L'Ingénu* who, brought up in feral innocence

> *...made rapid progress in the sciences. ...The cause of the rapid development of his mind was due to his savage education almost as much as to the quality of his intellect; for, having learned nothing in his infancy, he had not developed any prejudices. ...He saw things as they are...*

Here again, the scientist-turned-historian must modify the customary approach to a discipline and consider the significance of the blind alleys of research, the premature discoveries, the mistaken interpretations, and the "erroneous" or supplanted theories of the past. Without these our history, while more concise, would lack some of those condiments that are so very important for its full flavor.

The final one of the cherished (but essentially implicit) beliefs of the scientist which requires re-examination concerns *the impetus for scientific development*. By this I mean those forces which act to determine not only the direction but also the velocity of scientific activity and discovery. Most scientists seem to feel that this impetus is inherent within their discipline – an imperative driving force that dictates at least the sequence, and perhaps even the rate of its development. Thus, the scientist is fond of the notion of the "idea (or experiment) whose time has come," and supports this with case-histories of simultaneous and independent discoveries. To a certain extent, of course, this concept is apt, especially within the context of the current paradigm, as we saw above. But even leaving aside those major discontinuous and unlogical advances already mentioned, we are still left with anomalous developments. How to explain, for instance, a "premature" discovery whose significance goes unrecognized at the time (Spallanzani's refutation of spontaneous generation in the eighteenth century; Mendel's genetics; the Koch phenomenon)?

More interesting yet are those extra-scientific forces which impose themselves upon the course of scientific discovery and development. All too familiar is the effect of war upon science – the development of radar, of nuclear energy theory and practice, of transplantation immunology, to name but a few. One need only recall the Church's view of the Galilean heresy; the serious economic plight of the French silk industry whose appeal helped to direct the course of Pasteur's future work; or the benevolent view of science by Bismarck in Prussia and by Congressman James Fogarty and Senator Lister Hill in America, that did much to establish the scientific leadership in their respective countries. The ability of the Prussian Minister Friedrich Althoff to recognize talent and to reward the Kochs, Ehrlichs, and Behrings with university professorships and with their own institutes was one of the chief factors in German pre-eminence in bacteriology and immunology in the late nineteenth century. By contrast, Pasteur in France was forced to build his institute himself through public subscription, and later the operating funds of the Institut Pasteur came in no small part from its herd of horses to be immunized and from the commercial sale of antitoxins. The development of a yellow fever vaccine certainly owes much to the American occupation of Cuba after the Spanish-American war, and to the building of the Panama Canal. Similarly, not the least contribution to the development of the polio vaccine was the affliction of Franklin D. Roosevelt, while the critical choice between a killed versus an attenuated virus vaccine was made for mainly political reasons by a non-scientist, Basil O'Connor, Director of the Polio Foundation. Finally, when an American President and Congress declare a "War on Cancer" or on AIDS, and appropriate massive funds in its support, all of science changes in both direction and velocity.

These are but a few of the well-explored and documented instances of profound socio-political influences upon the course of scientific development, but there are many others deserving of the attention of the historian (and scientist), and some will be found in the text that follows. No history of a science would be complete or even fully comprehensible without their inclusion, and they add spice to what might otherwise be a rather dull and tasteless fare.

Notes

1. Butterfield, H., *The Whig Interpretation of History*, New York, W.W. Norton, 1965.
2. Julius H. Comroe's essay "Tell it like it was" speaks well to this point: Comroe, J.H., *Retrospectoscope: Insights into Medical Discovery*, Menlo Park, Von Gehr, 1977, pp. 89–98.
3. Kuhn, T., *The Structure of Scientific Revolution*, 2nd edn, Chicago, University of Chicago Press, 1970.
4. See, for example, G.W. Stocking's editorial "On the limits of presentism and historicism ...," *J. Hist. Behav. Sci.* 1:211, 1965.
5. Popper, K., *The Logic of Scientific Discovery*, London, Hutchinson, 1959.
6. Medawar, P.B., *Induction and Intuition in Scientific Thought*, Philadelphia, American Philosophical Society, 1969.

Preface to the second edition

It's this way, replied Samson; it is one thing to write as a poet and another to write as a historian: the poet can story-tell or sing about things, not as they were, but as they should have been; and the historian has to write of them, not as they should have been, but as they were, without adding or subtracting a single thing.

Don Quixote Part II, Ch. 3

Some twenty years ago, I wrote a book entitled *A History of Immunology*. It did not attempt to tell the day-by-day story of the early years as the discipline developed. Rather, it dealt with those aspects of the conceptual development of the field, and those major conflicts of ideas that interested me most. At the time, I felt that I had done full justice to "what had actually happened," by telling it as I was sure that it was.

But then I ran across Cervantes' few lines quoted above, and I began to reconsider what I had written earlier. I saw that much of what I had written was slanted by my own interests and priorities, the products of my own lifetime of experiences and responses. Some events might not have been given the weight that they deserved; others had perhaps been given too much emphasis. This situation became increasingly clear as I compared "my" history of immunology with other more recent writings in this field – by Pauline Mazumdar, Alfred Tauber, Anne-Marie Moulin, Gilberto Corbellini, and others. Each would stake out a somewhat different approach to immunology's early history, and no two might agree to the significance of the same phenomenon, the same interplay of ideas or personalities, or the same set of techniques. Each might well offer a different interpretation of any given event.

Here was the key word – *interpret*. If there were no differing interpretations, then each field would need only one historian, and there would only be one history! And this history would probably be pretty boring. It is thus clear that the historian must be to at least some extent a poet. He should interpret the "things as they were," not necessarily by making up a "things as they should have been" but at least by giving them his own version of a life, an inner vitality, and the importance that they might well have enjoyed in the only partly definable past.

In this new edition I have expanded the account in two different directions. On the one hand I have added a number of new chapters to clarify further the conceptual developments in the field. But since the initial publication of the book I have become increasingly conscious of the important contributions of more sociological factors to the development of a science – the role of government, specialty groups and societies, technological inputs to progress, and

subdiscipline formation. Even the blind alleys down which a field may sometimes wander deserve to be recorded, since they may contribute not only to an
interesting history but also to a rich cultural heritage. The book is therefore
divided into two sections, one devoted primarily to the intellectual history of the
field of immunology and the other to some of the more sociological factors that
have affected its progress. It is clear, however, that the two areas may overlap
considerably, as will be seen as early as the discussion in Chapter 2.

As with the earlier chapters, all new material reflects my own interests as
colored by my own set of prejudices. As before, I have given references to studies,
solutions, and reviews as close to the events as possible, in order to provide the
reader with a feel for the contemporary directions of progress and the various
viewpoints engaged; this, rather than later summaries that might be tainted with
the historical revisionism that so often accompanies later progress.

Finally, in addition to the acknowledgments made in the Preface to the first
edition, I wish also to thank Professor Thomas Söderqvist of Copenhagen, with
whom I published a study (*Cell. Immunol.* 158:1–28, 1994) that has been
modified to form Chapter 18 of this volume.

Woods Hole, Massachusetts
February, 2008

Preface to the first edition

I have always derived great enjoyment from reading the old literature. More specifically, I wanted to know not only what the earlier giants of immunology had said in their publications, but also how they thought and by whom they were influenced. These ventures into the past were only an innocent hobby at first, but became something more once I came into contact with students. The Johns Hopkins University has a very active interdepartmental immunology program, which includes a regular Tuesday evening informal seminar attended by faculty, graduate students, and interested postdoctoral fellows from various clinical and basic science departments. As, week after week, I listened to and participated in discussions that ranged over all aspects of current immunologic thought and practice, I slowly became aware of a troubling fact – most of the young scientists (and not a few of their elders!) appeared to believe that the entire history of immunology could be found within the last five years' issues of the most widely-read journals. Little that went before this was cited, and one might have concluded that each current line of work or current theoretical interpretation had arisen *de novo*, and without antecedents. But perhaps the single event that triggered my serious entry into the study of the history of immunology was the receipt of a manuscript for review from one of our leading journals. This was an elegant study of an important problem, using up-to-date techniques, but one that Paul Ehrlich had reported on eighty years earlier! Not only was the author unaware of Ehrlich's work, but he was also unaware that his data and conclusions differed little from Ehrlich's, despite the marvels of our newer technologies.

I began then to spend part-time in Hopkins' Institute of the History of Medicine, exploring in a more consistent fashion the treasures housed in its Welch Medical Library. These historical excursions led to a series of presentations at the Annual Johns Hopkins Immunology Council Weekend Retreat, and were in fact labeled "The Lady Mary Wortley Montagu Memorial Lectures" by a feminist colleague then in charge of the program committee. These same lectures also have served as the basis of many of the chapters in the present volume.

This book, then, is primarily directed to young immunologists, hopefully to provide both a better understanding of where immunology is today and how it got there, as well as an introduction to the many social, political, and interpersonal factors that have influenced over a century of progress in immunology. Of course, the book is not forbidden to more senior investigators, who may enjoy being reminded of some of the twists and turns that our science has taken, and of some of the grand debates and personalities that have so spiced its history.

If, in addition, the book should serve to interest professional historians of medicine in this important branch of twentieth-century biomedical science (to correct with further research some of my more egregious errors), then it will have more than served its purpose.

The reader will note in the title that this is *a* history of immunology and not *the* history of immunology. Each author will view and interpret the past differently, will be guided by a different background and different values in assigning importance to past events, and will emphasize some aspects more and others less. For my part, I have chosen to deal with the history of immunology in terms of what I consider its most important conceptual threads (whether or not they proved "useful" to future progress), attempting to trace each of these longitudinally in time, rather than to present a year-by-year list of the minutiae of its progress. It is hoped, in utilizing this approach, that most of the important events and discoveries and most of the important names will appear at one place or another. The price of this approach, however, is that some significant technical advances may receive short shrift, if they did not contribute significantly to the advance of a concept (a deficit that I attempt to redress in a final chapter devoted primarily to technologies). This approach also involves a certain amount of repetition, since certain discoveries or theories may have played a major role in the development of more than one important immunologic idea. In this sense, each of the chapters is meant to be internally self-sufficient and may be read independent of the others, but taken together they should present a fairly complete intellectual history of the discipline of immunology.

The reader of this book should be aware of a final caveat. No history of a discipline as active as immunology is today can hope to be completely up-to-date; otherwise the arrival of each new number of a journal would require immediate revision of the text. This is especially true in dealing with the history of ideas in such a field, since so many of our modern concepts and even phenomenologies are still the subject of debate, of verification, and of the test of time. I have therefore drawn an arbitrary line at the early 1960s, being the time when "modern" immunology entered the present biomedical revolution. Classical immunochemistry gave way then to modern immunobiology, a phase shift announced by Burnet's clonal selection theory. If later events and discoveries are mentioned, it is only to provide a context for the evaluation of or comparison with earlier events, or to provide an endpoint to illustrate the further consequences of those earlier developments.

I would like to express my deep thanks to the many individuals who helped me along the way. I owe a debt to Philip Gell for having helped to get me started on the historical path; to Noel Rose and Byron Waksman for their many helpful suggestions on the history of autoimmunity and immunopathology; to Fred Karush for helping to clarify many aspects of molecular immunology; and to Robert Prendergast, Rupert Billingham, and Leslie Brent for many valuable suggestions. I am indebted also to Anne-Marie Moulin for many interesting discussions, and for having permitted me to read and benefit from the manuscript of her thesis on the history of immunology. Chapter 1 was originally written in collaboration with Alexander Bialasiewicz, and Chapter 2 [now Chapter 13] in collaboration with Genevieve Miller, both of whom have given permission to include their important contributions in this book. The appendix containing the biographical dictionary would not have been as complete or as useful but for the generous assistance and encouragement of Mrs Dorothy Whitcomb, Librarian at the Middleton Library of the University of Wisconsin, and of her assistant, Terrence Fischer.

The faculty of the Johns Hopkins Institute of the History of Medicine have been especially helpful, not only in providing space, but also in giving me an informal training in certain aspects of historiography, and in putting up with my many questions of fact or technique. Among these are Drs Lloyd Stevenson, Owsei Temkin, Gert Brieger, Jerome Bylebyl, Caroline Hannaway, and Daniel Todes. Dr John Parascandola, Chief of the Division of the History of Medicine at the National Library of Medicine in Bethesda, contributed significantly to my understanding of Paul Ehrlich's work, and also made available to me the facilities and collections of the National Library. I heartily thank Irene Skop and Liddian Lindenmuth for their superb secretarial and editorial assistance. Finally, I thank Academic Press and H. Sherwood Lawrence, editor of *Cellular Immunology*, for permission to adapt for this book some chapters previously published in that journal.

A.M.S., 1989

The text of this volume was set in Sabon, designed by Jan Tschichold in 1964. It is named after the French typographer Jakob Sabon [1535-1580], a student of the famous Claude Garamond. The roman face was inspired by one of Garamond's fonts, and the italic follows one by Robert Granjon. The headings fonts are Officina Sans, designed in 1990 by Erik Spiekermann.

Part One

Intellectual History

1 Theories of acquired immunity

Blut ist ein ganz besonderer Saft.

Goethe

The Latin words *immunitas* and *immunis* have their origin in the legal concept of an exemption: initially in ancient Rome they described the exemption of an individual from service or duty, and later in the Middle Ages the exemption of the Church and its properties and personnel from civil control. In her impressive review of the "History of Concepts of Infection and Defense,"[1] Antoinette Stettler traces the first use of this term in the context of disease to the fourteenth century, when Colle wrote "*Equibus Dei gratia ego immunis evasi*" in referring to his escape from a plague epidemic.[2] However, long before that, poetic license permitted the Roman Marcus Annaeus Lucanus (39–65 AD) to use the word *immunes* in his epic poem "Pharsalia," to describe the famous resistance to snakebite of the Psylli tribe of North Africa. While the term was employed intermittently thereafter, it did not attain great currency until the nineteenth century, following the rapid spread of Edward Jenner's smallpox vaccination. Immunity was thus an available and apt term to employ during the 1880s and 1890s regarding the phenomena described by Pasteur, Koch, Metchnikoff, von Behring, Ehrlich, and other investigators. But long before any specific term such as immunity was applied, and some 1500 years before an explanation of it would be advanced, the phenomenon of acquired immunity was described.

Throughout recorded history, two of the most fearful causes of death were pestilence and poison. With great frequency, deadly epidemics and pandemics visited upon cities and nations, with enormous economic, social, and political consequences.[3] Despite a lack of knowledge of their origin, their nature, or even their nosologic relationship to one another, the keen observer could not help but notice that often those who by good fortune had survived the disease once might be "exempt" from further involvement upon its return. Thus the historian Thucydides, in his contemporary description of the plague of Athens of 430 BC, could say:[4]

> Yet it was with those who had recovered from the disease that the sick and the dying found most compassion. These knew what it was from experience, and had now no fear for themselves; for the same man was never attacked twice – never at least fatally.

The identity of this "plague" which killed Pericles and perhaps one-quarter of the population of Athens has been much disputed, and it is uncertain that it was due to *Pasteurella pestis*. However, some thousand years later, a pandemic of what is more likely to have been bubonic plague occurred in 541 AD, and is known as

A History of Immunology, Second Edition
ISBN: 978-0-12-370586-0

Copyright © 2009, Elsevier Inc.
All rights reserved

the Plague of Justinian after the Byzantine emperor of that time. In his history, Procopius said of the plague:[5]

> ...it left neither island nor cave nor mountain ridge which had human inhabitants; and if it had passed by any land, either not affecting the men there or touching them in indifferent fashion, still at a later time it came back; then those who dwelt roundabout this land, whom formerly it had afflicted most sorely, it did not touch at all. ...

And, after a further millenium, Fracastoro (1483–1553) felt free to offer the following tantalizing comment in his book *On Contagion*:[6]

> Moreover, I have known certain persons who were regularly immune, though surrounded by the plague-stricken, and I shall have something to say about this in its place, and shall inquire whether it is impossible for us to immunize ourselves against pestilential fevers.[7]

Unfortunately Fracastoro, despite his promise to return to this intriguing suggestion, failed to do so later in the book.

Man's continuous experience with poisons has also had a far-reaching influence on the development of concepts of disease and immunity.[8] During Roman times, Mithridates VI, King of Pontus, described in his medical commentaries (which his conqueror Pompey thought worthy of translation) the taking of increasing daily doses of poisons to render himself safe from attempts on his life. This immunity (or adaptation) had far-reaching influence throughout the Middle Ages, when complicated mixtures for this purpose were universally known as the *Mithridaticum* or *theriac*. Indeed, as we shall see below, its influence was felt as late as the 1890s, when an adaptation theory of immunity was advanced, based upon Mithridatic principles.

Even more important was the centuries-long belief that many diseases were due to poison, known universally by its Latin name *virus*. (The Greek word *pharmakeia* still means poisoning, witchcraft, or medicine.) In the absence of knowledge of etiology or pathogenesis, the causative agent was long considered to be the *virus*, connoting not only poison but also the slime and miasma from which the poison was thought to originate. Even into the early twentieth century, the term "virus" was used almost interchangeably with "bacterium" to describe the etiologic agent of an infectious disease. When, in 1888, Roux and Yersin isolated diphtheria toxin,[9] and in 1890 von Behring and Kitasato described antitoxic immunity to diphtheria and tetanus,[10] it appeared for a brief period that almost 2,000 years of interest in poison as the proximate cause of disease and in antidotes (German: *Gegengifte*) had been vindicated. However, the discovery soon thereafter of numerous diseases whose pathogenesis was based neither upon an exotoxin nor endotoxin led to an early correction of this over-generalization, although not before Paul Ehrlich had done his classical studies on immunity to the plant poisons abrin and ricin.[11]

Most textbooks of immunology begin with a short historical review, mention variolation and Jenner's vaccination against smallpox, but imply that theories of acquired immunity had to await Pasteur's germ theory of disease[12] and his first demonstration in 1880 of acquired immunity in the etiologically well-defined bacterial infection of chicken cholera.[13] This may be due in part to the surprising failure to find any hint of speculation in Jenner's writing on what he thought was the mechanism of vaccination in providing immunity to smallpox. Le Fanu suggests[14] that Jenner might have been influenced by the belief of his famous teacher John Hunter[15] that two diseases cannot coexist in an individual, or perhaps he took seriously Hunter's advice in an earlier letter to Jenner on another subject: "I think your solution is just, but why think? Why not try the experiment?"[16]

A "modern" theory of acquired immunity would seem to require, as minimal prerequisites:

1.   The concept of an etiologic agent
2.   A concept of transmission of this agent
3.   An understanding of the specificity and general reproducibility of a disease
4.   Some concept of host–parasite interaction.

However, as Stettler points out, there were earlier theories of acquired immunity. These appear to have required an awareness of only two factors: a recognition of the phenomenon of inability to succumb twice during the course of a pestilence, and some concept, however primitive, of disease pathogenesis (plus, of course, a speculative mind). We shall, in this chapter, expand upon Stettler's list, and examine these imaginative theories within the context of their times.

Magic and theurgic origin of disease

As Sigerist points out in his *A History of Medicine*,[17] there is only a nebulous border between magic and religion among primitive peoples. In the most primitive societies, both man and nature are thought to operate under the control of magical influences governed by spirits and demons.[18] These become formalized into sets of taboos and totems, followed often by the development of complex pantheons, and occasionally by a monotheistic unification. It is only natural, then, as Temkin[19] indicates, that in such ancient civilizations as Egypt, India, Israel, and Mesopotamia disease came to be considered a punishment for trespass or sin, ranging from the involuntary infraction of some taboo to a willful crime against gods or men. The wearing of amulets, the chanting of incantations, and the offering of sacrifice were common measures to neutralize "black" magic, to ward off demonic disease, or to propitiate the gods, and such practices persist to the present time, even among "advanced" peoples.

Throughout recorded history, every civilization has recognized the theurgic origin of disease. The Babylonian epic of Gilgamesh, about 2000 BC, records visitations of the god of pestilence, while in Egypt the fear of Pharaoh was compared with the fear of the god of disease during a year of severe epidemics. Throughout the Old Testament, God visits disease upon those who deserve punishment, including both His own people and those who oppose them. Thus, God through Moses smote the Egyptians (Exodus 9:9), the Philistines for their seizure of the Ark of the Covenant (I Samuel 5:6), and the Assyrians under King Sennacherib for invading Judea (Isaiah 37:36), but God equally brought down a pestilence that killed 70,000 people as punishment of David's sin of numbering the people (II Samuel 24). In ancient Greece, Sophocles records in "Oedipus the King" that the Sun god Phoebus Apollo caused the plague of Thebes because it had been polluted by the misdeeds of Oedipus, while the historians record that Apollo fired plague arrows upon the Greek host before Troy because their leader Agamemnon had abducted the daughter of his priest. Among Hindus also, sin, the breaking of a norm, the wanton cursing of a fellow man, and similar transgressions result in illness, for the gods – and particularly Varuna, guardian of law and order – punish the offender.

With the concept of a vengeful deity, and especially with the rise of a belief in a hereafter in which a life of earthly suffering might be followed by everlasting peace, the view of the nature of disease and of resistance to it underwent a significant change in early Christian times. While the opening of Pandora's Box might only have released disease-as-punishment into the world, Eve's eating of the forbidden fruit did more: it permitted redemption. Now not only did God punish the sins of man with disease, but He could also employ it to purge and cleanse man of his sins. Thus St Cyprian, Bishop of Carthage (?200–258 AD) could write of the plague then raging:[20]

> *Many of us are dying in this mortality, that is many of us are being freed from the world. …To the servants of God it is a salutary departure. How suitable, how necessary it is that this plague, which seems horrible and deadly, searches out the justice of each and every one …*

A theurgic view of disease has interesting implications for the immunologist. If, throughout early history, disease was considered as a punishment by the spirits or demons or gods for vice and sin, then being spared the initial effects of a raging pestilence or other disease (i.e., *natural immunity*) should automatically have been viewed as the inevitable result of having led a clean and pious life. Moreover, once disease came to be viewed as an expiation and purgative, the recovery from a deadly plague would imply not only that the sins of that individual had been minor, but further that he had been cleansed of those sins and thus did not merit further punishing disease when the plague returned (*acquired immunity*). Such concepts may have been so implicit in the religiosity of the times as not to warrant explicit statement.

It is true of course, as Edelstein[21] and others point out, that despite the common tendency among the ancients to consider a magical or religious origin of disease, physicians (and especially the Greeks) were in general rational and empirical, rejecting magic and any religious mysticism. Rationalism and empiricism were the greatest of the contributions of the Hippocratic school of Greece, a tradition that was maintained in the East by Islamic physicians and in the West throughout the Middle Ages and Renaissance until modern times. But it is difficult to know how such a rational approach might influence the thinking of physicians about so arcane a subject as infectious disease and resistance. While the officials of many cities were instituting the important public-health measure of quarantine (French, 40 days) against infectious disease, influenza was ascribed to the *influence* of the stars and *mal-aria* to bad air. Again, the same Fracastoro who devised such "modern" theories of contagion could in the same book ascribe the appearance of syphilis in Europe to an earlier evil conjunction of Mars, Saturn, and Jupiter. Two hundred years earlier, these same planets had been universally held responsible for the Black Death that ravaged Europe and the East.

The belief in astrological and theurgic bases for disease was not, however, confined to "less advanced" times – it persists even today. In his description of the cholera epidemics of the nineteenth century, Rosenberg[22] points out that few medical men then believed that cholera was a contagious disease, but rather thought (with Sydenham) that its cause lay in some change in the atmosphere. During the early days of the 1832 epidemic, the New York Special Medical Council announced "that the disease in the city is confined to the imprudent and intemperate," while the Governor of New York proclaimed that "an infinitely wise and just God had seen fit to employ pestilence as one means of scourging the human race for their sins," and he found support for this stand in a newspaper report that of 1400 "lewd women" in Paris, 1300 had died of cholera!

Expulsion theories of acquired immunity

From the time of the Hippocratic school in ancient Greece until its challenge by the rise of scientific medicine in the nineteenth century, most disease (whatever its provenance) was thought to reflect a disturbance of the four humors – blood, phlegm, yellow bile, and black bile – whence the terms sanguine, phlegmatic, choleric, and melancholic. During the earlier period, it was supposed that disease was due to a quantitative imbalance among the humors, and this led to widespread use of therapies which included bleeding, cupping, leeches, and purgatives and expectorants of many types. A further refinement (due in part to Galen, 130–?200 AD) held that disease might also be caused by qualitative changes in the humors, involving changes in their temperature, their consistency, or even their fermentation or putrefaction. For example, smallpox was long considered to have a special affinity for the blood, and to involve its fermentation. Given such a pathogenetic mechanism for this disease, and an increasing understanding of its symptomatology and course between the fifth and tenth centuries AD, it is not surprising that most early theories of acquired immunity would be formulated in the context of smallpox.

Rhazes

One of the most famous of the Arab physicians was Abu Bekr Mohammed ibn Zakariya al-Razi (880–932 AD), known in the West as Rhazes. In his *A Treatise on the Small-Pox and Measles*[23] he not only gave the first modern clinical description of smallpox, but also indicated very clearly that he knew that survival from smallpox infection conferred lasting immunity (although he did not employ this term). More than that, he provided a remarkable explanation for why smallpox does not occur twice in the same individual – the first such theory of acquired immunity that we have been able to find in the literature.

Like his contemporaries, Rhazes believed that smallpox affects the blood, and is due more specifically to a fermentation of the blood which is permitted by its "excess moisture." He considered the pustules which form on the skin and break to release fluid as the mechanism by which the body expels the excess moisture contained in the blood. Drawing a parallel between the change in the blood during the development of man and the change in wine from its initial production by the fermentation of grape juice (must) to its spoiling, he wrote:

> I say then that every man, from the time of his birth till he arrives at old age, is continually tending to dryness; and for this reason the blood of children and infants is much moister than the blood of young men, and still more so than that of old men. ...Now the smallpox arises when the blood putrefies and ferments, so that superfluous vapors are thrown out of it and it is changed from the blood of infants, which is like must, into the blood of young men, which is like wine perfectly ripened; as to the blood of old men, it may be compared to wine which has now lost its strength and is beginning to grow vapid and sour; and the smallpox itself may be compared to the fermentation and the hissing noise which takes place in must at that time. And this is the reason children, especially males, rarely escape being seized with this disease, because it is impossible to prevent the blood's changing from this state into its second state, just as it is impossible to prevent must...from changing.

This remarkable theory accounted satisfactorily for everything that Rhazes knew about smallpox. First, it affects virtually everyone, and that during youth, since youths have very moist blood. Next, he pointed out that smallpox was then seldom seen in young adults, and almost never in the aged, presumably because all had undergone the natural drying of the blood that accompanied the aging process. Finally, lasting immunity would follow from earlier infection, and a second experience of this disease would be impossible, since the "excess moisture" of the blood required to support the disease would have been expelled from the body during the first attack.

But there is another almost more interesting aspect of smallpox implicit in Rhazes' theory of pathogenesis and acquired immunity. He presented smallpox as an almost benign childhood disease, and as a salutary process which he apparently felt assisted the maturation from infancy to adulthood. Certainly no such theory could have been advanced by so astute an observer as Rhazes to explain the deadly disease which we know smallpox to be in modern times. Yet, this benign view of smallpox persisted into the seventeenth century in Europe, despite the ravages which it was observed to inflict upon virgin Amerindian populations in the New World immediately following the Spanish conquests. One must wonder where virulent smallpox was in the tenth century, or whether the pathogen underwent some subsequent change in its virulence.[24]

Girolamo Fracastoro: 1546

It was Fracastoro who first gave formal currency to the idea that not only was disease caused by small seeds (*seminaria*), but also that the contagion might spread directly from person to person, indirectly by means of infected clothing, etc., or even at a distance. Although Fracastoro thought that these seminaria might arise spontaneously within an individual or from air or earth or water, he still believed that they would reproduce truly and transmit the same disease from one person to another. He thought that all seminaria had specific affinities for certain things – some for plants, some for certain specific animals – and in this way he explained "natural immunity" to certain diseases. Some seminaria had an affinity for certain organs or tissues, or for one or another of the humors. The seminaria of smallpox, he felt, not only had an affinity for blood as Rhazes had suggested but also, more specifically, had an affinity for that trace of menstrual blood with which each of us was supposed to be tainted *in utero*, and which thenceforth contaminates our own blood. In this, Fracastoro picked up and expanded upon an idea advanced early in the eleventh century by Avicenna (Abu Ali al-Husein ibn Sina, 980–1037[25]). Fracastoro held that, following infection by smallpox seminaria, the menstrual blood would putrefy, rise to the surface beneath the skin, and force its way out via the smallpox pustules. In his own words:

> ...the pustules presently fill up with a thin sort of pituita and matter, and the malady is relieved by these very means...for this ebullition is a kind of purification of the blood; nor should we scorn those who assert that infection contracted by the child from the menstrual blood of the mother's womb is localized by means of this sort of ebullition and its putrefaction, and the blood is thus purified by a sort of crisis provided by nature. That is why almost all of us suffer from this malady, since we all carry in us that menstrual infection from our mother's womb. Hence this fever is of itself seldom fatal, but is rather a purgation. ....Hence when this process has taken place, the malady usually does not recur because the infection has already been secreted in the previous attack.

As with Rhazes' theory, that of Fracastoro appeared to explain all of the known phenomena associated with smallpox, with acquired immunity in this case resulting from the expulsion during the first illness of the menstrual blood contaminant, without which clinical disease could not recur. Again, it is worth noting that some six centuries after Rhazes, Fracastoro could still refer to smallpox in Italy as an essentially benign and almost beneficial process, apparently ignorant of its lethal effects upon the Mayan and Incan civilizations from 1518 onwards.[26] But Fracastoro's menstrual blood theory did not long survive critical evaluation, since Girolamo Mercuriali (1530–1606) certainly did know about the effects of smallpox on Amerindian populations. Mercuriali pointed out[27] that if the menstrual blood theory were correct and universally applicable, then smallpox should have pre-existed in America rather than being carried over by ship-laden miasmas, and that indeed Cain and Abel should have suffered the disease, rather than its first appearance being recorded about the time of the Arabs. He also questioned why smallpox was restricted to mankind, since all other mammalian young should also possess a menstrual contaminant and thus be subject to the disease. But most interesting was his objection that if smallpox, measles, and leprosy were all due to menstrual blood, as many physicians maintained, then affliction with one of these diseases should protect against the others, since their common substrate would have been expelled. Such cross-immunity was, of course, contrary to observed fact.

We may add to the list of theories of acquired immunity in smallpox several minor variants on the menstrual blood expulsion theme. Thus, Antonius Portus[28] maintained that it was not menstrual blood but rather amniotic fluid that contaminated the fetus *in utero*, and served after birth as the target for attacks by smallpox. In typical humoralist terms, amniotic fluid was supposed to undergo putrefaction, to rise to the surface, and to be expelled from the body of the smallpox victim by way of the pustules. Here too, recurrence of the disease was held to be impossible because the host no longer possesses the amniotic fluid substrate which would permit infection to manifest itself in typical clinical symptoms. Similarly, the theory was held by some Chinese physicians that it was the contaminating remnants of umbilical blood in the newborn rather than menstrual blood or amniotic fluid that was responsible for the development of smallpox, and that it was the expulsion of putrefied umbilical blood upon which lasting immunity depended.[29] Indeed, they recommended the careful squeezing out of the blood from the umbilicus prior to ligation as a means of preventing smallpox.

A distension theory: iatrophysics

The Renaissance that had so great an effect on the arts and literature during the fourteenth and fifteenth centuries did not significantly affect the sciences until some 200 years later. Thus, during the sixteenth and seventeenth centuries, physics and astronomy came alive in the hands of Copernicus, Galileo, Brahe, Kepler, and Newton; a new mathematics was developed by Napier, Descartes, Newton, and Leibnitz; the beginnings of modern chemistry could be seen rising from the occult practices of medieval alchemists, stimulated in great measure by

Robert Boyle; and great contributions to medicine were made by Paracelsus, Vesalius, Paré, Fallopio, Harvey, and Sydenham.

The new physical sciences had important implications for contemporary medical thought, and affected the manner in which diseases were viewed and their therapies formulated. Two new schools of medicine arose as a result of the scientific advances, each vying to apply its theories and its therapeutic regimens to the diseased patient.[30] On the one hand there was the iatrochemical school, which interpreted all of physiology as the product of chemical reactions. This approach originated with Paracelsus and was developed and strongly espoused by van Helmont, who in the early seventeenth century could make the very modern-sounding comment about acquired immunity to reinfection: "He who recovers from this disease possesses thenceforth a balsamic blood, which makes him secure from this disease in the future."[31] What van Helmont meant by "balsam" is unclear, but he seems to imply a chemical-physiological rather than a vitalistic interpretation. As may be seen, most theories of acquired immunity conform, more or less, to iatrochemical ideas.

One theory of acquired immunity was advanced, however, that was not based upon iatrochemical ideas, but rather upon the foundations of the second major school of medical thought – that of iatrophysics. These iatrophysical (or iatro-mechanical) concepts stemmed from Descartes' teaching that all bodily processes are mechanical in nature. The body was held to be a machine, and disease explicable in purely physical terms.

James Drake: 1707

The English physician, James Drake, was of the iatrophysical persuasion. In his book *Anthropologia Nova: Or, a New System of Anatomy*, he suggested that smallpox was caused by a "feverish disposition of the blood," whereby "peccant matter was concocted" and could only escape by forcing its way through the skin with the formation of pustules:[32]

> *I conceive therefore that the Alteration made in the Skin by the* Small-Pox, *at whatever Age it comes, is the true Reason why the Distemper never comes again. For the distention, which the* Glands *and* Pores *of the Skin suffer at that time, is so great that they scarce ever recover their* Tone *again, so as to be able any more to arrest the Matter in its Course outward long enough, or in such quantity, as to create those* Ulcerous Pustules *which are the very* Diagnosticks *of the* Small-Pox. *For tho' the same* Feverish Disposition *shou'd, and may again arise in the Blood, yet, the Passages thro' the Skin being more free and open, the Matter will never be stopt so there, as to make that appearance, from whence we denominate the* Small-Pox. ….*What has been said of the* Small-Pox, *will suffice to solve the* Phaenomena *of the* Measles, Scarlet Fever, *and* Erysipelatous Inflammation…*

Thus Drake stays well within the humoralist boundaries of his time but, by superimposing his mechanistic approach, is able to come up with a quite remarkable theory of acquired immunity. Unlike earlier expulsion theories or later depletion theories, Drake would permit smallpox infection to recur in the same individual and indeed to "concoct new peccant material" from the blood. But in an interesting and not uncommon identification of the symptoms with the disease itself, he maintained that the morbid matter would escape through the now-distended pores and glands of the skin as fast as it was formed, so that the symptoms (and thus the disease) could not appear a second time in the same individual. Again, Drake's theory implied a cross-immunity between smallpox and other exanthematous diseases, in apparent ignorance of Mercuriali's objections of 100 years earlier.

In fairness to Drake, we should point out that he advanced this interesting theory with great modesty and diffidence, writing:

> *Why the* Small-Pox *seldom visits any Person more than once in his Lifetime, has been a famous* Problem *much agitated with very little Success; & therefore if I succeed in my Attempt to resolve this no better than others have done before me, I shall not think it any Loss of Reputation, but shall freely wish others more Happy in theirs, when they undertake to reform my Notions."*

Drake's iatrophysical theory of smallpox immunity was taken up by Clifton Wintringham some years later.[33] Wintringham proposed that the "contagious matter" causes a coagulation of the blood, which "increases the Bulk of its constituent Particles," thus obstructing "the ultimate and perspirable vessels," leading to pustule formation. These vessels are left dilated, so that new disease (symptoms) cannot reappear.

Depletion theories

By the end of the seventeenth century, smallpox had become the serious disease in Western Europe that it was to remain until modern times. However, new attention was directed not only at smallpox but also at acquired immunity to this disease by a series of letters to the Royal Society of London in 1714 from two Greek-Italian physicians, Emanuele Timoni and Jacob Pylarini. For the first time, they brought to the official attention of Western medicine the Eastern practice of variolation, then currently popular in Constantinople. This involved the establishment of a mild infection by the insertion of crusts derived from the pustules of "favorable" cases of active smallpox. This had apparently become a very widespread part of the folk-medicine of many peoples, since reports soon emerged of its use not only in the Middle East but also in other parts of Asia, in Africa, in rural parts of Western Europe, and even in England. The practice was almost universally known as "buying the smallpox." Indeed the Chinese, who may have originated the practice, refined it by blowing the infected matter into the nose through a silver tube, employing the left nostril for males and the right for females.[34]

As Genevieve Miller so well describes, smallpox inoculation very rapidly became popular in England, thanks in part to the efforts of Lady Mary Wortley Montagu, wife of the British Ambassador to Constantinople. However, Miller suggests elsewhere[35] that the role of Lady Mary, given great prominence by Voltaire in his *Lettres sur les Anglais* was exaggerated, and that more credit is due to the Royal Experiment, conducted in 1721–1722 and followed avidly by the entire populace (see Chapter 13). This involved nothing less than the first clinical trial in immunity, in which the efficacy of inoculation was tested first upon condemned prisoners and then upon a group of orphans, in order that the Prince and Princess of Wales might be reassured and permit the inoculation of their children, which in fact took place in 1722 following the successful clinical trial. It is thus not surprising that the eighteenth century would be rich in both interest in and speculation about smallpox, inoculation, and the mechanism of the acquired immunity which inoculation furnished.

One of the most interesting examples of the general popularity of inoculation practices is furnished by Dühren in his diverting book *The Marquis de Sade and His Time*.[36] In a section entitled "The Bawdy House of Madame Gourdan," he describes the medical (and other) practices of that most famous of eighteenth-century Paris bordellos. Madame Gourdan apparently retained the services of a Dr Guilbert de Préval, one of France's most notorious charlatans, who possessed a most remarkable *spécifique* that was a true wonder drug. When injected into the skin it was held not only to immunize the recipient against syphilis, but also even to effect the cure of pre-existing disease. Further, Madame Gourdan herself injected it into newly-arrived girls as a diagnostic, to assure that they were free of syphilis. As Dühren exclaims, "Imagine, a sexual tuberculin in the 18th century. There is nothing new under the sun!"

Cotton Mather: 1724

Cotton Mather (1663–1728) was one of the remarkable figures of Colonial America. A man of great religiosity, he had played an active part in the Massachusetts witchcraft trials, but also found time to pursue an impressive range of other interests. He regularly received the Proceedings of the Royal Society of London, and thus quickly became aware of the communications of Timoni and Pylarini about inoculation. When, in 1721, a smallpox epidemic descended upon Boston, Mather was alone in urging the practice of variolation upon the Boston physicians, and finally convinced his friend Dr Zabdiel Boyleston to undertake this practice. Mather transmitted the Boston results to the Royal Society in several quite scholarly communications, and in 1724 published his *Angel of Bethesda*, the first medical book published in the American colonies.[37] In this remarkable book is a lengthy chapter entitled "Variolae Triumphatae, or the Small-Pox Encountred," in which Mather not only advanced a theory of acquired immunity in natural smallpox infection, but also explained (in florid prose) why variolation is effective in inducing lasting immunity:

> Behold, the Enemy [smallpox] at once gott into the very Center of the Citadel:
> And the Invaded Party must be very Strong indeed, if it can struggle with him,
> and after all Entirely Expel and Conquer him. Whereas, the Miasms of the
> Small-Pox being admitted in the Way of Inoculation, their Approaches are made
> only by the Outerworks of the Citadel, and at a Considerable Distance from the
> Center of it. The Enemy, tis true, getts in so far as to make Some Spoil, yea, so
> much as to satisfy him, and leaves no Prey in the Body of the Patient, for him
> ever afterwards to seize upon. But the Vital Powers are kept so clear from his
> Assaults, that they can manage the Combats bravely and, tho' not without
> a Surrender of those Humours in the Blood, which the Invader makes a Seizure
> on, they oblige him to march out the same way he came in, and are sure of never
> being troubled with him any more.

Thus, Mather does not view the inoculated material as being in any sense attenuated, but rather considers that the milder disease results only from a peripheral infection, in contrast to the natural infection which gains deadly access to "the very Center of the Citadel." But in both cases, he views the infection as acting upon some type of substrate (unidentified) which is depleted in the process, thus leaving "no Prey in the Body of the Patient" upon which subsequent infection can act. The similarity between this and other depletion theories, and those described above as expulsion theories, will be evident. In the one case the target or substrate of the infection is used up in the process, while in the other it is expelled from the body.

Thomas Fuller: 1730. The innate seed

The seventeenth and eighteenth centuries saw the development of many interesting notions about the etiology and pathogenesis of infectious disease. Perhaps none was quite as fanciful as the concept of the "innate seed," whose fertilization was thought to give birth to the disease process itself. In the context of smallpox and of acquired immunity, it was presented most elegantly by Thomas Fuller as follows:[38]

> Nature, in the first compounding and forming of us, hath laid into the Substance
> and constitution of each something equivalent to Ovula, of various distinct
> Kinds, productive of all the contagious, venomous Fevers we can possibly have as
> long as we live. Because these Ovula are of distinct Kinds,...as Eggs of different
> Fowls are from one another; therefore every sort of these Ovula can produce only
> its own proper Foetus... and therefore the Pestilence can never breed the Small
> Pox, nor the Small Pox the Measles. ...All Men have in them those specific Sorts of
> Ovula which bring forth Small Pox and Measles, and therefore we say that all
> Men are liable to them. ...The Ovula always lie quiet and unprolific, till
> impregnated, and therefore these Distempers seldom come without Infection,
> which is as it were the Male, and the active Cause. The Ovula of each particular

Fever, are all, and every individual one of them, usually impregnated at once....
And when these have been impregnated, and delivered of their morbid Foetus,
there is an End of them; ...Upon this Account no Man can possibly...be infected
with any of the respective Distempers any more than once.

Fuller's argument speaks elegantly for itself, and would appear to explain all of the known phenomenology of smallpox. Contagion and a specific etiologic agent are represented by the male element that comes from without, and specifically fertilizes the female elements (the ovula) that reside innately within each of us. As with all seeds, once germinated and sprouted the body suffers a depletion of the specific seeds of that disease, so that thenceforth new etiologic agents will fall upon sterile soil.

James Kirkpatrick: 1754

Kirkpatrick was a physician from Charleston, South Carolina, who, after an early experience with variolation in America, went to London where he became one of the principal proponents of the practice. He too espoused a theory that something was depleted from the blood during the course of smallpox infection, whose absence thenceforth prevented a recurrence of the disease.[39] He postulated the existence of a "pabulum" in the blood, with which contagious variolous "primordia" from the outside united. By the time the disease had run its course, the pabulum had been used up, and thus both natural infection and that following variolation were followed by longstanding acquired immunity. As Kirkpatrick said of reinfection, "Its Seeds were sown in an exhausted Soil."

Elsewhere in the same book, Kirkpatrick was guilty of a curious but prescient inconsistency. He suggested, without further amplification, that smallpox "left some positive and material quality in the constitution" which was responsible for prolonged immunity to reinfection. In this he may only have been parroting an earlier suggestion by the famous Boerhaave (1668–1738), who made the casual suggestion that "people who have smallpox must have something remaining in their body which overcomes subsequent contagious infection."[40] In any event, such suggestions had been made often, and were all but ignored during the eighteenth century, except for the occasional sarcastic reference such as was made by the anti-inoculationist Legard Sparham:[41]

> *Unless we could suppose some singular Virtue to remain in the Blood as a proper*
> *Antagonist, it would be absurd to think them secure from a second Infection,*
> *any more than that the Transfusion of the Blood or Matter of a venereal pocky*
> *Person into a sound Habit, should secure him from any future Amour with*
> *Impunity.*

The view that acquired immunity is due to the depletion of a substrate necessary to the action of the pathogen was repeated often during the eighteenth century, and became popular in France, following the English lead. Thus, the famous physician de la Condamine favored it in his communications to the Royal

Academy of Sciences,[42] and his translator Maty injected the following personal footnote (p. 32) into de la Condamine's book: "I lately tried this experiment (inoculation) upon myself,... and it had no effect upon my blood, as it had been sufficiently *defecated* 15 years before."

A similar view was repeated in 1764 by the remarkable Italian physician Angelo Gatti,[43] who for a time joined the *philosophes* in Paris to become one of the chief proponents of inoculation in France. In a book notably in advance of its times for its view of infection, resistance, and disease, and in its attempt to cut through the often meaningless jargon of contemporary medicine, Gatti compared smallpox infection and acquired immunity to a body which a single spark can set afire, but which has thenceforth become "incombustible" although surrounded by flames. As he says:

> *In like manner, when you have seen the smallest variolous atom, by its bare application, infecting a human body, and afterwards behold the same body covered with the same kind of matter, and not in the least affected by it, will you not conclude that it is no longer susceptible of infection, and, if I may so say, that it is become invariolable?*

Louis Pasteur: 1880

Plate 1 Louis Pasteur (1822–1895). Pasteur was honoured at the Sorbonne at the Jubliee celebration of his seventieth birthday in 1892

Plate 2 Louis Pasteur. Pasteur was so popular, *Vanity Fair* published this caricature with the legend "Hydrophobia"

The rise of modern bacteriology in the 1870s, thanks principally to the studies of Pasteur and Robert Koch, provided for the first time a well-established etiologic agent for infectious diseases, which could be studied both *in vivo* and *in vitro*. No sooner had he announced his epoch-making results on the induction of acquired immunity to fowl cholera using attenuated organisms than the exuberant Pasteur, never at a loss for ideas, theories, or biting repartee, advanced a theory to explain this phenomenon to the Academy of Sciences.[44] He pointed out that it was a frequent observation that bacteria grown in culture would initially multiply in great numbers, but that within days the growth would slow down and finally cease. When these cultures were filtered, then it was often found that while reseeding with unrelated bacteria might result in appreciable growth, reintroduction of the same bacteria would almost invariably lead to no new growth at all. Pasteur suggested that this phenomenon was due to the very highly specialized nutritional requirements of each species of organism, such that so long as the nutrients peculiar to a given organism remained in the solution growth could proceed, but upon depletion of these special nutrients growth would cease and could not resume thereafter. Pasteur likened the body to an artificial culture medium in which there were present only limited quantities of these special nutrients. Following natural infection, or artificial inoculation with attenuated organisms, the pre-existing supply of these nutrients would be depleted so that the body could not support renewed growth following reinfection. Thus, prolonged immunity could be induced with great specificity, given the highly specialized nutritional requirements of each pathogen.

Pasteur's theory of depletion did not long survive the rapid advance of bacteriology that took place in the 1880s, and Pasteur, ever the realist, quickly dropped it. But the theory was taken up and pursued for a very long time by no less a figure than Paul Ehrlich. Ehrlich very early developed a keen interest in cancer, and as the result of experimental studies on the inability of certain tumors to grow in some animal species, and of the regression of tumors, he formulated a theory of tumor immunity to which he applied the term *atrepsie*. He argued this theory elegantly and forcefully as late as 1907, in his Harben Lecture before the Royal Institute of Public Health in London.[45] Paying due respect to Pasteur's depletion theory (which the Germans called *Erschöpfung* – exhaustion), Ehrlich suggested that just as bacteria might have special nutritional requirements, so also might different cancers. Thus, he thought that a tumor would fail to grow in a host lacking those special nutrients that it required, or would regress when it had depleted the host of them. Being still much involved in elaborations of his side-chain receptor theory of antibody formation, he suggested that both bacteria and tumor cells might possess specific "chemoreceptors" which enable them to bind and then ingest those nutrients necessary to their growth. Ehrlich suggested that Pasteur need not have insisted upon complete depletion of a vital nutrient in the host – this he thought improbable – but that it may suffice that either the nutrient is reduced below a critical level, or more possibly that the pathogen has lost the ability (receptors) to utilize that nutrient – a sort of atrophy of specific receptors!

The retention theory and other concepts

In the ten years between Pasteur's first experimental demonstration of active acquired immunity and the discovery of antibody and of passive immunity by von Behring and Kitasato, rapid advances in the young field of bacteriology and the nascent field of immunology were matched only by the creativity of the investigators seeking explanations for their observations. All of these were, like Pasteur's depletion theory, couched in terms of the action of bacterial pathogens. However, they, like all earlier theories, were classified by Sauerbeck in his 1909 book on *The Crisis in Immunity Research*[46] as "passive" theories, in which the pathogen acts by itself to produce immunity in an otherwise inert host. With the exception of Metchnikoff's cellular (phagocytic) theory, originating in a zoological rather than a human disease context,[47] "active" theories of immunity involving host response awaited the discovery of antibody and complement.

The retention theory

Just as early experiments on the growth of pure cultures of the newly discovered bacteria led to Pasteur's depletion theory, so they also provided information upon which a diametrically opposite theory was formulated. Observations were made by numerous investigators that the growth of bacteria was accompanied by the formation of a variety of substances, such as phenol, phenylacetate, skatol, and other aromatic compounds. It was von Nencki who apparently first noticed that the growth of bacteria in culture might be inhibited by these and other products of their own metabolism. This led him to formulate the so-called retention theory of acquired immunity,[48] in which it was postulated that during the course of an infection, the initial bacterial growth in the body would result in the build-up of high concentrations of these chemical inhibitors. This would not only lead to cessation of growth during the initial infection, but retention of these inhibitors in the host would also confer lasting immunity. The specificity of this immunity was explained by assuming that each species of pathogen produces substances peculiar to its own metabolism, and to whose inhibitory effect they alone are sensitive. This theory was taken up and championed before the French Academy of Sciences by Chauveau, Director of the Veterinary School at Lyon.[49] In studies of anthrax infection of Algerian sheep, Chauveau observed that the offspring of ewes infected during pregnancy, and especially shortly before parturition, showed an increased resistance to anthrax infection. Chauveau suggested that this increased immunity was due to the retention of inhibitory substances within the body of the infected mother, and their transmission across the placenta to the fetus *in utero*. Little more was heard of the retention theory following the discovery of antitoxic and other antibacterial antibodies in the early 1890s.

Osmotic and alkalinity theories

The rapid progress made in physical chemistry toward the end of the nineteenth century had a strong influence on contemporary medical thought and practice. This was reflected in the famous dispute[50] between Paul Ehrlich on the one side and Jules Bordet and Karl Landsteiner[51] on the other, about whether antigen–antibody–complement reactions more closely resembled firm chemical unions, or weaker "colloidal" interactions. Similarly, the new physico-chemical concepts found their way into several early theories of acquired immunity.

Two years before his discovery of antitoxic antibodies, von Behring drew a parallel between blood alkalinity and bactericidal action.[52] He supposed that bacterial growth and the tissue changes that accompany it resulted in an increase in the alkalinity of the body to the point where bacterial growth was suppressed, and presumably could not be later reinitiated. This is, in a sense, analogous to the retention theory described above, and did not long survive further experimental work: indeed, von Behring himself helped to lay this theory to rest.

The osmotic theory was advanced by the prominent pathologist Baumgarten,[53] and was based on the suggestion that bacteria were destroyed in the body by osmotic rupture of their membranes. Again, it was supposed that bacterial growth resulted in the production of an increasingly less favorable osmotic environment which would presumably persist even after the initial infection had been cleared. Baumgarten maintained this view for many years and through many editions of his *Textbook of Pathogenic Microorganisms*, and held that the only function of antibodies was to render bacteria more susceptible to osmotic shock.

Adaptation theory

As pointed out above, disease was long associated with the action of poisonous miasmata. Among adherents of the concept of contagion, many followed the lead of Boerhaave in ascribing disease to "venomous corpuscles" which were not only transmissible but could also reproduce their own kind and thus poison the humors of the infected individual to induce putrefaction, inflammation, and disease. Since Mithridatic adaptation to various poisons was common knowledge, it is not surprising to find hints of an adaptation theory of acquired immunity to infectious disease throughout these times. However, it remained for von Behring to state this theory explicitly in his second paper on diphtheria immunity, if only to disprove it.[54] After recounting his elegant experiments demonstrating immunity to diphtheria toxin, he says:

> One might at first think that the resistance to poison described here depends upon an adaptation to that poison (Giftgewöhnung) in the sense that it is employed among alcoholics or morphine- and arsenic-eaters ...in short, that it is essentially a question of training or inurement.

Von Behring then goes on to show that such an explanation is impossible in the present instance, since normal mice who have never encountered diphtheria toxin can be protected against lethal doses of it by passive immunization: indeed, the toxin can be neutralized *in vitro* with the serum of immune animals. Although the word *virus* continued to be applied nonspecifically to all pathogens for many years until its usage was restricted to the ultrafilterable and ultra-microscopic agents, and although it might even have retained its connotation as "poison" to some, the advances of the late nineteenth and early twentieth centuries largely demystified and even detoxified many diseases. Concepts such as those outlined above could not long survive the new knowledge derived from bacteriology, immunology, and experimental pathology.

* * *

I have concluded this review of theories of acquired immunity just short of those "modern" concepts which guide investigators today. In contrast to the theories described above, in which the infected host was generally portrayed as a passive receptacle in which disease ran its course and immunity might be established, current theories involving antibodies, complement, macrophages, and lymphocytes speak of host–parasite *inter*actions to which the infected or immunized individual makes an active contribution. I shall review in the next chapter the early history of modern humoral and cellular theories of immunity, and the nature and implications of the early controversy that raged in the late nineteenth century between protagonists of these two concepts.

It will be apparent that throughout history there has been at least a rough consistency in the evolution of the concept of immunity, such as is found in the history of most ideas. At each stage, the contemporary understanding of the nature of immunity was very much a product both of its previous history as well as of contemporary developments in medicine in particular, and in the sciences and philosophy in general. Thus, no matter how improbable or inadequate these theories might appear today, whether derived from magic-theurgic principles, from post-Hippocratic humoralist doctrines, from later iatrochemical or iatrophysical teachings, or even from the early insights of modern bacteriology, they were all very much the product of their times. Each of them, if only transiently, appeared to explain satisfactorily the known phenomena of its day.

Notes and references

1. Stettler, A., *Gesnerus* **29**:255, 1972.
2. Colle, Dionysius Secundus, quoted in Stettler, note 1.
3. McNeill, W.H., *Plagues and Peoples*, New York, Doubleday, 1976.
4. Thucydides, *The Peloponnesian War*, Crawley translation, New York, Modern Library, 1934, p. 112.
5. Procopius, *The Persian War*, H.B. Dewing translation, Vol. I, London, Heinemann, 1914.
6. Fracastoro, Girolamo, *De Contagione et Contagiosis Morbis et Eorum Curatione*, 1546, W.C. Wright translation (with Latin text), New York, Putnam, 1930, pp. 60–63. See also Singer, C. and Singer, D., *Ann. Med. Hist.* **1**:1, 1917.
7. The translator has perhaps been too modern in his rendition. Fracastoro nowhere employs the word *immunis*, but rather "*et utrum consuescere pestilentiis possimus*," which is perhaps better translated "whether it is possible to accustom ourselves to pestilences."
8. Stevenson, L.G., *The Meaning of Poison*, Lawrence, University of Kansas Press, 1959; Hopf, L., *Immunität und Immunisirung; eine medicinische-historische Studie*, Tübingen, Pietzker, 1902.
9. Roux, E., and Yersin, A., *Ann. Inst. Pasteur Paris* **2**:629, 1888.

10. von Behring, E., and Kitasato, S., *Deutsch. Med. Wochenschr.* **16**:1113, 1890.
11. Ehrlich, P., *Deutsch. Med. Wochenschr.* **17**:976, 1218, 1891.
12. Pasteur, L., Joubert, J., and Chamberland, C., *C. R. Acad. Sci.* **86**:1037, 1878.
13. Pasteur, L., *C. R. Acad. Sci.* **90**:239, 952, 1880.
14. Le Fanu, W.R., Personal communication, 1977.
15. Hunter, J., *A Treatise on the Blood, Inflammation, and Gun-shot Wounds*, Philadelphia, Webster, 1823.
16. Baron, J., *Life of Edward Jenner*, London, Colburn, 1838. See also Le Fanu, W.R., *Edward Jenner*, London, Harvey & Blythe, 1951.
17. Sigerist, H.E., *A History of Medicine*, Vol. I, Chapter 2, New York, Oxford University Press, 1951. See also Procope, J., *Medicine, Magic, and Mythology*, London, Heinemann, 1954.
18. Thorndike, L., *A History of Magic and Experimental Science*, Vol. I, Chapter 1, New York, Macmillan, 1923. See also Rivers, W.H.R., *Medicine, Magic, and Religion*, New York, Harcourt Brace, 1927.
19. Temkin, O., "Health and Disease," in Wiener, P.P., ed., *Dictionary of the History of Ideas*, New York, Scribners, 1973.
20. Cyprian, *De Mortalitate*, M.L. Hannon translation, Washington DC, Catholic University, 1933, pp. 15–16.
21. Edelstein, L., *Bull. Inst. Hist. Med.* **5**:201, 1937.
22. Rosenberg, C., *The Cholera Years*, Chicago, University of Chicago Press, 1962. See also Ackerknecht, E.H., *Bull. Hist. Med.* **22**:562, 1948.
23. Rhazes, *A Treatise on the Small-Pox and Measles*, W.A. Greenhill translation, London, Sydenham Soc., 1848.
24. Carmichael, A.E., and Silverstein, A.M. (*J. Hist. Med. Allied Sci.* **42**:147, 1987) have speculated that in fact only a mild form of smallpox, akin to *Variola minor*, existed in Europe prior to the seventeenth century.
25. *Avicennae Arabum Medicorum Principis*, Latin translation by Gerard of Cremona, Venice, 1608, Vol. II, pp. 72–73.
26. McNeill, W.H., *Plagues and Peoples*, New York, Doubleday, 1976.
27. Mercurialis, Hieronymus, *De Morbis Puerorum*, Basel, 1584, Book I, pp. 17–21.
28. Antonius Portus, quoted in Miller, G., *The Adoption of Inoculation for Smallpox in England and France*, Philadelphia, University of Pennsylvania Press, 1959, p. 244.
29. Kirkpatrick, J., *The Analysis of Inoculation*, London, 1754, p. 41.
30. Garrison, F.H., *An Introduction to the History of Medicine*, Philadelphia, Saunders, 1917. See also Castiglioni, A., *A History of Medicine*, New York, Knopf, 1947.
31. van Helmont, J.B., 'De Magnetica Vulnerum Curatione', in Sylvestre Rattray, *Theatrum Sympatheticum Auctum*, Nürnberg, 1662, p. 477.
32. Drake, J., *Anthropologia Nova: Or, a New System of Anatomy*, Vol. I, London, 1707, p. 25.
33. Wintringham, C., *An Essay on Contagious Diseases*, York, 1721.
34. Another method of immunization employed by primitive peoples is that against pleuropneumonia of cattle. De Rochebrun (*C. R. Acad. Sci.* **100**:659, 1885) mentions that the Moors and the Pouls of Senegambia in Africa have a custom "whose origins are lost in the obscurity of history," in which a knife is plunged into the lung of an animal that died of the disease, and then used to make an incision in the skin of healthy animals. "Experience has demonstrated the success of this protective operation."

35. Miller, G., "Putting Lady Mary in her place: a discussion of historical causation," *Bull. Hist. Med.* 55:2, 1981.
36. Dühren, E. (pseudonym of Iwan Bloch), *Der Marquis de Sade und seine Zeit*, Berlin, Barsdorf, 1900, pp. 123, 213.
37. Mather, C., *The Angel of Bethesda*, G.W. Jones, ed., Barre, American Antiquarian Soc., 1972. See also Beall, O.T. and Shryock, R.H., *Cotton Mather: First Significant Figure in American Medicine*, Baltimore, Johns Hopkins University Press, 1954.
38. Fuller, T., *Exanthematologia: or, an Attempt to give a Rational Account of Eruptive Fevers*, London, 1730, p. 175 ff.
39. Kirkpatrick, note 29, p. 37.
40. Boerhaave, H., "*Praxis Medica*," Vol. V, Petavii, 1728, p. 308.
41. Sparham, L., *Reasons Against the Practice of Inoculating the Small Pox*, London, 1722, p. 20.
42. La Condamine, C.M., *A Discourse on Inoculation*, M. Maty translation, London, Vaillant, 1755.
43. Gatti, A., *Réflexions sur les préjugés qui s'opposent aux progrès et à la perfection de l'inoculation*, Bruxelles, Musier, 1764. See also Gatti, A., *New Observations on Inoculations*, M. Maty translation, London, Vaillant, 1768.
44. Pasteur, L., Chamberland, C., and Roux, E., *C. R. Acad. Sci.* 90:239, 1880.
45. Ehrlich, P., *Experimental Researches on Specific Therapeutics*, London, Lewis, 1908.
46. Sauerbeck, E., *Die Krise in der Immunitätsforschung*, Leipzig, Klinkhardt, 1909.
47. Metchnikoff, E., *Lectures on the Comparative Pathology of Inflammation*, London, Keegan, Paul, Trench, Trübner, 1893; reprinted by Dover, New York, 1968.
48. von Nencki, M., *J. Prakt. Chem.*, May, 1879, cited by Sirotinin, *Z. Hyg.* 4:262, 1888.
49. Chauveau, A., *C. R. Acad. Sci.* 89:498, 1880; 90:1526, 1880; 91:148, 1880.
50. See Chapter 6, and Zinsser, H., *Infection and Resistance*, New York, Macmillan, 1914.
51. Mazumdar, P.M.H., *Bull. Hist. Med.* 48:1, 1974. See also Mazumdar, P.M.H., *Species and Specificity: An Interpretation of the History of Immunology*, Cambridge, Cambridge University Press, 1995.
52. von Behring, E., *Centralbl. Klin. Med.* 9:681, 1888.
53. Baumgarten, P., *Berl. Klin. Wochenschr.* 37:615, 1900.
54. von Behring, E., *Deutsch. Med. Wochenschr.* 16:1145, 1890. For an earlier version, see Grawitz, P., *Virchow's Arch.* 84:87, 1881.

2 Cellular vs humoral immunity

....one of Metchnikoff's most suggestive biological romances...
George Bernard Shaw, The Doctor's Dilemma

In a major address to the congress of the British Medical Association in 1896, Lord Lister suggested that if ever there had been a romantic chapter in the history of pathology, it was certainly that concerned with theories of immunity. Lister's reference was to two epic but interrelated battles that had occupied pathologists, bacteriologists, and immunologists over the course of several decades – battles which saw opposing schools engage in passionate debate and a degree of vilification almost unknown in present-day science. When Lister spoke in 1896, the first of these great disputes was nearing its resolution. This involved the question of the basic nature of the inflammatory reaction – whether inflammation is an abnormal response harmful to the host, or a normal and beneficial component of its defensive armamentarium. However, the second of these battles had not yet been resolved, and was still being fought in every journal and at every congress relating to the subject. Its focus was on the question of whether innate and acquired immunity to infection could be best explained by cellular or by humoral mechanisms. In that exciting decade when remarkable discoveries crowded close on one another's heels, when new mechanisms, new organisms, new diseases appeared with almost every issue of the journals, the protagonists from one or the other camp grasped each new item eagerly to bolster their own theory, or to cast doubt upon that of the opposition.

The lines that divided the two camps were fairly sharply drawn. Conceptually, the cellularists argued that the chief defense of the body against infection resided in the phagocytic and digestive powers of the macrophage and the microphage (the polymorphonuclear leukocyte), while the humoralists claimed that only the soluble substances of the blood and other body fluids could immobilize and destroy invading pathogens. Geographically, the cellularists were predominantly French and rallied round Elie Metchnikoff at the Pasteur Institute in Paris, while the humoralists were predominantly German and followed the leadership of Robert Koch and his disciples at Koch's Institute in Berlin.

An examination of the history of the cellular–humoral dispute illustrates several interesting points. First, it provides the historian of ideas with yet another example of how earlier and even outmoded concepts help to determine the structure and content of future thought, and how the intransigent commitment to a scientific dogma often prevents timely and rational compromise. Secondly, it provides to the sociologist of science yet another striking example of the way in which non-scientific events may contribute importantly to both the direction and the velocity of scientific development. Finally, it provides to the modern immunologist the sobering caution that the triumph of one concept in such

A History of Immunology, Second Edition
ISBN: 978-0-12-370586-0

Copyright © 2009, Elsevier Inc.
All rights reserved

Plate 3 Robert Koch (1843–1910)

a dispute may for many decades stifle developments dependent upon the other concept, to the detriment of the scientific discipline.

Background to the conflict

As is true of most conceptual advances in science, the theory of a cellular basis for immunity and the violent opposition which it engendered arose not in a scientific and cultural vacuum, but in an environment which largely defined the nature and direction of the subsequent debate. Among the determinants of the battle over the central nature of immunity, some may be traced back over 2,000 years to the ideas of Hippocrates, some appeared only with the development of a true medical science and of Virchow's new pathology in the mid-nineteenth century, while some, surprisingly, were founded on the international politics and nationalistic rivalries of the contemporary era. It is only in the context of these background elements that the directions taken by this epic struggle, its intensity, and the full flavor of this "romantic chapter" may be fully appreciated.

The nature of disease

For over 2,000 years, following the teachings of Hippocrates, Celsus, and Galen,[1] disease was considered to be a maladjustment of the normal ratios of the four vital humors: the blood (*sanguis*), the phlegm (*pituita*), the yellow bile (*chole*), and the black bile (*melaine chole*). This humoral tetrad, almost a mystical part of a larger system which included Aristotle's four basic essences (earth, water, fire, and air) and the four primary qualities (hot, cold, wet, and dry), influenced medical theory and medical practice as late as the nineteenth century,[2] despite the long-growing appreciation that some diseases were contagious, and the early recognition that prior exposure to a "plague" might protect the individual from the current contagion. Even into the early nineteenth century, cupping, purgatives, phlebotomy, and the application of leeches were still common practices applied to all types of disease, to restore the ill-humored to healthier proportions. Given a humoral theory of disease going back over twenty centuries, coupled with a humoral approach to prophylaxis and therapy of similar ancestry, it is not surprising that the mere name "humoral" as applied to a theory of immunity would carry with it much traditional respect and prestige. This was true despite the serious criticism leveled at more modern humoralist offshoots, such as the hematohumoral theory of Rokitansky.[3]

It was only in 1858, some twenty-five years before Metchnikoff's first publication on the phagocytic theory, that Rudolph Virchow issued a comprehensive challenge to the remnants of the humoral theory of disease, in the form of a claim that all pathology is based upon the malfunction of cells rather than the maladjustment of humors.[4] While Virchow's cellular pathology was widely acclaimed and respected, even a quarter-century later in the 1880s humoralism had not yet fully given way to Virchow's cellular concepts, upon which Metchnikoff based his theory of immunity.

Another factor important for an understanding of the nature of disease, and another example that ancient concepts die hard, was that involving etiology. Any precise concept of immunity had necessarily to be based upon the acceptance that infectious diseases are specific and reproducible. However, only slowly did the medieval notion of ill-humors and miasmas give way to the recognition that each infectious disease is produced by its own specific pathogenic microorganism. The first barrier to be overcome in this acceptance was the old belief that the variety of microorganisms seen since the time of Leeuwenhoeck were spontaneously generated and almost infinitely mutable. The concept of spontaneous generation should have been destroyed in the eighteenth century by the work of Spallanzani (1768) and others, but it persisted and was defended by prominent scientists even up to the 1860s and 1870s. It finally gave way, at least in France, before the brilliant experimentalism and, more decisively, the forceful argumentation of Louis Pasteur.[5] Again, it was not until the late 1870s, barely five years before Metchnikoff advanced his new theory of immunity, that the germ theory of disease finally gained wide acceptance, due in part to its proclamation by Pasteur[6] and to Robert Koch's elegant description of the etiology of wound infections.[7]

One may thus conclude that the cellular theory of immunity advanced by Elie Metchnikoff in 1884[8] did not constitute just one further acceptable step in a well-established tradition, but rather represented a significant component of a conceptual revolution with which contemporary science had not yet fully learned to cope.

Plate 4 Ilya (Elie) Metchnikoff (1845–1916)

The nature of inflammation

It is important that the modern reader appreciate that at the time it was advanced, Metchnikoff's theory of phagocytosis was less a contribution to immunologic thought than to the field of general pathology, which for some thirty to forty years had been debating the nature of the inflammatory response. It will be recalled that at this point cellular pathology was only twenty-five years old, that a formal germ theory of disease was scarcely five years old, and that the demonstration by Louis Pasteur of a vaccine prophylaxis for chicken cholera (the first carefully designed scientific study that was to serve as the foundation for the new science of immunology) had appeared only four years earlier.[9]

Thus, there was little or no context of immunologic thought in which to fit the Metchnikovian theory: neither Edward Jenner with smallpox nor Louis Pasteur with chicken cholera had understood the mechanism responsible for the immunity which they were able to induce.[10] But there was a broad context in the general pathology of inflammation against which Metchnikoff's new theory could be measured, and here the phagocytic theory constituted a strong challenge to accepted dogma. The inflammatory reaction which accompanied infectious diseases and especially traumatic wounds had usually been considered deleterious to the host. This was perhaps understandable in the days before the concept of antisepsis, when the inflammatory response presented most often as a purulent and violent accompaniment of a wound, most often rendering an unfavorable prognosis. Even the repeated reference to a "laudable pus," most notably by that remarkable poet and scientist Erasmus Darwin (1731–1802) in his *Zoonomia* of 1801, did not seriously challenge the belief in the noxious contribution of the inflammatory response. With the rise of microscopy and anatomic pathology, purulent discharges were early associated with those inflammatory cells later named macrophages and microphages, and thus these cells were identified as the most obvious component of the deleterious inflammatory reaction. Moreover, it was generally thought by most pathologists of the era that phagocytic cells actually provided an admirable means of transport for infectious organisms in their dissemination throughout the body.

It was in this context that Metchnikoff dared to suggest that the phagocytic cells, far from being harmful, in fact constitute a first line of defense in their ability to ingest and digest invading organisms. It is not surprising, therefore, that when Rudolph Virchow visited Metchnikoff in his laboratory in Messina in 1883, he advised Metchnikoff to proceed with great caution in advancing his theories, since "most pathologists do not believe in the protective role of inflammation."[11]

Metchnikoff's challenge to contemporary pathological thought, however, was not limited to his iconoclastic view of the significance of the inflammatory response. Indeed, he dared to challenge the current concepts and authorities on the very nature of inflammation. Virchow himself had formulated a concept of parenchymatous inflammation, involving a disturbed nutrition with intensified local proliferation of parenchymal cells due to injury by the pathologic agent, thus leading to the *tumor* which he considered the most significant component of the process. Julius Cohnheim, on the other hand, concluded from his famous experiments that inflammation was due primarily to lesions of the walls of blood vessels, permitting passive leakage of all of the components (primarily humoral) then recognized in the inflammatory response.[12] Thus, Cohnheim considered that the *rubor* was the most significant sign of the inflammatory reaction. While the field of general pathology was divided on which of these two mechanisms was most important for the inflammatory reaction, almost all agreed with these great pathologists that inflammation was a deleterious reaction of no benefit to the host – a purely passive response on the part of the insulted organism. It is understandable that Metchnikoff's radical views would find difficulty in acceptance, since not only did they challenge the very foundation of then-current dogma, but they were also advanced by an individual who was (1) not a member of the confraternity of pathologists (Metchnikoff was a zoologist); (2) not even a physician (the chemist Pasteur had encountered similar problems); and (3) a Russian (a people then traditionally considered somewhat backward by many western Europeans).

International politics

In 1888, the itinerant expatriate Metchnikoff took up permanent residence in Paris as *Chef de Service* at the Pasteur Institute. His natural inclinations, reinforced by the fervent patriotism of Pasteur, engendered in Metchnikoff a strong passion for his adopted homeland. Proponents of the cellular theory thus naturally looked to Paris and the Pasteur Institute for their leadership, and the cellularists were drawn, for the most part, from among French scientists. In Germany, however, Metchnikoff's theory came under severe attack at an early date, first by Baumgarten in Berlin and then by other German pathologists. This curious geographic partisanship was even more sharply defined by a division of sentiment within recently unified Germany itself. The most vocal opponents of the Metchnikovian theory were Baumgarten, Bitter, Christmas-Dirckinck, Ziegler, Gaffky, and Emmerich, all of Berlin, Flügge of Göttingen, Weigert of Breslau, and Frank of Friedrichsheim, all from within Prussia. Of those Germans who spoke out on behalf of Metchnikoff, Hess was in Heidelberg, Ribbert in Bonn, and Buchner in Munich – all from regions of Germany that had historically resented Prussian power and hegemony. Elsewhere, Gamelaia and Banti in Italy and Calus in Vienna voiced their support of Metchnikoff.[13] British workers were in general neutral on the issue, with the notable exception of the Francophile Lord Lister, who repeatedly acknowledged the debt that his antiseptic theories owed to Pasteur and Metchnikoff.

However, the principal division in the cellularist–humoralist battle was between France and Germany – a division that reflected the overall nationalistic tendencies of the time. England had for over 600 years been the traditional enemy of France, dating from the time that the descendants of the Norman conquerors of England laid claim to their old lands and even to the throne of

France. Even after the English were expelled from the continent, the principal element of their foreign policy was to subsidize coalitions of the smaller states in Europe, including the German states, to neutralize a powerful France. The continuing French policy, on the other hand, was to prevent the development of a powerful continental opponent among the German-speaking peoples. The Germans, in their turn, had long resented the power of France, and felt that French Alsace should be a part of a greater Germany. A tradition of enmity between France and the German states (and especially Prussia) thus matured over a long period of time, and culminated in the ignominious defeat of the French and the loss of Alsace in 1870–1871 at the hands of a Germany now unified under Prussian rule.

The phagocytic theory of immunity was not the only dispute in which objective science appeared to have been compromised by the after-effects of the Franco-Prussian War. In the aftermath of the siege of Paris, Louis Pasteur, who in 1868 had received an honorary MD degree from the University of Bonn, returned his honors in anger. He wrote to the head of the Faculty of Medicine at Bonn that,

> Now the sight of that parchment is odious to me, and I feel offended at seeing my name, with the qualification of virum clarissimum that you have given it, placed under a name which is henceforth an object of execration to my country, that of Rex Gulielmus. ...I am called upon by my conscience to ask you to efface my name from the archives of your faculty, and to take back that diploma, as a sign of the indignation inspired in a French scientist by the barbarity and hypocrisy of him who, in order to satisfy his criminal pride, persists in the massacre of two great nations.

In response, Pasteur received a reply from the Principal of the Faculty of Medicine of Bonn, who "is requested to answer the insult which you have dared to offer to the German nation in the sacred person of its august Emperor, King Wilhelm of Prussia, by sending you the expression of its entire contempt." Ten years later found Pasteur and Robert Koch in violent debate about the etiology, pathogenesis, and prophylaxis of anthrax and other diseases, with unseemly and vituperative statements being issued from both sides. Indeed, the first volume of the Reports of the German Imperial Health Office in 1881 could almost have been subtitled "anti-Pasteur," containing as it did scathing criticisms of Pasteur's work by Koch and his students Löffler and Gaffky.[14] These authors declared that Pasteur was incapable of cultivating microbes in a state of purity, that he did not know how to recognize the septic vibrio (although he himself had discovered it!), and that many of his experiments were "meaningless." Pasteur, on his side, pursued the debate with his customary vigor, even going so far as to challenge Koch to face-to-face debate at the International Congress of Hygiene at Geneva in 1882. When, in the end, Pasteur's demonstration of the efficacy of anthrax vaccination was fully vindicated by the famous experiments at Pouilly-Le-Fort,[15] Pasteur rejoiced aloud that this great discovery had been a French one, and it is not difficult to define the alternative that he might have feared. It is interesting that yet another great immunologic debate, concerning the nature of the antigen–antibody interaction, was carried on between Jules Bordet at the Pasteur Institute in France and Paul Ehrlich in Germany.[16]

It was in this environment, then, that the cellular-versus-humoral debate proceeded. As was appropriate for the halls and journals of science, few overt hints appeared that anything other than pure objective science determined this debate. One such instance, however, appeared in a paper by Abel[17] which was highly critical of Metchnikoff, with a statement about "interpretations which we on the German side cannot share…." Metchnikoff was highly incensed by this statement, and in a later book called Abel to task for such unscientific nationalism.[18]

We may only wonder whether this debate would have been as vitriolic or protracted, had the international political setting been different during the latter half of the nineteenth century. Paul de Kruif goes too far perhaps in suggesting in his book *Microbe Hunters*[19] that this epic struggle in immunology contributed to the start of World War I, but it does seem probable that, in a minor way at least, it did represent one of the continuing reverberations of the Franco-Prussian War of 1870.

Cellular vs humoral immunity

The early debate

Ilya (later Elie) Metchnikoff was born in the Steppe region of Little Russia in 1845. He studied invertebrate zoology in both Russia and Germany, and developed a keen interest in invertebrate embryology while working at Naples with the great Russian embryologist Kovalevsky. His work in this field was extremely productive, so that by the late 1870s he had established for himself a significant reputation in zoological circles. It was during this period that he developed an interest in the digestive processes of invertebrates, and especially in the intracellular digestion exhibited by the wandering mesodermal cells of metazoans. This interest in digestion was to remain with Metchnikoff throughout his life, and accounts for his zeal in the popularization of yogurt in western Europe, and for the prominent place that digestive disorders play in so many of his writings.[20]

It was while working in the Marine Biology Laboratory on the straits of Messina that Metchnikoff first conceived of the phagocytic theory. In one of those conceptual leaps that occur in a science, an investigator may look at an old phenomenon and suddenly gain a new insight. In his own words:[21]

> *One day when the whole family had gone to the circus to see some extraordinary performing apes, I remained alone with my microscope, observing the life in the mobile cells of a transparent starfish larva, when a new thought suddenly flashed across my brain. It struck me that similar cells might serve in the defense of the organism against intruders. Feeling that there was in this something of*

surpassing interest, I felt so excited that I began striding up and down the room and even went to the seashore in order to collect my thoughts. I said to myself that, if my supposition was true, a splinter introduced into the body of a starfish larva, devoid of blood vessels or of a nervous system, should soon be surrounded by mobile cells as is to be observed in the man who runs a splinter into his finger. This was no sooner said than done. ...I was too excited to sleep that night in the expectation of the results of my experiment, and very early the next morning I ascertained that it had fully succeeded. That experiment formed the basis of the phagocytic theory, to the development of which I devoted the next twenty-five years of my life.

While continuing his studies of phagocytic cells in invertebrates, Metchnikoff immediately realized the significance of his theory for human disease, and as early as 1884 published a paper on the relationship of phagocytes to anthrax. He quickly followed this with studies on erysipelas, typhus, tuberculosis, and numerous other bacterial infections. In his book on the *Comparative Pathology of Inflammation* in 1891, Metchnikoff formalized the statement of the phagocytic theory and demonstrated in detail its Darwinian evolutionary development.[22]

No sooner had the phagocytic theory appeared in the literature than it came under severe and protracted attack.[23] At the outset these objections were of a quite general nature, as befitted a theory that flew in the face of so many conventional wisdoms. As the debate proceeded and new experiments flooded the literature, both the claims for the phagocytic theory and the counter-claims against it assumed a more precise form. In retrospect, the objections to Metchnikoff's theory may be catalogued under the following headings:

1. The phagocytes fail to ingest one or another pathogenic organism
2. Even where organisms are ingested by phagocytes, those organisms are either not destroyed, or had already come under humoral attack
3. Even when phagocytes can be shown to be effective components of the immune response, their role is secondary to the earlier action of some humoral factor.

(We will omit here a discussion of the criticisms of some of Metchnikoff's more exuberant claims, such as those that the phagocytes are the chief agents of the aging process, wherein active phagocytosis of neurons was claimed to contribute to senility, and the phagocytosis of hair pigment to graying.)

It was not until 1888 that the opponents of Metchnikoff's cellular theory found a proper banner around which they could rally, and a phenomenology upon which to base a humoral alternative to the phagocyte. In that year, Nuttall, during the course of experiments designed to put Metchnikoff's theory to the test, observed that the serum of normal animals possesses a natural toxicity for certain microorganisms.[24] This observation was quickly seized upon by many investigators, most notably by Buchner,[25] who was not only the first of the theoreticians of the humoral concept of immunity (without becoming anti-cellularist), but also named the active bactericidal factor *alexin* (protective substance; Ehrlich later renamed it complement). It is interesting that this observation had almost been foretold a century earlier by John Hunter, the famous surgeon, naturalist, and teacher of Edward Jenner in his *Treatise on the Blood, Inflammation, and Gunshot Wounds*, in which he noted that blood did not decompose as readily as other putrescible materials.[26] For a number of years after 1888, the scientific journals were filled with reports that the cell-free fluids of normal and especially of immunized animals could kill bacteria, so that no recourse was necessary to Metchnikoff's phagocytic cells in order to explain both natural and acquired immunity. This view received perhaps its most powerful support from the observation of Richard Pfeiffer, who found that the injection of cholera vibrios into the peritoneal cavity of immune guinea pigs was followed by their rapid destruction.[27] This Pfeiffer phenomenon involved an early granular change and swelling of the organism, followed soon after by complete dissolution and disappearance – i.e., the process of bacteriolysis. Moreover, it was soon shown that the Pfeiffer phenomenon could be passively transferred by injecting serum from an immunized guinea pig into the peritoneal cavity of a normal guinea pig, and even that bacteriolysis would proceed *in vitro*. The humoralists claimed that on those occasions when microorganisms could actually be found within phagocytic cells, it was probably part of the clean-up operation of damaged bacteria.

In response to these strong attacks by the humoralist school, Metchnikoff and his students at the Pasteur Institute were by no means silent. In paper after paper, these investigators demonstrated that there is often no relationship between the natural bactericidal powers of the serum of different species and their susceptibility to infection by a given organism. Rather, as in the case of anthrax, the resistance of a species could often be directly correlated with the ability of its phagocytes to ingest this organism. (It is interesting that so many of Metchnikoff's telling experiments were performed using the anthrax bacillus. As Zinsser later pointed out, this was a highly fortuitous choice, since the resistance of this bacillus to immune lysis is especially well marked and phagocytosis seems indeed to be the chief mode of bacterial destruction.)

A further method of investigation employed by Metchnikoff in endeavoring to prove his point was the attempt to demonstrate that virulent bacteria could be protected from destruction in the body of a resistant animal if the function of the leukocytes were inhibited. This resulted in a number of ingenious experiments, such as the one performed by Trapeznikoff on anthrax infection of frogs.[28] Whereas anthrax spores injected subcutaneously were rapidly phagocytosed and destroyed, those introduced in little sacks of filter paper were protected from phagocytes and remained virulent, although bathed in the tissue fluids. (This experiment is very reminiscent of the Algire chamber employed over sixty years later, to demonstrate that allograft rejection is based upon cellular rather than humoral mechanisms.[29]) Another of the interesting experiments of the era was that of Cantacuzène,[30] who showed that animals treated with opium are much more susceptible to infection than are normal controls, as the presumed consequence of the inhibition of motility of the drugged phagocytic cells. Finally, Metchnikoff showed repeatedly that the nonspecific creation of a macrophage-rich peritoneal exudate, with the attendant activation of those macrophages, would protect that host against intraperitoneal injection of lethal doses of bacteria. This was an early forerunner of another modern practice, that of nonspecific immunotherapy.[31]

The debate thus ebbed and flowed during the first decade following 1884, with the cellularists appearing at times to carry the day, while at other times it was the humoralists who claimed victory. But slowly the tide appeared to turn against the phagocytic theory, forcing Metchnikoff, in his zealous defense of it, to formulate rather extreme *ad hoc* hypotheses. When faced with increasingly convincing evidence of the bactericidal properties of immune serum, Metchnikoff postulated that immunization led to the formation of substances which he termed "stimulins," that acted directly upon phagocytes to enhance their activity. Again, when evidence mounted on the important role of serum complement (thanks in no small measure to the work of Jules Bordet in Metchnikoff's own laboratory), the cellularists were forced back to the position that complement probably originates in any event from the destruction of blood macrophages during the clotting process. Still later, Metchnikoff felt forced to go to great lengths to show that his theory was not inconsistent with Ehrlich's side-chain concept.

The growing humoral tide

The most telling blow to the cellular theory of immunity came in 1890 with the discovery by von Behring and Kitasato that immunity to diphtheria and tetanus is due to antibodies against their exotoxins.[32] When, shortly thereafter, it was demonstrated that passive transfer of immune serum would protect the naive recipient from diphtheria with no obvious intercession by any cellular elements,[33] the humoralists felt that they had been vindicated, and Koch felt free to proclaim the demise of the phagocytic theory at a congress in 1891. The discovery of antibodies against these exotoxins, and even against toxins of non-bacterial origin such as ricin and abrin,[34] supported the earlier view that most infectious diseases were toxic in nature, and thus it could be claimed that protection was due in large part to humoral antitoxic antibodies. Although the discovery of the Pfeiffer phenomenon quickly corrected this generalization by showing that circulating antibody could induce direct bacteriolysis of cholera organisms, even this observation provided yet another strong support for the humoral concept. Bordet's demonstration[35] that even the erythrocyte could be lysed with antibody in the absence of phagocytes demonstrated the generality of this phenomenon, and further reinforced the humoralists' claims.

As the decade of the 1890s progressed, new observations lent further weight to claims for the supremacy of the role of humoral antibody in the mediation of immunity. New antibodies against different microorganisms were reported regularly, and their specificities demonstrated. The discovery of the precipitin reaction[36] and Ehrlich's classical work on the titration of anti-diphtheria antibodies and diphtheria toxin[37] (which did much to found the field of immunochemistry)

Plate 5 Emil von Behring (1854–1917)

demonstrated that antibody was more than a concept: it was a substance that one could see and feel and study in the test tube, and with which most immunologists were much more comfortable than they were with the difficult phagocyte. Discovery of bacterial agglutination[38] provided still another convincing demonstration of the importance of humoral antibody in defense against infection. It only remained for Ehrlich to provide a theoretical formulation in his side-chain concept of the functions of antibody, antigen, and complement[39] to make the antibody the principal object of interest to almost all immunologists. This was helped in no small measure by the pictures which Ehrlich published to illustrate his side-chain theory – pictures that made it easier to believe that antibodies and complement were "real substances" with comprehensible receptors and simple modes of action.

By the turn of the century, then, it would appear safe to conclude that most active investigators favored one or another modification of a humoral theory to explain natural immunity and, certainly, acquired immunity. Metchnikoff was correct in receiving the impression at a congress in 1900 that his theory was not well understood, but was perhaps too late in his attempt to rectify this situation by publication a year later of his famous book *Immunity in the Infectious*

Diseases. Of course, the phagocytic theory continued to be referred to, and was covered extensively in the textbooks of the day, but more as a general phenomenon of great biological interest and as a tribute to a tireless and personally highly respected worker than as a serious competitor among general theories of immunity. Even the discoveries of anaphylaxis,[40] of the Arthus phenomenon,[41] and of serum sickness[42] provided indirect support for the humoralists' viewpoint. While no one at the time quite knew what the relationship was between immunity and these manifestations of allergy, yet it was clear that these were somehow immunologic phenomena dependent upon humoral antibody for their attainment.

As is most often the case in scientific disputes such as that between the cellularists and humoralists, the triumph of one theory over another is not proclaimed by some impartial arbiter, to be followed by general public acquiescence. Rather, the best measure of outcome is usually to assess the influence of these theories on the active members of that scientific community, and especially to determine the subjects of interest among the younger scientists. A review of the literature of the early twentieth century shows that although the cellularist–humoralist debate appeared still to be continuing, based upon what the older generation was *saying* in the literature, most scientists, old and young, were *doing* work on antibodies and complement rather than on cells. (Even later, as Brieger[43] points out, Metchnikoff's work was more revisited than extended.)

Two events occurred over the next few years to make it appear that the result of this silent vote in favor of the humoral theory of immunity was not widely appreciated. In 1908, the Swedish Academy conferred the Nobel Prize in Physiology or Medicine jointly on Metchnikoff, the current champion of cellularism, and Ehrlich, the then-leading exponent of humoralist doctrines," in recognition of their work in immunity." While it is dangerous to speculate upon motives, one cannot help but feel that this joint recognition was an attempt to arbitrate the dispute between cellularists and humoralists. But to judge again from the literature, the decision came too late, since active research on the participation of cells in immunity continued to decline.

Another apparently belated attempt to mediate the cellularist–humoralist dispute, and to rationalize their differences, followed the naming and description of the mode of action of opsonin by Wright and Douglas in England.[44] These investigators claimed that both humoral and cellular functions were equally important and interdependent, in that humoral antibody interacts with its target microorganism to render it more susceptible to phagocytosis by macrophages. Upon this simple structure, Wright constructed an elaborate and extremely complicated therapeutic scheme, involving the determination of "opsonic indexes" and the administration of autovaccines at certain critical periods during the course of the infectious process. This approach became so popular in early twentieth-century England that Bernard Shaw, a close friend of Sir Almroth Wright, used it as the subject of his play *The Doctors' Dilemma*. In his otherwise scathing castigation of the medical profession, Shaw, the skeptic and therapeutic nihilist, summarizes Wright's approach in his *Preface on Doctors*:

...Sir Almroth Wright, following up one of Metchnikoff's most suggestive
biological romances, discovered that the white corpuscles or phagocytes, which
attack and devour disease germs for us, do their work only when we butter the
disease germs appetizingly for them with a natural sauce which Sir Almroth
named opsonin. ...The dramatic possibilities of this discovery and invention will
be found in my play. But it is one thing to invent a technique: it is quite another
to persuade the medical profession to acquire it. Our general practitioners, I
gather, simply declined to acquire it...

However, Shaw was wrong and Wright was too optimistic: many tried, but the techniques proved so difficult and unreproducible in practice as to become unfashionable within a decade. In partial consequence of this he came to be referred to (out of his hearing) as Sir Almost Right, but while Wright acquired a new nickname, the cellular theory of immunity lost its last opportunity for revival for many years to come.

Consequences of the humoralist victory

It is the central thesis of this chapter that the fall from favor of Metchnikoff's cellular (phagocytic) theory of immunity carried with it profound implications for future developments in the young discipline of immunology. The most imaginative and productive investigators working on the cutting edge of a science tend to choose their problems based upon what they (or their teachers) feel is most significant, rather than what is technically the easiest. Behind them come the less imaginative, content to follow the fashions of the day. During the early decades of the twentieth century, it was clear to most workers that antibody held the key to an understanding of immunity, and thus it constituted the natural choice for investigative work. Moreover, the direction of even antibody research underwent a significant change detrimental to cellular studies, due to the decline of what has been called the Golden Age of bacteriology. As the discovery of new pathogens and new phenomena slowed around the turn of the century, and as those infectious diseases amenable to immunologic prophylaxis or therapy were satisfied, nascent immunology more and more turned away from medicine and biology and toward chemistry. This was initiated early on by the studies and theories of the ever chemically-oriented Ehrlich, and given a strong push by the famous chemist Svante Arrhenius.[45] This new direction was more than adequately reinforced during the 1920s and 1930s by the elegant work of Landsteiner on serological specificity[46] and of Heidelberger and his students on quantitative immunochemistry.[47] Leaving aside Pasteur himself (who, though trained as a chemist, was the quintessential biologist), it is interesting to note the number of workers trained in chemistry who became interested in immunology, including Arrhenius, Haurowitz, Heidelberger, Linus Pauling, and many others.

But the failure of the cellularist doctrine to gain adherents in the scientific community meant also that many approachable problems in cellular immunology were neglected as being "uninteresting," in the humoralist context of the times. This is not to suggest that all of the important problems could have been solved, or even that many of the important questions could immediately have been posed: rather, one might reasonably have expected slow but substantial progress in cellular immunology over the next forty to fifty years. Thus, instead of endless searches for circulating antibody associated with tuberculosis, the tuberculin reaction, and contact dermatitis, histopathologic studies and their resultant conclusions might have been obtained many decades earlier rather than awaiting the important descriptions of Gell and Hinde,[48] Turk,[49] and Waksman[50] in the 1950s. Such studies might have pointed up much earlier the importance of the lymphocyte in immunologic phenomena. Again, the phenomenologic demonstrations of Mackaness,[51] Rowley,[52] and others on the critical role of cellular immunity in certain bacterial infections required few advances over the techniques available to Metchnikoff, and could certainly have been pursued fifty years earlier, had the cellularists still held sway. Finally (but by no means exhausting the potential list), the pioneering cell transfer experiments of Landsteiner and Chase,[53] establishing the critical role of mononuclear cells in cellular immunity, were well within the technical competence of investigators earlier in the century. However, the notion of cellular immunity was out of favor, and few investigators in that environment were stimulated to pose the questions that might have led to such studies.

For a period of almost fifty years, few questions about cells in immunity were asked within a discipline comfortable with the dogma that circulating antibody would provide all essential answers to the problems of immunity and immunopathology. Only rarely during the first half of the twentieth century did an investigator think it worthwhile to study the role of cells in immunologic phenomena, or to explore the basis of "bacterial allergy" or "delayed hypersensitivity," as it was variously termed. In the 1920s and early 1930s Zinsser[54] studied bacterial allergy, and Dienes and his coworkers[55] studied delayed hypersensitivity to simple protein antigens injected into tubercles (the forerunner of the adjuvant), a possibility extended by Jones and Mote[56] and by Simon and Rackemann.[57] In tissue culture experiments, Rich and Lewis[58] showed the importance of inflammatory cells in tuberculin allergy, and Harris, Ehrich, and coworkers carried on extensive studies of the role of the lymphocyte in antibody formation.[59] But these were isolated excursions out of the mainstream, which made little impression upon immunologic thought at the time. As late as 1951, in his classic book on *The Pathogenesis of Tuberculosis*,[60] Arnold Rich could conclude that little was known about the nature of bacterial allergy, its relationship to immunity, or even the extent to which the familiar macrophage and the ubiquitous but mysterious lymphocyte were involved in its development.

The study of delayed hypersensitivity only attained respectability and became an appropriate topic for immunologic symposia[61] and books in the early 1960s, in conjunction with a shift in immunology from a chemical to a more biological approach. This radical change in emphasis can be traced directly to the development of the type of crisis in immunology that Thomas Kuhn has suggested is often responsible for major conceptual changes within a scientific discipline.[62] The new questions posed about the mechanism of allograft rejection, of immunologic tolerance, of immunity in certain viral infections, of the pathogenesis of autoallergic diseases, and of the phenomena associated with immunologic deficiency diseases could no longer be answered within the framework of a classical theory based solely upon the functions of humoral antibody. Despite the hiatus of over fifty years, the explosion of activity in cellular immunology during the 1960s was such that many of the gaps in our knowledge about cell functions in immunity were rapidly filled. A new journal bearing the title *Cellular Immunology* could appropriately be started in 1970, and at least partial vindication could be claimed for Metchnikoff's cellular theory of immunity. If Metchnikoff's "cellular [phagocyte] immunology" is not quite the same as modern "cellular [lymphocyte] immunology," yet his important contributions to the founding of the field cannot be gainsaid.[63]

It is still permissible, however, for one to wonder whether cellular immunology would not have achieved its modern successes even decades earlier, had not the humoral theory of immunity so strikingly overshadowed the cellular theory in the late nineteenth century.

Notes and references

1. Sigerist, H.E., *A History of Medicine*, Vol. II, New York, Oxford University Press, 1961.
2. Ackerknecht, E.H., *A Short History of Medicine*, New York, Ronald Press, 1955. See also Castiglioni, A., *A History of Medicine*, 2nd edn, New York, Knopf, 1947.
3. Miciotto, R.J., *Bull. Hist. Med.* 52:183, 1978.
4. Virchow, R., *Die Cellularpathologie in ihrer Begründung auf physiologische und pathologische Gewebelehre*, Berlin, Hirschwald, 1858; English edition, *Cellular Pathology*, New York, Dover, 1971.
5. Vallery-Radot, R., *La Vie de Pasteur*, Paris, Hachette, 1901; English edition, New York, Dover, 1960. For a comprehensive history of the spontaneous generation controversy, see Farley, J., *The Spontaneous Generation Controversy from Descartes to Oparin*, Baltimore, Johns Hopkins University Press, 1974.
6. Pasteur, L., Joubert, J., and Chamberland, C., *C. R. Acad. Sci.* 86:1037, 1878.
7. Koch, R., *Untersuchungen über die Aetiologie der Wundinfectionskrankheiten*, Leipzig, Vogel, 1878.
8. Metchnikov, E., *Virchows Archiv.* 96:177, 1884.
9. Pasteur, L., *C. R. Acad. Sci.* 90:239, 952, 1880.
10. Bulloch, W., *The History of Bacteriology*, London, Oxford University Press, 1938; see also Chapter 1.
11. Metchnikoff, O., *Life of Elie Metchnikoff*, Boston, Houghton Mifflin, 1921; see also the biography by his devoted student and coworker Besredka, A., *Histoire d'une idée, L'oeuvre de E. Metchnikoff*, Paris, Masson, 1921.
12. Cohnheim, J., *Neue Untersuchungen über die Entzundung*, Berlin, Hirschwald, 1873.

13. The order of battle in this dispute is discussed by Ziegler, E. (*Beitr. Pathol. Anat. Jena* **5**:419, 1889).
14. *Mitt. Kaiserlichen Gesundheitsamte* **1**:4, 80, 134, 1881.
15. Pasteur, L., Chamberland, C. and Roux, E., *C. R. Acad. Sci.* **92**:1378, 1881.
16. Zinsser, H., *Infection and Resistance*, New York, Macmillan, 1914; see also Mazumdar, P.H., *Bull. Hist. Med.* **48**:1, 1974.
17. Abel, R., *Zentralbl. Bakteriol. Parasitenk. Jena* **20**:760, 1896.
18. Metchnikoff, E., *L'Immunité dans les Maladies Infectieuses*, Paris, Masson, 1901. (English translation in 1905 by Macmillan, New York; reprinted by Johnson Reprint Corp., New York, 1968).
19. De Kruif, P., *Microbe Hunters*, New York, Harcourt Brace, 1926.
20. Metchnikoff, E., *The Nature of Man*, New York, Putnam, 1903.
21. Metchnikoff, note 11, pp. 116–117.
22. Metchnikoff, E., *Lectures on the Comparative Pathology of Inflammation*, London, Keegan Paul, Trench, Trübner, 1893. Reprinted by Dover, New York, 1968.
23. The most comprehensive criticism was by Baumgarten, P., *Z. klin. Med.* **15**:1, 1889.
24. Nuttall, G., *Z. Hyg.* **4**:353, 1888.
25. Buchner, H., *Zentralbl. Bakt.* **6**:561, 1889.
26. Hunter, J., *Treatise on the Blood, Inflammation, and Gunshot Wounds*, Philadelphia, Webster, 1823.
27. Pfeiffer, R., *Z. Hyg.* **18**:1, 1894.
28. Trapeznikoff, *Ann. Inst. Pasteur* **5**:362, 1891.
29. Algire, G.H., *J. Natl Cancer Inst.* **15**:483, 1954.
30. Cantacuzène, J., *Ann. Inst. Pasteur* **12**:273, 1898.
31. "Symposium on Immunotherapy of Tumours," in *Progress in Immunology*, III, New York, Elsevier, 1977, pp. 559–605.
32. von Behring, E., and Kitasato, S., *Deutsch. med. Wochenschr.* **16**:1113, 1890.
33. von Behring, E., and Wernicke, E., *Z. Hyg.* **12**:10, 45, 1892.
34. Ehrlich, P., *Deutsch. med. Wochenschr.* **17**:976, 1218, 1891.
35. Bordet, J., *Ann. Inst. Pasteur* **12**:688, 1899.
36. Kraus, R., *Wien. klin. Wochenschr.* **10**:736, 1897.
37. Ehrlich, P., *Klin. Jahrb.* **6**:299, 1897.
38. Gruber, M., and Durham, H.E., *Münch. med. Wochenschr.* **43**:285, 1896.
39. Ehrlich, P., *Proc. R. Soc. Ser. B* **66**:424, 1900.
40. Portier, P., and Richet, C., *C. R. Soc. Biol.* **54**:170, 1902.
41. Arthus, M., *C. R. Soc. Biol.* **55**:817, 1903.
42. von Pirquet, C., and Schick, B., *Die Serum Krankheit*, Vienna, Deuticke, 1906; English edition, *Serum Sickness*, Baltimore, Williams & Wilkins, 1951.
43. Brieger, G.H., "Introduction" to *Immunity in Infective Diseases*, by E. Metchnikoff, New York, Johnson Reprint Corp, 1968.
44. Wright, A.E., and Douglas, S.R., *Proc. R. Soc. Ser. B* **72**:364, 1903.
45. Arrhenius, S., *Immunochemistry*, New York, Macmillan, 1907.
46. Landsteiner, K., *The Specificity of Serological Reactions*, Boston, Harvard University Press, 1945; reprinted by Dover, New York, 1962.
47. Kabat, E.A., and Mayer, M.M., *Experimental Immunochemistry*, 2nd edn, Springfield, Charles C. Thomas, 1961.
48. Gell, P.G.H., and Hinde, I.T., *Br. J. Exp. Pathol.* **32**:516, 1951; *Int. Arch. Allergy* **5**:23, 1954.
49. Turk, J.L., *Delayed Hypersensitivity*, New York, Wiley, 1967.

50. Waksman, B.H., Suppl. to *Intl Arch. Allergy Appl. Immunol.*, **14**, 1959; also Waksman, B.H., in *Cellular Aspects of Immunity*, Wolstenholme, G.E.W., and O'Connor, M. eds., London, Churchill, 1960, p. 280.
51. Mackaness, G.B., and Blanden, R.V., *Progr. Allergy* **11**:89, 1967.
52. Rowley, D., *Adv. Immunol.* **2**:241, 1962.
53. Landsteiner, K., and Chase, M.W., *Proc. Soc. Exp. Biol. Med.* **49**:688, 1942.
54. Zinsser, H., *J. Exp. Med.* **34**:495, 1921; **41**:159, 1925.
55. Dienes, L., and Schoenheit, E.W., *Am. Rev. Tuberc.* **20**:92, 1929.
56. Jones, T.D., and Mote, J.R., *N. Engl. J. Med.* **210**:120, 1934.
57. Simon, F.A., and Rackemann, F.M., *J. Allergy* **5**:439, 451, 1934.
58. Rich, A.R., and Lewis, M.R., *Bull. Johns Hopkins Hosp.* **50**:115, 1932.
59. Harris, T.N., Grimm, E., Mertens, E., and Ehrich, W.E., *J. Exp. Med.* **81**:73, 1945.
60. Rich, A.R., *The Pathogenesis of Tuberculosis*, 2nd edn, Springfield, Charles C. Thomas, 1951.
61. Wolstenholme, G.E.W., and O'Connor, M., eds., *Cellular Aspects of Immunity*, Ciba Foundation Symposium, London, Churchill, 1960.
62. Kuhn, T.S., *The Structure of Scientific Revolutions*, 2nd edn, Chicago, University of Chicago Press, 1970.
63. Tauber, A.I., and Chernyak, L., *Metchnikoff and the Origins of Immunology: From Metaphor to Theory*, New York, Oxford University Press, 1991.

3 Theories of antibody formation

*...a cis-immunologist will sometimes speak to a trans-immunologist; but the
latter rarely answers.*

Niels Jerne

The discovery of humoral antitoxic antibodies in the early 1890s[1] exerted
a profound influence upon the future development of both immunologic practice
and immunologic thought. On the practical level, the demonstration of the
presence of specific agents in the serum of immunized animals opened the way for
the prevention or cure of infectious diseases by passive-transfer serum therapy.
Yet another direct consequence of this discovery was the development of such
serologic tests as agglutination, the precipitin reaction, and complement fixation,
all of which contributed to a veritable revolution in infectious disease diagnosis
during the following two decades. On the theoretical level, the discovery of
circulating antibody provided a new and almost impregnable rallying point for
those who argued that humoral factors rather than cellular mechanisms were all-
important in explaining natural and acquired immunity. This was a battle whose
outcome would direct the course of immunology for several generations.[2]
However, the discovery of antibody opened another theoretical doorway which
would entrance immunologists for the next eighty years or so. Where and how
were these antibodies formed within the immunized host, and how did they
acquire the exquisite specificity so characteristic of the immune response?

Throughout the long and meandering history of this particular set of immunologic ideas, several curious phenomena appear that deserve the attention of
both the historian and the philosopher of science. These are probably not unique
to immunology, but may rather provide more general hints about how science
and scientists operate.

1. *Cis- and trans-immunology-problems in scientific communication.* Niels Jerne has
   pointed out[3] that two competing schools of thought, each reflecting a different type of
   training and indeed a different world-view, long dominated immunologic speculation. On the one hand were the *cis*-immunologists (the biologists) who attempted to
   define immunology by working forward from the first interaction of antigen with cell,
   and worried much about the implications of such biological phenomena as the
   booster antibody response, changes in the "quality" of antibody with repeated
   immunization, and the problem of immunologic tolerance. On the other hand were
   the *trans*-immunologists (the chemists), who worked backward from the antibody
   molecule itself, and concerned themselves principally with quantitative relationships,
   the size of the antibody repertoire, and the structural basis of immunologic specificity.
   As this chapter will demonstrate, these two groups might sometimes ask the same
   question, but would invariably weigh the answer using different criteria, based upon
   the different aspects of the immune response which each felt to be critical. Thus, they

A History of Immunology, Second Edition
ISBN: 978-0-12-370586-0

Copyright © 2009, Elsevier Inc.
All rights reserved

would often argue not *with* one another, but *past* one another. Communication gaps such as this appear to have existed throughout science almost from its beginning. Among the more famous examples of this was the attempt to explain the basis of species evolution.[4] The field paleontologists and systematists studied populations and the phenotype, and argued ultimate causes backwards from the existing diversity of species, while the laboratory geneticists studied individuals and the genotype, and argued proximate causes forward from the still-hypothetical gene. For decades, the two schools did not appreciate the importance of one another's work. Again, a similar situation developed over the famous controversy about the age of the Earth.[5] Neglecting the calculations of biblical fundamentalists, one saw the geologists and paleontologists of the late nineteenth century demanding immense spans of time for the gradual attainment of present conditions, while the physicists, led by Lord Kelvin, showed with forceful thermodynamic argument that the Earth must have cooled from its initial high temperature in a far shorter time. It required the discovery of a new phenomenon, radioactivity, to resolve the issue.

2. *The perseveration of ideas.* In immunology, as in other scientific disciplines, we may note with interest how frequently an old concept, thought to have been rendered obsolete by the weight of countervailing data and/or a more satisfactory hypothesis, was revived. It is suggestive that the revival is often advanced without adequate acknowledgment of its predecessors, and almost invariably with a total disregard for the facts which contributed to the demise of that predecessor. Someone has said that good ideas must be rediscovered at least once in each generation. Is the timing similar, then, for bad ideas?

3. *The idea advanced "before its time."* The history of science is replete with instances of the publication of a new important concept, which passed unnoticed at the time. This happens occasionally because the idea may be "hidden away" in an obscure journal (like the genetic work of Gregor Mendel[6]), only to be discovered much later. More often, this occurs because the idea cannot be readily integrated into the governing rules and paradigms of the scientific discipline, and thus passes unnoticed. Only when the time is right will the "new" theory be acclaimed, often with little credit to its predecessors. Thus, in immunology, the antigen-instruction theory of antibody formation is universally credited to Breinl and Haurowitz, Mudd, and Alexander, and was rapidly and widely accepted in the early 1930s, although numerous instruction theories had been advanced previously, as we shall see. Again, Jerne's natural selection theory of antibody formation struck a sympathetic chord in 1955, although similar theories had been advanced at least twice in the preceding sixty years.

It will be the aim of this chapter on the history of theories of antibody formation to call attention to many long-forgotten contributions to its progress. We shall, however, also examine the general scientific contexts in which all these speculations were advanced, and the mind-set of the speculators themselves, hoping to learn something from them about how science itself functions.

Antigen-incorporation theories

Buchner: 1893

The noted German bacteriologist Hans Buchner was the first to confront the new conceptual problem posed by the discovery of antibodies. As early as mid-1893, in an address to the Medical Society of Munich on bacterial toxins and anti-toxins, Buchner offered a simple solution to the problem. He proposed that the antitoxin was formed directly from the toxin itself, hinting at some fairly simple transformation. As he put it:[7]

> *Everything speaks also for the fact that the antitoxin contains a cell plasma substance of the specific bacteria, that accumulates in the immunized animal's body. We saw already earlier that toxalbumin must be considered as a specific product of the bacterial plasma. Toxalbumin and antitoxin should be, by their nature, very closely related, and even substances of the same specific kind, or perhaps they may even be different modifications of one in the same substance. That one of these acts as a poison and the other not, I see therein no contradiction against this assumption, since we know that from almost non-poisonous choline, the very poisonous neurin arises by mere decomposition in water, or that poisonous peptone develops by simple digestion from fibrin derived from the circulatory system. On the other hand, however, a common origin of both substances from the bacterial plasma, the poisonous and the protective, opens directly a sure understanding of the specific nature of this protection.*

In an era when nothing was known about the chemical nature of toxins or antitoxins, and little was known about the chemistry of biological macromol-ecules in general, it is not surprising that Buchner's hypothesis found such ready favor. It appeared to explain the mechanism whereby the antibody was endowed with specificity for the immunizing antigen, and consigned to the antigen rather than to the host the primary role in antibody formation. Any requirement that the host contribute the new product would, as we shall see, raise more questions than it answered.

Even so, objections to Buchner's hypothesis were not long in coming. In the same year, Emile Roux, who had already made notable contributions to the study of toxins and antitoxins, showed with Vaillard that the continuous bleeding of a horse immunized with tetanus toxin did not diminish the antibody titer, even after the equivalent of its entire original blood volume had been removed.[8] How could antibody formation continue, without fresh supplies of antigen, if Buchner's view were valid? An even more telling blow against the antigen transformation (or incorporation) theory came with the work of Knorr, who showed that the injection of one unit of tetanus toxin into a horse might result in the production of as much as 100,000 units of circulating antitoxin.[9] The numbers appeared to argue too strongly against the theory and, despite a somewhat belated support for the theory provided by Metchnikoff in 1900,[10] Buchner's concept appeared already to have succumbed in the face of such strongly contradictory evidence.

Hertzfeld and Klinger: 1918

In a lengthy review entitled *The Reactions of Immunity*,[11] Hertzfeld and Klinger employed most of its forty-four pages to muster support for their own theory of antibody formation. With scarcely a nod to Buchner, and with no mention of the data that had doomed his proposal several decades earlier, they refer repeatedly to "our theory," but explain it in words similar to those that Buchner had employed:

> We shall explain, in what follows, that the essential in all immunization events depends upon the antigen being split up to a certain degree in the organism, from whose origin composite, yet still specific breakdown products are absorbed on the surface of appropriate colloidal proteins, and in this form represent "antibody."

And somewhat later, they go on to say:

> In order to make this type of specificity possible, the split products of the antigen still possess a characteristic chemical composition of their own, for were this not the case, then it would be impossible to understand why the antibody in question should react precisely only with this antigen, and not with a large number of others.

It is interesting that this elaborately propounded theory, published as it was in so widely read a journal, should receive so little attention either then or later. It was scarcely mentioned at all, except somewhat obscurely, and when the antigen incorporation concept was revived a decade later, any credit that was given was to Buchner, and not to Hertzfeld and Klinger.

Manwaring; Ramon; Locke, Main, and Hirsch: 1926–1930

In his presidential address to the American Association of Immunologists in 1926, W. H. Manwaring complained that Paul Ehrlich's immunology "constitutes today our most serious handicap to immunological progress, both in theoretical and in practical lines."[12] He insisted that the field was in desperate need of a new and consistent theory "to unravel the mystery of the origin and nature of antibodies." He did not have long to wait, for at the very same meeting a paper was presented by Locke, Main, and Hirsch,[13] proposing that specific antibody was nothing more than a derivative of antigen:

> It is postulated that antibodies are composed of clusters – of relatively large dimension – in which an elementary, naturally occurring, protein substance is absorbed in preponderating amount on nuclei of a binding substance derived from the injected antigen, and that they owe their individual properties to the proportion and character of this binding substance.

Three years later a similar theory was advanced, without reference to the others, by Gustave Ramon,[14] who was to contribute so much to the immunology of diphtheria. According to Ramon,

> *...antitoxin and antibodies in general will find their origin in the formation of humoral complexes constituted of materials originating in the organism and of elements derived from the specific antigen, this by a physicochemical process in the case of antitoxins and, more simply, a physical process for the other antibodies.*

In each case, the earlier rise and the reason for the fall of Buchner's original suggestion were either entirely neglected, or else given only slight attention.

While the others supported the antigen-incorporation theory of antibody formation with data, Manwaring propounded it only with fervor. After a brief flirtation with an instruction theory involving enzymes,[15] antigen-incorporation became a crusade in his hands. In flowery language not often matched in scientific journals, he celebrated the "Renaissance of Pre-Ehrlich Immunology,"[16] implying that Buchnerian immunology had been revived. He attributed the current parlous state of immunology to outdated physiologic concepts, and said:

> *Theoretical immunologists were soon convinced that there must be something radically wrong in their logic, but few of them dreamed that the error was not theirs, but in the basic mid-Victorian religiophysiology in which they placed such implicit faith.*

He saw salvation in a new immunology, based on the notion "that specific antibodies might not be hereditary specific antidotes, but might be retained, modified alien entities or partially dehumanized human proteins–hybridization products between toxic or infectious agents and host tissues."

If Buchner's original hypothesis had stimulated much experimental work to disprove it, then this reappearance of the same idea stimulated even more. The new wave of investigators seemed to be unaware of the earlier, similar studies. Now Heidelberger and Kendall[17] and Topley[18] showed, as had Knorr thirty-two years earlier, that the amount of antibody formed in the immunized animal was far greater than the amount of antigen utilized. Indeed, Hooker and Boyd pointed out, using Topley's data, that a single molecule of antigen might induce enough antibody to agglutinate 600 bacteria.[19] The same authors showed that anti-arsanilic acid antibody contained no arsenic,[20] and Berger and Erlenmeyer showed similarly the absence of arsenic in antibody against the atoxyl hapten.[21] Once again, the weight of all of this evidence was overwhelming that antigen could not possibly be incorporated in whole or in part into the antibody molecule. The theory was laid to rest once again, this time apparently for good.

The first selection theory

Paul Ehrlich: 1897

In addition to his medical studies, Paul Ehrlich spent time in the laboratories of the famous organic chemist and enzymologist Emil Fischer. He brought from this experience a lifelong interest in the relationship between chemical structure and biological function.[22] This was reflected in all of his subsequent work as one of the founders of modern immunology, and of chemical pharmacology as well.[23] Ehrlich's debt to structural chemistry is perhaps nowhere better illustrated among his immunologic publications than in his famous paper of 1897, describing how diphtheria toxin and antitoxin interact, and the method of their measurement.[24] Not only did he postulate that immunologic specificity is due to a unique stereochemical relationship between the active sites on antigen and antibody, but he also introduced the concepts of affinity and of functional domains on the antibody molecule. This work provided the taproot from which the field of immunochemistry later grew; Ehrlich would have been famous for this contribution alone; however, he also appended to this study a theory of the basis for antibody formation, which assured the report a unique position in the history of immunology.

Like Elie Metchnikoff's earlier phagocytic theory of immunity,[25] Paul Ehrlich's theory of antibody formation was based upon a Darwinian evolution of the process of intracellular digestion. He pointed out that many different types of nutrients were utilized, apparently specifically, in the metabolism of the cell, and suggested that these could interact and be absorbed by the cell only if structurally-specific receptors exist on the cell surface with which the nutrient molecules can react chemically. As Ehrlich put it, "The reactions of immunity, after all, represent only a repetition of the processes of normal metabolism, and their apparently wonderful adjustment to new conditions is only another phase of *uralte protoplasma Weisheit* [the ancient wisdom of the protoplasm]." Since certain toxins have a greater affinity for one organ than another, Ehrlich suggested that specific receptors for these toxin molecules also exist on the surface of certain cells. Like a nutrient, the toxin would bind to its specific receptor and thus be assimilated, following which the receptor would either be freed for renewed function or else be regenerated by the cell. When, however, large amounts or repeated doses of toxin were administered, then the cell would overcompensate for the loss of these side-chain receptors, producing so many that some would be released into the blood. Since they possessed complementary sites specific for the given antigen, these side-chains would now function as circulating antibody. In this formulation Ehrlich followed the lead of his cousin, the pathologist Carl Weigert, who had formulated a "law of overcompensation" to explain a variety of phenomena observed in general pathology.[26]

Ehrlich's side-chain theory contained all of the necessary elements to qualify as a true natural selection concept. Antibodies were *natural* constituents of the cell surface, formed within the cell. They possessed from the start the structural configuration that determined their specificity for a given antigen. The purpose of antigen was to *select*, from among all of the side-chains available, only those able to interact specifically, and the cell was then caused to produce more of these specific molecules for export into the blood, requiring only the triggering effect of antigen. In 1897, as we have seen, Ehrlich's ideas were not yet bothered by the problem later posed by an overly large repertoire of antibodies.

Further, his suggestion that the specific side-chain represented the shedding of a portion of some giant protoplasmic molecule was probably all that was permitted by the current state of knowledge of cells and macromolecules. But his prediction that the basis of immunologic specificity resides in a unique three-dimensional configuration of the antibody molecule would later be verified; and his inspired suggestion that antibody formation is the cellular response to the interaction of antigen with cell-surface receptors would not be improved upon for over sixty years.

Instruction theories

Despite the immediate and widespread success of Ehrlich's side-chain theory, some workers, following the lead of Jules Bordet, paid little attention to its explanation of *how antibodies are formed*. They attacked it, rather, because they considered it a too-complicated and erroneous explanation of *how antibodies function*. As Bordet said,[27]

> Ehrlich's theory has exerted a quite grievous influence, in engendering a series of
> artificial conceptions…relating notably to the constitution and classification of
> antibodies, to the mechanism of fixation of complement, etc. By the abuse which
> it has made of graphic representations which translate the outer aspect of the
> phenomena without penetrating to their inner meaning, it has spread the
> acceptance of facile and premature interpretations.

Other workers, however, voiced objections to the very basis of the theory. In 1897, when the theory was formulated, only a limited number of antibodies was recognized, specific for a variety of pathogenic organisms and for a group of plant toxins. Thus, the known antibody repertoire was quite limited and defensive antibody receptors on cells seemed a likely, if teleologic, Darwinian explanation for their function. The picture changed completely within only a few years, with the demonstration of antibody formation against isologous and heterologous erythrocytes, against spermatozoa and other cellular constituents, and against a wide variety of bland proteins.[28] These findings brought into question implicitly the *need* for such receptors, and explicitly raised the doubt that the Ehrlich theory was tenable, in view of the growing size of the antibody repertoire. If, however, the information for so large a repertoire could not possibly arise from within the host, then it surely must be carried in from the outside. What else was there but antigen? As early as 1905, Karl Landsteiner (an avowed opponent of Ehrlich's ideas) could say with M. Reich "that the activity of cells producing normal serum components…is altered following the stimulus of immunization, and so form differently constituted products."[29] This was the first time that anyone had suggested that antibody was a completely new substance.

Direct template theories

Bail and coworkers: 1909–1914

Oskar Bail spent most of his career at the German University in Prague, where he became Chairman of the Department of Hygiene and Bacteriology. In a series of papers published before World War I, he advanced an instruction theory of antibody formation that would be little improved upon over the next forty years. This historically important contribution passed almost completely unnoticed by later theorists.

Bail and Tsuda[30] were troubled, as were others, by the great number of different side-chains demanded by Ehrlich's theory. They suggested that antigen is not destroyed after its interaction with specific antibody, but may also release the latter and continue its function, which is to bind "natural antibodies" from normal blood, leaving the impress of its specificity upon the latter molecules. Working primarily with cholera vibrios, they pointed out that

> The reaction product between cholera substance and serum is able itself to function further as antigen, so that the quantitative relationship between the amount of immunizing antigen employed and the amount of antibody finally contained, will also be better understood...which of necessity must lead to a new view of antibody formation.

Bail and Tsuda were the first of many believers in the antigen-template theory of antibody formation to draw the obvious conclusion – that the process should work also *in vitro*, under appropriate conditions, and thus that antibody might be synthesized outside of the body.[31] Indeed, they claimed to have synthesized anti-cholera antibodies *in vitro*. As they summarized, "the principal result, the obtaining of a solution specifically active for cholera from a nonspecific normal serum with the help of cholera vibrios, is secure."

The theory was further supported and extended by Bail and Rotky in 1913,[32] and by Bail alone in 1914.[33] They held that immunizing antigen persists in the body, interacts with and impresses its specificity on normal human substances, and then gives these up into the circulation to continue its action, further enhancing the titer of specific antibody. In this way they accounted for the disparity in the amount of antibody produced by small amounts of antigen.

The complement theory of Thiele and Embleton: 1914

Thiele and Embleton were primarily interested in immune hemolysis and hemolytic antibody, and in how antibody participates with complement in the destruction of erythrocytes. In a paper on "The Evolution of Antibody"[34] they offered an instruction theory, suggesting that hemolytic antibody derives from complement by a series of "differentiation steps," under the influence of antigen. This contribution, quite out of the mainstream of immunologic thought, received little attention, and is included here only to illustrate the widely ranging formulations of early speculators.

Ostromuislensky: 1915

In the midst of World War I, there appeared two remarkable papers in the (obscure to Western immunologists) *Journal of the Russian Physicochemical Society.*[35] These publications not only reported the *in vitro* production of specific antitoxins, but also based this approach upon a theory of antibody formation that claimed that immunologic specificity was not due to a particular chemical constitution of the molecule, but to a special physical state of the colloidal antitoxin molecules which distinguishes antibodies from ordinary globulins. This special state is impressed upon an "ordinary" globulin by contact with the antigen molecule, which could then split off and repeat its function. Although he showed a wide familiarity with the Western literature, Ostromuislensky seems to have been unaware of the earlier work by Bail and coworkers, and thus we may credit him with an independent contribution to immunologic theory.

Haurowitz and Breinl; Topley; Mudd; Alexander: 1930–1932

By 1930, it was understood that antibodies were globular proteins, and that proteins were somehow built up of random arrangements of twenty-odd different amino acids – the so-called "building blocks of life." It was unclear, however, whether there was any regularity or reproducibility in the amino acid sequence of the polypeptide chain, or where and how the information for any particular sequence might be stored and retrieved. There were thus few restrictions during this period on the direction in which speculation might be carried in seeking an explanation of the basis for immunologic specificity. Even so, the increasing knowledge of the chemistry of proteins demanded that henceforth any theory of antibody formation involve the basic mechanism of protein formation.

Based upon the then reasonable assumption that the information for the universe of different antigenic determinants could not possibly be incorporated in the vertebrate genome, a wave of new instruction theories was proposed, since logic appeared to demand that each antigen must carry with it the information for its own immunologic specificity. Instruction, however, must now be on the level of protein synthesis, and for the first time the notion of antigen-as-template became explicit.

The new theory was advanced almost simultaneously by Topley[36] and by Breinl and Haurowitz,[37] and independently also by Mudd[38] and by Alexander.[39] It took its most definitive form in the hands of Breinl and Haurowitz, who proposed that an antigen would be carried in the body to the site of protein formation, where it would serve as a template upon which the nascent antibody molecule might be constructed. Since the antibody molecule was to be synthesized upon the surface of the antigen, it seemed reasonable to propose a mechanism whereby the stereochemical structure of the antigenic site would determine a unique amino acid sequence on the antibody, thus accounting for the complementary and specific fit of antibody for antigen. Implicit in this theory, of course, was the requirement that the antigen persist throughout the course of antibody formation.

This instruction template concept of antibody formation found broad acceptance in the chemically-oriented immunology of the day, since it appeared to dispose of several conceptual dilemmas that had worried earlier immunologists. First, the new theory answered the objection of those who claimed that the body could not possibly have accumulated in evolution the information required to produce antibodies against the thousands of synthetic haptenic determinants which Landsteiner and others had shown to be immunogenic (i.e., it solved the antibody repertoire size problem). Secondly, the template theory accounted well for the repeated observation that many thousands of molecules of antibody might appear in the blood for each molecule of antigen injected. Finally, of course, it accounted in structural terms for immunologic specificity. But one of the major shortcomings of the new theory, not even mentioned by its proponents, was its inability to explain why a second exposure to antigen should result in a much enhanced and more rapid booster antibody response.

It is worthy of note that none of these authors made any mention of the earlier instruction theories of antibody formation. This is especially interesting in the case of Breinl and Haurowitz, since they were then working in the department in Prague of the very same Oskar Bail who twenty years earlier had advanced an instruction theory of antibody formation substantially equivalent to their own.[40]

Pauling: 1940

The theory of interatomic and intermolecular forces which gained Linus Pauling his first Nobel Prize had important implications for an understanding of the specificity of many biological interactions. It was Pauling himself who applied these concepts to the antigen–antibody interaction, an interest which was stimulated by earlier conversations with Karl Landsteiner.[41] Pauling and his students, most notably David Pressman, showed formally that the specificity of the antibody–hapten interaction was due to the interaction of complementary three-dimensional configurations of atoms, as Paul Ehrlich had so long ago suggested. Their binding energy could also be well explained by a combination of ionic, hydrogen-bonding, and van der Waal's interactions.

Always on the lookout for the important scientific challenges of the day, the imaginative Pauling speculated on how the antibody protein molecule could possibly acquire and maintain the unique three-dimensional configuration that would endow it with specificity for a given antigen.[42] The answer was typical of the Pauling approach; antibody specificity must be due to the unique tertiary structure of a given antibody molecule, achieved through a unique folding of its peptide chain. (In 1940, the individuality of the primary amino acid sequence of proteins was unknown, as was the influence of this sequence on tertiary structure.) Since he did not favor a mysterious process whereby antigen instructed the amino acid sequence of the antibody molecule, as postulated by Breinl and Haurowitz, Pauling presented a simpler and more chemically justifiable theory. Antigen would serve as the template for the final step of protein formation, in which the coiling of the nascent polypeptide chain of the antibody molecule would conform more or less precisely to the template offered by the surface determinant of the antigen molecule.[43] Once the appropriate configuration had been attained, it would be stabilized by familiar interatomic bonds, and thus satisfy all of the requirements of specific antibody. Pauling's picture of this process is illustrated in Figure 3.1.

A further elaboration of Pauling's concept was provided by Karush,[44] who pointed out a critical defect in the Breinl–Haurowitz formulation. Any template which determines primary amino acid sequence must be *linear*, and thus cannot also "provide directly information for the development of noncovalent [tertiary] structure." The antigenic template must therefore act on the preformed chain, to permit it to fold uniquely into the antigen-specific complementary region required. Karush also proposed that this unique folding is stabilized thenceforth by multiple disulfide bridge cross-linkages, and that antibody heterogeneity is determined by the extent of such cross-linking.

Like other chemically-oriented instructionists before him, Pauling's chief concern was to explain specificity and repertoire size. The former area was his

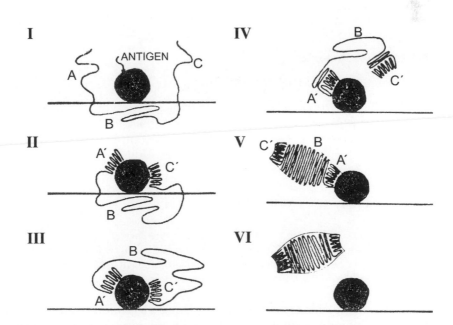

Figure 3.1 Pauling's direct template scheme of antibody formation.
From Pauling, L., *J. Am. Chem. Soc.* **62**:2643, 1940; *Science* **92**:77, 1940.

forte, and he could say of the latter: "The number of configurations accessible to the polypeptide chain is so great as to provide an explanation of the ability of an animal to form antibodies with considerable specificity for an apparently unlimited number of different antigens." He, like the others, did not demand of his theory an explanation of the more biological phenomenology of the antibody response.

Although Pauling's template theory of antibody production appeared to accord more than its predecessor with contemporary scientific theories, in fact all of these instruction theories shared much the same advantages and disadvantages. Indeed, Pauling's theory gained an additional defect which was pointed up by newer data. While it had not yet been firmly demonstrated that the antibody molecule was multivalent, the very fact of the antigen–antibody precipitin reaction and the form of the precipitin curve led Marrack to suggest in 1934 that antibodies were at least divalent and possibly multivalent, so that an antigen–antibody precipitin lattice could be established.[45] Further, the repeated demonstration by Landsteiner that multiply-substituted proteins could engage in precipitation with anti-hapten antibody suggested strongly that the antibody must not only be multivalent, but must also have each of its active sites specific for the same grouping. In contrast, Pauling's theory implied that the specific sites on the antibody molecule were formed and stabilized at different areas on the surface of the antigen, thus implying that they ought in general to be heteroligating (i.e., to exhibit a different specificity at each of their reactive sites).

Indirect template theories

The adaptive enzyme theory of Burnet: 1941

F. Macfarlane Burnet brought to his interest in immunology a broad background in virology and experimental pathology. However, of overriding importance for his future immunologic theories, he was, like Bordet, an unabashed biologist who had little use for the purely molecular concepts of his chemically-oriented predecessors. Thus, he faulted the template theory of Breinl and Haurowitz in his 1941 book on *The Production of Antibodies*.[46] In words that Jules Bordet might have used forty years earlier, Burnet could say that

> *in the circumstances, it would seem preferable to couch any general interpretation of the phenomenon of antibody production in biological terms which can be related to general conceptions in other biological fields, rather than to conceal ignorance by a pseudochemical formulation.*

Burnet treated the Pauling template theory somewhat more charitably, as providing a more impressive physical picture of the antibody molecule, but questioned also the biological basis and biological implications of the Pauling theory.

Like his predecessors, Burnet acknowledged that the information for antibody specificity must be carried by the antigen molecule. His criticisms of the earlier template theories, significantly, did not question the *chemical* basis for the specificity of the antibody combining site, but rather the *biological* basis for the production of the entire antibody molecule. First, claimed Burnet, direct template theories paid no attention to the modern knowledge of the importance of enzymes in the mechanisms of intracellular metabolism and synthesis. He pointed out, secondly, that these theories demanded the long-term persistence of antigen throughout the course of antibody formation – an event that Burnet claimed had not only not been formally demonstrated, but was even probably untrue. And finally and most significantly, Burnet the biologist made his most profound contribution in claiming that "antibody production is a function not only of the cells originally stimulated, *but of their descendants* [my italics]." Here was the key to the problem, which he would utilize so effectively some eighteen years later.

Burnet's instructionist theory of antibody formation was very much in line with contemporary biological thought. All proteins (including antibodies) are both broken down and synthesized by special proteinase enzymes. However, in addition to the normal complement of enzymes within a cell, recent work on bacteria had suggested that, under special circumstances, "adaptive" enzymes might appear in response to special modifications or requirements of the bacterial organism.[47] From this point of departure, Burnet postulated that once introduced into the body, antigen would find its way into the cells of the reticuloendothelial system, where contact with local proteinases would result in adaptive modification of the enzymes during the dissolution of the antigen molecule. These newly adapted enzymes would then be able to synthesize a globulin molecule specific for the antigen in question. Moreover, these adaptive enzymes would not only replicate within the antibody-forming cell itself, but the information for antibody specificity which they carried would be perpetuated also within any daughter cells that might result from proliferative activity. Such an expanded population of specifically adapted antibody-forming cells (later to be called a clone) would account well for the heightened secondary or booster antibody response upon subsequent re-exposure to antigen.

In a delightful extension of his theory, Burnet advanced a plausible explanation for the recent observation that when booster injections of antigen are administered, not only is the quantity of antibody increased but so also is its quality (i.e., its affinity for antigen). Adaptive enzymes, explained Burnet, are not unalterable structures, so that over time the adaptive enzymes for any given antibody will slowly deteriorate, thus producing lower-grade specific antibody and ultimately nonspecific "normal" globulin. However, further contact with the same antigen will intensify and make more perfect the adaptation of the enzyme to specific antibody formation, thus resulting in the production of an increasingly higher-grade antibody with each booster immunization. Burnet pointed out that it was precisely the ability of his theory to explain the qualitative changes in antibody which accompany prolonged immunization that provided its chief advantage over the earlier template theories.

The indirect template theory of Burnet and Fenner: 1949

The notion that enzymes might be adaptively modified, so popular early in the decade, had begun to lose favor near its close. The formation of protein was now held to be under the master control of an information-laden "genome" of uncertain composition. In a second edition of his book on antibody formation, written now with Frank Fenner, Burnet advanced a new indirect template hypothesis.[48] Each antigen, according to this theory, is able to impress the information for its specific determinant upon the (?RNA) genome, against which indirect template antibody specificity might be endowed during protein formation. The new theory continued to stress Burnet's concern with the importance of cellular dynamics in the immune response, since the new genomic copy would not only persist within the cell, but would be reproduced from mother to daughter cells during proliferation. It was these two factors, according to Burnet, that explained the persistence of antibody formation and the accentuated booster response. The genocopy might also deteriorate with time, or be sharpened by re-exposure to antigen, thus explaining later changes in the quality of antibody.

Medawar's early work[49] demonstrating that tissue homografts are routinely rejected while autografts are not, refocused attention on the ability of the immunologic apparatus to distinguish between one's own tissues and those of others. When Ray Owen then showed that non-identical cattle twins whose circulatory systems were connected *in utero* become antigenic mosaics, unable to respond immunologically to one-another's antigens,[50] it became apparent that the distinction between "self" and "other" was a learned rather than a genetically programmed phenomenon. Burnet and Fenner were the first to recognize the importance of these observations, and the first to insist that an adequate theory of antibody formation must encompass this important biological fact, later to be called "immunological tolerance." (Burnet would later share the 1960 Nobel Prize for this prescience.) They therefore suggested that body components acquire "self-markers" at some point in ontogeny, whose presence would thenceforth deflect self-components from participation in the immune process.

Here was the beginning of Burnet's longstanding preoccupation with the dichotomy of "self/nonself,"[51] a formulation that would become a central and almost defining metaphor for many immunologists. We shall examine the great influence of this metaphor in a later chapter.

The immunocatalysis theory of Sevag: 1951

We include here, in the interest of completeness, an instruction theory advanced by M. G. Sevag in his book *Immunocatalysis*.[52] Sevag believed that the chemical process of catalysis is important in many aspects of immunology, perhaps nowhere more than in the formation of specific antibodies. He suggested that "the specificity of an antibody molecule is the consequence of specific cellular synthetic processes catalytically modified by an antigen to conform with the configuration of certain active groups of the antigen molecule." Thus, in the production of "normal globulins," the catalytic effect of antigen is to change the configurational pattern of the protein, due to the special structure of the antigenic determinant. There appears to be a certain illogic in this theory. Sevag seems to overextend the accepted definition of catalysis, since catalysts do not make new reactions possible but merely accelerate pre-existing ones.

The template-inducer theory of Schweet and Owen: 1957

It was clear by 1957 that the repository of genetic information lay in DNA. As Schweet and Owen pointed out, information for protein synthesis could no longer be attributed to the *direct* function of other proteins (antigens). They proposed, therefore, a two-phase mechanism of antibody formation.[53] Antigen would first so modify the DNA of the globulin gene as to furnish somatically heritable information for the formation of a new RNA template, producing cells "primed" for specific antibody formation. Antigen would then act further on such cells as an inducer, stimulating the formation of many templates, and the exuberant production of antibody. This "biochemical model" of antibody formation not only seemed to accord better with contemporary knowledge, but also appeared to furnish a plausible explanation of the difference between primary and booster immunization.

The implications of instruction theories

It may be worth pausing for a moment to examine the broader biological implications of the theories that we have examined thus far. When Paul Ehrlich suggested in 1897 that antibodies were natural cell products, the only immunologic responses then recognized were to pathogenic organisms and toxic substances. It was thus not unreasonable to suppose that a Darwinian selection pressure had endowed the vertebrate host with the antibody specificities apparently so necessary to its survival. With the expansion of the antibody repertoire to almost unmanageable proportions, including a long list of the unnatural products of the synthetic organic chemist's imagination, a Darwinian explanation no longer seemed possible. In this situation, an instruction theory involving a direct template would be evolutionarily neutral, and thus appeared more acceptable. But the indirect template theories, involving the transmission of antigen-induced information from mother to daughter somatic cells, introduced a slightly Lamarckian flavor to the proceedings. The selection theories to which we now turn appeared to restore Darwin to favor among immunologists,[54] but simultaneously posed some of the most interesting and complicated evolutionary problems of all, as we shall point out in Chapter 21.

Selection theories

The influence of World War II on the tempo of scientific discovery cannot be overestimated; in its aftermath new information and new techniques commanded the attention of all biologists. From outside of immunology, perhaps the most significant advances concerned the structure and function of genes. From within immunology, the discovery of allograft rejection, of immunologic tolerance,[55] and of immunologic deficiency diseases[56] heralded a major shift in emphasis from immunochemistry to immunobiology, in which selectionist theories of antibody formation would find a more suitable environment. Within the context of the new biology, a plausible theory of antibody formation would now have to address these more biological aspects of the immune response. But even before the biologists led the return to selection theories, one such was advanced by the physicist Pascual Jordan.

The quantum-mechanical resonance theory of Jordan: 1940

Jordan attempted to apply quantum-mechanical arguments to biological systems, most notably to an explanation of the perplexing problem of the reproduction of biologically specific molecules such as enzymes and antibodies. His theory of antibody formation[57] was, in fact, the first of the post-Ehrlich natural selection theories, but has largely been forgotten since. It is presented here in part for completeness, but also to contrast its reception and influence with that of the natural selection theory of Jerne, with which it shared many important features.

Jordan held that injected antigen is first subjected to partial digestion within the host, after which its split products might combine preferentially with certain naturally occurring molecules, the antibodies. This antigen–antibody complex would then be capable, in special tissues in which the milieu was appropriate, of inducing an autocatalytic reproduction of the antibody moiety. Jordan suggested that quantum-mechanical resonance phenomena would lead to an attraction between molecules containing identical groups, and thus to self-reproduction of the antibody molecule. According to this concept, antigen would *select* from a pool of *naturally* occurring antibodies those with which it could specifically interact, and serve as a suitable carrier for the antibody during its autocatalytic phase of reproduction. Jordan even accounted for the Landsteiner observation of graded cross-reactions by suggesting that in many cases the reproductive process might result in daughter molecules whose structure, and therefore specificity, might differ slightly from that of the mother molecule.

Jordan's concept was substantially identical to Ehrlich's, with the substitution of a more "modern" mechanism for the reproduction of the specific antibody molecules. It was even more closely the equivalent of Jerne's natural selection theory, to be discussed below, but failed completely to attract the attention of biologists. It did, however, come to the attention of Linus Pauling, then promoting his own theory of antibody formation, who lost no time in attacking the Jordan formulation. The nature and role of intermolecular forces was Pauling's special domain, and he was quick to point out,[58] in Jordan's own quantum-mechanical notation, that resonance attractions were less likely between identical molecules than between complementary molecules, as Pauling's own theory had suggested. It is of some interest that although Pauling's attack was limited to Jordan's proposed mechanism for the reduplication of antibody molecules, it served also to eclipse the natural selection aspect of the argument.

The natural selection theory of Jerne: 1955

One of the early observations that led to the concept of a humorally-mediated immunity was the existence in "normal" blood and serum of specific antibacterial substances whose presence could not be accounted for by any known prior exposure to the antigens with which they reacted. These were termed *natural antibodies*, in contradistinction to those acquired after infection or immunization. So long as Ehrlich's side-chain theory of antibody formation was accepted, the presence of spontaneously produced antibodies excited no particular conceptual concern among immunologists. However, with the fall into desuetude of Ehrlich's theory, and the rise of instructionist theories of antibody formation, natural antibodies could no longer be accounted for, and interest in them waned. With the post-World War II burst of activity in all fields of biology, attention was once again directed to the nature and significance of natural antibodies, thanks in no small measure to a group at the Danish State Serum Institute in Copenhagen, to which Niels Jerne belonged.

In his landmark paper in 1955,[59] Jerne revived the old Ehrlich concept that antibodies of all possible specificities were normally formed by the vertebrate host and delivered in small amounts to the blood. Any antigen that chanced to enter the circulation would then react with those antibodies present that were specific for the antigenic determinants. Once the antigen–antibody interaction was complete, the role of antigen assigned by Jerne was to carry the antibody to specialized cells capable of reproducing this antibody. When the antigen had fulfilled its task as "selective carrier"[60] of antibody, it had no further role to play, and the internal mechanisms of the antibody-producing cell would then respond somehow to the signal provided by the selected globulins, initiating the synthesis of molecules identical to those introduced – i.e., of specific antibody. In view of the then-recent demonstration of the importance of ribosomal RNA for the assembly of protein molecules, Jerne suggested that the antibody prototype might initiate the synthesis of a specific RNA, or even modify the structure of a pre-existing RNA, upon which further specific antibody molecules might be synthesized.

Jerne's theory appeared, for the moment, to explain satisfactorily most of the biological phenomena associated with the immune response. The heightened booster response was attributable to the presence of increasing amounts of circulating antibody, thus providing a more efficient stimulus to a greater number of antibody-forming cells than was possible during primary immunization. Similarly, the presence of larger amounts of circulating antibody during the booster response would favor the binding to antigen of higher-affinity antibodies, thus accounting for the increase in the quality of antibodies with repeated immunization. Finally, immunologic tolerance was accounted for by postulating that the first natural antibodies produced against self-antigens during embryogenesis would immediately be absorbed by the tissues of the body, and thus would be unavailable to serve as stimuli for subsequent autoantibody formation. The demand of an immense number of pre-existing antibodies, earlier viewed as the chief objection to Ehrlich's theory, was not even mentioned as a possible drawback to the natural selection theory.

It is curious that although Jerne's theory of antibody formation appears to be the logical equivalent of both Ehrlich's side-chain theory (as Talmage[61] was quick to point out) and Jordan's resonance theory, he referred to neither of them in his paper. All three theories held antibodies to be naturally occurring substances which are selected for by antigens on the basis of their ability to interact specifically. Whereas in the Ehrlich theory this interaction was assumed to occur on the cell surface, signaling antibody formation, in both the Jerne and Jordan formulations the interaction was thought to occur in the blood. In all three theories, the actual antibody formation would then proceed in some sort of intracellular black box, the speculation about mechanisms being governed by the state of knowledge of the day. While future developments would show that the Ehrlich theory was closest to the truth, yet the times were apparently so ripe for this type of biological formulation that it was Jerne's theory which had a seminal influence on further immunologic speculation.

It was also pointed out by Talmage that the natural selection theory came close to sharing a critical defect with the earlier instruction theories. By the mid-1950s, it was becoming increasingly evident that the information governing the structure of proteins could flow in only one direction. This was formalized by Francis Crick as the "central dogma" of genetics, which held that information on protein structure flowed from DNA to RNA to protein and, once in the protein, could not escape.[62] Thus, neither antigen nor antibody could carry with it into the cell the information to program the production of specific antibody; it could at best only provide a signal for a pre-existing program – a concept that Ehrlich had originally advanced and to which Burnet and others now returned.

The clonal selection theory of Burnet, Talmage, and Lederberg: 1957–1959

The clonal selection theory of antibody formation was first advanced in somewhat vague and general terms by Talmage and Burnet,[63] and was then fleshed out in much more specific terms by Burnet,[64] Talmage,[65] and Lederberg.[66] The conceptual torch had now clearly passed from the chemists to the biologists.

The opening chapters of Burnet's book *The Clonal Selection Theory of Acquired Immunity* point up well the difference in approach between the chemically-oriented immunologists who had dominated the field prior to the

1950s, and the biologically oriented immunologists who came to the fore during the 1960s. In considering mechanisms of antibody formation, the former group had always placed great emphasis upon two principal characteristics of antibodies: the stereochemical requirement for immunologic specificity, and the almost incomprehensibly great size of the antibody specificity repertoire. Burnet, on the other hand, scarcely mentioned these factors. Rather, he placed great emphasis upon such questions as the difference between primary and secondary responses; the phenomena of immunologic tolerance and congenital agamma-globulinemia; and the population dynamics of differentiating cells. These factors, he felt, provide the key to the solution of the mechanism of antibody formation, and these factors also must be addressed and adequately explained by any suitable theory. The examples which he used to illustrate and support his theory came not from structural organic chemistry or chemical physics, but instead from bacterial genetics, from influenza and myxomatosis virus infections, and from general pathology.

Burnet acknowledged the power of Jerne's suggestion of the existence of pre-existing antibodies as the targets of antigen selection, but found fault with the subsequent steps leading to antibody synthesis. He felt that cells should somehow be more intimately involved in the process – not only single cells but *clones* of cells all devoted to the same function, just as one was accustomed to see in any specialized organ of the body or in tumor formation. Burnet therefore suggested, as Ehrlich had before him, that the "natural antibody" should more logically be placed on the surface of a lymphoid cell, phenotypically restricted to one or at most a very few types of specific receptor. The interaction of antigen with these receptors would then trigger (by some mechanism unknown) a signal for cellular differentiation to antibody production, as well as a signal for proliferation to form a clone of daughter cells possessing identical receptors and capable of identical immunologic responses. Antigen would thus serve to select and activate specifically the appropriate clonal precursor from a much larger population, thus accounting well for continued antibody formation, for enhanced secondary responses, and for changes in the quality of antibodies. The latter might also benefit from contributions by minor somatic mutations during the course of immunization, yielding closer-fitting antibodies. To explain the usual absence of response to self-antigens and acquired tolerance, Burnet postulated that clonal precursor cells might be especially susceptible to the lethal action of their respective antigens early in ontogeny, leading to the deletion of those clones which might in the future be embarrassing to the host. Should autoimmune disease develop in later life, it might be accounted for either because the antigen in question had been "sequestered" like lens antigen and not available for clonal deletion, or because a somatic mutation might occur, leading to the development of a "forbidden clone."

Both Talmage and Lederberg contributed importantly to the elaboration of the clonal selection theory of antibody formation. In addition to expanding on the role of antigen selection and antigen-induced cell differentiation, Talmage alone paid special attention to the question of specificity and the size of the antibody repertoire. He pointed out the important distinction between an antiserum composed of many different specificities, and the individual anti-bodies which it might contain. Based upon the type of graded cross-reactions so elegantly demonstrated by Landsteiner, Talmage suggested that varying mixtures of a limited number of different antibody specificities may be capable of distinguishing a far greater number of different antigenic determinants, because each combination of cross-reacting antibodies would appear as a distinct specificity. Thus, Talmage made a plausible case for the existence not of hundreds of thousands or millions of different antibody specificities, but rather something of the order of only 5,000 molecular types – not an unreasonable number to have stored in the genome.

Lederberg, on the other hand, specifically addressed some of the genetic implications of the clonal selection theory, contributing to it also the prestige of a Nobel Prize-winning molecular geneticist. He claimed that immunologic specificity is determined by a unique primary amino acid sequence, the information for which is incorporated in a unique sequence of nucleotides in a "gene for globulin synthesis." To account for antibody diversity, Lederberg suggested the existence in precursor cells of a high rate of spontaneous *and random* mutation of the DNA of the immunoglobulin gene. Such somatic mutation, according to Lederberg, might continue throughout life, rather than being restricted to fetal life as Burnet had suggested. This notion about the somatic generation of antibody diversity would serve as the focus of an extensive subsequent debate between germline and somatic theories.[67]

The molecular-genetic theory of Szilard: 1960

During the 1950s, the famous nuclear physicist Leo Szilard developed a strong interest in the genetic basis of protein formation in general, and of antibody formation in particular. He was, for a number of years, a familiar figure on the boardwalk at Atlantic City, immediately outside the annual meeting hall of the American Association of Immunologists. He would "hold court," and skillfully and closely cross-examine a selected list of immunologic witnesses whose experiments he had decided were important to the formulation of his concepts. Those immunologists whose data he could not completely extract at the boardwalk sessions were later invited to dinner at his apartment in Washington, where they would be cross-examined closely and drained dry of useful information. The result of this exercise was a molecular theory of antibody formation, based upon the latest information from the new field of molecular genetics, and which sought to explain the latest phenomenologic observations of the immunologists.[68] As nearly as can be determined, the theory exerted absolutely no influence on the subsequent course of immunologic speculation. It is summarized briefly here not only to complete our review of theories of antibody formation, but also because its attention to detail and its elegance of inner logic cannot fail to excite the interest and even the admiration of the reader.

Szilard postulated that the large variety of enzymes which governed the steps in normal metabolic pathways are encoded by germline genes whose duplication and modification result in the formation of proteins possessing similar specific binding sites, but which lack the catalytic activity of the enzyme. These are the antibodies. Immunization was held by Szilard to involve penetration of the antigenic determinant into the cell, where it would combine specifically with the site on a "coupling" enzyme responsible for the formation of a repressor of gene activity. Precipitation of the coupling enzyme would then reduce the rate of repressor formation, so that the de-repressed antibody gene might engage in the production of large amounts of specific protein. Since the antibody itself could also bind to the repressor, the cell would then lock in, to continue production of the given antibody. The secondary booster antibody response was explained by the existence of an enzyme postulated to inhibit cell division, and which possessed many of the properties of serum complement. The injection of new antigen would then lead to intracellular precipitation of immune complexes in those cells already producing antibody, so that the enzymatic inhibitor of cell proliferation would be bound to the precipitate, freeing the cell for proliferation to yield daughter cells still restricted to the production of the given antibody.

Szilard attempted to explain the development of immunologic tolerance along much the same lines. Now, the presence of excessive amounts of antigen within the cells of the newborn animal might permit it to initiate specific antibody formation, but would prevent the cell from locking in on that production, since excess antigen would then precipitate the antibody formed, and prevent it from further neutralization of repressor molecules.

Szilard's notions of the basis for antibody formation derived from the contemporary view of the importance of enzyme induction and repression that emerged from the study of bacterial systems. But progress in the field of molecular genetics was so rapid that new approaches had almost superseded the old ones by the time that Szilard published his theory, and thus it attracted little attention.

Conclusions

Within a decade of its introduction, the clonal selection theory of antibody formation had won general acceptance for its principal features. It is well to remember, however, that the theory of clonal selection – indeed the *law* of clonal selection – is based upon two principal concepts advanced sixty years apart: Paul Ehrlich's suggestion that the trigger for the immune response is based upon the interaction of antigen with cell membrane antibody receptors, and Macfarlane Burnet's suggestion that the consequences of the triggering event involve the cellular dynamics of differentiation and proliferation.

In following the course of the history of theories of antibody formation, several interesting aspects of the manner in which a science progresses have been well illustrated. Not long after emerging from its purely bacteriological beginnings, immunology suffered a division into two main traditions or schools of thought: those investigators with a chemical orientation who dominated the field for almost half a century, and those with a biological orientation, who only became a strong force in the field in the 1960s. Chemically-oriented immunologists demanded of a theory of antibody formation primarily that it explain the structural basis of immunologic specificity and the overwhelming size of the antibody repertoire, while generally neglecting the more biological aspects of the immune response. Biologically oriented immunologists, on the other hand, paid scant attention to specificity and repertoire size, and required only that an acceptable theory of antibody formation explain adequately such aspects of the immune response as the difference between primary and booster immunization, changes in the quality of antibody, and immunologic tolerance. For decades, *cis-* and *trans-* immunologists spoke different languages, and communicated incompletely to one another their views, priorities, and criteria. Only at the very end did it become apparent that a useful theory would have to satisfy the demands of both schools of thought, and indeed the modern synthesis appears to have done this. It has, at the same time, almost completely blurred the boundaries that earlier divided the two camps, although the *cis* or *trans* orientation of at least the older generation of investigators can still be discerned.

If Burnet's theory of clonal selection and its biological implications carried the day, it should not be thought that it had settled all outstanding conceptual problems. With the acceptance of a genetic basis for the production of antibodies, the specter of a possibly exorbitantly large repertoire once again raised its head. There then began a strenuous debate about whether this repertoire was germline encoded, or the result of a hyperactive mechanism of somatic mutation. Again, because Burnet had so forcefully suggested that his clonal selection theory was dependent on self–nonself discrimination, a number of dissidents came forth to challenge the very basis of the theory itself. These two issues will be the subjects of the next chapters.

Notes and references

1. Behring, E., and Kitasato, S., *Deutsch. med. Wochenschr.* **16**:1113, 1980; Behring, E., and Wernicke, E., *Z. Hyg.* **12**:10, 45, 1892; Ehrlich, P., *Deutsch. med. Wochenschr.* **17**:976, 1218, 1891.
2. The details of this important conflict are discussed in Chapter 2.
3. Jerne, N.K., *Cold Spring Harbor Symp. Quant. Biol.* **32**:591, 1967.
4. See Ernst Mayr's introductory summary of the historical positions of the opposing factions, in Mayr, E., and Provine, W.B., eds, *The Evolutionary Synthesis: Perspectives on the Unification of Biology*, Cambridge, Harvard University Press, 1980, pp. 1–48.
5. Badash, L., *Proc. Am. Phil. Soc.* **112**:157, 1968; Brush, S.C., *The Temperature of the Earth*, New York, Burt Franklyn, 1978, pp. 29–44.
6. There is now reason to believe, however, that prior to its "rediscovery" in 1900 by de Vries, Correns, and Tschermak, Mendel's work had in fact received fairly wide

dissemination; see, for example, Olby, R., *Origins of Mendelism*, 2nd edn, Chicago, University of Chicago Press, 1985.

7. Buchner, H., *Münch. med. Wochenschr.* **40**:449, 480, 482, 1893.
8. Roux, E., and Vaillard, L., *Ann. Inst. Pasteur*, **7**:65, 1893.
9. Knorr, A., *Münch. med. Wochenschr.* **45**:321, 362, 1898.
10. Metchnikoff, E., *Weyl's Handbuch Hyg.* **9**(1):48, 1900.
11. Hertzfeld, E., and Klinger, R., *Biochem. Zeitschr.* **85**:1, 1918.
12. Manwaring, W.H., *J. Immunol.* **12**:177, 1926.
13. Locke, A.L., Main, E.R., and Hirsch, E.F., *Arch. Pathol.* (abstract) **2**:437, 1926.
14. Ramon, G., *C. R. Soc. Biol.* **102**:287, 379, 381, 1929.
15. Manwaring, W.H., *J. Immunol.* **12**:177, 1926; see also Manwaring, W.H., in Jordan, E.D., and Falk, I.S., eds, *The Newer Knowledge of Bacteriology and Immunology*, Chicago, University of Chicago Press, 1928.
16. Manwaring, W.H., *J. Immunol.* **19**:155, 1930; *Science*, **72**:23, 1930.
17. Heidelberger, M., and Kendall, F.E., *Science* **72**:252, 1930.
18. Topley, W.W.C., *J. Pathol. Bacteriol.* **33**:339, 1930.
19. Hooker, S.B., and Boyd, W.C., *J. Immunol.* **21**:113, 1931.
20. Hooker, S.B., and Boyd, W.C., *J. Immunol.* **23**:465, 1932.
21. Berger, E., and Erlenmeyer H., *Z. Hyq. Infektionskr.* **113**:79, 1931.
22. Marquardt, M., *Paul Ehrlich*, New York, Schuman, 1957.
23. The full scope of Ehrlich's preoccupation with structure and function is detailed in Silverstein, A.M., *Paul Ehrlich's Receptor Immunology: The Magnificent Obsession*, San Diego, Academic Press, 2002.
24. Ehrlich, P., "Die Wertbemessung des Diphtherieheilserums," *Klin. Jahrb.* **60**:299, 1897 (English translation in *The Collected Papers of Paul Ehrlich*, Vol. 2, London, Pergamon, 1956, pp. 107–125); the theory was further amplified in Ehrlich's Croonian Lecture to the Royal Society, *Proc. R. Soc. Lond.* **66**:424, 1900.
25. Metchnikoff, E., *Lectures on the Comparative Pathology of Inflammation*, London, Keegan, Paul, Trench, Trübner, 1893. Reprinted by Dover, New York, 1968.
26. Weigert, C., *Verh. Ges. deutsch. Naturforsch. Aerzte* **68**:121, 1896.
27. Bordet, J., *Traité de l'Immunité dans les Maladies Infectieuses*, 2nd edn, Paris, Masson, 1939. In the first edition, Bordet had also called the illustrations "puerile." For a fuller account of the Ehrlich–Bordet debates, see Chapter 6.
28. These studies are well summarized in Karl Landsteiner's book *The Specificity of Serological Reactions*, New York, Dover, 1962.
29. Landsteiner, K., and Reich, M., *Centralbl. f. Bakt.* **39**:712, 1905.
30. Bail, O., and Tsuda, K., *Z. Immunitätsf.* **1**:546, 772, 1909.
31. The list of attempts to synthesize antibody artificially stretches from Bail to Pauling; Manwaring's collection of such reports (note 16), up to 1929, includes ten items. We know of no efforts in this direction after those of Pauling, L., and Campbell, D.H., *Science* **95**:440, 1942; *J. Exp. Med.* **76**:211, 1942. A patent (No. 392055) for the production of artificial diphtheria antitoxin was issued in Germany in the late 1920s!
32. Bail, O., and Rotky, H., *Z. Immunitätsf.* **17**:378, 1913.
33. Bail, O., *Z. Immunitätsf.* **21**:202, 1914.
34. Thiele, F.H., and Embleton, D., *Z. Immunitätsf.* **20**:1, 1914.
35. Ostromuislensky, I.I., *J. Russ. Phys. Chem. Soc.* **47**:263, 1915; Ostromuislensky, I.I., and Petrov, D.I., *ibid*, **47**:301, 1915. An English abstract of these papers appeared in *Chem. Abstr.* **10**:214, 1916.

36. Topley, W.W.C., *J. Pathol. Bacteriol.* **33**:341, 1930.
37. Breinl, F., and Haurowitz, F., *Z. Physiol. Chem.* **192**:45, 1930; see also Haurowitz, F., *Nature* **205**: 847, 1965.
38. Mudd, S., *J. Immunol.* **23**:423, 1932.
39. Alexander, J., *Protoplasma* **14**:296, 1931.
40. Haurowitz informed me (personal communication, 1982) that there was little interaction between Bail and the younger members or students in the department. He said that Breinl in fact worked under the senior virologist, Weil. Breinl had become interested in immunology through Landsteiner's work while at the Rockefeller Institute in 1928, and on his return invited Haurowitz to join him. Haurowitz recalled that even Breinl "considered my chemical aspects of immunology as fantastic," and refused to submit them to the immunology journal.
41. Pauling, L., personal communication, 1984.
42. Pauling, L., *J. Am. Chem. Soc.* **62**:2643, 1940; *Science* **92**:77, 1940.
43. It is interesting that the same suggestion had been made earlier by Rothen, A., and Landsteiner, K., *Science* **90**:65, 1939.
44. Karush, F., *Trans. NY Acad. Sci.* **20**:581, 1958.
45. Marrack, J.R., *The Chemistry of Antigens and Antibodies*, London, HMSO, 1934.
46. Burnet, F.M., *The Production of Antibodies*, Melbourne, Macmillan, 1941.
47. For a review of the adaptive enzyme story, see Monod, J., and Cohn, M., *Adv. Enzymol.* **13**:67, 1952.
48. Burnet, F.M., and Fenner, F., *The Production of Antibodies*, 2nd edn, Melbourne, Macmillan, 1949.
49. Medawar, P.B., *J. Anat. Lond.* **78**:176, 1944; **79**:157, 1945; *Harvey Lect.* **52**:144, 1958.
50. Owen, R.D., *Science* **102**:400, 1945.
51. Burnet seemed to feel that this dichotomy almost defined the philosophical basis of immunologic thought, and used it as the title of a book directed at general biologists, *Self and Not-Self*, Cambridge, Cambridge University Press, 1969. See also Chapter 5, and Tauber, A.I., *The Immune Self: Theory or Metaphor?*, New York, Cambridge University Press, 1994.
52. Sevag, M.G., *Immunocatalysis*, 2nd edn, Springfield, Charles C. Thomas, 1951.
53. Schweet, R.S., and Owen, R.D., *J. Cell. Compar. Physiol.* **50**(Suppl. 1):199, 1957.
54. See Alain Bussard's discussion of "Darwinisme et Immunologie," *Bull. Soc. Franç. Phil.* **77**:1, 1983. See also Chapter 21.
55. Billingham, R.E., Brent, L., and Medawar, P.B., *Nature (Lond.)* **172**:603, 1953.
56. Bruton, O.C., *Pediatrics* **9**:722, 1952; Bruton, O.C., Apt, L., Gitlin, D., and Janeway, C.A., *Am. J. Dis. Child.* **84**:632, 1952.
57. Jordan, P., *Z. Immunitätsf.* **97**:330, 1940.
58. Pauling, L., *Science* **92**:77, 1940.
59. Jerne, N.K., *Proc. Natl Acad. Sci. USA* **41**:849, 1955.
60. Rather than antigen serving as carrier for antibody, the concept of *antibody-as-carrier* was espoused for several decades by Pierre Grabar, *Clin. Immunol. Immunopath.* **4**:453, 1975; *Med. Hypotheses* **1**:172, 1975. Grabar was less interested in *how* antibodies are formed than in *why*. In an approach reminiscent of Metchnikoff and especially of Ehrlich, he viewed the antibody as part of a broader, normal physiologic system for the transport and assimilation of both foreign and native substances.
61. Talmage, D.W., *Annu. Rev. Med.* **8**:239, 1957.

62. Crick, F.H.C., *Symp. Soc. Exp. Biol.* **12**:138, 1958.
63. Talmage, D.W., *Annu. Rev. Med.* **8**:239, 1957; Burnet, F.M., *Austral. J. Sci.* **20**:67, 1957.
64. Burnet, F.M., *The Clonal Selection Theory of Acquired Immunity*, Cambridge, Cambridge University Press, 1959.
65. Talmage, D.W., *Science* **129**:1643, 1959.
66. Lederberg, J., *Science* **129**:1649, 1959.
67. The somatic theories are best summarized by Cohn, M., *Progr. Immunol.* **2**(2):261, 1974, and germline theories by Hood, L., and Talmage, D.W., *Science* **168**:325, 1970; see also Cunningham, A.J., ed., *The Generation of Immunologic Diversity*, San Diego, Academic Press, 1976. See also Chapter 4.
68. Szilard, L., *Proc. Natl Acad. Sci. USA* **46**:293, 1960.

4 The generation of antibody diversity: the germline/somatic mutation debate

...the ad hoc assumptions required under each construct begin to strain the imagination.

J.D. Capra, 1976[1]

It is one of the curious phenomena of science that substantive debates about mechanism often engage opponents who take extreme positions on either side of the issue. Then, as additional data emerge the positions are modified, so that the final solution often shows that both sides were partially correct and agreement is found somewhere between the extremes. In immunology, we have seen that such was the case in the debate between those who thought that the immune response depends solely upon the action either of cells or of circulating antibodies, or among those who argued for one or another of the various mechanisms advanced to explain the establishment and maintenance of immunological tolerance. Resolution of the debate about the mechanism for the generation of diversity (referred to as GOD in the whimsical cartoons of the ever-imaginative Richard Gershon) between paucigene and multigene proponents (somaticists vs germliners) witnessed a similar splitting of the difference.

The background: the ever-enlarging repertoire

We saw previously that when Louis Pasteur discovered how to induce acquired immunity by immunizing with attenuated pathogens,[2] it was generally thought that all infectious diseases were caused by toxins associated with the microorganisms involved. Diseases such as chicken cholera, anthrax, and rabies yielded to vaccine therapy, and immunity was demonstrated to such bacterium-free preparations as diphtheria and tetanus toxins, and even to the plant toxins ricin and abrin. It could thus reasonably be concluded that: (1) the earlier view that disease results from the action of toxins was correct (Pasteur had named these organisms *virus*, meaning toxin or poison); and (2) the immune response is a Darwinian adaptation directed specifically to counter the toxic threat posed by these pathogens. This latter view found strong support in Elie Metchnikoff's theory of the evolution of vertebrate phagocytic functions[3] and in Paul Ehrlich's suggestion[4] that antibody formation depends upon the presence of *preformed* specific antibody receptors with which antigen reacts to induce exuberant antibody production.

A History of Immunology, Second Edition
ISBN: 978-0-12-370586-0

Copyright © 2009, Elsevier Inc.
All rights reserved

It did not take very long before data accumulated to challenge both of these assumptions. First, many dangerous pathogens were found (e.g., typhus, treponemes, mycobacteria, tropical parasites, etc.) against which the immune system appeared incapable of furnishing protection. Not only did these organisms lack obvious toxins to mediate the diseases that they caused; they also represented such major threats to mankind that one would have expected that a system evolved to protect against dangerous infection should have included these too. Secondly, and even more conceptually disturbing, a variety of bland and innocuous proteins and even cells were found able to stimulate the formation of specific antibodies demonstrable by the formation of immune precipitates, agglutinates, or hemolysates. Where, one asked, was the selective advantage in being able to "protect" oneself against egg albumin, bovine serum globulin, or sheep red cells? The immunological repertoire was growing.

But worse was to come! In 1906, Obermayer and Pick reported that the addition to a protein of simple chemical groupings (later termed *haptens*) would redirect the immune response to the formation of antibodies specific for these chemical structures.[5] In the hands of Pick[6] and especially of Karl Landsteiner,[7] the *repertoire* of possible antibodies was suddenly increased by many orders of magnitude. Again, it appeared unreasonable to suppose that evolution had prepared the rabbit, the guinea pig, or man to form antibodies against synthetic organic chemicals hitherto unknown to Nature. Even more unreasonable in this context seemed Ehrlich's suggestion that specific receptors pre-exist in the body for perhaps millions of different molecular structures. Here was the conceptual rock upon which Ehrlich's side-chain theory foundered during the early years of the twentieth century.[8]

If the information for the formation of these many antibodies could not possibly reside within the host, then logically it could only be introduced by the antigen itself. During the next several decades, a number of suggestions were advanced to explain how antigen might direct the formation of specific antibody – the widely-accepted *instruction* theories of Breinl and Haurowitz[9] and of Pauling.[10] Only as part of the shift to a more biomedical approach to immunology in the 1950s, and with the support of modern genetic concepts such as Francis Crick's *Central Dogma* that information flows only in the direction DNA to RNA to protein, would Darwinian concepts return to influence speculation about the origin and workings of the immune system. (We shall return to the role of Darwinian concepts in immunological theory in Chapter 21.)

The cornerstones of the opposing positions

In 1955, Niels Jerne revived Ehrlich's notion of preformed antibodies,[11] stimulating the imaginative Macfarlane Burnet to advance his clonal selection theory of antibody formation.[12] This was predicated, like Ehrlich's, on the notion that all antibodies are naturally occurring and, in the modern view, encoded in the DNA of genes. In an elaboration of the theory,[13] Burnet proposed that only a very few such genes pre-exist in the organism, and that the specificity repertoire is expanded by somatic mutation of these genes. In a further discussion of the genetics of antibody formation, Joshua Lederberg pursued this notion and indeed spoke of "an immunoglobulin gene"[14] susceptible to such rapid mutation that the full repertoire could be generated in a timely fashion. Herein lay the foundation of the paucigene position.

In a companion paper in support of the clonal selection theory, David Talmage addressed the question of specificity and repertoire size.[15] Were there really so many *different* antibodies (the numbers bandied about ran from 10^5 to more than 10^7)? Taking his cue from Landsteiner's demonstration of the extensive degeneracy of the immune response (graded cross-reactions among related haptenic structures), Talmage suggested that one must distinguish between the functional specificity of an antiserum composed of many different antibodies and the specificity of its individual components. Different combinations selected from among a more limited set of antibody specificities would result in an appreciably wider apparent repertoire. A plausible case could thus be made, not for millions of different specificities, but only for thousands of molecular types expanded combinatorially. Here was the seed of the germline approach – a few thousand immunoglobulin genes were not too much to ask of so important a biological function as acquired immunity.

A similar argument would later emerge from the demonstration that the antibody molecule is formed from a combination of light and heavy chains. If every light chain may combine with every heavy chain, then the so-called $p \times q$ hypothesis would allow perhaps 3,000 light chains and 3,000 heavy chains to provide some 10^7 specificities.[16] (Proponents of this argument would later be embarrassed by the demonstration that the antibody response to a single haptenic determinant such as the dinitrophenyl or the iodo-nitrobenzoyl group might comprise over 5,000 *different* clonotypes.[17] Assuming, reasonably, that each clonotype is determined by unique DNA, there would hardly be sufficient genes to constitute a full repertoire.)

The question of whether diversity is generated by many genes or by few may be put another way; had immunological diversity developed over evolutionary time,[18] or does it arise *de novo* (somatically) during the maturational time of each individual?

The data that initially addressed the problem of the generation of diversity came from three different methodological approaches: the ontogenetic (the study of the fetal and neonatal development of the repertoire); the biochemical (the study of the structure and amino acid sequence of immunoglobulin molecules); and the serological (the study of genetic markers on various parts of immunoglobulin chains). Each of these approaches furnished important data, sometimes interpretable in support of and sometimes in contradiction to one or another theory. These research areas overlap in time, and only the highlights will be considered here. A more detailed discussion can be found in Kindt and Capra's *The Antibody Enigma*.[19]

As we view the developing data, we should keep in mind the prevailing biases of the two camps, both of which realized that they were dealing with a mechanism unique in biology. The somaticists assumed that any solution other than a paucigene one would require the commitment of too much DNA, and that there was no way that any type of selective pressure could conserve those genes for specificities rarely if ever utilized.[20] The germliners, on the other hand, assumed that there is not enough time in ontogeny to fully expand the repertoire from only a few genes,[21] and that one ought not rely on pure chance to assure the appearance of those antibody specificities critical to survival.

These, then, were the polar hypotheses. As data accumulated, some investigators would advance variations on one or the other themes, usually in the context of their own methodological approaches and results. Thus, there was the DNA repair error model of Brenner and Milstein,[22] the paucigene crossing-over model of Smithies,[23] and the gene duplication–somatic recombination model of Edelman and Gally,[24] among others.

The ontogenetic data

The initial description of immunological tolerance and the formulation of the clonal selection theory implied that the mammalian fetus is incapable of an immune response for most of its gestational time. Further, the Burnet–Lederberg concept of somatic generation of diversity seemed to call for a random process, where chance alone would determine the precise time and order of appearance of a given antibody specificity. Thus, when preliminary experiments showed that an immune response might be elicited quite early during the gestation of the fetal opossum and lamb and in young tadpoles,[25] a test of these theories seemed to be at hand.

The first suggestive finding was that some developing animals might be capable of manifesting an extensive repertoire of antibody specificities despite having only very limited numbers of lymphocytes.[26] There hardly seemed adequate time to have generated this diversity by a somatic process. More to the point, it was found that fetuses and neonates develop immunological competence to different antigens at very precise stages of fetal or neonatal development.[27] There seems to be a species-specific program in which all young animals mature their antibody responses in precise order – a timetable apparently incompatible with a random mutational process.

Perhaps the most significant data along these lines emerged from the experiments in Norman Klinman's laboratory, where they studied the development of the clonotype repertoire in the neonatal mouse. These investigators found that there is an ordered maturation from fetal to adult life of the different clonotypes formed against a hapten such as the nitrophenyl group.[28] Again, the data appeared to argue strongly against a random somatic process, and in fact these authors proposed a mechanism for the generation of diversity that they termed *predetermined permutation*. They postulated a basic germline mechanism further enhanced by additional well-ordered intrachain permutations, insertions and cross-overs, and even by mutations.

The ontogenetic data developed in both the fetal lamb and neonatal mouse carried with it a further implication that seemed to favor a germline approach. Each variant of a somatic mutation model required that mutations occur during cell division; thus, it was assumed that the proliferative component of earlier responses to antigenic stimulus would accelerate the somatic expansion of the repertoire.[29] However, the ontogenetic data showed clearly that prior non-specific expansion of lymphocyte numbers did not affect the developmental program of the individual, neither hastening the maturational event nor enhancing the quality and quantity of the response. In addition, the fact that germ-free animals with retarded lymphoid development do not suffer a parallel defect in their immunological maturation was taken as further evidence against a somatic process.

The biochemical data

The splitting of immunoglobulins by enzymes[30] and then by reductive cleavage of disulfide bonds[31] led to the conclusion that the molecule is a heterodimer composed of light (L) and heavy (H) chains. With the finding that myeloma proteins are homogeneous populations of immunoglobulin molecules[32] and that the Bence-Jones proteins found in the urine of such patients are free light chains, the determination of their amino acid sequences provided a key to their genetic origins.[33] It quickly became evident that the amino terminal half of light chains are quite variable in their amino acid sequences, whereas the carboxy terminal portions have quite constant sequences.[34] This, taken with the demonstration that human Inv allotypes located in the constant region of the light chains are inherited as simple Mendelian alleles, led to the postulate by Dreyer and Bennet that two genes are used to form a single light chain[35] – one common to all constant regions, and a (?large) set of separate and independent genes which encode for the variable regions. A similar two gene–one polypeptide chain formulation would soon be advanced to explain immunoglobulin heavy chains, in this case the variable region comprising only about one-fourth of the chain length, rather than the half seen for light chains. Needless to say, the Dreyer–Bennet hypothesis was quickly adopted as supporting evidence by the germliners.

Comparison of the amino acid sequences of variable regions of light chains by Wu and Kabat[36] and of heavy chains by Kehoe and Capra[37] showed that they contain three to four hypervariable regions, the combinations of which would be shown to comprise the antibody-specific site. Far in advance of their times, Wu and Kabat suggested that the immunoglobulin V regions are composed of the products of multiple "mini-genes," in which the short segments coding for the hypervariable regions are episomally inserted into the stable "framework" of the V-region gene.[38] This suggestion seemed to be compatible with both theoretical extremes; variability would be germline-encoded, but diversity would be accomplished somatically by variable insertions, presumably during ontogeny.

Data from structural studies of the immunoglobulin molecule continued to point first in one direction and then in the other. Appearing to support the multigene position was the finding that there are many different heavy chain C regions (ultimately defining the Ig classes IgM, IgG, IgA, IgE, and IgD, with multiple IgG subgroups), each requiring a separate gene. Similarly, kappa and lambda light chains were discovered, and then an increasing number of light chain subgroups. Indeed, there came a time when multiple subgroups and thus multiple germline genes had to be acknowledged even by the somaticists, and the argument turned on what in fact constitutes a subgroup. In 1970, Hood and Talmage constructed a "phylogenetic tree" of forty-one human kappa and twenty-three human lambda proteins.[39] They not only showed the branching from a common origin of kappa and lambda chains, but also suggested that each of many further branchings reflected the evolution of yet additional subgroups, and thus of additional genes. The somaticists countered that the definition of subgroups employed in this analysis was far too liberal, but had to acknowledge the need for increasing numbers of V region genes.

Some aspects of the amino acid sequence studies, however, appeared to favor the somaticist position. Comparison of the Ig sequences of many different species revealed that the V regions of each possess unique residues not shared by other species. These were termed species-specific or phylogenetically-associated residues. If each of the many putative germline genes evolves independently, as do other proteins, argued the somaticists, then how can they all develop *and conserve* these same species markers, and how can these all change in concert during the process of speciation? These data seemed strongly to favor the mutation of only one or a few genes that carry the species-specific residues. Somewhat embarrassed, the germliners proposed two explanations, neither really satisfying. In the one, a gene expansion/contraction model, it was proposed that a set of genes on a chromosome might be expanded by homologous but unequal crossing-over, where a given sequence might dominate in one species and a different one in another species. Alternatively, a gene conversion model was proposed, where gene duplication would be followed by rectification (partial in this case) against a master gene to account for the phylogenetically identical residues. The need to appeal to these complicated *ad hoc* concepts weakened the position of the germliners, one of whose main advantages had been the simplicity of their original theory.

Another strong support of the somatic view was found in the analysis of a large number of mouse V lambda chains.[40] Two-thirds of them had identical sequences, and amino acid substitutions in the rest lay within the hypervariable regions, explicable as mutations in a single lambda subgroup gene. Unfortunately, expression of the lambda chain in mouse immunoglobulins is rare, so that advancing it as representative and proof of a somatic mechanism lacked force, especially in view of the large number of subgroups found in other immunoglobulin chains.

The serological data

In 1956, Jacques Oudin discovered immunoglobulin allotypes by immunizing rabbits with antibodies produced by other rabbits.[41] It quickly became apparent that these serological entities represent structural differences, and are inherited according to Mendelian principles. Thus, the study of allotypes might provide a key to the genetics of the immune system. Soon two *unlinked* allotype groups were found, each with three alleles, no animal forming more than two of the three. One allotype group (a) was localized to the heavy chain V region, and the other group (b) was restricted to the light chain. The same heavy chain allotypes were found on different Ig classes;[42] these findings actually represented the first proper challenge to the one gene–one polypeptide chain dogma. More perplexing, however, was the observation that only one allele is utilized by a single antibody-forming cell[43] – a phenomenon hitherto unknown apart from the functions of the X chromosome. The demonstration that a given V region allotype, especially one located within the framework region, might be shared by antibodies of different specificities raised a problem similar to that later encountered for the case of species-specific residues; it suggested a common origin (?single gene), absent some sort of complex gene conversion mechanism to maintain genetic purity among the many different germline representatives.

The discovery of idiotypes in the early 1960s was made independently in three laboratories – those of Jacques Oudin, Henry Kunkel, and Philip Gell.[44] When these were shown to represent antigenic sites corresponding to the combining sites (hypervariable regions) of the antibody molecule, it seemed that the serological use of anti-idiotypes might provide the most direct approach to the genetic basis of repertoire generation. If diversity resides in many germline genes, then closely related animals should share the same idiotypes; alternatively, if random mutation determined each hypervariable region, then it would be unlikely that different animals would share the same idiotypes. Studies of murine antibodies against such antigenic determinants as arsonate (designated the Ars idiotype),[45] streptococcal carbohydrate (the A5A idiotype),[46] and phosphorylcholine (the T15 idiotype)[47] and others showed that whereas many of these V regions are inherited, presumably as intact germline genes,[48] a significant number show cross-reactions and a variability suggesting that they may be the products of somatic changes. Protagonists on both sides could take some solace from these data.

Resolution of the debate

There was, for almost twenty years, an ebb and flow of support for one or the other extreme position in the somaticist–germliner debate; such concessions as were forced from either side were made reluctantly. As often happens in such situations, debaters on each side would appeal to those data and methods that supported their position, while questioning closely the techniques and results which favored their opponents. We saw this happen in the cellularist–humoralist debate, where Metchnikoff studied pathogens susceptible to phagocytosis and the humoralists studied pathogens that could be lysed or neutralized by antibodies. Again, in arguing the basis of immunological tolerance, those in favor of central mechanisms emphasized the functions of the thymus, while those who espoused peripheral mechanisms studied cytokines, suppressor cells, and networks, and spoke of "regulation."[49]

The Cold Spring Harbor meeting of 1967 seemed already to point to the direction from which the solution of the problem of diversity would come; it was due to the presence of so many molecular biologists and to the early results obtained with their new methods. Their approaches of estimating numbers of genes by liquid hybridization kinetics[50] and then of actually counting genes by DNA cloning and hybridization would produce numbers far greater than would please the somaticists, but far fewer than the germliners had insisted upon. But it would be the actual nucleotide sequencing of DNA that would soon tell the whole story.[51]

Thus, Tonegawa first verified the two gene–one polypeptide theory of Dreyer and Bennet,[52] and it was shown that the mouse has in the germline about fifty V kappa, two V lambda, and some fifty V_H functional genes, as well as nonfunctional pseudogenes; the human has somewhat fewer germline genes. Surprisingly, however, the variable region of both light and heavy chains has the additional contributions of other DNA segments: J (for joining) segments in all light chains and both J and D (for diversity) segments for all heavy chains. In the human, for example, the four to five J segments and the twenty-three D segments, which lie between the twenty-seven to thirty-nine V segments and the C-region genes, contribute importantly to the combinatorial diversity potential.[53] In addition, there are superimposed further diversities in each species. These may involve combinations of junctional variations between gene segments, one or another mechanism of gene conversion, and point mutations in each gene segment. Taken all together, the molecular biological solution to the problem of the immunologic diversity provides for the generation of a quite adequate number of different antibody specificities.[54] Even allelic exclusion found a reasonable explanation, in that activation of all alleles by a pathogen might produce destructive V_L–V_H anti-self combinations.[55]

The solution of most scientific debates usually involves at least the partial validation of the basic assumptions of both sides; in this instance, the solution also exposed mechanisms undreamed of earlier. The paucigene proponents had to acknowledge the presence of far more germline genes than originally proposed, but their chief argument for a somatic mechanism was validated, although in a quite unexpected manner. The multigene proponents, for their part, while forced to acknowledge the importance of somatic mechanisms, found some vindication in the demonstration of multiple germline genes, although in far fewer numbers than initially predicted, and in a quite unexpected form.

If molecular biology provided the solution of the mechanisms by which immunological specificity is encoded and accessed, it left open several ancillary conceptual problems relating to the provenance of this elegant system.

The evolutionary paradox

During the decade-long debate on the genetic basis of immunologic diversity, one of the most telling arguments employed by the proponents of a paucigene model expanded by somatic variation, against those who espoused the idea that all specificities were encoded in the germline, focused on the problem of Darwinian evolution. How, they asked, could the gene pool be maintained when any given organism was likely to employ such a small proportion of its specificity repertoire during its lifetime, and when so many of the specificities that it did employ were against antigens that posed little threat to survival? In the absence of positive selective pressures, it would not take long for such unused or "unimportant" genes to lose their identity. Thus, the evolutionary question still remains with us.

There are, in fact, three different questions to be asked about the evolution of the specificity repertoire in immunology:

1. What are the specificities encoded for by the germline genes, such that Darwinian selective pressures might function to maintain their integrity?
2. How has the complicated overall mechanism evolved, which includes multiple V_L and V_H genes and an elaborate mechanism for the somatic expansion of their specificity potential and for their splicing to J_L, J_H, D_H, and the constant region sequences of DNA, including even intracodon recombination?
3. How can speciation of these *linear sets* of immunoglobulin genes be explained?

We are still far from understanding the answers to any of these questions, and may not even have phrased the questions correctly.

What is encoded by germline V region genes?

Whatever may be the basis for the further somatic expansion of the immunological repertoire, it appears necessary to invoke Darwinian selective forces for the maintenance intact of the set of variable region genes with which we are endowed in the germline. But the single gene does not, as we have seen, define a specificity – this is a function of the V_H and V_L combination. Fortunately, selection does not act upon the genotype but rather upon the phenotype, so that an individual would presumably be deselected should he suffer functional loss of a single V region gene whose light or heavy chain product was critically important for survival.

What, then, are the germline specificity phenotypes? Jerne, impressed by the large number of T cells that show specificity for alloantigens shared within a species,[56] proposed that the germline V genes code for receptor specificities which recognize the full range of the species' polymorphic histocompatibility units.[57] He cites the importance in the ontogeny even of invertebrates of cell-to-cell recognition, to enable differentiation and histogenesis to take place, and suggests that the parallel evolution of a set of histocompatibility units and V gene combinations may mediate these important interactions. Pointing to the tremendous lymphocyte proliferation in the thymus (and bursa of Fabricius),

Jerne suggests that these organs may in fact function as mutant-breeding sites, where the immunologic repertoire is somatically expanded by stepwise mutational deviations from the histocompatibility-determinant starting point.

On the other hand, Cohn and colleagues have pointed out that alloaggression and allospecificities appear to be significant only with respect to the T cell repertoire, and seem not so prominent in the B cell repertoire. They support this view by noting also that whereas B cells appear to recognize only antigen, T cells usually recognize the combination antigen-and-self – the gene products encoded in the major histocompatibility complex. Thus, while willing to concede that the specificity of the germline T cell repertoire *may* be for alloantigens, they insist that the specificity of the germline B cell repertoire must be devoted to the important infectious pathogens, to assure their selective survival.[58]

It is possible that the stepwise maturation of immunological competence in the fetus represents the initial utilization of the proximal germline gene combinations (what has been termed by Langman and Cohn the STAGE I repertoire[59]), but the earliest immune responses in different species do not seem to include antibodies against the species' most important pathogens. If, in fact, the adult-type repertoire is seamless and is achieved fairly rapidly during late fetal and early neonatal life, then it might be argued that the precise composition of the germline set might not matter, since almost any set of gene segments might generate a full repertoire.

Evolution of the immunoglobulin mechanism

It must be recognized that immunology is not unique in presenting a problem of the Darwinian evolution of complex biological systems, often involving multiple independent constituents acting in sequence to produce a complicated physiologic result. As Ernst Mayr points out, the self-reproduction of complex biological systems which are based upon the trials and errors of several thousand million years of evolution is what distinguishes the biological from the physical sciences.[60]

In tracing the evolution of a complex biological system, it may not always be necessary to posit a step-by-step *forward* development from the first reactant. Thus, the complicated vertebrate blood-clotting system, involving multiple factors and co-factors, pro-enzymes and enzymes acting in sequence, might have started in evolution at the end result – the selective advantage of a fairly simple clotting protein in metazoan invertebrates (Limulus, for example) – and then evolved elaborate and more efficient mechanisms by working *backward* to what we now consider the initiating factors in clotting. Again, the complicated cascade reactions seen in the complement system, involving a dozen or more components acting sequentially and along at least two different pathways,[61] might have started somewhere in the middle, perhaps with the physiologically important activities associated with the third or fifth components of complement. In this instance, one can conceive of evolution working in both directions: backward, to select the earlier components which render the production of chemotactic factors and anaphylotoxins more efficient, and forward to extend the utility range of the complement system to additional biological functions.

In defining the molecular evolution of the immunoglobulins, one is impressed by the amino acid sequence homology between the variable and constant regions of the light chains, among the different domains on the heavy chains, and among the light and heavy chains themselves,[62] thus suggesting an evolution through gene duplication.[63] But what is the molecular starting point for such an evolution? Here, the sequence homology of immunoglobulins with β_2 microglobulin is impressive.[64] The immunoglobulin Urpeptide (and its immediate evolutionary descendents) may well have functioned as cell-membrane recognition or adhesion molecules, whose selective value in the differentiation and maintenance of integrity of *all* multicellular organisms is well recognized.[65]

It is difficult, however, to see how selection of the phenotype can conserve such specific combinations, given that each specific site is composed of three V_H and two V_L gene segments and that each response is probably degenerate and composed of many clonotypes. Ohno has addressed this question in a most imaginative way, pointing out that the answer may be as applicable to the function diversity of the nervous system and human intelligence as it is to immunity.[66] By analogy with the Greek myth of the Titan brothers, foresighted Prometheus and hindsighted Epimetheus, he suggests that there may in fact be two types of evolution – an Epimethean process based upon *past* adaptations (corresponding to classical Darwinian principles), and a Promethean process which may prepare the organism advantageously for *future* adaptations. Given that the generation time of viral and bacterial pathogens is several orders of magnitude less than that of vertebrate hosts, Ohno suggests that Epimethean natural selection might not afford adequate time to catch up with the rapid adaptive changes which parasites may manifest, and thus there may be much selective advantage in the development of a new evolutionary mechanism based upon Promethean principles.

The problem of speciation

In dealing with the evolution of single genes, it is easy to understand that mutations which do not impair the physiologic function of the gene product may introduce species-specific DNA sequences, or a polymorphism associated with the presence in a population of multiple alleles at a single locus.[67] But if one considers the effect of speciation on tandemly-arranged sets of genes of related function, such as exist in the immunoglobulin system, then the acquisition of shared species characteristics by *all* members of such gene families becomes more difficult to explain. In considering the evolution of immunoglobulin chains, the question of speciation may be posed at two different levels.

The first problem is to explain how species-specific substitutions, including allotypes, on the constant regions of the immunoglobulin chains or on the framework regions of the variable portions of these chains, can be achieved simultaneously by tandemly-arranged gene families during the evolution of a species. While a number of allotypic markers in such species as rabbit and man appear confined to one or another of the heavy chain isotypes, and thus to a single gene, some allotypic markers appear to be shared by multiple genes (e.g., the several V_H allotypes of the rabbit, and the light chain INV marker in the human). In addition, other nonpolymorphic species marker sequences appear to exist elsewhere in immunoglobulin chains.[68]

It may not be necessary, however, to postulate some novel genetic mechanism that would insure that speciation be accompanied from the outset by an abrupt *and concerted* shift of species markers by all members of a given gene family. Edelman and Gally originally proposed a mechanism for the conservation of homology among the members of immunoglobulin gene families, which they termed "democratic gene conversion"[69] – a suggestion that Baltimore revived.[70] These authors suggest that gene conversion (the transfer of DNA sequence information from gene to gene) may have played the most significant role in immunoglobulin evolution, in insuring the uniform acquisition (or, rather, the uniform spread) of species markers along the linear array of a given family of immunoglobulin genes.

The problem of speciation becomes more difficult, however, when we consider the evolution of the set of germline V region genes – if in fact the specificities for which they code are species-related. Assuming, with Jerne, that the germline variable region genes of T cells encode for receptor specificities which recognize species-specific histocompatibility alloantigens, speciation would require a concerted redirection of the entire family of V genes to include now a new library of allospecificities. Such a genetic shift would appear to impose a greater conceptual problem than does the suggestion that the germline V genes encode for the antigens carried by the major pathogens, for, in addition to the obvious selective value which such immunity would confer, the susceptibility of related species to similar sets of pathogens would obviate the requirements for a major shift in V gene-coded specificities.

Finally, a consideration of the large size of the vertebrate immune repertoire raises the question about how small animals survive. If in fact we need a mature repertoire of 10^6–10^7 specificities, then the human with some 10^{12} lymphocytes, the mouse with 10^8 lymphocytes, and even the 1-g pygmy shrew (*Suncus etruscus*) with some 10^7 lymphocytes should have no problem. Indeed, Cohn and Langman[71] have postulated that the shrew (and hummingbird) possess the minimal immunological requirement in their lymphoid mass which they have termed the "protecton." But the pygmy gobi fish and other very small species, weighing less than 20 mg and presumably with proportionately fewer lymphocytes, should have had a difficult time of it. Yet the individuals survive, and some of these species do not produce the very numerous progeny nor do they live in the protected environments often pointed to as the facile explanations for the survival of such species. Of course, invertebrates do well without any adaptive immune system at all, but no vertebrate is known to survive *normal* conditions with a grossly impaired adaptive immune apparatus.

Notes and references

1. Capra, J.D., in Cunningham, A.J., ed., *The Generation of Antibody Diversity*, New York, Academic Press, 1976, p. 75
2. Pasteur, L., *C. R. Acad. Sci.* **90**:239, 1880.
3. Metchnikoff, E., *Lectures on the Comparative Pathology of Inflammation*, London, Keegan, Paul, Trench, Trübner, 1893; see also Metchnikoff's *Immunity in the Infectious Diseases*, New York, Macmillan, 1905.
4. Ehrlich, P., *Proc. R. Soc. Lond.* **66**:424, 1900.
5. Obermayer, E., and Pick, E.P., *Wien. klin. Wochenschr.* **19**:327, 1906.
6. Pick, E.P., in A. v. Wassermann, ed., *Handbuch der pathogenen Mikroorganismen*, 2nd edn, pp. 685–868, Jena, Fischer, 1912.
7. Landsteiner, K., *The Specificity of Serological Reactions*, New York, Dover, 1962, a reprint of the 1945 2nd edn.
8. Max von Gruber had earlier challenged Ehrlich on the size of the repertoire (Gruber, M., *Münch. med. Wochenschr.* **48**: 1214, 1901; *Wien. klin. Wochenschr.* **16**, 791, 1903). A sign that Ehrlich's theory was in decline was the way that it was treated, as early as 1914, as "of historical interest" by Hans Zinsser in his *Infection and Resistance*, New York, Macmillan, 1914, and similarly in W.W.C. Topley and G.S. Wilson's *Principles of Bacteriology and Immunity*, 2nd edn, Baltimore, William Wood, 1938.
9. Breinl, F., and Haurowitz, F., *Z. Physiol. Chem.* **192**:45, 1930.
10. Pauling, L., *J. Am. Chem. Soc.* **62**:2643, 1940.
11. Jerne, N.K., *Proc. Natl Acad. Sci. USA* **41**:849, 1955.
12. Burnet, F.M., *Austr. J. Sci.* **20**:67, 1957. David Talmage had arrived independently at a similar notion of selection, *Ann. Rev. Med.* **8**:239, 1957.
13. Burnet, F.M., *The Clonal Selection Theory of Antibody Formation*, London, Cambridge University Press, 1959.
14. Lederberg, J., *Science* **129**:1649, 1959.
15. Talmage, D.W., *Science* **129**:1643, 1959. See also Titani, K., Whitley, E., Avogardo, L., and Putnam, F.W., *Science* **152**:1513–1516, 1965.
16. This argument was advanced by Hood, L., and Talmage, D.W., *Science* **168**:325, 1970, in the context of two genes, not for entire light or heavy chains, but for their respective variable regions. They would calculate that no more than 0.2 percent of the total DNA in the genome would suffice.
17. Klinman showed that as many as 5,000 different clonotypes reactive with the dinitrophenyl group could be found in the mouse (Klinman, N.R., *J. Exp. Med.* **136**:241, 1972; Sigal, N.H., and Klinman, N.R., *Adv. Immunol.* **26**:255, 1978), and Kreth and Williamson calculated that the mouse can produce some 8–15,000 individual clones reactive with the *o*-nitro-*p*-iodophenyl (NIP) hapten (Kreth, H.W., and Williamson, A.R., *Eur. J. Immunol.* **3**:141, 1973. See also Pink, J.R.L., and Askonas, B., *Eur. J. Immunol.* **4**:426, 1974.
18. According to Hood, L., and Prahl, J., *Advances Immunol.* **14**:291, 1971, a new antibody gene develops by a slow process of mutation and selection in evolutionary time.
19. Kindt, T.J., and Capra, J.D., *The Antibody Enigma*, New York, Plenum, 1984. These authors present in detail all of the data (except for the ontogenetic) that contributed to the original debate and then to the ultimate molecular biological solution to the problem of the generation of diversity.

20. See, for example, Cohn, M., *Cell. Immunol.* 1:461, 1970; Jerne, N.K., *Ann. Inst. Pasteur* 125C:373, 1974.
21. However, Melvin Cohn (*Cell. Immunol.* 1:461, 1970) calculated that there might indeed be enough time.
22. Brenner, S., and Milstein, C., *Nature* 211:242, 1966.
23. Smithies, O., *Science* 157:267, 1967; *Nature* 199:1231, 1963. See also Smithies, *Cold Spring Harbor Symp. Quant. Biol.* 32:161, 1967.
24. Edelman, G.M., and Gally, J.A., *Proc. Natl Acad. Sci. USA* 57:353, 1967.
25. Reviewed in Šterzl, J., and Silverstein, A.M., *Adv. Immunol.* 6:337, 1967 and Solomon, J.B., *Foetal and Neonatal Immunology*, Amsterdam, North Holland, 1971.
26. Thus, the tadpole does reasonably well with fewer than one million lymphocytes (Du Pasquier, L., *Immunology* 19:353, 1970; Du Pasquier, L., and Wabl, M.R., in Cunningham, note 1, pp. 151–164). The very young fetal lamb does equally well (Silverstein, A.M., and Prendergast, R.A., in J. Šterzl and I. Řiha, eds, *Developmental Aspects of Antibody Formation and Structure*, Vol. I, pp. 69–77, Prague, Czech Academy of Sciences, 1970.
27. Thus, the fetal lamb, Silverstein and Prendergast, note 26; the mouse, Yung, L., et al., *Eur. J. Immunol.* 3:224, 1973; the opposum, Sherwin, W.K., and Rowlands, D.T., *J. Immunol.* 113:1353, 1974. These data were used to argue in favor of the germline side of the multigene/paucigene debate by Silverstein, A.M., in P. Liacopoulos, and J. Panijel, eds., *Phylogenic and Ontogenic Study of the Immune Response and its Contribution to the Immunological Theory*, pp. 221–227, Paris, INSERM, 1973.
28. Klinman, N.R., Press, J.L., Sigal, N.H., and Gearhart, P.J., in Cunningham, note 1, pp. 127–149; Klinman, N.R., et al., *Cold Spring Harbor Symp. Quant. Biol.* 41:165, 1976.
29. Such a theory of antigen-enhanced generation of diversity was advanced by Cunningham, A.J., and Pilarsky, L.M., *Scand. J. Immunol.* 3:5, 1974.
30. Porter, R.R., *Biochem. J.* 46:479, 1950; Fried, M., and Putnam, F.W., *J. Biol. Chem.* 193:1086, 1960; Nisonoff, A., Wissler, F.C., and Woernley, D.L., *Biochem. Biophys. Res. Commun.* 1:318, 1959.
31. Edelman, G.M., *J. Am. Chem. Soc.* 81:3155, 1959; Fraňek, F., *Biochem. Biophys. Res. Commun.* 4:28, 1961.
32. Slater, R.J., Ward, S.M., and Kunkel, H.G., *J. Exp. Med.* 101:85, 1955; Kunkel, H.G., *Harvey Lect.* 59:219, 1965. Potter's discovery that myelomas can be induced in mice by intraperitoneal injection of mineral oil provided extensive material for sequencing, Potter, M., and Boyce, C., *Nature* 193:1086, 1962; Potter, M., *Physiol. Rev.* 52:631, 1972.
33. Edelman, G.M., and Gally, J.A., *J. Exp. Med.* 116:207, 1962; see also Putnam, F.W., and Hardy, S., *J. Biol. Chem.* 212:361, 1955.
34. Hilschmann, N., and Craig, L.C., *Proc. Natl Acad. Sci. USA* 53:1403, 1965; Putnam, F.W., et al., *Cold Spring Harbor Symp. Quant. Biol.* 32:9, 1967.
35. Dreyer, W.J., and Bennet, J.C., *Proc. Natl Acad. Sci. USA* 54:864–869, 1965.
36. Wu, T.T., and Kabat, E.A., *J. Exp. Med.* 132:211, 1970.
37. Kehoe, J.M., and Capra, J.D., *Proc. Natl Acad. Sci. USA* 68:2019, 1971.
38. A variant of this gene segment idea was taken up by Capra, J.D., and Kindt, T.J., *Immunogenetics* 1:417, 1975.
39. Hood, L., and Talmage, D.W., *Science* 168:325, 1970.
40. See Weigert, M.G., et al., *Nature* 228:1045, 1970.

41. Oudin, J., *C. R. Acad. Sci.* **242**:2606, 1956. See also Grubb, R., *Acta Pathol. Microbiol. Scand.* **39**:195, 1956.
42. Todd, C.W., *Biochem. Biophys. Res. Commun.* **11**:170, 1963; Feinstein, A., *Nature* **199**:1197, 1963.
43. Pernis, B., et al., *J. Exp. Med.* **122**:853, 1965; Weiler, E., *Proc. Natl Acad. Sci. USA* **54**:1765, 1965.
44. Oudin, J., and Michel, M., *C. R. Acad. Sci.* **257**:805, 1963; Kunkel, H.G., Mannick, M., and Williams, R.C., *Science* **140**:1218, 1963; Gell, P.G.H., and Kelus, A., *Nature* **201**:687, 1964.
45. Kuettner, M.G., Wang, A., and Nisonoff, A., *J. Exp. Med.* **135**:579, 1972.
46. Eichmann, K., *Eur. J. Immunol.* **2**:301, 1972.
47. Lieberman, R., et al., *J. Exp. Med.* **139**:983, 1974.
48. See, for example, Eichmann, K., and Kindt, T.J., *J. Exp. Med.* **134**:532, 1971, and Eichmann, K., *Immunogenetics* **2**:491, 1975.
49. See, for example, Cantor, H., Chess, L., and Sercarz, E., eds, *Regulation of the Immune System*, New York, Alan Liss, 1984; Bock, G.R., and Goode, J.A., eds, *Immunological Tolerance*, New York, Wiley, 1997. In his introduction to the latter volume, Avrion Mitchison concludes, "The problem posed by Ehrlich turns out to be far, far more complex than the pioneers had imagined" (p. 4).
50. See, for example, Delovitch, T.L., and Baglioni, C., *Proc. Natl Acad. Sci. USA* **70**:173, 1973; Premkumar, E., Shoyab, M., and Williamson, A.R., *Proc. Natl Acad. Sci. USA* **71**:99, 1974; Leder, P., et al., *Proc. Natl Acad. Sci. USA* **71**:5109, 1974; Tonegawa, S., et al., *FEBS Letts* **40**:92, 1974.
51. Leder, P., *Sci. Am.* **246**:102, 1982.
52. Tonegawa, S., et al., *Cold Spring Harbor Symp. Quant. Biol.* **41**:877, 1976.
53. Weigert, M., et al., *Nature* **276**:785, 1978.
54. For a full review of the molecular genetics of immunoglobulin formation, see Max, E.E., in *Fundamental Immunology*, 4th edn, Paul, W.E., ed., pp. 111–182, New York, Lippincott-Raven, 1998.
55. See Cohn, M., and Langman, R.E., eds, *Sem. Immunol.* **14**:153, 2002.
56. Simonsen, M., *Acta Pathol. Microbiol. Scand.* **40**:480, 1967; Wilson, D.B., and Nowell, P.C., *J. Exp. Med.* **131**:391, 1970.
57. Jerne, N.K., *Eur. J. Immunol.* **1**:1, 1971.
58. The suggestion has also been made (Thomas, L., in Lawrence, H.S., ed., *Cellular and Humoral Aspects of Hypersensitivity States*, New York, Hoeber-Harper, 1959) that the evolutionary significance of the vertebrate immunologic apparatus is not so much to protect against *exo*genous pathogens as to mount a surveillance against *endo*genous tumor formation. This proposal was explored at length in Smith, R.T., and Landy, M., eds, *Immune Surveillance*, New York, Academic Press, 1970; and in *Transplant. Rev.* **28**, 1976.
59. Langman, R.E., and Cohn, M., *Molecular Immunol.* **24**:675, 1987. They suggested that the problem of natural selection is simpler, in that the critical germline specificities maintained by selection are formed by *unique* pairing of germline V_H and V_L segments.
60. Mayr, E., *The Growth of Biological Thought*, Cambridge, Belknap, 1982.
61. Fearon, D.T., *Crit. Rev. Immunol.* **1**:1, 1979; Müller-Eberhard, H.J., and Schreiber, R.D., *Adv. Immunol.* **29**:1, 1980.
62. Kabat, E.A., Wu, T.T., and Bilofsky, H., *Variable Regions of Immunoglobulin Chains. Tabulations and Analyses of Amino Acid Sequences*, Medical Computer Systems,

Cambridge, Bolt, Baranek, and Newman, 1976; idem, *Sequences of Immunoglobulin Chains. Tabulation and Analysis of Amino Acid Sequences of Precursors, V-regions, C-regions, J-chain, and β_2 Microglobulins*, US Dept of Health, Education and Welfare Publication 80-2008, Bethesda, 1980; Dayhoff, M.O., ed., *Atlas of Protein Sequence and Structure*, (multiple volumes and supplements), Washington, DC, National Biomedical Research Foundation.

63. Hill, R.L., Lebowitz, H.E., Fellows, R.E., and Delaney, R., in Killander, J., ed., *Gamma Globulin Structure and Control of Biosynthesis, Proceedings of the Third Nobel Symposium*, Stockholm, Almqvist and Wiksells, 1967; Grey, H.M., *Adv. Immunol.* **10**:51, 1969; Hood, L., Campbell, J.H., and Elgin, S.C.R., *Ann. Rev. Genetics* **9**:305, 1975. For a more general discussion of the gene duplication mechanism, see Ohno, S., *Evolution by Gene Duplication*, New York, Springer, 1970.

64. Peterson, P.A., Cunningham, B.A., Berggard, I., and Edelman, G.M., *Proc. Natl Acad. Sci. USA* **69**:1697, 1972; Poulik, M.D., *Progr. Clin. Biol. Res.* **5**:155, 1976.

65. Katz, D.H., *Adv. Immunol.* **29**:137, 1980. A possible evolutionary precursor of the vertebrate major histocompatibility complex is discussed by Scofield, V.L., Schlumpberger, J.M., West, L.A., and Weissman, I.L., *Nature* **295**:499, 1982. It is suggested that Igs may derive from primitive cell adhesion molecules: Cunningham, B.A., et al., *Science* **236**:799, 1987; Edelman, G.M., *Immunol. Rev.* **100**:11, 1987.

66. Ohno, S., *Perspect. Biol. Med.* **19**:527, 1976; *Progr. Immunol.* **4**:577, 1980; see also Cohn, M., Langman, R., and Geckeler, W., *Progr. Immunol.* **4**:153, 1980.

67. Among the many remarkable genetic mechanisms associated with the immune response, we have thus far not mentioned yet another which is important to an understanding of immunologic specificity. The utilization of immunoglobulin genes is characterized also by allelic exclusion, which permits only one (paternal or maternal) chromosomal allele to be transcribed within a given cell. Whatever the molecular basis for this phenomenon (see Leder, et al., *Progr. Immunol.* **4**:34, 1980), without it some antibodies might be heteroligating and inefficient, with two different specificities on a single Ig molecule. Alternatively, the activated B cell might produce an undesirable anti-self antibody.

68. Mage, R., *Contemp. Topics Mol. Immunol.* **8**:89, 1981; see also *Ann. Immunol.* (Paris) **130C**, 1979.

69. Edelman, G.M., and Gally, J.A., in Schmitt, F.O., ed., *The Neurosciences: Second Study Program*, New York, Rockefeller University Press, 1971, p. 962.

70. Baltimore, D., *Cell* **24**:592, 1981. See also Egel, R., *Nature* **290**:191, 1981.

71. Cohn, M., and Langman, R.E., *Immunol. Rev.* **115**:1, 1990.

5 The clonal selection theory challenged: the "immunological self"

Like every theoretical statement...the 1957 theory was made in terms of contemporary knowledge... [and is] incomplete... [and] expressed in terms that have now become meaningless.

F.M. Burnet, 1967[1]

In Chapter 3 we outlined the general features of the clonal selection theory (CST) of Macfarlane Burnet and David Talmage.[2] It may now be appropriate to examine more closely what exactly is central to the theory and what is peripheral, by attempting to differentiate its basic postulates from the secondary inferences that may flow from them. The reason for this is that the theory has come under attack from several different directions, in each of which one or more of Burnet's original assumptions have been challenged. However, even Burnet admitted, only ten years after advancing the theory, that "some of the terms...have now become meaningless." We will now examine the nature and validity of the principal challenges to the theory. An analysis of the components of the theory appears to show that it is only its secondary postulates that are under attack, while the central core of the clonal selection theory survives intact.

Challenges to clonal selection

The suggested alternatives to Burnet's concepts have taken different forms; some have proposed only subtle variations to the underpinnings of clonal selection theory proper, while others have boldly asserted a challenge to the central concept itself, suggesting that "the ruling paradigm" of modern immunology is no longer valid.

Niels Jerne: idiotypic networks

The first theory to be examined was not presented as an explicit challenge to clonal selection, but rather merely as a mechanism concerned with the regulation of the immune response. This was the idiotype–anti-idiotype network theory of Niels Jerne.[3] However, this theory assumes greater importance in the present context because it seems to have served as the intellectual basis for an overt and boldly explicit challenge to clonal selection – that of Irun Cohen, outlined below.

Jerne proposed that even in the absence of antigens, the first spontaneous antibody products of the immunological repertoire would induce the formation

A History of Immunology, Second Edition
ISBN: 978-0-12-370586-0

Copyright © 2009, Elsevier Inc.
All rights reserved

of auto-anti-idiotypic antibodies (an anti-antibody), since each antibody combining site would represent a new self-antigen (idiotope). This new combining site would, in turn, stimulate the formation of yet another level of anti-antibodies, until a stable network of multi-level id–anti-id antibodies was formed that would not only define the "self" but would also regulate all future immune responses. (A more complete account of early suggestions that anti-antibodies might be formed will be found in Chapter 10.)

Irun Cohen: the immunological homunculus

In considering the role of autoimmunity in the economy of the body, Irun Cohen has suggested that: [4]

> Progress in immunology appears to have rendered the clonal selection paradigm incomplete, if not obsolete; true it accounts for the importance of clonal activation, but it fails to encompass, require, or explain most of the subjects being studied by immunologists today...

However, Cohen does acknowledge the validity of three components of Burnet's theory:

1. The existence of a pre-established diversity of receptors
2. One cell–one specificity
3. Antigen selection and activation of clonal precursors for specified antibody formation (and implicity for proliferation).

Cohen suggests that the CST does not explain regulation – what he calls the "patterns of response" involving multiple possibilities among the many components of the immune response: the selection from among the array of specificities due to the degeneracy of the response and from among the array of cytokines that may mediate this response. Thus, he claims, CST does not provide for the regulation of the response repertoire.

Cohen suggests further that Burnet's idea of clonal purging during the time of immunological immaturity is wrong. He postulates that there exists a "physiological autoimmunity" comprising the immune response to a critical set of self-antigens and to the anti-idiotypic T and B cells that, in their turn, recognize the receptors on these autoimmune cells themselves. This network, he claims, constitutes an "immunological representation of self," what he calls "the immunological homunculus" that helps regulate immune responses, and in fact serves more generally to protect the body and heal its defects. In this context, autoimmune *disease* would be the result of a "dysregulation" of the homuncular network.

Note the use of the same general concepts and terminology employed a century earlier by Paul Ehrlich in discussing his concept of *Horror Autotoxicus*. Ehrlich implicitly allowed for the formation of autoantibodies, but suggested that disease was prevented by "certain regulatory contrivances." When autoimmune disease does take place (as with paroxysmal cold hemoglobinuria described in 1904[5]), then a "dysregulation" must have occurred.

Polly Matzinger: the danger signal

Matzinger has taken a somewhat different approach to the attack on Burnet's clonal selection theory.[6] While still arguing in the context of immunoregulation and the basis for tolerance, she has suggested that Burnet was wrong in thinking that the simple interaction of antigen with immunocyte would lead to an active immune response. Rather, she proposes that pathogens that infect the host induce tissue damage and cell death, and it is this process that releases signaling substances that shout "danger"! It is this signal that stimulates an immune response and immunopathological processes. Normal, programmed cell death (apoptosis) and other normal tissue housekeeping processes will not release such "danger signals," but abnormal tissue damage does, and autoimmune disease may result. From these data Matzinger concludes that Burnet's idea of a self/nonself divide cannot explain why some stimuli elicit immune responses while others do not.

In later studies, Matzinger and colleagues attacked another of Burnet's proposals – his suggestion that the fetus and neonate are immunologically immature, thus permitting clonal elimination of anti-self to take place. In a series of papers,[7] these authors showed that immune responses could be elicited in newborn mice, and that the decision on immunity vs tolerance depends only on the manner of presentation of the immunogen – i.e., whether a "danger signal" is present. Arguing from these data, Matzinger suggested that the entire clonal selection paradigm that had ruled immunology for some thirty-five years had been overthrown! Given Burnet's international prestige and the hint that a scientific revolution might be at hand, these claims received wide popular attention in the press.[8]

Matzinger's thesis received at least indirect support from a group led by Charles Janeway at Yale. From studies of the response of the innate immune system, they concluded that foreign pathogens carry markers that identify them to the immune system as "strangers."[9] Thus, the body appears to be more attuned to the differentiation of "infectious non-self" from "non-infectious self" than to Burnet's classical "self/nonself" discrimination. Herein, the "stranger signal" was something akin to Matzinger's "danger signal." Moreover, here too was a similar challenge to one of Burnet's favorite positions.

Now let us examine Burnet's theory more closely, to see whether these challenges stand up to close analysis.

The clonal selection theory

Burnet published his theory in, as he later put it with unaccustomed modesty, "an exceedingly obscure journal...."[10] If the concept proved to be important, he would have priority; if it were wrong, then "very few people in England or

America would see it." It had developed from "what might be called a "clonal" point of view."[11] Its core hypotheses may be put succinctly:[12]

1. The entire immunological repertoire develops spontaneously in the host (i.e., there is no information furnished by antigen)
2. Each [antibody] pattern is the specific product of a cell, and that product is presented on the cell surface (as an Ehrlich-type receptor)
3. Antigen reacts with any cells that carry appropriate specific receptors, to induce the activation of these cells to proliferation and differentiation
4. Some of these cells and their daughters differentiate (become plasmacytoid) to form clones of antibody-forming cells, while others survive as clones of [undifferentiated] memory cells.

This, then, is the essence of the clonal selection theory (henceforth "CST"). It is illustrated in its simplicity by Burnet in his 1957 elaboration of the theory[13] as in Figure 5.1.

It is a theory of *selection* (hypotheses 1–3), involving the selective interaction of antigen with preformed antibody on the cell surface, and of *clonality* (hypotheses 3–4), involving the cellular dynamics of proliferation and differentiation to yield clones of cells and clones of their product. (Although T cells were not at the time even on the horizon, we may note in passing how reasonably well these hypotheses hold for T cells! Even many of the subsidiary questions will be the same for both systems.)

The secondary implications of CST

Each of the core hypotheses raises obvious questions which must ultimately be answered, although some of them did not become obvious until later. Speculation about each of these questions would then lead to the formulation of

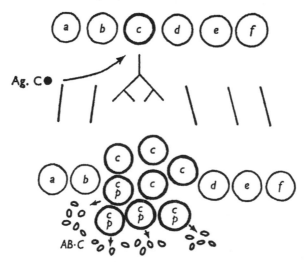

Figure 5.1 Burnet's illustration of the clonal selection theory (from note 13).

subsidiary hypotheses to be tested. Thus, some of the more obvious questions relating to the core hypotheses (and some contemporary answers) are as follows.

- *Question 1A.* If a "Landsteiner-size" repertoire[14] arises spontaneously, what is the mechanism for its generation?

    Sub-hypothesis 1A1. Burnet suggested "in some stage in early gestation a genetic process for which there is no available precedent,"[15] i.e., a rapid somatic muta-tion. This was taken up and expanded elegantly by Joshua Lederberg,[16] who spoke of the rapid somatic mutation of "an immunoglobulin gene." This would come to be known as the *paucigene model.*[17]

    Sub-hypothesis 1A2 (a somewhat later arrival). Why not a gene available *de novo* for each specificity? This would come to be known as the *multigene (or germline) model.*[18] Talmage would anticipate this side of the future debate in his suggestion that the total repertoire is limited, since any Landsteinerian specificity may be determined by a unique combination of selected representatives from a relatively modest repertoire of antibodies.[19] (In the end, both germliners and somaticists would be proved partly right.[20])

- *Question 1B.* If a repertoire is generated randomly and somatically, why are destructive autoantibodies not formed against native antigens to engender immediate autoimmune disease?

    Sub-hypothesis 1B. Deserting his earlier "self-marker" explanation,[21] Burnet assumed ("again following Jerne") a special susceptibility of immature cells *in utero,* such that any antigen then present would abort that clonal precursor. Lederberg would extend this notion of a susceptible stage to cover the life of the individual,[22] since there is no reason why somatic mutation to expand the repertoire should be restricted to fetal life. (Note that nowhere in his initial formulation does Burnet mention the terms "self" or "self–nonself discrimina-tion." However, Burnet would later become much preoccupied with the question of self/nonself discrimination, as we shall see below.)

- *Question 2.* Is there more than one receptor specificity on a single cell?

    Sub-hypothesis 2. Burnet's hypothesis of clonal deletion (tolerance induction) implied the potential loss of desired specificities if a cell produces many different receptors. Alternatively, the cell might produce undesired (autoimmune) anti-bodies when activated – what Burnet would term *forbidden clones.* Burnet thus suggested that a cell could have reactive sites corresponding to only "one (or possibly a small number of) potential antigenic determinants."[23] (Given the above, and the diploid genome, the question of one cell–one antibody would engage the field for a time.)

- *Question 3A.* If somatic mutation persists after clonal expansion, how can clonal specificity be maintained?

    Sub-hypothesis 3A. Burnet does not address this question, but Lederberg suggests in his Proposition A8 that the expanding clone somehow becomes "genetically stable."[24]

- *Question 3B.* How can interaction of antigen with a surface antibody receptor induce cell proliferation and differentiation?

    Sub-hypothesis 3B. It was far too early for Burnet even to ask this question; it would be decades before the complicated biochemical mechanisms of signal transduction could be attacked.

- *Question 4.* What determines and regulates which clonal daughters differentiate to form antibody and which survive as memory cells?
  Sub-hypothesis 4A. Burnet does not raise this question. Again, it is too early to envision cytokine effects, costimulatory molecules, feedback controls, etc.

Here, then, is Burnet's clonal selection theory broken down to its four essential propositions and then to the subsidiary hypotheses that stem from the implications of the core components. It will be immediately apparent that the central theory need not fall just because its promulgator was wrong in failing to address, or in proposing a mechanism to answer correctly, one (or more) of its subsidiary questions. We shall now test the several challenges to CST in the light of these criteria.

Evaluation of the challenges

A critical examination of Macfarlane Burnet's clonal selection theory highlights the difference between its core hypotheses and the ancillary hypotheses advanced to address the many secondary implications of the theory. It seems safe to conclude that the two major tenets that comprise the theory remain unchallenged: (1) that antigen *selects* (more-or-less specifically) from among a population of spontaneously-formed receptor-bearing cells, to stimulate those bearing the homologous receptors; and (2) that this interaction results in the *clonal* proliferation and differentiation of these cells. None of those who purport to "overthrow" clonal selection challenge these two central tenets; indeed, as we saw above, Irun Cohen explicitly acknowledges their validity. Rather, these challenges question Burnet's proposals to answer what I have termed above the "subsidiary" or "secondary" questions that stem from the core hypotheses. These challenges deal, in the main, with mechanisms of immunoregulation – with the nature of tolerance and autoimmunity – and thus they question whether there exists a functioning "self/nonself discrimination." It is also curious that while clonal selection has been challenged based upon Burnet's error in explaining self–nonself discrimination and tolerance, no one has suggested that CST might be challenged because Burnet was wrong about the *mechanism* for the generation of diversity, although these are hierarchically equal hypotheses.

Even Joshua Lederberg recognized early the precise nature and limits of Burnet's theory. In his genetic elaboration of clonal selection, he presented nine propositions (hypotheses), of which four refer to genetics, one to tolerance, and three to antibody formation and memory cells. As he says, "*Of the nine propositions given here, only number 5 is central to the elective theory* [my italics]." [25] This is the one that supposes the spontaneous production by a cell of antibody "corresponding to its own genotype." Note that geneticist Lederberg failed to recognize the centrality of the second main component of the theory, cellular dynamics (i.e., clonal expansion). Lederberg even suggested that his elaboration of Burnet's explanation for tolerance (Proposition 6) is not vital to CST, and is "equally applicable to instructive theories." Again, in one of the most critical surveys of the data that tested the validity of the clonal selection theory,

Sigal and Klinman[26] evaluated only the central components of the theory as defined above. (They did, however, include tests of the validity of the one cell–one specificity question (sub-hypothesis 2 above), because it was a critical component of their clonotype repertoire review.)

Even philosophers of science have occasionally blurred the important distinction between the *core* hypotheses of CST and the *ancillary* hypotheses that may stem from it. In his book *The Immune Self: Theory or Metaphor?*, Alfred Tauber calls Burnet's view of tolerance "a cornerstone of his later immune theory [CST],"[27] and throughout appears to accept the various challenges to Burnet's ideas on tolerance and self–nonself as challenges to the central meaning of CST. Again, in their book *The Generation of Diversity*, Scott Podolsky and Tauber discuss the several challenges to Burnet's notion of self–nonself, and conclude that "Specifically, we must ponder whether CST, as constructed by Burnet, Talmage, and Lederberg[28]...is now being seriously challenged." [29] Kenneth Schaffner, in his elegant discussion of the philosophical bases of CST, *Discovery and Explanation in Biology and Medicine*,[30] formally defines three levels of hypothesis in clonal selection, and actually assigns Burnet's tolerance hypothesis to a secondary level. But even he sometimes seems to suggest that tolerance experiments may serve as serious tests of CST.

The fact that Burnet was (at least partially) mistaken in his subsidiary hypothesis about the mechanism of tolerance induction (clonal deletion *in utero*) influences not at all the validity of the central theory. One might as well suggest that Darwin's Theory of Evolution was overthrown by the demonstration that its author was shown to be wrong about one of its important but subsidiary mechanisms – how variation is inherited; Darwin suggested soft inheritance (the inheritance of acquired characteristics), an idea that was dispensed with in the modern "Neo-Darwinian" period. We might point out also, to return to Irun Cohen's criticism, that like Burnet's CST, Darwin's theory also does not today "encompass, require, or explain" most of what evolutionists study today! Such is the nature of scientific progress.

In the end, then, we must not lose sight of the fact that the clonal selection theory is only a theory of *how* antibodies are formed, not a theory of *why* they are formed.

Burnet's "immunological self"

It would appear from the above that Burnet's theory of "selection" and of "clonality" may safely continue as the governing paradigm as concerns antibody formation. But since attacks on Burnet's ideas have involved his views on regulation, tolerance, and autoimmunity – in brief, his preoccupation with "the immunological self" – it might be well to explore how this notion has become so pervasive in modern immunology.

As early as 1949, in analyzing Ray Owens' observations on red cell chimerism in twin cattle,[31] Burnet suggested that somehow a foreign antigen had failed to be recognized as such, and had been accepted as a part of "self." This would come to be known as immunological tolerance, and his prediction of it as an intrauterine process would gain him a share in the 1960 Nobel Prize. Burnet went on to speak and write about it extensively, including such books as *Self and Notself* and *The Integrity of the Body*.[32] With Burnet's help, the borders of tolerance and "self/nonself discrimination" quickly expanded from the simple explanation of tolerance mechanisms to a metaphor with evolutionary and even philosophical implications.[33] Since then, immunology has more than once been called "the science of self–nonself discrimination."[34]

Burnet became so involved with the question of self and with his explanation of the mechanism of tolerance that even he began to view it as an integral part and even a test of CST, rather than as merely a subsidiary question to be approached by trial and error. In discussing the foundations of CST, Burnet admitted that if immunologists are correct in doubting that "tolerance is wholly a matter of the *absence* of the immunocyte...an extensive reorientation [of CST] will become necessary."[35] No wonder that others might feel the same! Further evidence that even Burnet confused his core postulates with their secondary implications can be seen in the fact that he came close to giving up on CST in 1962[36] when reports came in of two and even four different antibody specificities formed by a single cell;[37] when Szenberg *et al.* found too many pocks on the chorioallantoic membrane of chick embryos injected with small numbers of allogeneic lymphocytes;[38] and when Trentin and Fahlberg, using the Till–McCulloch spleen colony technique, found that a single clone of cells used to reconstitute a lethally irradiated mouse seemed able to form antibodies of different specificities.[39] Burnet, perhaps conceding prematurely before all the returns were in, would say, "This blows out the *original* clonal selection theory. I've said before that I don't believe the original clonal selection theory..."

There are two further reasons why modern immunologists might concentrate so much on "the immunological self." Following the discovery of T cell functions and of T cell receptors, it was found by Zinkernagel and Doherty[40] that these receptors react with a polypeptide attached to a native MHC molecule. Here was "recognition in the context of self," appearing to reinforce the notion of the sharp divide between the self and the other.

The second reason for the prevailing interest in self–nonself discrimination is perhaps more important; for many, the phrase *self–nonself discrimination* has come to epitomize one of the major unsolved problems facing the discipline today. Most of the other subsidiary questions raised by CST have been clarified fully or in great measure – the mechanism for the generation of diversity; the nature and role of T and B cell subsets and their markers; immunoglobulin class switching; the mechanism of allelic exclusion; the mechanisms of signal transduction; and the nature and role of cytokines and other pharmacological participants. Still to be defined clearly, however, are the complex regulatory mechanisms that control the events that follow the interaction of an antigenic determinant with its T or B cell receptor – those that determine whether the response will be positive or negative, activation or tolerance.

Given this wide-open theoretical terrain, it is no wonder that debate continues on such questions as a "big bang" versus the continuous generation of diversity; the relative roles of central versus peripheral mechanisms of tolerance; the number and type of signals required for one or the other response;[41] whether autoimmunity is dangerous or beneficial; and whether the immune apparatus evolved to recognize infectious pathogens,[42] "danger,"[43] or, following Jerne, an internally-modeled "self."[44]

In view of this preoccupation with "the immunological self," it may be appropriate to point out that views on this subject cover the spectrum from true believers to agnostics. A recent extended discussion of the topic revealed that at least four groups think that self–nonself discrimination (*sensu strictu*) is **not** central to the problem of immunoregulation and tolerance.[45]

Notes and references

1. Burnet, F.M., *Cold Spring Harbor Symp.*, **32**:1, 1967.
2. Burnet, F.M., *Austral. J. Sci.*, **20**:67, 1957; Talmage, D.W., *Ann. Rev. Med.* **8**:239, 1957.
3. This theory was first hinted at in Jerne, N.K., *in Ontogeny of Acquired Immunity*, Elsevier, New York, 1972, and then formalized in Jerne, N.K., *Ann. Immunol. (Paris)* 125C:373, 1974. See also Jerne, N.K., *Harvey Lect.* 70:93, 1974.
4. Cohen, I.R., "The cognitive principle challenges clonal selection," *Immunol. Today*, **13**:441, 490, 1992. See also Cohen's *Tending Adam's Garden*, San Diego, Academic Press, 2000.
5. Donath, J., and Landsteiner, K., *Münch. med. Wochenschr.* 51:1590, 1904.
6. The concept first appeared in Matzinger, P., "Tolerance, danger, and the extended family," *Ann. Rev. Immunol.* **12**:991, 1994, but Burnet's theory was not yet seriously threatened.
7. Ridge, J.P., Fuchs, E.J., and Matzinger, P., *Science* 271:1723, 1996; Sarzotti, M., Robbins, D.S., and Hoffman, P.M., *Science* 271:1726, 1996; Forsthuber, T., Yip, H.C., and Lehmann, P.V., *Science* 271:1728, 1996.
8. See, for example, Pennisi, E., *Science* 271:1665, 1996; Johnson, G., *The New York Times* March 26, 1996, p. C1; and Dreifus, C., *The New York Times*, June 16, 1998, p. F4. The Matzinger challenge did not go unanswered, however. See, for example, Silverstein, A.M., *Science* 272:1405, 1996, and Stockinger, B., *Immunol. Today* 17:241, 1996.
9. Janeway, C.A., *Immunol. Today* **13**:11, 1992; Janeway, C.A., Goodnow, C.C., and Medzhitov, R., *Curr. Biol.* **6**:519, 1996.
10. Burnet, F.M., *Austral. J. Sci.* 20:67, 1957.
11. Burnet, F.M., *Changing Patterns: An Atypical Autobiography*. Melbourne, Heinemann, 1968, p. 206.
12. These are not exactly the same as the five "slightly modernized and simplified" principles given in his 1968 autobiography (Burnet, note 11, p. 213), which now included the one cell–one antibody requirement.
13. Burnet, F.M., *The Clonal Selection Theory of Acquired Immunity*, Cambridge, Cambridge University Press, 1959, p. 59.
14. A "Landsteiner-size repertoire" refers to the ability of the host to make an antibody against almost any chemical structure that may be attached as a hapten to

a carrier protein, i.e., an extremely large repertoire characterized by great degeneracy.

15. Burnet, note 13, p. 68.
16. Lederberg, J., *Science* **129**:1649, 1959.
17. See, e.g., Melvin Cohn's discussion, "A rationale for ordering the data on antibody diversity," *Progr. Immunol.* **2**(2):261, 1974.
18. See, e.g., Dreyer, W.J., and Bennett, J.C., *Proc. Natl Acad. Sci. USA* **54**:864, 1965; Hood, L., and Talmage, D.W., *Science* **168**:325, 1970. See also Cunningham, A.J., ed., *The Generation of Antibody Diversity*, New York, Academic Press, 1976.
19. Talmage, D.W., "Immunological specificity," *Science* **129**:1643, 1959.
20. See Kindt, T.J., and Capra, J.D., *The Antibody Enigma*, New York, Plenum, 1984, and Chapter 4 of this volume.
21. Burnet, F.M., and Fenner, F., *The Production of Antibodies*, 2nd edn, New York, Macmillan, 1949.
22. Lederberg, note 16, p. 1651.
23. Burnet, note 13, p. 54.
24. Lederberg, note 16, p.1652.
25. Lederberg, note 16, p. 1649.
26. Sigal, N.H., and Klinman, N.R., *Advances Immunol.* **26**:255, 1978.
27. Tauber, A.I., *The Immune Self: Theory or Metaphor?*, New York, Cambridge University Press, 1994, p. 93. See also Moulin, A.-M., in Bernard, J., Bessis, M., and Debru, C., eds, *Soi et Non-Soi*, Paris, Seuil, 1990, pp. 55–68.
28. Lederberg's name has been closely associated with CST because of his paper "Genes and Antibodies" (note 16). In this, however, while giving genetic substance to Burnet's notions, little is added to the core hypotheses of clonal selection.
29. Podolsky, S.H., and Tauber, A.I., *The Generation of Diversity: Clonal Selection Theory and the Rise of Modern Immunology*, Cambridge, Harvard University Press, 1997, p. 369.
30. Schaffner, K.F., *Discovery and Explanation in Biology and Medicine*, Chicago, University of Chicago Press, 1993. See also Schaffner's discussion of CST in *Theoretical Med.* **13**:175, 1992.
31. Owens, R.D., *Science* **102**:400, 1945.
32. Burnet, F.M., "Immunological recognition of self," *Science* **133**:307, 1961; Burnet, F.M., *The Integrity of the Body*, Cambridge, Harvard University Press, 1962; Burnet, F.M., *Self and Not-Self*, Cambridge, Cambridge University Press, 1969.
33. Tauber, note 27.
34. Wilson D., *The Science of Self: A Report of the New Immunology*, Harlow, Longman, 1971; Klein, J., *Immunology: The Science of Self–Nonself Discrimination*, New York, Wiley, 1982. See also Carosella D., et al., eds, *L'Identité? Soi et Non–Soi, Individu et Personne*, Paris, Presses Universitaires, 2006.
35. Burnet, *Self and Not-self*, note 32, p. 30.
36. Burnet, F.M., in *Conceptual Advances in Immunology and Oncology*, Hoeber-Harper, New York, 1963, pp. 7–21.
37. Nossal, G.J.V., and Mäkelä, O., *Annu. Rev. Microbiol.* **16**:53, 1962. See also Melvin Cohn's retrospective review of this debate, "The wisdom of hindsight," *Annu. Rev. Immunol.* **12**:1, 1994, p. 16ff, and Kindt and Capra, note 20.
38. Szenberg, A., et al., *Br. J. Exp. Pathol.* **43**:129, 1962.
39. Trentin, J., and Fahlberg, W.J., in *Conceptual Advances*, note 36, pp. 66–74.
40. Zinkernagel, R.M., and Doherty, P.C., *Adv. Immunol,* **27**:51, 1979.

41. Bretscher, P.M., and Cohn, M., *Science* **169**:1042, 1970; Langman, R.E., and Cohn, M., *Scand. J. Immunol.* **44**:544, 1996. See also Langman, R. *The Immune System*, San Diego, Academic Press, 1989.
42. Cohn, M., Langman, R., and Geckeler, W., "Diversity 1980," *Prog. Immunol.* **4**:153–201, 1980; Janeway, note 9.
43. See Matzinger, note 6, and Matzinger, P., "An innate sense of danger," *Seminars Immunol.* **10**:399, 1998.
44. Cohen, note 4. The original network concept of Jerne has been elaborated on by, among others, Coutinho, A., et al., *Immunol. Rev.* **79**:151, 1984; Varela, F.J., and Coutinho, A., *Immunol. Today* **12**, 159, 1991; and Coutinho, A., Kazatchkine, M.D., and Avrameas, S., *Curr. Opinion Immunol.* **7**:812, 1995.
45. These positions are elaborated in Langman, R., ed., "Self–nonself discrimination revisited," *Seminars Immunol.*, Vol. **12**:159–344, 2000: Silverstein, A.M., and Rose, N.R., pp. 173–178; Grossman, Z., and Paul, W.E., pp. 197–203; Coutinho, A., pp. 205–213; and Cohen, I.R., pp. 215–219. See also Silverstein, A.M., and Rose, N.R., *Immunol. Rev.* **159**:197–206, 1997.

6 The concept of immunologic specificity

behind the diversity of discoveries moved a unity, a constant direction of change…the development of the concept of biological specificity.

Horace Freeland Judson[1]

This quotation reflects Judson's insightful perception of the *leitmotiv* that guided the discovery of the genetic code and the molecular biology of the storage and retrieval of genetic information. He recounts this exciting story in his book *The Eighth Day of Creation*. In a similar way, the 125 years that have elapsed since Pasteur first established immunology as an experimental science can also be viewed as the continuing quest for the meaning and basis of immunologic specificity. In surveying the history of the discipline, this search appears logically to be divisible into three somewhat overlapping phases.

The first centered on the questions of how immunologic specificity is expressed, and what its biological implications are. This involved the study of the major phenomena of immunology – precipitation, agglutination, hemolysis, and the reactions and cross-reactions of immunity and allergy – as well as the pathological, diagnostic, and therapeutic consequences of these phenomena. These were, in general, the principal preoccupations of the bacteriologic era of immunology, which saw its peak about the turn of the century but spilled over into the early years of the immunochemical era, which encompassed the decades of the 1920s to 1950s.

The second phase in the development of the discipline addressed the question of how immunologic specificity is structurally determined. It started with studies of the nature of antigens, following the lead that Karl Landsteiner established in the 1920s and 1930s, but soon turned to the chemistry and anatomy of the immunoglobulin molecule, chain sequences, molecular domains, isotypes and idiotypes. It then expanded to include the study of the specific and nonspecific receptors and factors that initiate and regulate immune responses. This latter area was launched in the mid-1950s by the pioneering work of Porter, Edelman, Putnam, and others[2] on the structure of the immunoglobulin molecule. It quickly spread from the molecular immunologists to the cellular immunologists, and soon demanded for the first time that *cis*- and *trans*-immunologists[3] speak to one another in mutually comprehensible terms.

Finally, the third phase of investigation has sought to answer the question of how the information for immunologic specificity is carried and retrieved. Every theory of antibody formation has been an attempt to answer this question, and it finds modern expression in studies of the molecular biology of the

A History of Immunology, Second Edition
ISBN: 978-0-12-370586-0

Copyright © 2009, Elsevier Inc.
All rights reserved

immunoglobulin and immune response genes, and of the origin, nature, and function of the receptors on T and B lymphocytes.

The background to biological specificity

The concept of specificity did not appear unheralded when the term was employed in an immunologic context by Pasteur and Koch in the 1880s, and by Ehrlich and Metchnikoff in the 1890s. It had, rather, a long history of usage in many fields of science in one form or another, and in philosophy as well. Indeed, as early as the sixteenth century, the physician Hieronymus Mercurialis showed an understanding of that most basic aspect of immunologic specificity when he pointed out,[4] in criticizing Girolamo Fracastoro's theory of acquired immunity,[5] that defense against reinfection by smallpox does not protect against measles or leprosy.

Medicine and pharmacy

The ancient Greek view that disease represents a disturbance of the humors prevailed in the Western world for some 2,000 years, and discouraged the view that individual illnesses might be specific and subject to differential diagnosis and classification, and to specific therapy. Thus, descriptions of the phenomenon of acquired immunity to plagues and pestilences from Thucydides down to the Middle Ages carried with them no implication of specificity, since it was usually not even clear then or later which of the major infectious diseases was involved in any given epidemic.[6] Only in the tenth century did the Islamic physician Rhazes first differentiate between smallpox and measles,[7] and by the sixteenth century it was generally understood that such diseases as plague, smallpox, syphilis, malaria, and many others had more or less specific identities which allowed their differentiation.[8]

In his famous 1546 book *On Contagion*,[9] Fracastoro held that disease "seeds" or "germs" (*seminaria*) had specific affinities for certain targets – some for plants and others for animals. Some germs were held to be specific for certain animal species, and not to affect other species (a primitive notion of natural immunity); and even within a given host, some germs were held to have a specific affinity for certain tissues, organs, or humors. Perhaps even more significant was the contemporary development of the iatrochemical approach to disease, originated by Paracelsus in the early sixteenth century and expanded a century later by van Helmont and many others. Paracelsus helped to replace the lack of specificity of the classical humoralist school by emphasizing the individuality of diseases, their cause by specific agents foreign to the body, and the possibility therefore of specific therapy.[10] This approach has been termed the *ontological concept of disease*, and is essentially the modern one.

Perhaps the greatest impetus to the consideration of diseases as specific entities was provided by Thomas Sydenham in the seventeenth century,[11] as Knud Faber points out in his book *Nosography*.[12] Sydenham's emphasis on clinical rather than theoretical medicine, and on careful bedside diagnosis, convinced him that diseases are specific, subject to uniform laws, and therefore classifiable. This concept led naturally to his later doctrine of specific remedies. Based on his knowledge that quinine exerts a specific effect on the ague, "Sydenham entertained no doubt that the Creator had provided specific remedies for the chief diseases...."[13] But the notion of therapeutic "specifics" was not new with Sydenham. It has probably always existed in folklore and even in professional practice, and persists in sometimes curious form even today. Thus, the physician of the Middle Ages often used ground precious stones as a specific against the plague, mercury was long employed as a specific for syphilis, and pulverized animal horn was and still is employed in many cultures to treat impotence.[14]

Sydenham's approach to medicine exerted a strong influence on his contemporaries, and many of his followers attempted to expand Sydenham's nosologic approach, with the idea that diseases could be described and classified in the same way that botanists describe and classify plants. Just as the eighteenth century saw the epitome of botanic taxonomy in the hands of Linnaeus, so did it witness a similar attempt at a grand systematization of disease, most notably by François Boissier de Sauvages. In 1763, Boissier published a *Nosologia*[15] which divided all diseases into ten classes, these into forty orders, the orders into genera, and the genera into species – 2,400 in all! In both botany and medicine, the aim was to distinguish the *specific*. In botany, where structure was easily studied and defined, this approach succeeded; in medicine, where pathologic anatomy and pathophysiology were still in their infancy, these classifications proved not very useful.

The eighteenth-century taxonomic excesses in medicine, and the obvious defects of all of the classification systems proposed, induced a reaction in the early nineteenth century to the notion of disease specificity. This was reinforced by the widely held view of spontaneous generation,[16] in which even microorganisms were not specifically fixed but could develop spontaneously and undergo transformations of both form and function. Thus it was that Pasteur experienced great difficulty at the outset in convincing the world of mid-nineteenth-century science of the specificity of action of yeast ferments, of the silkworm parasite, and of the specificity of the agents responsible for the diseases of wine and beer. Pasteur and Koch would later experience the same difficulty when they advanced the germ theory of disease, including as it did the claim that these agents of disease were morphologically specific, were capable of reproduction to type, and were the etiologic agents of specific and reproducible diseases. But their notions finally gained broad acceptance, and it was the developments in bacteriology that helped to give true meaning to the term specificity as employed in medicine.

Chemistry

By at least the fifteenth century, the concept of the specificity of action of poisons and of certain drugs was spreading, along with the alchemical understanding that nitric acid, antimony, mercury, sulfur, and many other substances were each able to react in *predictable* fashion with certain other chemicals.[17] Paracelsian iatrochemistry was firmly based on alchemy and, with overtones of medieval mysticism, held chemical specificity to be the central issue. Not only did the Paracelsians postulate that every individual physiological action was specific in nature and every disease a distinct entity specific in location; they also maintained that the best form of therapy lay in the finding of a chemical able specifically to combat the given disease.

The ability of certain substances to react preferentially with other substances was termed *affinity*. With the discovery that a given chemical might react more strongly with one substance than with another, the late eighteenth and early nineteenth centuries saw extensive "tables of affinity" developed in an attempt to codify the many reactions and cross-reactions observed in the chemistry laboratory. In an age when most sciences were dominated by the popularity of Newtonian mechanics, the speculation of such scientists as Bergman and Berthollet viewed chemical affinity as a force similar in nature to gravity. Only in the nineteenth century did the concept of chemical affinity take on a more significant meaning, with the increasing knowledge of substitution reactions in organic chemistry, and with the development of the fields of thermochemistry, electrochemistry, and stereochemistry. All of these trends were crystallized in 1867 in Guldberg and Waage's law of mass action, which showed that the affinity of chemical reactants for one another not only had meaning, but could also be quantified.

Developments in biochemistry during the nineteenth century also pointed in the direction of a significant role for affinity and specificity. This was nowhere more apparent than in the study of enzymes and of fermentation.[18] In the 1820s, Justus Liebig, perhaps the foremost chemist of his day, conceived of disease as the specific consequence of infection by external inanimate ferments, and the disease process itself as one of specific fermentation. It was his opposition to this view that preoccupied Pasteur during his early years, and perhaps contributed to the transition of his interests from the fermentation of wine and beer to the study of the agents of animal and human disease. As information developed about the multiplicity and mode of action of enzymes, it became quite clear that there exists a remarkable specificity in the interaction of an enzyme with its substrate. This led, in 1894, to the famous metaphor of the noted biochemist Emil Fischer: "To use a picture, I will say that enzyme and glucoside must join one another as lock and key, in order to be able to exert a chemical effect."[19] Fischer's metaphor, as well as the chemical emphasis he gave it, were to exert a profound effect upon Paul Ehrlich, and thereby upon the nascent field of immunology.

Thus, by the 1890s, significant progress had been made toward an understanding of the general nature of the specificity of chemical interactions. Problems persisted in two areas of chemistry that would cause great difficulties for immunologic theoreticians in the decades to come. The first difficulty involved a lack of understanding of the central difference between the ionic bonds of the electrochemist and what would later be called the covalent bond of the organic chemist. This problem would not be resolved until G. N. Lewis established in

1916 the basis of the strong covalent bond,[20] and Linus Pauling showed in 1939 how weak bonds are formed.[21] The second difficulty faced in the 1890s, and for some fifty years thereafter, involved the difference between chemical and "physical" interactions. Aided in part by the physicist Ernst Mach and the physical chemist Wolfgang Pauli, and attended by a certain degree of mysticism, it was conceived that "colloidal" interactions were all-important in biological systems, and involved adsorption processes and reactions with electrolytes that were predominantly nonspecific.[22] Such colloids included proteins which, at least until the 1930s, were assumed to be nonspecific aggregates of smaller molecules rather than discrete molecular entities themselves.[23]

It was the difference between the *strong* bonds of organic chemistry and the *weak* ionic bonds of electrochemistry that highlighted the conceptual differences between Paul Ehrlich and Svante Arrhenius. Again, Ehrlich's emphasis on firm *chemical* union between antigen and antibody clashed with Jules Bordet's view that the antigen–antibody union represented a weaker, *physical* (colloidal) interaction. The problem of strong vs weak bonds and of chemical vs physical interactions would serve as the *leitmotif* during the ensuing debates on immunologic specificity and its phenomenologic consequences.

Philosophy

Among the philosophical questions that have interested mankind, one has exerted a profound influence in modern times on almost every field of biology. This is the question of whether nature is continuous in all respects, or divisible into discrete components and types essentially unrelated to one another. Plato's concept of *essentialism* held that the vast observed variabilities of the world represent complete discontinuities between types (*eidos*), and the Aristotelian classification of animals and plants into separate groups accorded with this notion, as did the later concept of creationism which maintains that each form of plant or animal life is unique and specific, and persists unchanged. In opposition to this view, others saw an overriding continuity in nature and only gradual and quantitative transitions between apparently discrete species of things.

The debates over whether nature is seamless or discontinuous persisted throughout the eighteenth and nineteenth centuries. On the one hand was the growing tendency to categorize plants and animals according to the sharply defined differences among them, and on the other was the familiar scholastic expression *natura non facit saltum*, and Leibnitz' claim that this was one of his greatest and most highly verified maxims, which he called the law of continuity.[24] Immanuel Kant, in his *Critique of Pure Reason*, suggested that "this logical law of *continuum speciarum*...presupposes however a transcendental *lex continui in natura*."[25] Nature, according to Kant, is herself discontinuous, and only the human mind imposes continuity. In opposition to this, the botanist Karl von Nägeli held, like his teacher Mathias Schleiden, that nature's continuity is real, and imposes itself upon the observer. von Nägeli believed in the continuity of change, and that species differ from one another by only gradual transition and quantitative gradation (*quantitative Abstufung*), rather than by qualitative absolutes. Once again, the "lumpers" and "splitters" were in dispute, the former searching for similarities and the latter for differences.

Not only did these divergent world views exert a marked effect upon theories of spontaneous generation, on concepts of the fixity of species, and on Darwinian evolution, as Ernst Mayr points out in *The Evolutionary* Synthesis;[26] they also exerted an interesting influence on immunologic thought during the first decades of this century. Their implications for immunology were pointed out by Pauline Mazumdar in her study *Species and Specificity: an Interpretation of the History of Immunology.*[27] In this impressive work, she calls attention to the fact that this difference in the approach to nature was responsible for a dispute that covered 130 years and four generations of scientists, of whom the most recent were immunologists. Thus, von Nägeli argued the question of continuity vs discontinuity in botany with the botanist Ferdinand Cohn. In turn, von Nägeli's student Max von Gruber entered into a longstanding dispute in bacteriology, first with Cohn's protégé Robert Koch and then even more violently in immunology with Koch's students Richard Pfeiffer and Paul Ehrlich. Ehrlich next had conceptual differences with Gruber's student Karl Landsteiner. Finally, in the modern era, Landsteiner's student Alexander Wiener engaged in long and bitter controversy in immunohematology about the interpretation and nomenclature of the rhesus blood group system. Wiener argued for unity and continuity against the intellectual descendants of the diversity–specificity school, R. A. Fisher and R. R. Race.

This running battle over the period of 130 years may have had its roots in an abstract philosophical difference, but it profoundly influenced the development of modern immunology, as it found expression in concepts of the nature and function of immunologic specificity.

Paul Ehrlich: specific receptors

Paul Ehrlich received his medical degree in 1878, for which he presented a thesis on the theory and practice of histologic staining.[28] He quickly saw that tissue staining depends upon the chemical nature of the substances used. For the next dozen years, most of his work was focused on demonstrations of the usefulness and specificity of chemical stains in histology, hematology, and bacteriology. These studies involved him deeply in problems of organic synthesis and chemical interaction, and he very early became aware that such interactions depend upon the presence of specific groups of atoms. Following the lead of Edward Pflüger, Ehrlich conceived of the physiologic functions of the living cell as dependent upon such specific atom groups, attached as distinct side-chains to the "chemical nucleus" of the cell.[29] He became interested in the problem of immunity during this period, and undertook experiments in this new field – first privately, and then part-time in Koch's Institute for Infectious Diseases.[30]

Plate 6 Paul Ehrlich (1854–1915)

Because of Ehrlich's interest in antitoxins, and his demonstrated knowledge of chemistry, he was charged in 1895 with finding a solution to the problem that had vexed workers throughout the world following Behring's discovery of diphtheria antitoxin – the measurement and standardization of both toxin and antitoxin. So elegant was his solution to this problem that publication of the results gained him almost instantaneous worldwide recognition and, shortly thereafter, his own institute, the Royal Prussian Institute for Experimental Therapy. Ehrlich's 1897 paper on the measurement of diphtheria antitoxin[31] is of great historical interest from several points of view. First, it laid the foundation for the field that would later be known as immunochemistry, and pointed the way for fifty years of quantitative studies of the antigen–antibody interaction. Next, it introduced Ehrlich's concept of the side-chain receptor theory of antibody formation, which would have important repercussions on immunologic speculation for many decades to come. Finally, it declared to the world that the reactions and specificity of immunity depend upon the laws of structural chemistry.[32]

The side-chain theory

While the practical applications of preventive vaccines and of antitoxin serum therapy were being widely exploited, the physiologic mechanisms responsible for the functions of immunity were still a mystery. As we saw in Chapter 1, early theories of acquired immunity pictured the host as a passive receptacle in which it was the pathogen itself that contributed to its own demise. Only Elie Metchnikoff, in his theory of cellular immunity,[33] considered that the host might mount an active immunologic defense against infection, but Metchnikoff's phagocytic theory was widely opposed.[34] The discovery of antibody in 1890 reinforced the trend toward humoral theories of immunity, and interest in cellular immunity declined, not to be seriously revived for almost sixty years. After the work of Emil Behring and his collaborators,[35] the central theoretical questions in immunology involved how antibodies are formed, and how they acquire and exercise their specificity. These were the questions that Paul Ehrlich addressed in his famous side-chain theory.

Ehrlich's side-chain theory was not excessively complicated in its initial formulation in 1897. The key postulates were that:

1. Antibodies are *normal cell products* which serve as cell membrane receptors, and are not "made to order" to fit a given antigen
2. Antibody *specificity* is the consequence of the interaction of chemically defined complementary molecular structures
3. The antigen–antibody interaction represents a firm (irreversible) chemical union.

Since most of the antibodies recognized at that time were antitoxins,[36] Ehrlich conceived of the antibody molecule as having a single binding site for the "haptophore" group on the antigen, which would simultaneously neutralize its "toxophore" group. Very quickly, however, the implications of the newly discovered phenomena of agglutination,[37] the precipitin reaction,[38] and immune hemolysis[39] made new demands upon the side-chain theory, and a complicated set of *ad hoc* hypotheses was required to bring the theory into conformity with the newer facts. Thus, the antitoxin molecule was called by Ehrlich a receptor or "haptine" of the first order, possessing only a binding site for toxic antigens. Receptors of the second order were postulated to contain two different domains; one responsible for interaction with antigen, and a separate portion to account for the secondary biological phenomena of agglutination and precipitation.[40] Ehrlich called this second site the "zymophore" group, believing that agglutination and precipitation were probably the result of some type of enzyme action. To account for complement-mediated hemolysis and bacteriolysis, Ehrlich postulated the existence of third-order receptors, which possess a site for the binding with antigen (the cytophile group) and a separate site (the complementophile group) to which complement was bound. This complement-fixing antibody, since it contained two receptors (one for antigen and one for complement), Ehrlich termed a *Zwischenkörper* (amboceptor). It is interesting that even this early, Ehrlich conceived of the lytic activity of complement as an enzymatic process – a speculation that would not be confirmed for some sixty to seventy years.[41]

Ehrlich's side-chain theory had a marked effect upon the biological and medical sciences in the decade or two that followed its formulation. This was due in part to the famous pictures used by Ehrlich and his adherents (Figure 6.1) to illustrate the structures and their reactions. It is interesting to note that immunologists even today employ the same assortment of geometric shapes to depict different antibody and antigenic specificities. Despite Ehrlich's caution that "Needless to say, these diagrams must be regarded as quite apart from all morphologic considerations..."[42] yet they seemed almost to define *and even to be* the molecules and

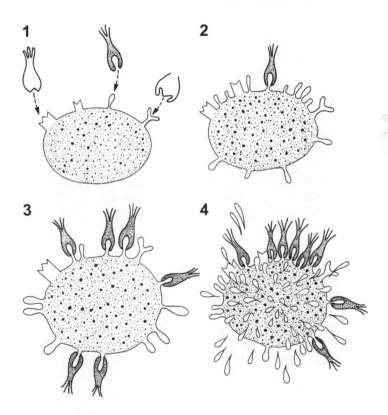

Figure 6.1 Paul Ehrlich's side-chain theory of antibody formation. Note the use of geometric shapes to indicate different specificities. Antigen selects for the production and release of the appropriate membrane receptors.
From Ehrlich, P., "Croonian Lecture – on Immunity...," *Proc. R. Soc. Lond.* **66**:424, 1900.

reactions depicted, so that one was almost convinced that one understood the reality from these pictures. This vexed Jules Bordet, who complained: [43]

> By the abuse that it [the Ehrlich theory] has made of quite puerile graphical
> representations which merely translate the exterior aspect of phenomena
> without in any way penetrating to their inner meaning, it has extended the
> deceptive use of explanations that are facile, but illusory.

Nevertheless, entire books were devoted to the theory, with chapters on "the side-chain theory in internal medicine," "the side-chain theory in obstetrics and gynecology," and so on.[44] More substantially, perhaps, Ehrlich's side-chain theory represented an important contribution to pharmacology and to Ehrlich's new concepts of chemotherapy,[45] to which he devoted most of his activities from the turn of the century until his death in 1915.

The antigen–antibody interaction

It is important to examine more closely Ehrlich's view of toxin–antitoxin and other antigen–antibody interactions, for they not only illustrate his view of the workings of immunologic specificity, but also introduce for the first time in immunology the equally important concept of affinity.

Ehrlich's studies on diphtheria toxin and antitoxin introduced many now-familiar terms to immunology. Antibody was initially a *receptor* on the cell surface which possessed a configuration specific for antigen. Ehrlich considered the antigen to be a distinct molecular entity, containing a complementary haptophore grouping (Greek *aptein* – to grasp or fasten) and a second toxophore group responsible for toxicity. He had early recognized that while the toxicity of a preparation of diphtheria toxin might decrease with time, its ability to bind to antibody remained undiminished. He suggested that this represented a deterioration only of the toxophore group, and named the resulting product a *toxoid*. To explain the discrepancy between L_+ and L_0 values,[46] Ehrlich postulated that the unit of antitoxin had a *valency* of about 200. He suggested further that preparations of diphtheria toxin contained not only toxin and toxoid, but also other related substances with different *affinities* for the antibody receptor. Those with the lowest affinity he called *toxons*, and soon he introduced a veritable congeries of substances, including α and β modifications of proto-, deutero-, and tritotoxins and toxoids. Ehrlich held that each of these components would react with antibody in sequence, the higher affinity components first and the lower affinity components last. In this manner, he sought to explain discrepancies in the titration curve of diphtheria toxin with antibody, and he established elaborate charts called "toxin spectra" which purported to show the composition of any preparation in terms of the various high- and low-affinity toxic and nontoxic constituents.[47]

As another example of the extent to which *ad hoc* hypothesis might be carried by Ehrlich and his adherents, we may cite the example of the Danysz (or Bordet–Danysz) phenomenon, discovered in 1902.[48] Danysz showed that the degree of neutralization of diphtheria toxin by antitoxin depends upon whether the toxin is added all at once, or stepwise. When the addition is made in steps, then the supernatant is found to be much more toxic than when the same quantities of toxin and antibody are mixed all at once. By this time Ehrlich had substantially left immunology to pursue his interest in chemotherapy, but, in the Ehrlich tradition, von Dungern hypothesized the existence of "epitoxinoids," which were claimed to have even lower affinity for diphtheria antitoxin than did toxon.[49] von Dungern suggested that the stepwise addition of toxin mixtures resulted in the irreversible binding of significant amounts of epitoxinoid, whereas mixture of the toxin broth all at once should result in preferential binding and neutralization of the higher-affinity toxin, with the epitoxinoid left substantially unbound.

In his studies on immune hemolysis, Ehrlich and others noted the existence of immunologic cross-reactions, such as the ability of rabbit anti-ox erythrocyte antibody to hemolyze goat erythrocytes.[50] Differential absorption studies showed that the goat cells would absorb only the cross-reacting antibody, whereas the homologous ox cells would absorb all hemolysins. Ehrlich explained these results, in the context of his side-chain theory, by assuming that any complex cell must necessarily contain a number of unique antigenic determinants, some of which might be shared among the cells of related species. Thus, immunologic specificity was presented as absolute, with cross–reactions resulting only from mixtures of antibodies whose specific antigens happen to be shared by different cells (Figure 6.2).

The complexity of Ehrlich's formulation of the diphtheria toxin–antitoxin interaction points up the extent to which his fertile imagination was grounded in structural chemistry and concepts of affinity; it was this same commitment, and a willingness to support his theories with extensive *ad hoc* assumptions, that would later carry him into a prolonged debate with Jules Bordet over the mechanism of immune hemolysis and the nature and interactions of complement. But, Ehrlich's complex theories aside, we must not lose sight of the important contributions of his work in this area. Apart from advancing for the first time a *practical* method for the standardization of diphtheria toxin and antitoxin, these studies introduced a number of important concepts to immunology. These include: the notion that antibody specificity is based upon structural-chemical complementarity of combining sites; the notion of heterogeneity of binding affinity (although in Ehrlich's formulation heterogeneity was restricted to antigenic variability – the antibody "side-chains" reactive with a given antigen were implicitly assumed to be homogeneous); and, finally, the observation that antigen–antibody interactions are temperature- and concentration-dependent like any other chemical interaction, and thus amenable to quantitative study.

The Ehrlich–Bordet debates

As we have seen, Ehrlich maintained that the antigen–antibody bond was the result of a firm and irreversible chemical union, with a specificity based on molecular structure. Moreover, Ehrlich held that the receptor on the antibody

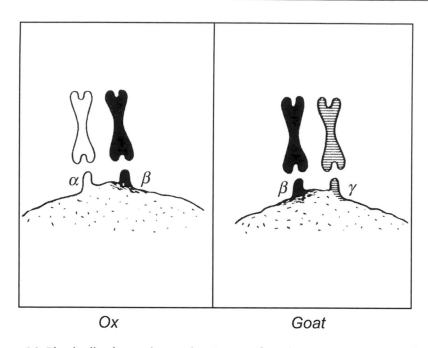

Figure 6.2 Blood cells of ox and goat, showing specific and common receptors. Different shadings are used to code for different specificities.
From Ehrlich P., and Morgenroth, J., *Berl. klin. Wochenschr.* **38**:569, 1901; English translation in Ehrlich, P., *Collected Studies in Immunity*, New York, Wiley, 1906, p. 88.

molecule that fixes complement follows the same rules of specificity and firmness of bonding. This forced the conclusion that complement must also be bound firmly by antibody, even in the absence of antigen, since even the side-chain pictures represented the antigen-binding site and the complement-binding site as separate entities on the antibody molecule. Ehrlich and his students (most notably Sachs) also concluded, partly on the basis of experiments but also by strong analogy with the antigen receptor, that just as there were different antigens so were there different complements, each capable of binding specifically to its appropriate receptor on antibody.[51] In both aspects of Ehrlich's side-chain theory he encountered strong opposition, most notably from the Belgian immunologist Jules Bordet.

The principal difference between Ehrlich and Bordet focused not upon the existence of immunologic specificity, but rather upon its basis (i.e., on the fundamental nature of the antigen–antibody interaction). Ehrlich insisted upon the chemical nature of immunologic specificity, which he thought required a firm and irreversible bond. When challenged by data suggesting that antigen–antibody reactions might not proceed to completion, or that dissociation was possible, as suggested by the diphtheria toxin titration curve and by the Danysz phenomenon, Ehrlich skillfully extended his theory by postulating sufficient new components and sufficient differences in affinity to account for the necessary stoichiometry.

Plate 7 Jules Bordet (1870–1961)

Bordet, on the other hand, approached these problems as a biologist rather than as a chemist. He was convinced from the outset that antigen and antibody might interact in multiple proportions, and that a firm chemical union between them was neither required nor indeed permitted by the pertinent data.[52] Bordet rejected the notion of a chemical interaction between antigen and antibody, and suggested rather that the binding was more comparable to the adsorption of dyes. He suggested that the combinations of antibody with antigen are in fact "colloidal" interactions – a term then very popular in many fields, and thought to represent an adequate explanation for many different types of phenomena.[53] The large and complex surface area presented by colloids was deemed adequate to encompass almost any sort of adsorptive process. As Bordet claimed, "The affinity of adsorption is sufficiently delicate, graduated, and elective, so that the notion of its participation in antigen–antibody reactions is compatible with that of specificity."[54]

Throughout his writings, Bordet repeatedly criticized Ehrlich for his insistence upon chemical notions, and especially for the complexity of his theories. Bordet insisted that he himself was not a theorist, and that the ideas that he had advanced were not even worthy to be called theories, but rather "merely represent a description of the true state of affairs." He says, however, that:[55]

> My predilection for the realities cannot deter me from considering briefly the hypotheses...one knows with what luxuriance they have been developed on the fertile ground of immunity, where so much of the unknown still stimulates the imagination and invites audaciously synthetic concepts from the schools desirous of affirming their superiority...conceptions that are defended with all of the partisanship that amour propre mixed with chauvinism so readily inspire.

If the driving force behind Ehrlich's side-chain theory was the desire to explain immunologic specificity and the origin of antibodies, the stimulus to Bordet's more modest speculations was certainly the desire to explain the phenomena of immunity – bacterial agglutination and neutralization, the precipitin reaction, and especially complement-mediated hemolysis. In this approach, the specificity of the primary antigen–antibody interaction seemed less important than the secondary physiological consequences. Bordet apparently did not feel impelled to question closely whether "colloidal interactions" or "physical adsorption processes" were indeed able adequately to explain the striking precision of immunological specificity that was apparent even in those early days. Ehrlich, however, did not hesitate to point out this defect in the "physical" theory of Bordet. It should be noted that this conflict between structural-chemical and physical theories was not restricted to immunology, but was then being argued in many fields of biology.[56]

The fundamental difference in the approaches of Ehrlich and Bordet is highlighted by their respective responses to the new theory of Arrhenius and Madsen,[57] elaborated in detail by Arrhenius in a book whose title, Immunochemistry, named the new field.[58] These authors suggested that antigen reacts with antibody like a weak acid with a weak base, setting up an equilibrium that obeys the law of mass action. Bordet welcomed this new theory, because it showed that the complicated Ehrlich-type theories were not really necessary. However, he maintained that the Arrhenius–Madsen theory could not be correct because it seemed to permit more dissociation of toxin–antitoxin complexes than the data supported. Ehrlich's school, on the other hand, rejected it because it did not adequately explain affinity and specificity.[59]

The second principal dispute between Ehrlich and Bordet centered on the mechanism of complement fixation and immune hemolysis. Ehrlich's chemical concept required two distinctly separate and specific receptors on the antibody molecule, one for antigen and one for complement, and he saw no good reason why complement receptors should not have the range of specificities for different complements that antigen receptors have for different antigens. Thus, for a long time Ehrlich and his coworkers argued that there was a multiplicity of specific complements in each serum, and that the firm complement–antibody interaction could occur independent of the antigen–antibody interaction. Bordet, for his part, was content to have antibody adsorbed onto antigen – an interaction that he suggested would result in a "change of configuration" appropriate for the nonspecific fixation of a single complement.

Even the terminology used by Ehrlich and Bordet typified their different approaches, and helped to prevent reconciliation. Ehrlich persisted in his use of the German term *Komplement*, and referred to hemolytic antibody as the *Zwischenkörper* or *Ambozeptor*, which suggested that the function of antibody was to provide a specific bridge between complement and antigen. Bordet preferred the more French-sounding term *alexine* (actually coined by Buchner), and called antibody *la substance sensibilisatrice*, implying that the function of the antibody was merely to "sensitize" the antigen-bearing cell so that complement might nonspecifically effect its lysis.

If the major elements of Ehrlich's structural-chemical approach to the antigen–antibody bond (save its irreversibility) were later vindicated, at the expense of Bordet's physical theory, it was Bordet whose general concept of the mode of action of complement was proved correct. From this latter dispute, only Ehrlich's nomenclature survived. However, the impressive contributions to immunology of both these scientific giants were subsequently recognized by Nobel Prizes. Ehrlich shared his in 1908 with Elie Metchnikoff, while Bordet received this honor in 1919.

The Ehrlich–Gruber debate

The decade that immediately followed Paul Ehrlich's side-chain theory of antibody formation and definition of the structural-chemical basis for antibody specificity witnessed a flood of new findings that extended the scope of the young field of immunology beyond its original narrow boundaries of protection against bacterial diseases. The discoveries of the precipitin reaction and of hemagglutination quickly demonstrated that antibodies could be raised against a wide variety of nontoxic and even nonbacterial substances. More than this, the phenomenon of complement-mediated immune cytolysis disclosed for the first time that antibodies were not restricted to the destruction of pathogenic bacteria, but might engage also in the destruction of other, non-noxious cell types such as erythrocytes, spermatozoa, etc.[60] The same decade also saw the description of most of the principal phenomena of allergy,[61] including anaphylaxis,[62] the Arthus reaction,[63] and serum sickness.[64] Each of these pathologic reactions, as well as the earlier-described Koch reaction in tuberculosis and the tuberculin skin test,[65] were quickly demonstrated to be based upon immunologic mechanisms and to obey the general rules of immunologic specificity. In yet another direction, these early developments led to the use of antibodies in the new field of serotaxonomy and forensic medicine,[66] in which the interrelationship of animal and plant species would be examined in what would be the first of many applications of antibody molecules as molecular probes in other fields of biology.

One of the important consequences of all of these new discoveries, though, was the growing realization that the repertoire of possible antibodies was very large indeed. Another consequence was the increasing difficulty in establishing the boundaries of immunologic specificity itself; certain *in vitro* reactions were found to be highly specific, whereas other reactions might show appreciable cross-reactivity among bacterial species or among different animal sera. It thus became increasingly more difficult to find overwhelming and irrefutable support in the literature for those who argued with Paul Ehrlich for a chemically-defined irreversible binding and a narrow specificity for antibody, or for those who argued with Jules Bordet for a physically-based reversible binding and a some-what broader specificity. It might perhaps be more appropriate to say that it had become easier to support *either* position, by an appropriate selection from the mass of data that was accumulating. In the event, the controversy continued.

Max von Gruber descended from a philosophical school of science (Schleiden and von Nägeli) that had long held basic differences with the tradition (Cohn, Koch, and Pfeiffer) to which Paul Ehrlich was heir. For the Ehrlich school, the password had always been discrete specificity – first in botanic classification, then in bacterial action, and currently in the interactions of antigen, antibody, and complement. In the tradition in which Gruber had been reared, the key words were *Kontinuität* and *quantitative Abstufung* (quantitative gradation). A traditional conflict between two differing world views had now shifted its field of battle to the young discipline of immunology.[67]

Ehrlich's controversy with Jules Bordet had always been marked by politeness and respect on both sides. Despite their differences, they continually referred to one another as "distinguished," and Ehrlich even allowed that, while Bordet might be in error, his opposition had served to stimulate much fruitful research that further strengthened the side-chain theory. But Ehrlich's battle with von Gruber was far different – it soon became quite vitriolic, as charges and countercharges appeared month after month in the leading journals. More than once, the dispute was reduced to personal attacks.

When Gruber and Clemens von Pirquet attacked Ehrlich on the question of toxin–antitoxin interactions,[68] Ehrlich responded that:[69]

> It is all the more remarkable that Gruber should choose the subject of toxins for the main portion of his attack on me, for according to his own admission that is the field which he knows merely from literary studies. Against such critics I am in the unpleasant position of a man who is compelled to discuss colors with the blind.

Gruber, in his turn, sent a satirical letter to the *Wiener klinische Wochenschrift* entitled "New Fruits of the Ehrlich Toxin Theory," purportedly addressed to Gruber and signed "Dr. Peter Phantasus, *by God's Grace Chemist*."[70] It starts:

The Ehrlich side-chain theory has made its triumphant way throughout the world. You, Herr Professor [Gruber], were literally the only one to have called this theory, with astonishing shortsightedness and almost incomprehensible arrogance "completely worthless, unbridled hypothesis-spinning, and dangerous numbing with words"...

Then, in apparent support of the Ehrlich thesis, it employs Ehrlich's approach to prove that distilled water (which also lyses erythrocytes) must contain a whole series of Ehrlich-type toxins, prototoxoids, etc. To this, Ehrlich replied, "I shall now take up Gruber's recent experiments. These were first published in the *Wiener klinische Wochenschrift*, in the form strongly suggestive of the comic supplement of a newspaper."[71] In the introduction to his collected works, Ehrlich takes to task "such authors as Gruber, who have absolutely no personal experience in the main questions, [but] wage a bitter war merely because they have made a few literary studies." And in final exasperation at the Gruber attacks, Ehrlich complains that "In a way, therefore, my position is like that of a chess player who, even though his game is won, is forced by the obstinacy of his opponent to carry on move by move until the final 'mate'."[72]

Gruber's attacks on Ehrlich are extremely interesting, not only because they question the very possibility in Nature of a set of discrete and absolute specificities such as Ehrlich postulated, but also because in the process Gruber raised questions that would continue to vex immunologists for several generations to come. How is it possible to explain even in chemical terms, asked Gruber, the astoundingly large number of different specificities demanded by Ehrlich's theory?[73] Again, Gruber asked how, from the point of view of Darwinian selection, can all of these antibody specificities have evolved to antigens not in the normal environment? For Gruber, the absence of suitable answers to such questions was sufficient cause to reject the Ehrlich hypothesis, and indeed these questions may have been significant in the fall from favor of Ehrlich's selection theory of antibody formation in the 1920s, to be replaced by instruction theories that suffered no such defect.

Gruber's central attack, however, was on the key to Ehrlich's theory – specificity – and on its principal implication, the firmness of the antigen–antibody union. Gruber argued against the existence of absolute and discrete specificities for antibodies, which in Ehrlich's hands would not permit extensive serologic cross-reactions. For Ehrlich, as for Richard Pfeiffer before him, an antibody could react only with its homologous antigen; any evidence of cross-reactions must imply different antibodies interacting with varying mixtures of antigens. For Gruber, however, a single antibody could interact, by *graded affinities*, against a number of different antigens. Pfeiffer, in his studies of bacterial agglutination and lysis, and Ehrlich, in his studies of hemolysis, preferred to use weak or diluted antisera whose reactions were narrowly specific. Gruber, on the other hand, always insisted upon "high-grade" antisera that permitted the study of nonspecific reactions. Each group argued that the other's approach was erroneous.

Throughout the long dispute, few of their differences were resolved, since each spoke a language incomprehensible to the other (see Chapter 14 for a further discussion of the semantic aspects of immunologic disputes). The Ehrlich–Bordet disputes are widely remembered in the English-speaking world, thanks in no small measure to the efforts of Bordet's student and translator-popularist, Frederick Gay.[74] The Ehrlich–Gruber disputes are largely unknown outside Germany and Austria, in part because Gruber had no Gay, but in greatest measure because his work was substantially eclipsed by that of his more famous student, Karl Landsteiner, who made the conflict his own.

Karl Landsteiner, The Compleat Immunochemist

No single individual contributed as importantly to so many different areas of immunology as did Karl Landsteiner. In a scientific career that spanned almost half a century, he involved himself in almost every significant area of

Plate 8 Karl Landsteiner (1868–1943)

immunology, and on each of them he left his mark.[75] Thus, he founded the field of immunohematology by first describing the ABO system in 1900,[76] and followed this up by discovering the M, N, and P erythrocyte isoantigens with Philip Levine in 1926,[77] and the Rh antigen system with Alexander Wiener in 1940.[78] He was among the first to produce experimental syphilis in the monkey in 1906,[79] and the first to produce experimental poliomyelitis in the monkey in 1909.[80] In the field of serodiagnosis, Landsteiner (with Donath in 1904) defined the antibody responsible for the first reported autoimmune disease, paroxysmal cold hemoglobinuria,[81] and later correctly identified the antigen involved in the Wasserman test for syphilis,[82] as well as advancing the first useful serological test for poliomyelitis in 1909.[83] Finally, in collaboration with Merrill Chase, he showed in 1942 that delayed hypersensitivity could only be transferred passively by sensitized leukocytes and not by serum antibody[84] – a critical step forward in our understanding of the nature of cellular immunity.

Landsteiner's most significant work, however, for the purposes of this discussion of immunologic specificity, lay in his studies of the immune response to artificial haptens, which he employed to gain a better understanding of the specificity of serologic reactions, as his famous book is entitled.[85] Landsteiner himself apparently considered that the study of hapten–antibody reactions represented his most significant scientific contribution, for, when awarded the Nobel Prize in Medicine in 1930 for his discovery of the blood group antigens, he is reported to have felt that he was recognized for the wrong thing![86]

The Vienna of the 1890s, in which the young Karl Landsteiner took his medical degree and postgraduate training, was both scientifically and intellectually one of Europe's most exciting centers.[87] Like Paul Ehrlich, Landsteiner developed a strong interest in synthetic organic chemistry, and spent two years at various laboratories, including that of the famous chemist Emil Fischer in Würzburg. This early training in structural chemistry ultimately influenced the direction of Landsteiner's work, but along pathways quite different from those toward which a similar background had pointed Paul Ehrlich. Following additional clinical experience at the University of Vienna, Landsteiner became assistant to Max von Gruber in 1896–1897, in the newly established Department of Hygiene. It was here that his interest in serology and immunology was stimulated, and it was here also, from Gruber, that he acquired his basic beliefs about the nature of immunologic specificity. Indeed, his first scientific paper in immunology, published in 1897, was on antigen–antibody cross-reactions,[88] and this set the tone for his entire life's work in this area.

From the very beginning, in agreement with Gruber, Landsteiner voiced his opposition to Ehrlich's view of a discrete and absolute immunologic specificity, and he rejected Ehrlich's structural-chemical approach to the antigen–antibody reaction. In a paper with Jagić in 1903,[89] Ehrlich adopted Jules Bordet's physical (colloid) explanation of the antigen–antibody interaction, and claimed that "immune specificity can be seen as in principal the sum of a number of individual, nonspecific reactions." So ardently did Landsteiner espouse the cause of a colloidal interpretation of the antigen–antibody interaction that for a time he, rather than Bordet, was the recognized champion of this viewpoint. However, Landsteiner's commitment to a colloidal interpretation did not survive the decade, and he quietly gave up this "physical" position in favor of a more chemical one, although he never ceased to argue against Ehrlich's absolute specificity and in favor of Gruber's concept of quantitative and almost continuously graded affinities. One who believed otherwise would hardly have devoted his life's work to a study of the *cross*-reactions of antibodies.

There appeared in 1912 a publication that redirected the course of Landsteiner's future research interests. In this, E. P. Pick published an encyclopedic review of the chemistry of antigens,[90] including an extensive section on the specificity of chemically altered antigens – a doorway that Obermeyer and Pick had opened with their initial study published in 1906.[91] This paper pointed out that not only were a number of different chemical treatments available, but that many of these would also confer new and unique specificities upon proteins, depending only upon the nature and extent of such chemical treatment. Even the normal rules of species specificity might be abrogated. Thus, an antibody raised against a chemically-treated egg albumin might now react specifically with a similarly modified horse-serum protein.

It did not take long for Landsteiner to recognize the power of this new approach, and in 1917 he published two important papers with H. Lampl which marked clearly the course of his future work. In the first of these[92] he presented data on the cross-reactions of antibodies against a homologous series of acyl-substituted proteins, and in the second study he employed for the first time protein antigens substituted with diazonium compounds. These papers are of singular importance for an understanding of Landsteiner's future work on immunologic specificity as well as the contemporary trend of immunologic thought. In his work with artificial haptens, Landsteiner saw a powerful tool with which to prove two important immunologic points, perhaps best summarized in a quotation from a follow-up paper published with Lampl in 1918:[93]

> We...did not think it was possible to explain the workings of normal or immune serum on innumerable cell types by the supposition of innumerable different antibodies and, by analogy, a similar fantastic number of receptors on each cell. We held, like Gruber, that the simpler concept was quite adequate, that an antibody can react with a variety of related but not necessarily identical antigens...The specificity of serum reactions appears to be the expression of grades of affinity which reach a maximum in certain combinations – those of antigen and homologous antibody...while the Ehrlich theory admits of only a single absolute specificity.

In this quotation, we can almost hear the death knell of Ehrlich's side-chain theory of antibody formation. For if the number of *naturally occurring* antigens was too embarrassingly large to be explained on the basis of pre-existing antibody side-chains, Pick in his review had shown and Landsteiner would soon amply confirm that the number of *synthetic* haptens which might serve as antigenic determinants was immeasurably larger yet. How could the mammalian host have *prior* knowledge to encompass antibody specificities for such unnatural structures, many of which were not yet even a gleam in the eye of the synthetic organic chemist? We shall see below the direction in which this aspect of Landsteiner's work forced the speculation of chemically-oriented immunologists.

The defeat and break-up of the Austro-Hungarian Empire following World War I had serious consequences for Viennese society and for Viennese science.[94] Among those who suffered the effects of this defeat was Landsteiner, who soon lost his position and was pensioned off while still only in his early fifties. He had no recourse but to look elsewhere, and, after two somewhat insecure years in Holland, finally received an invitation from Simon Flexner to take up a position at the Rockefeller Institute in New York. Here, his work on antibody cross-reactions with artificial haptens flourished, most notably in collaboration with his colleague James van der Scheer. In much of his work with homologous series of related chemical compounds, he showed repeatedly the nature and extent of the cross-reactions that one might expect from an antibody made against any given member of the chemical series. And yet, despite his continuing emphasis upon *cross*-reactions, it was Landsteiner's work that helped convince the world of the elegance and narrow precision of immunologic specificity, in which an antiserum could distinguish clearly among subtle differences in antigenic (haptenic) structure. Landsteiner might argue for graded affinities, but his demonstrations that appropriate antisera could distinguish between optical isomers or between ortho-, meta-, and para-substituted azobenzoates seemed to argue for a sharper specificity.

It is interesting that Landsteiner rarely succumbed to the common practice of contemporary immunologists of drawing Ehrlich-type geometric pictures to designate immunologic specificities. (He may have remembered Gruber's distaste of such pictures, and Bordet's comment that these were "quite puerile graphical representations which…extended the deceptive use of explanations that are facile, but illusory.") What Landsteiner did publish to display his feelings about the nature of immunologic specificity were data of the type illustrated in Table 6.1. It is almost as though Landsteiner saw, in the diagonal sweep of + and ± cross-reactions, all of the pictorialization that his concept required. For Landsteiner, each of these tables of cross-reactions represented a reconfirmation of his basic belief in how immunologic specificity functions. Thus he could say in the last edition of his famous monograph *The Specificity of Serologic Reactions* substantially what he had said twenty-five years earlier: [95]

> *The high specificity of many serum reactions led Ehrlich to the view that each antibody is sharply adjusted to one particular structure (receptor), and accordingly that overlapping reactions of antigen must depend upon the presence in each of them of identical substances or chemical groupings…From numerous observations on artificial conjugated antigens, however, this notion is seen to be inadmissible, and it is certain that antibodies react most strongly upon*

Table 6.1 Serologic cross-reactions of a homologous series of anilic acids[a]

Antigen[b]	Reactions with antisera against			
	p-Amino-oxanilic	p-Amino-succinanilic	p-Amino-adipanilic	p-Amino-suberanilic
p-Amino-oxanilic acid ($n = 0$)	++++	0	0	
p-Amino-malonanilic acid ($n = 1$)	0	±	0	0
p-Amino-succinanilic acid ($n = 2$)	0	++++	+±	+
p-Amino-glutaranilic acid ($n = 3$)	0	+	++	++
p-Amino-adipanilic acid ($n = 4$)	0	)	++++	++++
p-Amino-pimelanilic acid ($n = 5$)	0	0	+++	++++
p-Amino-suberanilic acid ($n = 6$)	0	0	++	++++

[a]After Landsteiner, K., and van der Scheer, J., *J. Exp. Med.* **59**:751, 1934.
[b]General formula of aminoanilic acids: $NH_2C_6H_4NHCO(CH_2)_nCOOH$.

> the homologous antigen, but also regularly with graded affinity, on chemically related structures. Or, as Haldane wrote regarding an enzyme reaction, "The key does not fit the lock quite perfectly, but exercises a certain strain upon it."

If the above discussion of Landsteiner's views on the nature of immunologic specificity implies a consistency of concept based upon an untroubled consistency in his experimental results, then we must point out that such was true only with respect to Landsteiner's work with the precipitin reaction. Throughout Landsteiner's career, he was beset by a paradox which he was never able fully and satisfactorily to resolve. On the one hand, all of his work on the precipitin reaction with naturally occurring proteins and hapten-substituted antigens suggested extensive cross-reactions and spectra of graded affinities of antibody for proteins of related species, or for haptens of related structure. In contrast, however, all of his work with hemagglutination, starting with the isoagglutinins of the ABO system and later with the M, N, and P antigens and the Rh blood groups, seems to show absolute specificities and few cross-reactions. It was his work on horse–donkey interspecies hybrids[96] which perhaps best pointed up this paradox. Here, antibodies against the serum proteins might show extensive cross-reactions between these related species using the precipitin test. In contrast, however, agglutinating antibodies could distinguish clearly between the erythrocytes of these closely related species. In this system, discrete inherited antigens seemed able to stimulate discrete antibodies more in line with Ehrlich's than with Gruber's concepts.

The lack of accord between the results obtained with agglutination reactions and isoagglutinins, and those observed in precipitin tests, led Landsteiner and van der Scheer to do a special comparative study[97] that concluded that there must be a difference in the specificities of these two reactions, which "suggest an essential difference in the chemical structures which determine the specificity of the two kinds of antigens (precipitinogens and agglutinogens)." This, for Landsteiner, must have been a disappointingly inconclusive result, but it was all that could be said at the time. He attempted to circumvent this inconsistency by pointing out that most of the cell antigens thus far identified (the heterogenetic antigen of Forssman and the pneumococcal polysaccharides of Heidelberger and Avery) were nonprotein haptens. Landsteiner therefore concluded that:[98]

> there exist two systems of species specificity in the animal kingdom, the specificity of proteins and that of cell haptens. The proteins, it would seem, undergo gradual variation in the course of evolution, while haptens are subject to sudden changes not linked by intermediate stages.

Thus, the paradox remained, not even causing Landsteiner to modify his earlier statement that Ehrlich's view of an antibody sharply adjusted to one particular structure "is seen to be inadmissible."

Specificity and theories of antibody formation

The term *Antikörper*, as it was first coined to describe the agent in the blood that Behring and Kitasato had shown capable of passively transferring immunity, was originally a noncommittal term. It only implied a recognition that there must exist a discrete entity or *body* capable of carrying immunologic specificity, and thus able to act against (*anti*) the offending toxin. But if a discrete physical entity rather than some vague physiologic process were held responsible for acquired immunity, then the nature and manner of formation of such a substance would furnish a valid topic for speculation. While theories of antibody formation have been dealt with in detail in Chapter 3, their formulation has been so intimately tied to considerations of antibody specificity (and its corollary, the repertoire size of these specificities) that they deserve at least a brief summary in the present context. Indeed, we have already seen that the principal basis for the attack on Ehrlich's theory by Gruber, Bordet, and Landsteiner was precisely their claim that the antigen–antibody interaction was colloidal and thus nonspecific.

At the outset, the protein nature of antibodies was unknown, but even after this was established no hints were yet available on the nature or mode of production in the body of *any* macromolecules.[99] Before the appreciation developed that the repertoire of antibodies might be large, only considerations of immunologic specificity drove theoreticians of antibody formation, and it appeared reasonable to Buchner that antigen itself carried the information for specificity by somehow being incorporated into the antibody molecule in such a manner that it would thenceforth react specifically with other similar antigen molecules.[100] However, such a theory could not long survive the rapidly developing quantitative studies which showed that much more antibody was formed than could be accounted for by the amount of antigen injected, and that antibody formation, once started, would continue without further administration of antigen.[101]

These objections were dealt with effectively by Paul Ehrlich in 1897, in his side-chain theory of antibody formation. In this anticipation of all subsequent natural selection theories, Ehrlich postulated that antibodies are naturally occurring cell products present as receptors on the cell surface, to be selected for specifically by an appropriate antigen. Such an antigen–receptor interaction would then lead to compensatory overproduction of these receptors, which would appear in the blood as circulating antibody. As we have already seen, Ehrlich postulated that the specificity of these antibody receptors was a function of certain stereochemical configurations whose complementarity with structures on the antigen permitted specific interaction. With this theory, Ehrlich dealt simultaneously with the origin of antibodies and with the basis of their specificity. He was, in 1897, untroubled by the problem of the size of the immunologic repertoire, since the only antibodies then known were thought to be antitoxins directed against a limited number of human and animal pathogens. Within a very few years, however, the scope of immune reactions was so far extended that it became apparent that the size of the specificity repertoire was almost too great to permit of their *natural* occurrence, and so the Ehrlich side-chain theory fell into disrepute.

Now specificity *and* repertoire size became the central issues in all theories of antibody formation – an approach reinforced by the changing character of the discipline from about 1910 onward. If the first decades in immunology may be called its Age of Bacteriology, then the period from the 1920s to the 1950s may appropriately be termed its Age of Immunochemistry, thanks in part to the stimuli provided by Landsteiner's work on haptens and Heidelberger's work on the pneumococcal polysaccharides and quantitative immunochemistry.[102] Immunology in this period was dominated by chemical approaches and chemical thinking.[103] This meant inevitably that contemporary theories of antibody formation would have to explain the elegant specificity and the extensive specificity repertoire of antibodies; they tended to neglect the more biological aspects of antibody formation, such as its persistence and its ability to undergo anamnestic boosting.

In the main, chemically-oriented speculation about the nature of antibody formation sought to explain specificity and its large repertoire by means of instruction theories, wherein the antigen itself somehow transmitted the information for its specificity to a nascent globulin molecule. The most notable of the instruction theories called upon the antigen to act as a template upon which specific antibody might be formed, either by directing the formation of a unique amino acid sequence on the polypeptide chain[104] or by molding a preformed polypeptide chain into an appropriate tertiary configuration in which stereochemical specificity was incorporated.[105] Such direct template theories enjoyed a broad popularity at the time, since they appeared to present the only reasonable explanation for the large number of antibodies that Landsteiner had shown could be formed by the vertebrate host.

In contrast, the developing Age of Immunobiology in the late 1950s and 1960s, arising from observations in such fields as tissue transplantation, immunologic tolerance, immunodeficiency diseases, and immunopathology, caused a shift toward more biological directions. The biologists pointed out that "chemical" theories failed to explain how antibody production could persist in the apparent absence of antigen, or why a second exposure to antigen should result in an enhanced booster response. Moreover, they provided no explanation for the newer data which suggested that repeated immunization produces changes in the *quality* of the antibody, in some instances sharpening the specificity and in others considerably broadening the potential for cross-reactions.[106] The demands upon a theory of antibody formation made by biologists now tended to emphasize exactly these and other biological phenomena, but in the process the structural-chemical aspects of specificity and the repertoire problem were neglected.

This substitution of biological considerations for chemical ones is well illustrated in the several theories that Macfarlane Burnet advanced to explain antibody formation. (Burnet was, at this time, almost unique in bringing a broad background in biology to bear upon questions of theoretical immunology.) In 1941, he proposed an instructionist alternative to the earlier theories, which held that the function of antigen was to stimulate an adaptive modification of those enzymes necessary for globulin synthesis, such that a unique protein molecule with the required specificity would result. But fashions in biology change fairly rapidly, and a decade later the concept of adaptive enzymes was somewhat out of style; now, protein formation was held to be encoded for in a "genome" of uncertain composition. Still impelled by essentially biological considerations, Burnet and Frank Fenner advanced an indirect template hypothesis,[107] in which each antigen was thought able to impress the information for its specific determinant upon the (?RNA) genome, against which indirect template a specific antibody might be formed. Not only would this new genocopy persist within the cell; it would also be reproduced from mother to daughter cells during proliferation, thus explaining persisting antibody formation and the heightened booster response.

From the earliest days of the study of immunity, immunologists had been perplexed by finding "natural antibodies" in the serum of normal animals, often present in modest amounts in the absence of overt exposure to the various organisms and toxins with which they reacted.[108] These could hardly be explained by any instruction theory that demanded the presence and even the persistence of antigen. For many decades, the specificity of these natural antibodies was questioned, their provenance remained mysterious, and their very existence was neglected in the main by the proponents of instruction theories of antibody formation. However, in 1955 Niels Jerne focused attention upon these natural antibodies by assigning to them the central role in a "natural selection" theory of antibody formation.[109] Jerne proposed, as had Paul Ehrlich before him, that the host could in fact synthesize small amounts of all possible antibody specificities as part of its normal physiologic processes. Jerne suggested that the function of antigen was merely to act as a "selective carrier" of natural antibody, transporting it to appropriate cells somewhere in the body, where it would signal the reproduction of molecules identical to those introduced – i.e., of specific antibody. Jerne's theory appeared to cope quite adequately with most of the biological aspects of antibody formation, and even included an explanation of the newly-discovered phenomenon of immunological tolerance. It neglected entirely, however, any discussion of a chemical basis for antibody specificity, or any consideration of the size of the antibody repertoire, which had doomed the similar Ehrlich side-chain theory some half-century earlier.

The historic importance of Jerne's theory lies in the fact that it appeared to provide for the first time a biologic alternative to the instruction theories of the immunochemists, and served as the stimulatory point of departure for biologically-oriented theoreticians. The seed that Jerne had planted was not long in germinating, and within three years Burnet, Talmage, and Lederberg had given birth to the clonal selection theory of antibody formation.[110] Central to this concept was Jerne's (and Ehrlich's) postulate that antibodies are natural products; they appear on the cell surface as receptors with which antigen can interact; such interaction signals clonal proliferation of a population of cells phenotypically restricted for the given antibody specificity; and some daughter cells of the clone differentiate into antibody-forming cells while others remain as immunologic memory cells, able to participate in an enhanced booster antibody response.

It is probably safe to say that the clonal selection theory took the increasingly biologically-oriented world of immunology by storm, save for a few instructionist diehards who were unhappy with its failure to deal adequately with the structural basis of immunologic specificity, and with the repertoire problem. Indeed, once the DNA control of antibody structure was accepted, clonal selection theory generated its own repertoire controversy, as described in Chapter 4.

The clonal selection theory appeared to accord well with developments in the new genetics,[111] and especially with Francis Crick's "Central Dogma," which held that information could only pass from nucleic acids to protein and not in the reverse direction,[112] and with the demonstration that the tertiary structure of proteins is under precise genetic control.[113] Moreover, the newer techniques of fluorescent antibody immunohistochemistry[114] and of hemolytic plaque assays,[115] which permitted for the first time the study of single immunocytes in large populations of cells, provided rapid confirmation of the principal aspects of clonal selection theory.

Notes and references

1. Judson, H.F., *The Eighth Day of Creation*, New York, Simon and Schuster, 1979. This book is the subject of a superb essay review by John T. Edsall (*J. Hist. Biol.*

13:141, 1980), which also emphasizes the importance of the search for biological specificity.

2. The early work in this field is reviewed by Porter, R.R., and Press, E.M., *Annu. Rev. Biochem.* **31**:625, 1962; Cohen, S., and Milstein, C., *Adv. Immunol.* **7**:1, 1962.

3. Niels Jerne, in his typically perceptive discussion of the status of contemporary immunology entitled "Waiting for the end" (*Cold Spring Harbor Symp. Quant. Biol.* **32**:591, 1967), first pointed out clearly the schism between *trans*-immunologists (the molecular immunochemists) and *cis*-immunologists (the cellular immunobiologists), and their difficulty in communicating with one another.

4. Mercurialis, Hieronymus, *De Morbis Puerorum*, Book I, Basel, 1584, pp. 17–21.

5. Fracastoro, Girolamo, *De Contagione et Contagiosis Morbis et Eorum Curatione*, 1546 (translation by W.C. Wright, with Latin text), New York, Putnam, 1930. Fracastoro's theory of acquired immunity is discussed in detail in Chapter 1.

6. Castiglioni, A., *A History of Medicine*, New York, Knopf, 1947, p. 243.

7. Rhazes, A., *Treatise on the Small-Pox and Measles* (translation by W.A. Greenhill), London, Sydenham Soc., 1848.

8. Castiglioni, note 6, p. 353 ff; p. 452 ff.

9. Fracastoro, note 5.

10. Pagel, W., "Paracelsus," *Dict. Sci. Biog.*, **10**:304, 1974. See also Pagel's *Paracelsus: An Introduction to Philosophical Medicine in the Era of the Renaissance*, Basel, Karger, 1958. Pagel also gives a summary of von Helmont's contributions in *Dict. Sci. Biog.*, **6**:253, 1972, and in *Bull. Hist. Med.* **30**:529, 1955.

11. Bates, D.G., "Thomas Sydenham," *Dict. Sci. Biog.* **13**:213, 1976.

12. Faber, K., *Nosography: The Evolution of Clinical Medicine in Modern Times*, 2nd edn, New York, Hoeber, 1930, pp. 1–27.

13. Faber, note 12, p. 16.

14. Leake, C.D., *A Historical Account of Pharmacology to the XX Century*, Springfield, Chas. Thomas, 1975.

15. Boissier de Sauvages, François, *Nosologia Methodica*, 1768.

16. Farley, J., *The Spontaneous Generation Controversy from Descartes to Oparin*, Baltimore, Johns Hopkins Press, 1974.

17. von Mayer, E., *A History of Chemistry* (translation by G. McGowan), London, Macmillan, 1898. A briefer and more readable account is in Leicester, H.M., *The Historical Background of Chemistry*, New York, Wiley, 1956.

18. Fruton, J., *Molecules and Life: Historical Essays on the Interplay of Chemistry and Biology*, New York, Wiley-Interscience, 1972, pp. 22–86. See also Leicester, H.M., *Development of Biochemical Concepts from Ancient to Modern Times*, Cambridge, Harvard University Press, 1974, pp. 176–188.

19. Fischer, E., *Ber. Deutsch. chem. Ges.* **27**:2992, 1894.

20. Lewis, G.N., "The atom and the molecule," *J. Am. Chem. Soc.* **38**:762, 1916.

21. Pauling, L., *The Nature of the Chemical Bond*, 2nd edn, Ithaca, Cornell University Press, 1944.

22. Partington, J.R., *A History of Chemistry*, Vol. IV, London, Macmillan, 1965, p. 729 ff. See also Mazumdar, P.M.H., *Species and Specificity: An Interpretation of the History of Immunology*, Cambridge, Cambridge University Press, 1995, pp. 214–236. Pauli's book of essays, *Physical Chemistry in the Service of Medicine* (translation by M.H. Fischer), New York, Wiley, 1907, intimately connects the colloidal state with life processes.

23. Fruton, J., note 18, pp. 87–179.

24. Leibnitz, G.W., "Nouveaux Essais sur l'Entendement," in Gerhardt, C.J., ed., *Die philosophischen Schriften von Gottfried Wilhelm Leibnitz*, Vol. 5, Berlin, Wiedmann, 1882, p. 49.

25. Kant, I., *Critique of Pure Reason* (translation by N.K. Smith), New York, Macmillan, 1965, p. 540 ff.

26. Mayr, E., and Provine, W.B., eds, *The Evolutionary Synthesis: Perspectives on the Unification of Biology*, Cambridge, Harvard University Press, 1980.

27. Mazumdar, P.M.H., note 22.

28. Ehrlich, P., *The Collected Papers of Paul Ehrlich*, Vol. 1, London, Pergamon, 1956, pp. 65–94.

29. The background to Ehrlich's interest in the physiological function of receptors and his later application of this concept in immunology and pharmacology is detailed in Silverstein, A.M., *Paul Ehrlich's Receptor Immunology: The Magnificent Obsession*, New York, Academic Press, 2002.

30. Marquardt, M., *Paul Ehrlich*, New York, Schuman, 1957; Bäumler, E., *Paul Ehrlich: Scientist for Life* (translation by G. Edwards), New York, Holmes and Meier, 1984.

31. Ehrlich, P., "Die Wertbemessung des Diphtherieheilserums," *Klin. Jahrb.* 60:299, 1897 (English translation in *Collected Papers*, Vol. 2, pp. 107–125).

32. For analyses of Ehrlich's toxin–antitoxin work, see Mazumdar, P.M.H., "The antigen–antibody reaction and the physics and chemistry of life," *Bull. Hist. Med.* 48:1, 1974; Rubin, L.P., *J. Hist. Med.* 35:397, 1980; and Silverstein, *Paul Ehrlich's Receptor Immunology*, note 29.

33. Metchnikoff, E., *Immunity in the Infectious Diseases* (1905), Johnson reprint, New York, 1968.

34. The battle over the phagocytic theory is detailed in Chapter 2.

35. Behring, E., and Kitasato, S., *Deutsch. med. Wochenschr.* 16:113, 1890. See also Behring, E., and Wernicke, E., *Z. Hyg Infektionskr.* 12:10, 45, 1892. Only some years later, after receiving his Nobel Prize, would Behring be ennobled by the Kaiser, and granted the coveted "von."

36. In addition to diphtheria and tetanus antitoxins, Ehrlich himself had shown that antibodies could be formed against the plant toxins ricin and abrin (Ehrlich, P., *Deutsch. med. Wochenschr.* 17:976, 1218, 1890).

37. von Gruber, M., and Durham, H.E., *Münch. med. Wochenschr.* 43:285, 1896.

38. Kraus, R., *Wien. klin. Wochenschr.* 10:736, 1897.

39. Bordet, J., *Ann. Inst. Pasteur* 12:688, 1899.

40. Ehrlich's concept of distinct domains on the antibody molecule would find striking confirmation three-quarters of a century later in the work of Edelman, G.M., *Biochem.* 9:3197, 1970.

41. Müller-Eberhard, H.J., *Adv. Immunol.* 8:1, 1968. See also Lachmann, P.J., "Complement before molecular biology," *Mol. Immunol.* 43:496, 2006.

42. Ehrlich, P., "Croonian Lecture – on Immunity...," *Proc. R. Soc. Lond.* 66:424, 1900. This is also the clearest exposition of his side-chain theory.

43. Bordet, J., *Traité de l'Immunité dans les Maladies Infectieuses*, Paris, Masson, 1920, p. 504.

44. See, for example, Aschoff, L., *Ehrlichs Seitenkettentheorie und ihre Anwendung auf die künstlichen Immunisierungsprozesse*, Jena, Gustav Fischer, 1902; Römer, P., *Die Ehrlichsche Seitenkettentheorie und ihre Bedeutung für die medizinischen Wissenschaften*, Vienna, Hölder, 1904.

45. See, for example, Parascandola, J., and Jasensky, R., *Bull. Hist. Med.* **48**:199, 1974. J. Parascandola's paper on "The theoretical basis of Paul Ehrlich's chemotherapy" (*J. Hist. Med.* **36**:19, 1981) is especially interesting in this regard.

46. The L_0 (*limes* = threshold) dose of toxin would just neutralize 1 unit of antitoxin, while the L_+ (*limes* death) dose would leave 1 lethal dose free, in the presence of 1 unit of antitoxin. Thus, L_+ minus L_0 should equal 1 minimum lethal dose (MLD), but in fact might equal 40–60 MLDs or more.

47. Ehrlich, P., *Berl. klin. Wochenschr.* **40**:793, 825, 848, 1903. Ehrlich's toxin spectra were elaborated upon by T. Madsen (*Z. Hyg. Infektionskr.* **32**:214, 1899), and figured significantly in every comprehensive text on immunology and microbiology for the next thirty years.

48. Danysz, J., *Ann. Inst. Pasteur* **16**:331, 1902. Jules Bordet had earlier reported a similar observation on immune hemolysis (*Ann. Inst. Pasteur* **14**:257, 1900).

49. von Dungern, E., *Deutsch. med. Wochenschr.* **30**:275, 310, 1904.

50. Ehrlich P., and Morgenroth, J., *Berl. klin. Wochenschr.* **38**:569, 1901. English translation in Ehrlich, P., *Collected Studies in Immunity*, New York, Wiley, 1906, p. 88.

51. Ehrlich, P., and Sachs, H., *Berl. klin. Wochenschr.* **39**:297, 335, 1902.

52. Jules Bordet's side of this argument is well summarized in a resumé chapter of his book *Studies in Immunity* (translation by F. Gay), New York, John Wiley & Sons, 1909, p. 496 ff, and in his famous book *Traité de l'Immunité*, (note 43), written while Bordet was isolated in Belgium by the war.

53. Mazumdar (note 32) discusses at length the importance attributed at that time to colloidal interactions, and the almost mystical role assigned to them in biological processes.

54. Bordet, *Traité de l'Immunité*, (note 43), p. 546.

55. Bordet, *Traité de l'Immunité*, note 43, p. vi ff. Bordet's use of the epithet "chauvinism" against his German colleagues is not surprising. We have commented in Chapter 2 on the influences of Franco-German enmity on immunologic disputes; see also Rubin, note 32.

56. See, for example, Parascandola, J., *Pharmacy in History* **16**:54, 1974; Kohler, R.E., *J. Hist. Biol.* **8**:275, 1975; and Fruton, J., *Molecules and Life*, note 18.

57. Arrhenius, S., and Madsen, T., "Physical chemistry as applied to toxins and antitoxins," in Salomonson, C.J., ed., *Festskrift ved Indvielsen af Statens Serum Institut*, Copenhagen, 1902.

58. Arrhenius, S., *Immunochemistry*, New York, Macmillan, 1907.

59. See, for example, Neisser, M., *Zentralbl. Bakt. Parasitkde* **36**(1):671, 1904. Rubin, (note 32) provides an interesting discussion of the substantive and stylistic differences between Ehrlich and Arrhenius, and touches on the broader debate between "physical chemists" and "biologists," an issue which Ehrlich himself addressed in his second Herter Lecture at Johns Hopkins in 1904 (*Collected Papers*, Vol. II, p. 414). In this lecture, curiously, he defended biology from the physical chemist Arrhenius, whereas earlier he had attacked the "biologist" Bordet from a chemical position.

60. See, for example, the several papers on antitissue antibodies (cytotoxins) in *Ann. Inst. Pasteur* **14**, 1900.

61. An encyclopedic four-volume review of the entire field of allergy, with much historical background, will be found in Schadewaldt, H., *Geschichte der Allergie*, München-Deisenhofen, Dustri, 1979.

62. Portier, P., and Richet, C., *C. R. Soc. Biol.* **54**:170, 1902.
63. Arthus, M., *C. R. Soc. Biol.* **55**:817, 1903.
64. von Pirquet, C., and Schick, B., *Die Serumkrankheit*, Leipzig, 1905. English translation, *Serum Sickness*, Baltimore, Williams & Wilkins, 1951.
65. Koch, R., *Deutsch. med. Wochenschr.* **17**:101, 1891.
66. Nuttall, G.H.F., *Blood Immunity and Blood Relationship*, Cambridge, The University Press, 1904.
67. Mazumdar (note 22) has discussed the philosophical differences between the school that saw Nature as composed of discretely different entities and the other that viewed Nature as composed of a seamless continuum of related components.
68. von Gruber, M., and von Pirquet, C., *Münch. med. Wochenschr.* **50**:1193, 1903.
69. Ehrlich, P., *Münch. med. Wochenschr.* **50**:2295, 1903; English translation in Ehrlich's *Collected Studies*, note 49, p. 514 ff.
70. von Gruber, M., *Wien. klin. Wochenschr.* **16**:791, 1903.
71. Ehrlich, P., *Collected Studies*, note 50, p. 525.
72. Ehrlich, P., *Collected Studies*, note 50, p. viii.
73. von Gruber, M., *Münch. med. Wochenschr.* **48**:1214, 1901. See also Hopf, L., *Immunität und Immunisierung*, Pietzker, Tübingen, 1902, p. 89. Hans Buchner had also raised this perplexing question earlier (*Münch. med. Wochenschr.* **47**:277, 1900).
74. Bordet, J., *Studies in Immunity* (translation by F. Gay), New York, John Wiley & Sons, 1909.
75. For further information on Landsteiner and his work, see: Lesky, E., *The Vienna Medical School in the 19th Century*, Baltimore, Johns Hopkins University Press, 1976; Speiser, P., in *Dict. Sci. Biog.* **7**: 622, 1973; Mazumdar, note 22; Mazumdar, P.M.H., *J. Hist. Biol.* **8**:115, 1975; and Speiser, P., and Smekal, F.G., *Karl Landsteiner: The Discoverer of Blood Groups and a Pioneer in the Field of Immunology* (translation by R. Rickett), Vienna, Brüder Hollinek, 1975.
76. Landsteiner, K., *Centralbl. Bakt. Orig.* **27**:357, 1900; *Wien. klin. Wochenschr.* **14**:1132, 1901.
77. Landsteiner, K., and Levine, P., *Proc. Soc. Exp. Biol. Med.* **24**:600, 941, 1926.
78. Landsteiner, K., and Wiener, A. S., *Proc. Soc. Exp. Biol. Med.* **43**:223, 1940.
79. Finger, E., and Landsteiner, K., *Arch. Dermatol. Syph.* **78**:335, 1906.
80. Landsteiner, K., and Levaditi, C., *C. R. Soc. Biol.* **67**:592, 789, 1909.
81. Donath, J., and Landsteiner, K., *Münch. med. Wochenschr.* **51**:1590, 1904.
82. Landsteiner, K., Müller, R., and Potzl, O., *Wien. klin. Wochenschr.* **20**:1565, 1907.
83. Landsteiner, K., and Popper, E., *Z. Immunitätsforsch.* **2**:377, 1909. Landsteiner's important contributions to the study of poliomyelitis led to his election to the Polio Hall of Fame.
84. Landsteiner, K., and Chase, M.W., *Proc. Soc. Exp. Biol. Med.* **49**:688, 1942.
85. Landsteiner, K., *The Specificity of Serological Reactions*, New York, Dover, 1962, a reprint of the 2nd (1945) edition with a complete bibliography. The original German version, *Die Spezifizität der serologischen Reaktionen*, was published in Berlin by Springer, 1933.
86. Chase, M.W., Personal communication cited in Corner, G.W., *A History of the Rockefeller Institute*, New York, Rockefeller Institute Press, 1964, p. 205; also in Speiser and Smekal, note 75.
87. Lesky, E., *The Vienna Medical School*, note 75.
88. Landsteiner, K., *Wien. klin. Wochenschr.* **10**:439, 1897.

89. Landsteiner, K., and Jagic, N., *Münch. med. Wochenschr.* 50:764, 1903.
90. Pick, E.P., in Kolle, W., and von Wassermann, A., *Handbuch der pathogenen Mikroorganismen*, 2nd edn, Vol. 1, pp. 685–868, Jena, Fischer, 1912.
91. Obermayer, F., and Pick, E.P., *Wien. klin. Wochenschr.* 19:327, 1906.
92. Landsteiner, K., and Lampl, H., *Z. Immunitätsforsch.* 26:258, 293, 1917.
93. Landsteiner, K., and Lampl, H., *Biochem. Z.* 86:343, 1918.
94. Lesky's book on the Vienna Medical School (note 75), while concentrating on its nineteenth century greatness, shows also how it and intellectual Vienna declined following the War.
95. Landsteiner, *The Specificity*, note 85, p. 266.
96. Landsteiner, K., and van der Scheer, J., *Proc. Soc. Exp. Biol. Med.* 21:252, 1924; *J. Immunol.* 9:213, 221, 1924.
97. Landsteiner, K., and van der Scheer, J., *J. Exp. Med.* 40:91, 1924.
98. Landsteiner, *The Specificity*, note 85, p. 76. See also Landsteiner's presidential address before the American Association of Immunologists (*J. Immunol.* 15:589, 1928).
99. Fruton, note 18, pp. 87–179.
100. Buchner, H., *Münch. med. Wochenschr.* 40:449, 1893.
101. See, for example: Roux, E., and Vaillard, L., *Ann. Inst. Pasteur* 7:65, 1893; Knorr, A., *Münch. med. Wochenschr.* 45:321, 362, 1898.
102. Heidelberger's important contributions are perhaps best summarized in the landmark book by his students: Kabat, E.A., and Mayer, M.M., *Quantitative Immunochemistry*, 2nd edn, Springfield, Charles C. Thomas, 1961.
103. The titles of the leading *basic science* books in immunology testify that it was indeed an "Age of Immunochemistry." Thus: Wells, H.G., *The Chemical Aspects of Immunity*, New York, Chemical Catalog Co., 1924; Marrack, J.R., *The Chemistry of Antigens and Antibodies*, London, HMSO, 1934; Kabat, E.A., and Mayer, M.M., *Quantitative Immunochemistry*, (note 102, 1st edn, 1949); Boyd, W.C., *Introduction to Immunochemical Specificity*, New York, Wiley-Interscience, 1962; Kabat, E.A., *Structural Concepts in Immunology and Immunochemistry*, New York, Holt, Rinehart, and Winston, 1968; Pressman, D., and Grossberg, A., *The Structural Basis of Antibody Specificity*, New York, Benjamin, 1968. The leading textbook in the field, Boyd, W.C., *Fundamentals of Immunology*, New York, Wiley-Interscience, 1943, was quite chemically oriented, and was only replaced in 1963 by a text aimed at biologists: Humphrey, J.H., and White, R.G., *Immunology for Students of Medicine*, Philadelphia, Davis, 1963.
104. Breinl, F., and Haurowitz, F., *Z. physiol. Chem.* 192:45, 1930; Alexander, J., *Protoplasma* 14:296, 1931; Mudd, S., *J. Immunol.* 23:423, 1932.
105. Pauling, L., *J. Am. Chem. Soc.* 62:2643, 1940.
106. The biologists' arguments against chemical template theories are best summarized in Burnet, F.M., *The Production of Antibodies*, Melbourne, Macmillan, 1941.
107. Burnet, F.M., and Fenner, F., *The Production of Antibodies*, 2nd edn, Melbourne, Macmillan, 1949.
108. The history of natural antibodies is discussed in Chapter 16.
109. Jerne, N.K., *Proc. Natl Acad. Sci. US* 41:849, 1955.
110. A clonal selection theory was first hinted at by Talmage, D.W., *Annu. Rev. Med.* 8:239, 1957, p. 247; outlined by Burnet, F.M., *Austral. J. Sci.* 20:67, 1957; and fleshed out in detail by Burnet, F.M., *The Clonal Selection Theory of Acquired Immunity*, Cambridge, Cambridge University Press, 1959. See also Talmage, D.W., *Science* 129:1643, 1959; and Lederberg, J., *Science* 129:1649, 1959.

111. Highly readable accounts may be found in: Olby, R., *The Path to the Double Helix*, Seattle, University of Washington Press, 1974; and Judson, H.F., *The Eighth Day of Creation*, note 1.
112. Crick, F.H.C., in *Symp. Soc. Exp. Biol.* **12**:138, 1958. See also Crick, F.H., *Nature Lond.*, **227**:561, 1970.
113. Epstein, C.J., Goldberger, R.F., and Anfinsen, C.B., *Cold Spring Harbor Symp. Quant. Biol.* **28**:439, 1963.
114. Coons, A.H., Leduc, E.H., and Connolly, J.M., *J. Exp. Med.* **102**:49, 1955.
115. Jerne, N.K., and Nordin, A.A., *Science* **140**:405, 1963.

7 Immunologic specificity: solutions

...enzyme and glucoside [read antibody and antigen] must join one another as lock and key...to exert a chemical effect.

Emil Fischer

It was evident by the early 1930s that if Paul Ehrlich's biologic theory of antibody formation was out of favor, his chemical concept of the basis for antibody specificity was very much in vogue.[1] The very data (primarily Landsteiner's work with synthetic haptens) that had made the antibody repertoire appear too large to be explicable in terms of naturally occurring antibodies almost demanded an immunologic specificity based upon a very precise stereochemical complementarity of configuration between antigen and the putative combining site of antibody. Indeed, as we have seen, such a precise structural "fit" between antigen and antibody was explicitly required by most instructionist theories of antibody formation, each of which postulated the existence of some form of template upon which specific configuration might be molded.[2] But the work of Landsteiner on artificial haptens and of Heidelberger and coworkers on polysaccharide antigens accomplished more than this; they signaled to a generation of immunologists that progress in understanding the functions of antibody and the nature of its specificity would only come from chemical approaches to the problem. Nor did biologically-oriented immunologists have much to offer at this time in competition with the new trend. Their startling and attractive advances in antitoxic and antibacterial immunity, in novel techniques of serodiagnosis and serotaxonomy, and their important contributions to forensic medicine were mostly a generation in the past, while the discovery of immunologic tolerance and deficiency diseases, of transplantation immunobiology and of cellular functions in the immune response would only come with the new biology of a future generation. The occasional development of a new vaccine or the finding of a new blood group thus had little effect upon the growing influence of immunochemistry within the larger field of immunology between about 1920 and 1960.[3]

The introduction of more chemical approaches to immunology – of quantitative methods and studies of the fine structure of antigens and antibodies – had profound implications for the science of immunology. Not only did it reorient the research goals of a generation of scientists; it also led to the production of impressive amounts of "hard" data that altered the very direction of immunologic conceptualization. It is typical of the development of a science that in its infancy, conceptual advances are often based primarily upon philosophical viewpoints, given the scarcity and uncertainty of the facts at hand. As the science matures, hypotheses tend to depend less upon the world view of the scientist, and more upon the imperatives contained within the growing body of evidence itself.

A History of Immunology, Second Edition
ISBN: 978-0-12-370586-0 Copyright © 2009, Elsevier Inc.
All rights reserved

This phenomenon has been no less true of the development of the concept of immunologic specificity than in other fields of biology – a source of potential hazard to the chronicler who attempts to trace the history of an idea through the entire timespan.

If the earlier period lends itself to more philosophical approaches, and furnishes interesting accounts of often vitriolic debates (which the times and the journals then permitted), modern developments tend to make for a drier and more factual presentation. Not only are there more facts to deal with, but the very training and background of the scientist concerned also becomes an important factor. During the last half of the nineteenth century, the scientist was more likely to have had a classical education that predisposed for a broad philosophical approach to his discipline, and could hope to comprehend his own as well as other related disciplines. In the mid-twentieth century, education for a more complex science was often at the expense of the humanities, and the scientist found it difficult to encompass fully even his own subspecialty. The gap between C. P. Snow's "Two Cultures" is thus reflected not only in the relationship between science and society, but to a degree also in the "generation gap" that develops within the science itself.

The structural basis of immunological specificity

By the 1930s, it was known that antibodies are protein in nature, that they belong to the class of proteins termed globulins, and that antibody activity can be found variously in both the euglobulin and pseudoglobulin classes, as defined by solubility in water and ammonium sulfate solutions. But apart from the knowledge that proteins were composed of chains of apparently randomly arranged amino acids of indeterminate length, little was known of the protein molecule.[4] A theory had been advanced that the precipitation of antigen and antibody was attained by means of a molecular lattice,[5] which implied at least bivalency of the antibody; other than this, the nature of antibody structure and specificity was as little known as that of enzymes. Any approach to the definition of the antibody specific combining site would thus of necessity have to rely upon chemical studies of the antigenic determinants with which they interacted.

Approaches to specificity via the antigen molecule

In his studies of the serologic cross-reactions among homologous series of structurally related haptens,[6] Karl Landsteiner provided a powerful tool which permitted the size and shape of the specific site on antibody to be estimated. His demonstration that the precipitation of antihapten antibodies by hapten–protein conjugates might be inhibited by free hapten[7] was further seized upon as a means of estimating the thermodynamic characteristics of the antigen–antibody interaction. These predominantly physicochemical approaches provided a wealth of new, if indirect, information on the structure and function of the specific combining site of antibody.

The shape of the specific antibody site

The strength of this new approach to the definition of antibody specificity was most forcefully provided by the studies of Linus Pauling and his scientific descendants. By combining quantitative hapten-inhibition studies with the newer knowledge of atomic size and of the orientation of interatomic bonds within and between molecules, Pauling and his students (most notably David Pressman) were able to define precisely, in terms of their van der Waal's radii, the configuration of various haptens, and therefore by inference the configuration of the "pockets" in the specific antibody site into which they fit. Such studies served also to provide a measure of the varying contributions to the antigen–antibody interaction of ionic interactions, of hydrogen bonding, and of van der Waal forces. These studies are summarized *in extenso* in Pressman and Grossberg's book on *The Structural Basis of Antibody Specificity*,[8] where even the difference in size between a chlorine and a bromine atom on a benzene ring is shown to influence the binding affinities of haptens, and where it is shown that even the influence of the water of hydration of a hapten molecule in solution can be measured.

As more information became available on the correspondence between the three-dimensional structure of haptens and their ability to combine with specific antibody, molecular diagrams of the type illustrated in Figure 7.1 were drawn,[9] leading to the suggestion by Hooker and Boyd, and then by Pauling, that these structures in fact define a cavity in the globulin molecule into which the hapten might fit more or less tightly,[10] representing thus an interaction of greater or lesser affinity. More careful measurements appeared to show, however, that antibody might not always react with the entire haptenic grouping, especially when the latter attained sizeable proportions – an observation that led Pressman

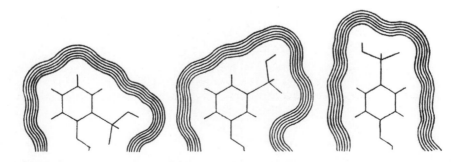

Figure 7.1 Scale drawing of the antibody cavities specific for ortho-, meta-, and para-azophenylarsonic acid groups.
From Pauling, L., and Pressman, D., *J. Am. Chem Soc.* **67**:1003, 1945.

to suggest[11] that the cavity in the globulin molecule that determined antibody specificity might sometimes form as an invagination, while in other instances it might be pictured as either a shallow trough or a slit trench,[12] as illustrated in Figure 7.2.

Antibody heterogeneity and thermodynamics

Hapten inhibition studies of immune precipitation quickly confirmed what had long been known – that an immune serum to even a well-defined haptenic grouping was apparently composed of a fairly heterogeneous mixture of antibodies of different affinities. With the revival in the late 1940s by Eisen and

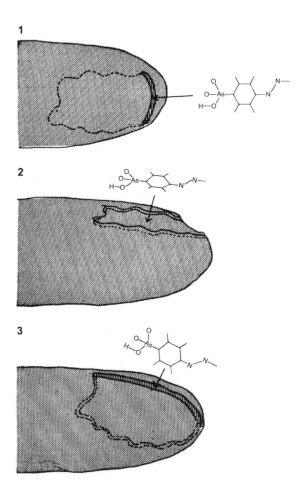

Figure 7.2 Speculation on the possible ways that the specific combining site might be arranged on the antibody molecule.
From Boyd, W.C., *Introduction to Immunochemical Specificity*, New York, Interscience, 1962.

Karush of the technique of equilibrium dialysis (involving direct measurements of hapten–antibody interactions free of the complications of such secondary phenomena as precipitation), these observations were elegantly extended and given a firm quantitative basis.[13] Now, for the first time, it was possible to obtain direct measurements of hapten–antibody interactions, and to measure these interactions in absolute rather than relative terms. By assessing the degree of hapten binding at different free hapten concentrations, and at different temperatures, measurements could now be made of the free energy of interaction of the hapten and antibody combining site, and of the enthalpy and entropy changes associated with these interactions.[14] Only then was the remarkable range of antibody affinities first appreciated, some reacting only weakly with the antigen with association constants of the order of 10^4 liters/mole, while others might react with their respective haptens with association constants of 10^8 to 10^{10} liters/mole or higher.

Equilibrium dialysis provided two additional types of information about antibodies that were invaluable. From the law of mass action, an appropriate plot of hapten binding data at different initial hapten concentrations should yield a straight-line isotherm. Any deviation of the curve from linearity is a measure of the heterogeneity of affinities in the antibody population measured, and so it was possible for the first time to obtain quantitative estimate of the heterogeneity of different antisera, and thus of the range of specific affinities present in the mixture.[15] The second advantage of this type of data plot lay in the fact that extrapolation of the curve to the abscissa would give a precise estimate of the valence of the antibody molecule. Immunologists had argued for many years about whether the antibody molecule had only one combining site or many, some suggesting that parsimony of hypothesis did not require more than univalency. Indeed, one prominent immunologist suggested that the single antibody combining site on the globulin molecule was itself so much a miracle that it would be too much to insist upon two such miracles on the same molecule![16] But equilibrium dialysis settled this question, since it showed that most antibodies were in fact divalent.

The growing notion that a specific antiserum might be composed of a mixture of many different antibodies with greater or lesser "fit," or affinity for the antigenic structure against which they were formed, had a further interesting implication for the concept of immunologic specificity. This was pointed out most clearly by Talmage in 1959,[17] although in a somewhat different context – that of an attempt to explain away the apparently large size of the immunologic repertoire, a subject to which we shall return below. Following an earlier suggestion by Landsteiner,[18] Talmage noted that if indeed the antibody response is degenerate, and results in a mixture of antibodies of overlapping specificities and varying affinities for antigen, then such a hetero-geneous mixture might appear to react more specifically and to discriminate more finely between related antigenic structures *than would any of its constituent antibodies*. An antigen might fit only partially into the combining site of any particular antibody present in relatively low concentrations, but would be well recognized by the totality of all combining sites in a heterogeneous antiserum. By postulating that an antiserum might manifest different specificities depending only upon variations in the relative concentrations of a *limited* number of different specific antibodies, Talmage suggested that the requirement for an unlimited repertoire of antibodies was sharply reduced. In addition, this concept accorded well with the clonal selection theory, which implied the existence of a discrete and discontinuous set of more-or-less specific receptors in the immune response, rather than a continuously changing spectrum of affinities such as Landsteiner, and Gruber before him, had suggested.[19]

The size of the antibody combining site

Numerous studies by Landsteiner and others had shown that a single terminal saccharide, a substituted benzene ring, or even a dipeptide might suffice to determine the specificity of an antibody combining site, and this appeared to set the lower limit on its size. But with the finding that the carrier molecules to which these "immunodominant" groups were attached might influence the antigen–antibody interaction, interest was focused on the maximum size that the antibody combining site might attain. An early and imaginative approach to this question was made by Landsteiner and van der Scheer,[20] who immunized animals with "two-headed haptens," synthesized by attaching symmetrically to a benzene ring two distinctive groupings (such as *sym*- aminoisophthalyl glycine-phenylalanine, or (3-amino, 5-succinylaminobenzoyl)-*p*-aminophenylarsenic acid). The antibodies formed against these large structures were invariably specific for one or the other of the two determinants on the molecule, and never appeared large enough to encompass both. Using these results, Campbell and Bulman calculated that the specific combining site of an antibody could not be larger than some 700 $\mathring{A}^2$.[21]

A more detailed study of the size of the antibody combining site was made by Kabat.[22] This investigation took advantage of the ability to prepare antibodies against dextran, composed of long chains of one to six linked glucose units, and of the availability of all of the oligosaccharides of glucose from the disaccharide isomaltose to the heptasaccharide isomaltoheptaose. By testing the ability of the various oligosaccharides to inhibit the precipitation of dextran by antidextran, it was possible to calculate that whereas simple glucose might contribute some 40 percent to the total binding energy of the interaction, the addition of further saccharide residues contributed successively less to the interaction, until no further effect could be found beyond isomaltohexaose. These results appeared to set an upper limit on the determinant size of $34 \times 12 \times 7 \mathring{A}$, if the unlikely assumptions were made that the molecule in solution interacted in its extended form. These findings were substantially confirmed by other investigators employing D-lysine oligopeptides, where it was found that oligomers beyond a chain length of five to six units made little or no additional contribution to the energy of interaction.[23]

Approaches to specificity via the antibody molecule

Paul Ehrlich's suggestion in 1897 that immunologic specificity is based upon a three-dimensional arrangement of atoms in the antibody molecule that permits a close complementarity to the antigen was a brilliant conceptual leap for his times, the verification of which would have to wait more than a half-century for the development of appropriate technologies. As we have seen, progress in understanding the protein molecule was almost nonexistent prior to World War II, and only indirect information could be obtained about antibodies by studying antigens and haptens. But starting in the late 1930s and for a quarter-century thereafter, a series of technical innovations initiated an explosive burst of progress that permitted the structure of the antibody molecule and the location and nature of its combining sites to be worked out in the finest detail.

The initial steps in defining the antibody molecule were due to the development of ultracentrifugation by Svedberg and of electrophoresis by Tiselius, and especially of the subsequent modification of the latter technique to permit immunoelectrophoresis in gels.[24] By allowing antibodies to be separated by weight and by electrical charge, it was established that some antibodies had a molecular weight of about 160,000, while others (macroglobulins) had a molecular weight of almost one million. Again, some antibodies were found to migrate slowly in an electrical field in the γ-globulin region, while others might migrate in the β- and even α-globulin regions. Most interesting was the observation that differences in biologic function (fixation of complement, passage across the placenta, involvement in allergic disorders, etc.) might be correlated with these physical differences. What emerged most forcefully from these early studies was an appreciation of the fact that antibodies, unlike most other serum proteins, constituted a distinctly heterogeneous population of molecules, related in some way to their heterogeneity of specificities and/or to their heterogeneity of biological function.[25]

With the increasing ability to characterize different molecular species, and especially with the use of antibody probes with which the *antigenic* character of antibody globulins could be tested, dissection of the immunoglobulin molecule (as it soon came to be known) could be undertaken. Two principal approaches were pursued, which strongly complemented one another. The first was the finding that the immunoglobulin molecule might be selectively split by such enzymes as papain and trypsin,[26] and the second was the finding that reductive cleavage of disulfide bonds of the immunoglobulin molecule would lead to a different set of products.[27] Edelman and Poulik[28] then showed that immunoglobulins were composed of two polypeptide chains with molecular weights of 20,000 and about 50,000 respectively (later termed the light (L) and heavy (H) chains). Taken together with the enzyme cleavage results, Porter was able to suggest a structure of the immunoglobulin molecule[29] as illustrated in Figure 7.3a, and it was now evident that the specific antibody combining site must somehow be formed utilizing portions of both the H and L chains. In addition, it soon became clear that it was a portion of the heavy chain that

defined the secondary biological functions of different immunoglobulin classes –
a finding that was formalized by Edelman[30] (Figure 7.3b) in his description of
the immunoglobulin molecule in terms of a combination of subunit chains so
assembled as to give rise to a set of different functional domains (thus validating
Ehrlich's speculation of some seventy years earlier).

All of these structural studies were aided immeasurably by the growing
appreciation that the abnormal proteins present in the serum of multiple
myeloma patients were fairly homogeneous populations of immunoglobulin

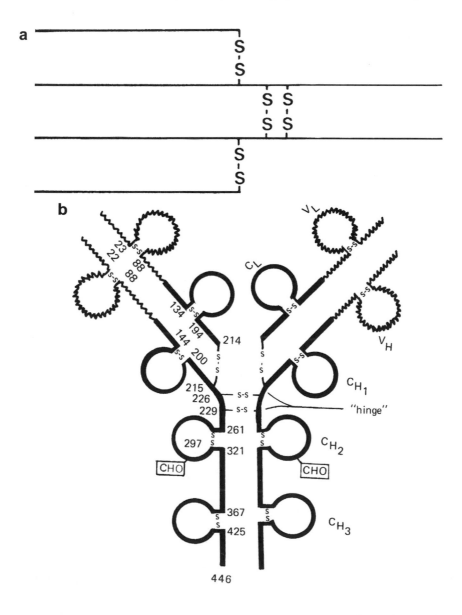

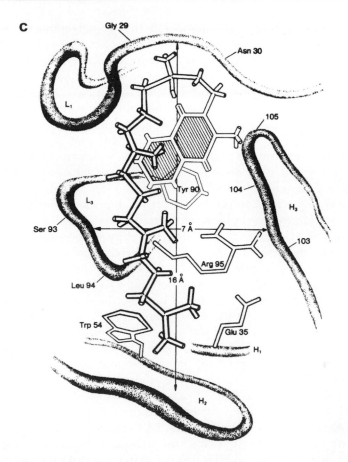

Figure 7.3 Changing concepts of the immunoglobulin molecule. **a**, the four-chain stick model of Porter in 1963; **b**, the heavy chain domains of Edelman in 1970; and **c**, the three-dimensional cavity specific for vitamin K, formed of L and H chain segments (from Poljak et al, 1974, note 35).

a, from Porter, R.R., *Br. Med. Bull.* **19**:197, 1963; b, from Edelman, G.M., *Biochemistry* **9**:3197, 1970; c, from Poljak, R.J., Amzel, L.M., Avey, H.P., Chen, B.L., Phizackerley, R.P., and Saul, F., *Proc. Natl Acad. Sci. US* **71**:1427, 1974.

molecules,[31] and that the Bence Jones proteins found in the urine of such patients were in fact free immunoglobulin light chains.[32] Based upon studies of this type, it was finally possible to define a set of immunoglobulin classes (isotypes) and subclasses which depend upon the presence of distinctive heavy chains, each with somewhat different biological properties: immunoglobulin G (IgG), composed of two gamma heavy chains and two kappa or lambda light chains with a molecular formula H_2L_2; the pentameric macroglobulin IgM with mu heavy chains; IgA (mono- or dimeric), involved in the secretory immune system, with alpha heavy chains; IgE, involved in allergic disease with epsilon heavy chains; and IgD, currently assumed to function as a lymphocyte membrane receptor, utilizing delta heavy chains.

With the increasing ability to split the light and heavy chains of immuno-globulins with a variety of enzymes, and with the development of improved techniques to fractionate and to establish the amino acid sequences of these fragments, it was finally possible to work out the complete primary structure of an entire immunoglobulin molecule, for which Edelman and Porter shared the Nobel Prize in 1972.

The ability to establish the primary amino acid sequence of immunoglobulin molecules did not of itself establish the location or structure of the specific binding site on the molecule, since secondary and tertiary configuration could not be directly inferred from primary structure. But such studies did open the way for a solution to this problem, along two interesting and complementary lines. The first of these derived from the new ability to compare the amino acid sequences of immunoglobulin chains from different antibody specificities and from different species. Not only were there homologies between light and heavy chains and among different segments of both light and heavy chains, suggesting an evolution through gene duplication,[33] but also the amino-terminal segments of both light and heavy chains showed impressive variations in amino acid sequence, especially in certain "hypervariable" regions, whereas the carboxy-terminal portions of these chains were much more constant in composition. A comparison of the sequences of many immunoglobulin chains permitted Wu and Kabat[34] to identify the precise locations of these hypervariable regions, and to suggest that these portions of the Ig chains were most likely to be involved in defining the specific combining site on the antibody molecule. It remained for high resolution X-ray crystallographic studies to establish the three-dimensional structure of the antibody molecule, to provide a physical picture of the combining site itself, and to confirm that it was in fact a sort of pocket formed by H and L chain hypervariable regions, into which the antigen determinant might fit with greater or lesser precision[35] (Figure 7.3c).

At long last, immunologic specificity had been provided with a firm structural basis. However, if a unique organization of amino acid sequences with a special spatial configuration were sufficient to determine an antibody binding site, it should also comprise an equally unique *antigenic* determinant on the immunoglobulin molecule, and this was soon confirmed and called an idiotype.[36] It has been possible to obtain antibodies specific for these special structures, whose use has contributed importantly to the analysis of immunologic specificities, and to a clarification of some of the mechanisms that may modulate the immune response.[37]

Specificity in cellular immunity

When Elie Metchnikoff introduced the notion of cellular immunity to infection in the 1880s, and when he defended and extended his theory over the next thirty years,[38] he identified macrophages (wandering monocytes and sessile histiocytes) and microphages (polymorphonuclear leukocytes) as the mediators of this protection. But Metchnikoff never really addressed in detail the question of the specificity of the phagocytes involved in protective immunity. This may have been due in part to his preoccupation with "natural immunity," where discrete specificity was not absolutely demanded. In his first comprehensive review of the subject, he speaks of the "sensitivity" of phagocytes to chemotactic factors released by pathogens and foreign bodies, which leads to enhanced diapedesis and the engulfing and intracellular enzymatic destruction of pathogens. The hallmark of *acquired* immunity, however, is specificity, and Metchnikoff had to concede that immunization might increase the sensitivity of phagocytes (by a mechanism unknown), thus enhancing diapedesis, immigration, and phagocytosis.[39]

When he wrote his famous *Immunity in the Infectious Diseases* in 1901,[40] Metchnikoff was forced to deal with the increasing evidence that not only did the finding of circulating antibody support the opposing theory of humoral immunity, but that these antibodies were themselves also highly specific. To counter opposition to this theory, he suggested that these antibodies were in fact merely "stimulins," serving to increase the sensitivity of phagocytes to foreign bacteria. Quoting his student Mesnil, he pointed out that "the effect of the [immune] serum is to stimulate the phagocyte...they ingest more quickly, they digest more quickly. The serum is, therefore, a stimulant of the cells charged with the defense of the animal."[41] Elsewhere, Metchnikoff hinted at a specificity of the phagocyte, and spoke of the immunized animals as possessing "leukocytes, impressed with a special sensitiveness...,"[42] but did not elaborate on this. But if Metchnikoff generally begged the question of phagocyte specificity in his writings, yet his theory of cellular immunity based upon phagocytic function imposed an implicit requirement for specificity, and the "specific phagocyte" would be a recurrent theme in any discussion of cellular immunity for the next seventy-five years.

For a time, it appeared that the question had been settled by the suggestion that circulating antibody might function as an opsonin (Greek *opsonein* – to render palatable) wherein antibodies were thought to coat the pathogen specifically, thus rendering it more susceptible to phagocytic action. This was based upon an observation by Denis and Leclef in 1895,[43] who found that the destruction of bacteria by phagocytosis was substantially increased by the addition of specific immune serum. These observations provided the basis for an extensive series of investigations by Wright and Douglas,[44] who sought to mediate the dispute between the cellular and humoral theories of immunity by showing that humoral antibody and phagocytic cells might *collaborate* in combating infectious disease. Indeed, the opsonic theory was broadly accepted for several decades, and appeared to explain quite satisfactorily the mechanism of acquired immunity in a number of infectious disease processes. But evidence slowly mounted that certain diseases, most notably tuberculosis, were not so readily explained.

From the very outset, macrophages had been observed to play an important role in the granulomatous inflammation associated with tubercle formation and with the Koch phenomenon, and to contribute importantly to the inflammatory infiltrate associated with positive tuberculin tests. However, it soon became clear that there was little or no correlation between circulating levels of anti-tubercle antibodies and immunity to this disease.[45] Moreover, immunity could not be passively transferred with serum antibody, so the intimate collaboration of opsonins and phagocytes in tuberculosis and many other significant diseases appeared questionable, and belief in a "knowledgeable" and specific macrophage was revived.

The concept of the immune macrophage was supported by extensive studies by Lurie in the 1930s and 1940s,[46] purporting to show with "purified" macrophage preparations that those from immune animals killed or inhibited the growth of tubercle bacilli better than those from normal donors. These findings received additional support from many investigators,[47] although some continued to insist that the protective function of these phagocytic cells was essentially nonspecific in nature.[48] One of the most suggestive observations supporting the notion of the immune macrophage was that of Rich and Lewis,[49] who showed that the normal migration of macrophages from *in vitro* explants of bits of spleen from tuberculous animals could be specifically inhibited by tuberculin.

It was first assumed that the macrophages were directly and specifically killed, which implied the presence on their surface membrane of receptors specific for the antigen involved. In due course, however, the weight of evidence forced the conclusion that while the macrophage might contribute importantly to cellular immunity and even to antibody formation, its functions were essentially nonspecific. The inhibition of macrophage migration proved to be due to the antigen-induced release of a soluble factor from extremely small numbers of contaminating specific lymphocytes.[50] The role of the macrophage in antibody formation was also shown to be a nonspecific one, involving the processing and/ or presentation of antigen to lymphocytes,[51] but not before it was suggested that antigen might induce in the macrophage the formation of a specific RNA which could transfer information to antibody-forming lymphocytes,[52] or that, less specifically, an antigen-macrophage RNA complex might serve as a "super antigen" in stimulating lymphocytes to antibody formation.[53]

Delayed-type hypersensitivity

Observations of two sorts helped slowly to define a dichotomy in the phenomenology of immunity and allergy. The first of these was the clinical finding that whereas the symptoms of local and systemic anaphylaxis, the skin test for allergy, and the hayfever–asthma group of allergies were all characterized by an almost immediate onset following antigenic challenge, the intradermal tuberculin test, the luetin test for syphilis, the lepromin test for leprosy, and the response to vaccination in the sensitized host all required twenty-four to forty-eight hours to develop. The former were grouped together under the heading *immediate* hypersensitivities, and were acknowledged to be due to the participation of humoral antibodies. The latter, on the other hand, were termed *delayed* hypersensitivities and, since circulating antibody could not be implicated in their pathogenesis, were ascribed to some type of cellular function, or to "cell-bound" antibodies.

The appropriateness of this division was made clear when Landsteiner and Chase demonstrated the ability to transfer these hypersensitivities passively with cells and not with serum antibody,[54] and when it was recognized that contact dermatitis, allograft rejection, and certain viral and autoallergic diseases somehow belonged to the same category of delayed hypersensitivity cellular responses. With the discovery that certain immunologic deficiency diseases might inhibit antibody formation and immediate hypersensitivities while others would impair delayed hypersensitivity and cellular immunity,[55] the stage was set for the establishment of a major division of lymphocytic function. Thymus-derived cells (T cells) were shown to function in cellular immunity, whereas avian bursal or mammalian bone-marrow derived cells (B cells) were shown to be responsible for antibody formation.[56]

The second line of investigation stemmed from the observation that somehow there was a major difference between immunologic responses to soluble exotoxins and innocuous antigens, and the body's response to infectious agents. This led to the early use of such terms as "immunity of infection" and "bacterial allergy." It was not until the late 1920s that Dienes and Schoenheit[57] demonstrated that this type of allergy could be induced even against bland antigens, by injecting them directly into the tubercles of infected animals – a procedure that was considerably simplified by the introduction of complete Freund's adjuvants containing dead mycobacteria. Now delayed-type hypersensitivity could be induced against purified proteins,[58] and it was not long before it was shown that hapten–protein conjugates would serve as well (analogous to Landsteiner's use of artificial antigens for the study of immediate hypersensitivities[59]). The parallelism was not exact, however, for while antibodies raised against a hapten coupled to carrier protein X would interact with the same hapten attached to protein Y, the delayed skin reaction against hapten–protein conjugates required that the carrier protein employed to elicit the response be the same as that used for sensitization.[60]

Here was a conundrum that taxed the ingenuity of investigators, since it appeared to call for a different order of immunologic specificity than that established by the study of serum antibodies. From one direction, Benacerraf and Gell suggested that the specific combining site that mediated delayed hypersensitivity might be larger in physical dimensions than that normally encountered on circulating antibody, and would thus encompass both the haptenic determinant and a portion of the adjacent carrier protein molecule.[61] From another direction (and somewhat neglectful of the "carrier-effect" while emphasizing the passive transfer experiments), Karush and Eisen suggested that these reactions were due to the participation of very small quantities of very high-affinity antibodies, with association constants of the order of 10^{10} or more.[62] But both of these hypotheses were eclipsed by a remarkable series of investigations that showed that in fact *two* different types of lymphocytes were required for responses to conjugated proteins; one a "helper" cell which recognizes the carrier protein, and the other an effector cell which recognizes the haptenic determinant.[63] In delayed hypersensitivity and other forms of cellular immunity the effector cells proved to be specialized subsets of T lymphocytes, whereas in antibody formation helper T cells were shown to collaborate with effector B cells (both sharing in specificity for antigen) by stimulating the latter to active antibody formation.[64] Evidence was soon forthcoming that antigen-specific T cells might serve other functions as well, such as participating in the feedback suppression of immune responses.[65]

But if lymphocytes are to engage in specific interactions, then they must have appropriate receptors on their surface membranes with the full repertorial range of immunologic specificities which immunocyte reactions have been shown capable of distinguishing. In the case of the B lymphocyte the demonstration of specific surface receptors proved fairly simple, thanks to such techniques as immunofluorescent analysis. These proved to be samples of the antibody specificities for which these cells were programmed, thus providing final vindication of Paul Ehrlich's side-chain theory of 1897, and of Macfarlane Burnet's clonal selection theory sixty years later.

The search for the T cell receptor, however, proved more elusive. As Marrack and Kappler point out, in their review of antigen-specific T cell receptors:[66]

> *Early attempts to isolate these proteins relied heavily on the idea that T cell receptors might be similar, if not identical, to immunoglobulin. In retrospect, although this idea was not unreasonable, it certainly created a good deal of confusion in the field.*

(It will be remembered that the extremely complicated mechanism for immunoglobulin formation was even then on the horizon; one was loathe at the time to predict that *two* such unique systems had evolved independently, even for so worthy a purpose as the immune response. This was a suggestion still apparently vulnerable to severe damage by Occam's razor.) In the event, immunofluorescence with anti-immunoglobulin sera usually failed to demonstrate these receptors, and even microchemistry of the surface membrane constituents of T lymphocytes led to very mixed results. Some authors – most notably Marchalonis and Cone[67] – claimed that the T cell receptor is monomeric IgM, a finding strongly contested by Vitetta and Uhr.[68]

Three findings appeared in fairly rapid succession that made it appear that the T cell receptor was indeed not an immunoglobulin. First, it became evident that T and B cell receptors do not recognize the same determinants on a given antigen.[69] Next, it was shown that T and B cells specific for a given antigen often cross-react differently with other antigens.[70] Finally, it was shown by several laboratories that immune response genes associated with the major histocompatibility complex affect T cell function, with little *direct* effect upon B cells.[71] A clearer picture of the difference between T and B cell receptors was obtained when Zinkernagel and Doherty showed that whereas B cells see antigen alone, T cells interact with antigen only in the presence of MHC products.[72] The demonstration that T cells do not contain mRNA for immunoglobulin chains[73] made it evident that a completely different system governs T cell specificity.

We need not explore *in extenso* the voluminous subsequent work on the molecular biology of the T cell receptor gene families.[74] It will suffice to indicate that the T cell receptor has been shown to be composed of heterodimeric glycoproteins, consisting of an α and a β chain and, more rarely, of a γ and a δ chain. Each of these chains, although apparently unrelated to any of the immunoglobulin chains, has both constant and variable regions. Of interest to this discussion is that specificity and a large repertoire are attained, just as in the case of Ig molecules, by the rearrangement of numerous gene segments (including multiple V, [D], and J exons). What is especially fascinating is that the T cell receptor shows only modest affinities for the antigen–MHC ligand, and that it appears not to interact appreciably with either component alone. (These two features may in fact go hand-in-hand, since interaction with and activation by either component might obviate the special functions played by T cells in the immune response.)

Transfer factor

No discussion of the basis for and functions of immunologic specificity would be complete without mention of the curious substance transfer factor, first described by Lawrence.[75] This material is obtained from extracts of the lymphocytes of delayed hypersensitive humans, and appears capable of transferring specific hypersensitivity to naive recipients. Unlike most other passive transfer systems, however, this one seems to work only in humans. Transfer factor has been applied clinically with some success in the protection of human immunodeficient patients from a variety of viral and mycotic diseases.[76] Although it has been known for over fifty years, the precise chemical composition and mode of action of transfer factor remain a mystery. It is apparently dialyzable and of relatively low molecular weight (<10,000) and unlikely to be either DNA or messenger RNA – the only substances known to be able to mediate the transfer of information. (Some have suggested that the extracts might contain sensitizing antigen.) Transfer factor represents one of those interesting examples in science of an observation that appears to be so mysterious, yet so unapproachable, that it is no longer mentioned, even as a curiosity.

Specific triggers and nonspecific amplifiers

We have thus far treated the phenomena of immunity and allergy which follow upon the interaction of antigen with antibody, or antigen with specific lymphocyte, as though these complex reactions were entirely specific from start to finish. But progress in sorting out the cellular dynamics and pathophysiology of antibody formation, or of delayed and immediate allergic inflammatory responses, shows that there have evolved a number of complicated mechanisms by which an oft-times exceedingly minor specific immunologic trigger may be amplified to a remarkable degree, depending upon the release of a variety of pharmacologic agents whose further function is usually nonspecific in nature.

The first of these nonspecific mediators to be described was complement. Jules Bordet showed originally that complement could mediate the hemolysis of erythrocytes that had been sensitized with specific antibody, and speculated that complement fixation and its consequences were the nonspecific byproducts of the antigen–antibody union.[77] The process of complement fixation proved in fact to be nonspecific, and unexpectedly complicated. Thus, complement is composed of a large number of individual components and factors which function in a sequential cascade, with the activation or release of a variety of enzymes which can damage cell membranes, and of a variety of split-products that may be chemotactic for polymorphonuclear leukocytes or that may exercise physiologic effects upon muscle or blood vessels.[78] It is mechanisms of this type, triggered by an initial antigen–antibody interaction, that contribute so importantly to, among others, the inflammatory reaction seen in the Arthus phenomenon and the glomerular damage seen in immune complex disease of the kidney.

Another example of the nonspecific enhancement of a modest immunologic interaction occurs in hayfever and asthma-type allergies. Here, exceedingly small amounts of allergen may interact with nanogram quantities of the specialized IgE antibody, but the initial site of interaction on the surface of mast cells leads to their degranulation, with the release of such active pharmacologic agents as histamine and serotonin.[79] These latter substances then incite nonspecific sequelae, such as dermal wheal and erythema reactions, inflammation of the ocular conjunctiva, and bronchiolar constriction.

Nonspecific mediators and enhancers are no less important in cell-mediated phenomena than they are in those triggered by circulating antibody. It was initially difficult to understand, in passive transfer experiments of delayed hypersensitivity or allograft rejection reactions, why the proportion of *specific* lymphocytes in the inflammatory infiltrate should be so low – often only a fraction of 1 percent of the lymphocytes present.[80] If these reactions were immunologically specific, as one in fact knew them to be, why then were so many "innocent bystander" cells present at the site? The answer emerged from studies that were stimulated by the phenomenon of antigen-induced inhibition of macrophage migration, to which we alluded above. It was found that the specific interaction of antigen and lymphocyte leads to the release of pharmacologically active substances (lymphokines) able nonspecifically to immobilize (and often activate) macrophages locally. In time, other agents were identified whose release could be triggered by specific interaction with antigen. Some of these act on T cells to attract them to a local site of inflammation and there to stimulate them to mitosis, while others appear to act on B cells to induce polyclonal activation and antibody formation.[81] Nor are lymphocytes the only source of nonspecific agents that may contribute to immunogenic reactions. Monocytes may also give rise to such active factors (monokines), now often two steps removed from the specific immunologic triggering event.[82] Indeed, it is now believed that many other cell types throughout the body, even those unrelated to the immunologic apparatus, may be excited by a variety of stimuli (hormones, etc.) to release pharmacologically active agents (cytokines) which may act upon lymphocytes, among other target cells.

It is currently clear that without this congeries of nonspecific factors, the workings of protective immunity (and of immunopathology) would be far more modest than those that we see in actual practice.

Specificity and repertoire size

From the earliest days of immunology, it has been an integral part of the received wisdom that antibody is endowed with a fine specificity for its inducing antigen. This view is reinforced by repeated demonstrations of stereochemical molecular complementarity between antigen and antibody and, from outside of immunology, by increasing knowledge of the specificity of enzyme action. The demonstrations by Landsteiner and by Pauling and Pressman of serologic cross-reactions among haptens of closely related structure modified this view only slightly, by permitting minor variations in the antibody combining site into which *very closely related* structures might fit with reduced binding energies. However, these same studies with artificial haptens also immeasurably expanded the universe of antigenic structures against which specific antibodies could be formed; and we have seen how the requirement of a large specificity repertoire affected the thinking of immunologists about the mechanism of antibody formation.[83] But does the modern immunogenetic synthesis even now provide sufficient clonal precursors (clonotypes) to encompass the full repertoire requirements of the immunologically active organism? Some investigators think not!

It has always been difficult to arrive at a reasonable estimate of the size of the specificity repertoire of the vertebrate. Some investigators have put the maximum number of *completely different* immunogenic structures as low as 50,000, while the usual number quoted is 10^5 to 10^6. Inman has suggested,[84] from an analysis of the known natural synthetic structures that have been catalogued, that as many as 10^{16} different antigenic structures may exist – a number appreciably larger than the total number of lymphocytes ($<10^9$) in the immunologically well-studied mouse. There is, however, another approach to the problem.

It has been possible to estimate the number of different clonotypes that may be produced by a mouse, and the precursor cell frequency for each clonotype, employing such techniques as isoelectric focusing, fine specificity analysis, idiotypy, and the transfer of limiting dilutions of clonal precursors to irradiated recipients or to *in vitro* cultures, to allow an actual count of responding clones.

These approaches have been extensively reviewed by Sigal and Klinman in their discussion of the B cell clonotype repertoire.[85] Employing such approaches, it has been estimated that there may be as many as 5,000 *different* clonotypes (i.e., specific antibodies of varying primary structure and varying affinity, but reactive nevertheless with the immunizing antigen) for the dinitrophenyl (DNP) or for the 3-iodo 5-nitrobenzoyl (NIP) haptens, and that each of these clonotypes is represented by some ten precursor cells per mouse.[86] With $2-3 \times 10^8$ B cells in the lymphoid system of a mouse, this would permit only some 6,000 *different* antigenic determinants against which the adult mouse might be capable of responding. This figure accords well with the finding that some 1 in 5,000 B cells in the mouse is specific for DNP, and between 1 in 7,000 and 1 in 15,000 B cells is specific for NIP.[87] Thus, while the total clonotype repertoire of the mouse may be quite large (10^7), the degeneracy of the immune response appears to allow an almost embarrassingly restricted coverage of the universe of potential stimuli.[88]

This paradox is further pointed up by observations in other areas. Whereas man, with his 10^{13} lymphocytes, might have little difficulty in expressing a suitably broad specificity repertoire, smaller vertebrate species such as the mammalian shrew at 1–2 g or certain species of fish at less than 100 mg of adult weight (with proportionately fewer lymphocytes) might experience greater difficulties. Yet these small animals appear to cope very well, and to survive attack by their respective pathogens with little sign of immunologic impairment. Indeed, du Pasquier has shown that the tadpole, with some 10^6 lymphocytes (and perhaps one-third as many B cells), can form an adequate immune response against a variety of antigens.[89]

The dilemma posed by these data has been countered by the suggestion that the antibody combining site may not be as tightly restricted to a small antigenic determinant as had earlier been supposed. The hypothesis has been advanced that the combining region on antibody might be "polyfunctional," in that it might be large enough to permit of the binding of two or more quite disparate molecular structures.[90] Thus, an immune response to antigenic determinant A might consist of different clonotypes, one binding determinants A and B, another binding determinants A and C, etc. The resulting immune serum would *appear* to be of anti-A specificity, because other specificities would be at very low concentration, but the universe of different antigens could be dealt with on such a basis. There is even some direct indication in the literature that such a general multispecificity of the antibody combining site may exist. Monoclonal myeloma proteins have been found which are able to bind such unrelated haptens as ε-DNP lysine (with an affinity constant of 10^5 l/M) and 2-methyl-1,4-naph-thoquinone (menadione) with an affinity constant of 2×10^4 l/M, and neither of these may represent the "best fit" hapten.[91] Similarly, a myeloma protein has been found[92] which reacts with three unrelated structures (dinitrophenyl, 5-acetouracil, and purines), while the homogeneous human immunoglobulin Wag has been shown to bind both ε-DNP lysine and an Fc fragment of IgG.[93] Finally, in another biological system whose specificity appears to have a basis similar to that of antibodies, enzymes have been found which bind a number of structurally unrelated compounds to their binding sites.[94]

A theory of receptor site multispecificity was first advanced by Talmage,[95] in an attempt to show that a heterogeneous immune serum might show a greater specificity for antigen than any of its constituent antibodies. This suggestion has been taken up and extended by Inman and by Richards and colleagues as "the only reasonable solution" to the continuing repertoire paradox.

Conclusions

We have attempted, in these chapters on the development of the concept of immunologic specificity, to trace the history of one of the most central ideas in immunology (and indeed in biology in general). The result must be viewed as preliminary and incomplete, and as an invitation to others to add and to amend. Nevertheless, several interesting conclusions may be drawn that reveal much about the workings of immunology in particular, and perhaps science in general.

First, the roots of any important scientific concept (such as that of immunologic specificity) do not grow in isolation; they draw nourishment from many other disciplines. Similarly, the growth of an important concept within a given discipline will have far-reaching implications and fruits for other fields of science. Secondly, we may note a marked change in the manner in which immunology is currently practiced, compared with that of 100 years ago. The quantum leaps forward in funding, in numbers of scientists, and in masses of crucial data have not been without a certain cost – the substantial reduction in elegant *personal style* that characterized so many of our scientific forebears, and that makes so interesting and even enjoyable the reading of their reports. Finally, we see again and again how much one's philosophical bases and disciplinary upbringing determine a scientist's approach, the questions asked, and the type of answers one will accept. Throughout much of immunology's history, as Jerne put it so well,[96] *cis-* and *trans-*immunologists hardly spoke to one another. Or rather, a *cis-*immunologist sometimes spoke to a *trans-*immunologist, but the latter rarely answered! Fortunately, one of the attributes of scientific progress is often a merging of these disparate languages, and eventual mutual comprehension.

Notes and references

1. Ehrlich, P., "Croonian Lecture – On Immunity..." *Proc. R. Soc. Lond.* **66**:424, 1900.
2. See Chapter 3.
3. That this was indeed the "Age of Immunochemistry" is discussed in Chapter 17 and in Appendix A1.
4. Fruton, J., *Molecules and Life: Historical Essays on the Interplay of Chemistry and Biology*, New York, Wiley-Interscience, 1972.
5. Marrack, J.R., *The Chemistry of Antigens and Antibodies*, London, HMSO, 1934.
6. Landsteiner, K., *The Specificity of Serological Reactions*, New York, Dover, 1962.

7. Landsteiner, K., *Biochem. Z.* **104**:280, 1920.
8. Pressman, D., and Grossberg, A., *The Structural Basis of Antibody Specificity*, New York, Benjamin, 1968.
9. Pauling, L., and Pressman, D., *J. Am. Chem Soc.* **67**:1003, 1945.
10. Hooker, S.B., and Boyd, W.C., *J. Immunol.* **42**:419, 1941; Pauling, L., and Itano, H.A., eds, *Molecular Structure and Biological Specificity*, Washington, DC, American Institute of Biological Sciences, 1957.
11. Pressman, D., in *Molecular Structure and Biological Specificity*, note 10.
12. These factors are reviewed by Boyd, W.C., *Introduction to Immunochemical Specificity*, New York, Interscience, 1962.
13. Eisen, H.N., and Karush, F., *J. Am. Chem. Soc.* **71**:363, 1949. A forerunner of this technique had first been employed by Marrack, J.R., and Smith, F.C., *J. Exp. Pathol.* **13**:394, 1932.
14. Karush, F., *Adv. Immunol.* **2**:1, 1962.
15. See, for example, Pauling, L., Pressman, D., and Grossberg, A., *J. Am. Chem. Soc.* **66**:784, 1944; Karush, F., and Sonnenberg, M., *J. Am. Chem. Soc.* **71**:1369, 1949; and Karush, F., *J. Am. Chem. Soc.* **78**:5519, 1956.
16. The author recalls having heard this comment at a meeting, perhaps in the mid-1960s, and believes it was made by Felix Haurowitz. Professor Haurowitz (personal communication, 1982), while not remembering it specifically, allowed that he could well have made it.
17. Talmage, D.W., *Science* **149**:1643, 1959.
18. Landsteiner, K., *Wien. klin. Wochenschr.* **22**:1623, 1909.
19. The philosophical background to the dispute about continuity or discontinuity in nature, and its implications for immunology, are discussed by Mazumdar, P.M.H., *Species and Specificity: An Interpretation of the History of Immunology*, Cambridge, Cambridge University Press, 1995, pp. 214–236.
20. Landsteiner, K., and van der Scheer, J., *J. Exp. Med.* **67**:709, 1938.
21. Campbell, D.H., and Bulman, N., *Fortschr. Chem. org. Naturstoffe* **9**:443, 1952.
22. Kabat, E.A., *J. Immunol.* **77**:377, 1956. See also Kabat, E.A., *J. Immunol.* **97**:1, 1966, and Kabat, E.A., *Structural Concepts in Immunology and Immunochemistry*, New York, Holt, Rinehart and Winston, 1968.
23. Arnon, R., Sela, M., Yaron, A., and Silber, H.A., *Biochemistry* **4**:948, 1965; van Vunakis, H., Kaplan, J., Lehrer, H., and Levine, L., *Immunochemistry* **3**:393, 1966.
24. Grabar, P., and Williams, C.A., Jr., *Biochem. Biophys. Acta* **10**:193, 1953; **17**:65, 1955. See also Grabar, P., and Burtin, P., *Immunoelectrophoretic Analysis*, Amsterdam, Elsevier, 1964. Also critical for the development of immunoelectrophoresis, and of great importance in its own right, was the introduction of immunoprecipitation in gels by Oudin, J., *Ann. Inst. Pasteur*, **75**:30, 1948, and by Ouchterlony, A., *Acta Pathol. Microbiol. Scand.* **26**:509, 1949.
25. Fahey, J.L., *Adv. Immunol.* **2**:41, 1962.
26. Porter, R.R., *Biochem. J.* **46**:479, 1950; Fried, M., and Putnam, F.W., *Fed. Proc.* **18**:230, 1959; Nisonoff, A., Wissler, F.C., and Woernley, D.L., *Biochem. Biophys. Res. Comm.* **1**:318, 1959. As early as the mid-1930s, I.A. Parfentiev had obtained patents on the pepsin cleavage of diphtheria antitoxin without altering its antibody activity (US Patents 2,065,196–1936; 2,123,198–1938).
27. Edelman, G.M., *J. Am. Chem. Soc.* **81**:3155, 1959; Franek, F., *Biochem. Biophys. Res. Comm.* **4**:28, 1961.
28. Edelman, G.M., and Poulik, M.D., *J. Exp. Med.* **113**:861, 1961.

29. Porter, R.R., *Br. Med. Bull.* **19**:197, 1963.
30. Edelman, G.M., *Biochemistry* **9**:3197, 1970.
31. Slater, R. J., Ward, S. M., and Kunkel, H. G., *Harvey Lect.* **59**:219, 1965.
32. Edelman, G.M., and Gally, J.A., *J. Exp. Med.* **116**:207, 1962; see also Putnam, F.W., and Hardy, S., *J. Biol. Chem.* **212**:261, 1955.
33. Hill, R.L., Lebowitz, H.E., Fellows, R.E., and Delaney R., in Killander, J., ed., *Gamma Globulin Structure and Control of Biosynthesis*, Proceedings of the Third Nobel Symposium, Stockholm, Almqvist and Wiksells, 1967; Grey, H.M., *Adv. Immunol.* **10**:51, 1969; Hood, L., Campbell, J.H., and Elgin, S.C.R., *Annu. Rev. Genetics* **9**:305, 1975. For a more general discussion of the gene duplication mechanism, see Ohno, S., *Evolution by Gene Duplication*, New York, Springer, 1970.
34. Wu, T.T., and Kabat, E.A., *J. Exp. Med.* **132**:211, 1970. The reader should be aware that much of the progress in understanding the relationship between Ig structure and function, Ig evolution, and even the clarification of genetic mechanisms involved has depended upon comparisons of the amino acid sequences of Ig chains, and thus upon such tabulations as those of Kabat, E.A., Wu, T.T., and Bilofsky, H., *Variable Regions of Immunoglobulin Chains. Tabulations and Analyses of Amino Acid Sequences*, Medical Computer Systems, Cambridge, Bolt, Baranek, and Newman, 1976; *idem, Sequences of Immunoglobulin Chains. Tabulation and Analysis of Amino Acid Sequences of Precursors, V-regions, C-regions, J-chain, and β_2 Microglobulins*, Bethesda, US Dept. of Health, Education and Welfare Publication 80–2008, 1980; and Dayhoff, M.O., ed., *Atlas of Protein Sequence and Structure* (multiple volumes and supplements), Washington, DC, National Biomedical Research Foundation.
35. Poljak, R.J., Amzel, L.M., Avey, H.P., Chen, B.L., Phizackerley, R.P., and Saul, F., *Proc. Natl Acad. Sci. USA* **70**:3305, 1973; **71**:1427, 1974.
36. Kunkel, H.G., Mannik, M., and Williams, R.C., *Science* **140**:1218, 1963; Oudin, J., and Michel, M., *C. R. Acad. Sci. (Paris)* **257**:805, 1963; Gell, P.G.H., and Kelus, A., *Nature* **201**:687, 1964. The early history of idiotypes and anti-idiotypes is discussed in Chapter 10.
37. Jerne, N.K., *Ann. Immunol.* **125C**:373, 1974; Nisonoff A., and Green, M.I., *Progr. Immunol.* **4**:57, 1980. The status of idiotypes and anti-idiotypic antibodies in the definition of immunologic specificity, the study of repertoire size, and the functions of what Jerne called *immune networks* of immunoregulation at the height of interest in the subject is reviewed in *International Conference on Immune Networks*, New York, NY Academy of Sciences, 1983.
38. See Chapters 2 and 3.
39. Metchnikoff, E., *Lectures on the Comparative Pathology of Inflammation*, London, Kegan Paul, Trench, Trübner, 1893. Reprinted by Dover, New York, 1968, pp. 154 ff.
40. Metchnikoff, E., *L'Immunité dans les Maladies Infectieuses*, Paris, Masson, 1901. (English translation, New York, Macmillan, 1905; reprinted by Johnson Reprint, New York, 1968.)
41. Metchnikoff, E., note 40 (English edition), pp. 270–274.
42. Metchnikoff, E., note 40 (English edition), p. 306.
43. Denis, J., and Leclef, J., *La Cellule* **11**:177, 1895. The observation was soon confirmed by Mennes, F., *Z. Hyg. Infektionskr.* **25**:413, 1897, and by Leishman, W.B., *Br. Med. J.* **1**:73, 1902.

44. Wright, A.E., and Douglas, S.R., *Proc. R. Soc. B.* **72**: 364, 1903; **73**:136, 1904; Wright, A.E., *Studies on Immunization and their Application to the Treatment of Bacterial Infection*, London, A. Constable, 1909.
45. The best summary of the early work on tuberculosis, and on the relationship of immunity and allergy to its pathogenesis, is still that of Rich, A.R., *The Pathogenesis of Tuberculosis*, 2nd edn, Springfield, Thomas, 1951; see also Chapter 9.
46. Lurie, M.B., *J. Exp. Med.* **57**:181, 1933; **75**:247, 1942.
47. Elberg, S.S., and Faunce, K., Jr., *J. Bacteriol.* **73**: 211, 1957; Elberg, S.S., *Bacteriol. Rev.* **24**:67, 1960; Suter, E., *J. Exp. Med.* **97**:235, 1953.
48. Mackaness, G.B., *Am. Rev. Tuberc.* **69**:495, 1954. For more general reviews, see Rowley, D., *Adv. Immunol.* **2**:241, 1962; Suter, E., and Ramseier, H., *Adv. Immunol.* **4**:117, 1964; and Nelson, D.S., *Macrophages and Immunity*, Amsterdam, North Holland, 1969.
49. Rich, A.R., and Lewis, M.R., *Bull. Johns Hopkins Hosp.* **50**:115, 1932. These authors thought that antigen killed the macrophages, but see Waksman, B.H., and Matoltsy, M., *J. Immunol.* **81**:220, 1958.
50. Bloom, B.R., and Bennett, B., *Science* **153**:80, 1966; David J.R., *Proc. Natl Acad. Sci. USA* **56**:72, 1966.
51. Möller, G., ed., "Role of macrophages in the immune response," *Immunol. Rev.* **40**:1978; Unanue, E.R., and Rosenthal, A.S., eds, *Macrophage Regulation of Immunity*, New York, Academic Press, 1980.
52. Fishman, M., and Adler, F.L., *J. Exp. Med.* **117**:595, 1963; Fong, J., Chin, D., and Elberg, S.S., *J. Exp. Med.* **118**:371, 1963.
53. Askonas, B.A., and Rhodes, J.M., *Nature Lond.* **205**:470, 1965. See also Plescia, O.J., and Braun, W., eds, *Nucleic Acids in Immunology*, New York, Springer, 1968.
54. Landsteiner, K., and Chase, M.W., *Proc. Soc. Exp. Biol. Med.* **49**:688, 1942.
55. Bergsma, D., *Immunologic Deficiency Diseases in Man*, New York, The National Foundation, 1968.
56. Miller, J.F.A.P., and Mitchell, G.F., *Transplant. Rev.* **1**:3, 1969; Davies, A.J.S., *ibid.* **1**:43, 1969; Claman, H.N., and Chaperon, E.A., *ibid.* **1**:92, 1969.
57. Dienes, L., and Schoenheit, E.W., *Am. Rev. Tuberc.* **20**:92, 1929.
58. See especially Uhr, J., Salvin, S.B., and Pappenheimer, A.M., Jr., *J. Exp. Med.* **105**:11, 1957, and Salvin, S.B., *J. Exp. Med.* **107**:109, 1958.
59. Landsteiner, K., *Kgl. Acad. Wet. Amsterdam* **31**:54, 1922; *J. Exp. Med.* **39**:631, 1924; Landsteiner, K., and van der Scheer, J., *J. Exp. Med.* **57**:633, 1933.
60. Benacerraf, B., and Gell, P.G.H., *Immunology* **2**:53, 1959; Benacerraf, B., and Levine, B.B., *J. Exp. Med.* **115**:1023, 1962.
61. Benacerraf, B., and Gell, P.G.H., *Immunology* **2**:219, 1959.
62. Karush, F., and Eisen, H.N., *Science* **136**:1032, 1962.
63. Mitchison, N.A., in *Immunological Tolerance*, Landy M., and Braun, W., eds, New York, Academic Press, 1969, p. 149.
64. Claman, H.N., Chaperon, E.A., and Triplett, R.F., *Proc. Soc. Exp. Biol. Med.* **122**:1167, 1966; Mitchell, G.F., and Miller, J.F.A.P., *J. Exp. Med.* **128**:801, 821, 1968.
65. Gershon, R.K., *Contemp. Top. Immunol.* **3**:1, 1974; Tada, T., and Okumura, K., *Adv. Immunol.* **28**:1, 1979.
66. Marrack, P., and Kappler, J., *Adv. Immunol.* **38**:1, 1986.
67. Marchalonis, J.J., Atwell, J.L., and Cone, R.E., *Nature Lond.* **235**:240, 1972; Feldmann, M., and Nossal, G.J.V., *Transplant. Rev.* **13**: 3, 1972; Cone, R.E., and Marchalonis, J.J., *Biochem. J.* **140**:345, 1974.

68. Vitetta, E.S., Bianco, C., Nussenzweig, V., and Uhr, J.W., *J. Exp. Med.* **136**:81, 1972; Vitetta, E.S., Uhr, J.W., and Boyse, E.A., *Proc. Natl Acad. Sci. USA* **70**:834, 1973; Uhr, J.W., in Landy, M., and McDevitt, H.O., eds, *Genetic Control of Immune Responsiveness*, New York, Academic Press, 1972, p. 228.

69. Parish, C.R., *J. Exp. Med.* **134**:21, 1971; Schirrmacher, V., and Wigzell, H., *J. Exp. Med.* **136**:1616, 1972.

70. Hoffmann, M., and Kappler, J.W., *J. Immunol.* **106**:261, 1972; Playfair, J.H.L., *Nature New Biol.* **235**:115, 1972.

71. Katz, D.H., Hamaoka, T., Dorf, M.E., Maurer, P.E., and Benacerraf, B., *J. Exp. Med.* **138**:734, 1973; Press, J.L., and McDevitt, H.O., *J. Exp. Med.* **146**:1815, 1977.

72. Zinkernagel, R.M., and Doherty, P.C., *J. Exp. Med.* **141**:1427, 1975; *Adv. Immunol.* **27**:51, 1979.

73. Kronenberg, M., et al., *J. Exp. Med.* **152**:1745, 1980; **158**:210, 1983.

74. See, for example, Marrack and Kappler, note 66, and Waldmann, T.A., *Adv. Immunol.* **40**:247, 1987. A feeling for the rapidity of progress in this field may be obtained by comparing the foregoing reviews with that of Tada and Okumura, note 65.

75. Lawrence, H.S., *J. Clin. Invest.* **33**:951, 1954; Lawrence, H.S., *Harvey Lect.* **68**:239, 1974.

76. Ascher, M.S., Gottlieb, A.A., and Kirkpatrick, C.H., eds, *Transfer Factor: Basic Properties and Clinical Applications*, New York, Academic Press, 1976.

77. Bordet, J., *Ann. Inst. Pasteur* **12**:688, 1899; see also Chapter 6.

78. Lichtenstein, L., et al., *Immunology* **16**:327, 1969; see also Müller-Eberhard, H.J., and Schreiber, R.D., *Adv. Immunol.* **29**:1, 1980; Lachmann, P.J., *Mol. Immunol.* **43**:496, 2006.

79. Plaut, M., and Lichtenstein, L.M., in Middleton, E., Ellis, E., and Reed, C.E., eds, *Allergy: Principles and Practice*, St Louis, C. V. Mosby Co., 1978, pp. 115–138.

80. McCluskey, R.T., Benacerraf, B., and McCluskey, J.W., *J. Immunol.* **90**:466, 1963; Prendergast, R.A., *J. Exp. Med.* **119**:377, 1964.

81. Cohn, S., Pick, E., and Oppenheim, J.J., eds, *Biology of Lymphokines*, New York, Academic Press, 1979.

82. Rocklin, R.E., Bendtzen, K., and Greineder, D., *Adv. Immunol.* **29**:55, 1980.

83. The role of repertoire size in the decline of Ehrlich's side-chain theory, and in the stimulation of instructionist theories, is discussed in Chapter 3.

84. Inman, J.K., in Bell, G.I., Prerlson, A.S., and Pimbly, G.H. Jr., eds, *Theoretical Immunology*, New York, Marcel Dekker, 1978, p. 243.

85. Sigal, N.H., and Klinman, N.R., *Adv. Immunol.* **26**:255, 1978.

86. Kreth, H.W., and Williamson, A.R., *Eur. J. Immunol.* **3**: 141, 1973; Pink, J.R.L., and Askonas, B., *Eur. J. Immunol.* **4**:426, 1974; see also Köhler, G., *Eur. J. Immunol.* **6**:340, 1976.

87. Klinman, N.R., *J. Exp. Med.* **136**:241, 1972; Press, J.L., and Klinman, N.R., *Eur. J. Immunol.* **4**:155, 1974; Nossal, G.J.V., Stocker, J.W., Pike, B., and Goding, J.W., *Cold Spring Harbor Symp. Quant. Biol.* **41**:237, 1977.

88. The degeneracy of the immune response is perhaps best pointed up by the reports of Sharon, J., Kabat, E.A., and Morrison, S.L. (*Mol. Immunol.* **18**:831, 1981; **19**:375, 389, 1982) on 12 hybridomas to 1-6 dextran, no two of which are identical with respect to idiotypic specificity, binding affinity, or moiety of the antigenic determinant with which they interact.

89. Du Pasquier, L., *Curr. Top. Microbiol. Immunol.* **61**:37, 1973; see also Eisen, H.N., *Progress in Immunology*, New York, Academic Press, 1971, p. 243.
90. See Karush, note 14; also Richards, F.F., and Konigsberg, W.H., *Immunochemistry* **10**:545, 1973; Richards, F.F., Konigsberg, W.H., Rosenstein, R.W., and Varga, J.M., *Science* **187**:130, 1975; see also Berzofsky, J.A., and Schechter, A.N., *Mol. Immunol.* **18**:751, 1981.
91. Eisen, H.N., Michaelides, M.C., Underdown, B.J., Schulenberg, E.P., and Simms, E.S., *Fed. Proc.* **29**:78, 1970.
92. Schubert, D., Jobe, A., and Cohn, M., *Nature* **200**:882, 1968.
93. Metzger, H., *Proc. Natl Acad. Sci. USA* **57**:1470, 1967; Otchin, N.S., and Metzger, H., *J. Biol. Chem.* **246**:7051, 1971.
94. Glazer, A.N., *Proc. Natl Acad. Sci. USA* **65**:1057, 1970.
95. Talmage, note 17.
96. Jerne, N.K., *Cold Spring Harbor Symp. Quant. Biol.* **32**:591, 1967.

8 Horror autotoxicus: the concept of autoimmunity[1]

It would be exceedingly dysteleologic, if in this situation self-poisons, autotoxins, were formed.

Paul Ehrlich

When Paul Ehrlich speculated in 1901 about whether an individual is able to produce toxic autoantibodies, and about the implications of such antibodies for disease,[2] it might almost have appeared that he was making one of those conceptual leaps into the unknown that occasionally accelerate the normally slow pace of science. A closer examination of contemporary ideas, however, reveals that this new concept was the eminently logical result of the convergence of three historically important trends: the 2,000-year-old tradition of Greek humoral medicine; the century-old developments in the new (but not yet so-named) pathophysiology; and more recent developments in the new sciences of bacteriology and immunology. If the implacably logical Ehrlich (see Chapter 10) was at all out of step with his times, it was with the concept of *horror auto-toxicus* itself, as we shall see below.

The teachings of Hippocrates and Galen held that disease results from dysfunctions of the four humors, usually instigated by external (and often demonic) factors.[3] Normal bodily functions might be disturbed, leading to quantitative changes in the humors (too much or too little) or to qualitative changes (a "sharp" humor), with resulting disease. With the advent of a more scientific medicine, these ancient concepts were translated in the nineteenth century by John Brown and François Broussais into a new physiologic concept of health and disease, in which disease was defined as a disturbance of normal (and now presumably identifiable) physiological processes.[4] Claude Bernard's famous 1865 book *Introduction to the Study of Experimental Medicine* became the classic exposition of this new pathophysiology,[5] which held that disease is essentially a *functio laesa* – i.e., one of the patient's inherent bodily processes in a state of disorder.

It was this pervasive nineteenth-century view of the close relationship of the normal and the pathological, strongly supported by August Compte's positivist philosophy of biology,[6] that lent support to the notion that just as altered normal bodily functions might cause disease, so they might be recruited to fight disease as well. Thus, Ilya Metchnikoff was able to invoke the normal digestive functions of phagocytes in his cellular theory of immunity to infectious diseases, while Paul Ehrlich proposed that antibodies are normal cell receptors with pre-assigned functions in the body's economy.

A History of Immunology, Second Edition
ISBN: 978-0-12-370586-0

Copyright © 2009, Elsevier Inc.
All rights reserved

The next step in this conceptual progression came from the young field of bacteriology. With the triumph of the germ theory of disease, thanks to Pasteur and Koch, it was held initially that bacterial toxins (rather than the organisms acting directly) were the major offenders – a view reinforced by the identification of diphtheria and tetanus toxins. Not only were such toxins elaborated directly by pathogenic organisms, but they might also result from the action of even saprophytic bacteria on normal bodily elements, leading to the formation of a variety of noxious ptomaines and so-called toxalbumins. This led Charles Jacques Bouchard to advance a theory of autointoxication in 1886.[7] It was held that toxic products arising most usually in the intestinal tract from otherwise normal digestive processes could produce a variety of different diseases. It is remarkable how popular the notion of autointoxication became in the twenty-five years prior to World War I. Hundreds of papers were written on the implications of autointoxication for one or another disease process or organ system, and extensive reviews were published on the implication of autointoxication for such medical specialties as ophthalmology, pediatrics, internal medicine, etc.[8] To cite but a single case, autointoxication from colonic stasis was deemed so important that great numbers of surgical procedures for colon bypass or colectomy were performed for indications ranging from lassitude to epilepsy![9] Even Metchnikoff developed a fascination for the intestinal tract and its imperfections, for the treatment of which he was instrumental in popularizing yogurt in the Western world.[10]

It was thus at almost the height of general interest in so-called autointoxication and after it had been shown, primarily at the Institut Pasteur in Paris, that toxic antibodies (cytotoxins) could be formed against a variety of cells in the body[11] that Paul Ehrlich considered the question of autotoxic antibodies. Given the prevailing views, it is not surprising that he would speculate on the possibilities that antibodies against self might account for yet another kind of auto-intoxication. What is surprising is that he would conclude, in the face of a general contemporary belief in so many other forms of autointoxication (of which he should have been aware), that the production of toxic autoantibodies was "dysteleologic in the extreme."[12] Why postulate an immunologic *horror autotoxicus* in quite absolute terms, when no such horror appeared to exist in other physiologic processes?

The real meaning of *horror autotoxicus*

As we saw in Chapter 3, Ehrlich's side-chain theory of antibody formation viewed the antibody not as a unique attribute of the immune apparatus, but as part of a larger physiologic system of cell receptors. While some of these receptors might function as antibodies to neutralize bacterial toxins, others served to promote drug action, to assimilate the nutrients required by cells, or even to aid in the breakdown and elimination of both foreign bacteria and native effete cells. When Bordet showed that anti-erythrocyte antibodies could mediate immune hemolysis,[13] and cytotoxic antibodies against a variety of other cell types were demonstrated, it was but a simple and logical step to imagine that *self*-produced hemolytic antibodies might assist in the normal destruction of worn-out erythrocytes. But an intensive search for such hemolytic autoantibodies by injecting animals with their own blood and that of other members of their species led only to the formation of isoantibodies, and never to autoantibodies. It was only then that Ehrlich concluded that either autoantibody formation does not occur because the appropriate receptors do not exist in the individual, or, more probably, that they may be formed but are inhibited in their toxic action. As Ehrlich put it:[14]

> ...the organism possesses certain contrivances by means of which the immunity reaction, so easily produced by all kinds of cells, is prevented from acting against the organism's own elements and so giving rise to autotoxins...so that we might be justified in speaking of a "horror autotoxicus" of the organism. These contrivances are naturally of the highest importance for the individual.

Here is the true meaning of Ehrlich's *horror autotoxicus*, as Dietlinde Goltz makes abundantly clear in her treatise on this subject.[15] Ehrlich's dictum of *horror autotoxicus* makes no claim that autoantibodies may not be formed; it only suggests that they are somehow prevented from acting. As Goltz pointed out, several generations of immunologists have misunderstood Ehrlich, to the detriment of progress in the science of autoimmune diseases. An interesting case in point is that of Ernest Witebsky, a "second-generation" Ehrlichite (by way of Ehrlich's student and Witebsky's teacher Hans Sachs). When Witebsky and his students discovered thyroid autoantibodies in experimental thyroiditis animals in the early 1950s, Witebsky (as a fervent adherent of the Ehrlich theories) refused for some time even to believe his own data.[16] He actually withheld publication of the results for some three years, while the experiments were repeated and re-examined to find the error that had produced data in such apparent contravention of Ehrlich's rule.[17]

We can be fairly certain that Ehrlich did not intend, with the phrase *horror autotoxicus*, to prohibit all autoantibody formation. When Serge Metalnikoff in Metchnikoff's laboratory produced autoantispermatozoa,[18] Ehrlich did not object to the antibodies themselves, but rather argued that they were not autocytotoxins "within our meaning," since they did not function to destroy spermatozoa in their normal *in vivo* location.[19] But how then did Ehrlich picture the putative "regulatory contrivances" that would inhibit the development of autoimmune diseases? For a while, when it seemed that autoanti-antibodies (what would later be called anti-idiotypes) were produced with great facility, Ehrlich (along with Besredka in Paris) conceived of a steady-state immunoregulation provided by the balanced production of autoantibodies and their neutralizing anti-antibodies. (This fascinating interlude, a long-forgotten forerunner of modern theories of idiotype–anti-idiotype network immunoregulation, will be discussed in greater detail in Chapter 10.) But belief in the existence of autoanti-antibodies was short-lived in the early twentieth century, and the search for the theoretical basis of such regulatory mechanisms was left to a later generation of immunologists.

The "classical period" of autoimmunity research

Despite the teleologic appeal of Ehrlich's *horror autotoxicus*, the first decade of the twentieth century witnessed an ever-increasing willingness to speculate that autoantibodies might contribute to the pathogenesis of certain diseases. This movement was especially notable among those such as Landsteiner in Vienna and Weil in Prague who did not accept Ehrlich's teachings as gospel, but it was detectable even among the faithful. It appears to have been based primarily upon observations made in two different experimental areas, each of which contributed importantly to the intellectual environment that favored such speculation.

The first set of observations, as we saw, involved the demonstration that hetero- and even isoantibodies (cytotoxins) could be obtained against almost any organ or cell type one chose to inject into the experimental animal. Many investigators turned to this diverting pastime during the next decade, and few were the tissue types that were not put to this test, as was witnessed by the many reviews that were published on this subject.[20] To many, it seemed but a short step from an isoantibody to an autoantibody, and even though Metalnikov's auto-antispermatozoa were only cytotoxic *in vitro* and did not cause obvious disease in the experimental subject, they appeared to point in the same direction.

The second and perhaps even more significant contribution to this speculative environment came with the succession of discoveries that antibodies (or something remarkably similar) could in fact produce disease. In 1902, Portier and Richet discovered anaphylaxis[21]; in 1903, Arthus discovered the phenomenon named after him[22]; and in 1906, von Pirquet and Schick described and analyzed serum sickness.[23] Even though there was a general disinclination to identify these reactions with the same mechanisms that produced protective antitoxins and antibacterial immunity – hence the special term *allergy*, or altered reactivity – yet there was more than a hint that the mechanisms that protected from a disease and those that led to a disease were somehow interrelated. Anaphylaxis became thenceforth a sort of passkey to the study of disease causation, especially among clinicians interested in explaining the pathogenesis of their particular subspecialty group of interesting diseases.[24] Where exogenous factors that might serve as inciting antigens were not immediately apparent, it was only a short step to the conclusion that endogenous antigens and "auto-anaphylactic responses" might hold the key.

Paroxysmal cold hemoglobinuria (PKH)

The details of the discovery in 1904 by Julius Donath and Karl Landsteiner[25] of the mechanism responsible for this rare hemolytic disease will be covered more fully in Chapter 14, since it illustrates other important sociological and linguistic aspects of our immunologic science. Suffice it to say here, in the context of autoimmunity, that these authors may be credited with the discovery of the first human disease based upon an autoimmune pathogenesis. In carefully controlled experiments, they showed that there exists in the blood of PKH patients an autoantibody of a special type, one that combines with its specific antigen on the surface of the patient's own erythrocytes only in the cold. It requires rewarming of the erythrocyte–antibody complex before complement is able to participate, to induce the immune hemolysis of the now-sensitized cell. All of the clinical symptoms were explicable in terms of these autoimmune events: the "cold" because of the special characteristics of this peculiar antibody; the "paroxysmal" because it occurred suddenly after exposure of the patient's extremities to the cold; and the "hemoglobinuria" as a consequence of the sudden release of so much hemoglobin following intravascular hemolysis. Here was a useful precedent for an autoimmune disease, which even Ehrlich and his followers could not gainsay, despite its incompatibility with the rule of *horror autotoxicus*. This finding in PKH would ease the way for later speculations on the autoimmune nature of other disease entities.

The "Wassermann" antibody

Not long after Bordet and Gengou showed that *any* antigen–antibody interaction could be measured by the nonspecific uptake of complement,[26] it occurred to numerous investigators that here was a useful method for the detection either of antigen with a known antibody,[27] or of specific antibody by utilizing an appropriate antigen. Wassermann and Bruck,[28] and Citron[29] independently, showed that bacterial extracts could be successfully substituted for whole bacteria in these reactions, and it was demonstrated that complement fixation could be applied to the serodiagnosis of tuberculosis, using various tuberculin preparations as antigen. The recent isolation of *Spirochaeta pallida*[30] had stimulated numerous studies on experimental syphilis, and Wassermann and his colleagues quickly realized that here was an important serodiagnostic application of the complement fixation test. However, since the spirochete could not be cultured, Wassermann, Neisser, and Bruck[31] utilized extracts of the organs of syphilitic humans as antigen, and showed that the sera of syphilitic monkeys would yield positive results. Shortly thereafter, the same authors, with Schucht[32] (and, independently, Detré[33]), extended this method to the diagnosis of syphilis in human beings. These and many other investigations very quickly showed that syphilis might be more-or-less reliably diagnosed by testing the blood and even the cerebrospinal fluid of infected individuals.

The history of the Wassermann reaction, and of its acceptance by the scientific community, was the object of a very detailed study by the Polish serologist Ludwik Fleck in his 1935 book *Genesis and Development of a Scientific Fact*.[34] Fleck's thesis, which has attracted much attention from sociologists and epistemologists of science,[35] was that the directions of research are generally determined by the body of contemporary views held by a *Denkkollektiv* – that group of intellectual leaders in the field whose views establish what Thomas Kuhn would later call the "reigning paradigm." Further, even a scientific "fact," according to Fleck, does not actually become one until it is accepted by and integrated within the normative science of the day. (In an interesting side-comment on scientific revisionism, Fleck called attention to a lecture by Wassermann in 1920 in which the latter claimed sole title to the discovery of the "Wassermann" reaction [called by the French the Bordet–Wassermann reaction], and egregiously revised the history of its development. These claims were quickly challenged by Wassermann's former student Carl Bruck, and by Wassermann's long-time opponent from Prague, E. Weil, and there ensued a series of exchanges among the three that was both vituperative and *ad hominem.*[36])

Many investigators were attracted by the new serodiagnostic test for syphilis, and the next few years saw many modifications and improvements, to render the test both more specific and more sensitive. But the most curious new observation revealed that extracts of syphilis-infected organs were not actually required for positive results – extracts of *normal* organs would function as combining antigen just as well.[37] This was an extremely perplexing finding, since all previous experience indicated that only *specific* antigen could interact with antibody to fix complement and yield a positive diagnostic test. If the antigen in these extracts was not of spirochetal origin, then what was it, and why should it have stimulated antibody formation in the syphilitic individual?

It was not long before Weil and Braun offered a possible explanation, entirely consistent with earlier speculations on the broad biological functions of antibody. Infection with *Treponema pallidum* induces tissue breakdown in the affected organs, claimed these authors, and the antibodies circulating in syphilitic patients are in fact autoantibodies specific for the breakdown products! This would account also, they held, for the "false positive" cross-reactions seen in such other diseases as malaria and leprosy, where analogous tissue breakdown also occurred. Indeed, they claimed that such autoantibodies would exacerbate the disease:[38]

> When the very tissue destruction which follows the first infection has led to antibody formation, so it will during the further course of its formation be generally directed not only against the material formed from damaged cells, but also against [normal] human cell substances…thus these autoantibodies attack the cells to liberate antigen, which is able to evoke further autoantibody formation. Should the newly formed protein possess toxicity for the organism, then will the antibody contribute to enhance this toxicity.

In an interesting extension of this provocative speculation, Weil and Braun also considered the possibility that paresis, one of the more prominent symptoms of tertiary syphilis, might be attributed to the same pathogenetic process. Should the infection spread to the brain and there cause analogous cell breakdown, then autoantibodies to neuroantigens might further attack and destroy the normal cells of the brain, thus accounting for the progression of the neurologic complications of syphilis. But even as they advanced this hypothesis, they cautiously suggested that it might be better to reject it as too hypothetical, "because specificity has not yet been demonstrated."[39]

Now, some 100 years later, an acceptable explanation for the presence of these serodiagnostic antibodies in syphilis (which have been shown to differ from those which react specifically with known treponemal antigens) is still awaited. The antigen active in complement fixation tests for syphilis was quickly shown to be a lipid, and later was purified and named cardiolipin, but the origin of the antibodies from syphilitic patients with which it reacts is still a mystery. Modern science has not gone much further in this respect than Weil in 1907, who concluded that "the facts seem rather to speak to the view that the complement-binding material [antibody] is the consequence and not the cause of [the disease]."

Autoimmunity to lens proteins

In his contribution to the Festschrift in 1903 honoring the sixtieth birthday of Robert Koch, Paul Uhlenhuth opened up a new chapter in immunologic research by demonstrating the organ specificity of the proteins of the lens of the eye.[40] This was the first clear demonstration, not only that unique antigens might exist within a single organ and nowhere else in the body, but also that these antigens might be shared from species to species. Here was a finding whose implications would fascinate clinicians interested in ocular diseases, as well as generations of immunopathologists searching for the underlying basis of autoimmune disease. Over the next few years, Uhlenhuth's report was confirmed in a number of different laboratories, and extended in several provocative directions. First, Kraus and co-workers[41] showed that these organ-specific lens antigens can induce both active and passive anaphylaxis in experimental animals – a finding quickly confirmed by Andrejew.[42] Then Uhlenhuth and Haendel[43] showed that a guinea pig could be rendered sensitive to its *own* lens protein, and sent into anaphylactic shock with the protein from any other lens. But these authors made no further comment on the possible clinical significance of this phenomenon, although the ophthalmologist Paul Römer had earlier speculated that the pathogenesis of senile cataract formation might possibly involve the production of autocytotoxins.[44] It remained for the ophthalmologist F. F. Krusius to perform the critical experiment relating lens anaphylaxis to an actual ocular disease, by showing that the experimental rupture of the lens capsule in a *normal* guinea pig could not only actively sensitize the animal but also function as the antigenic challenge of the "anaphylactic" state, with resulting ocular disease.[45] All of these data were re-evaluated, and their implications considered in a lengthy review of the field by Römer and Gebb in 1912. In a separate section "on the question of the formation of autoanaphylactic antibodies by means of lens proteins," these authors asked whether an autologous lens can actually be characterized as *foreign* in the guinea pig, or "whether the 'law of immunity research,' which

Ehrlich has popularly termed *horror autotoxicus*, does not rather apply to the lens."[46]

The first part of this question is an interesting one. Here, as early as 1912, is a preview of what would later be called the "sequestered antigen" theory. If indeed the body cannot respond immunologically to "self," then, *ipso facto*, such antigens as do stimulate the immune response must be foreign – i.e., somehow sequestered from the immunologic apparatus of the host. But Römer and Gebb did not yet misunderstand Ehrlich's rule of *horror autotoxicus* as future investigators would; this is made clear by the way that they interpreted Ehrlich's law. In a further elaboration of this understanding, they state that:[47]

> *we shall by no means assert that the homologous [read autologous] lens protein fails in all circumstances with respect to the formation of these anaphylactic autoantibodies. We are rather convinced that the regulatory mechanism of the organism can and will refuse to serve under special conditions. And the investigation of these situations will further promote our understanding of pathological states.*

This statement clearly advances Ehrlich's original proposition by a significant step forward. Ehrlich was willing to permit the formation of autoantibodies, but invoked an immunoregulatory process to inhibit their deleterious reactions. Now, in the light of these more recent experiments, Römer and Gebb can conceive of a *failure* of these regulatory mechanisms, with consequent autoimmune disease.[48]

Sympathetic ophthalmia

Sympathetic ophthalmia is a blinding disease that has long fascinated ophthalmologists, due to its curious sequence of clinical events. Even long after a penetrating injury to the eye, that eye may suddenly become inflamed, accompanied by a spontaneous involvement of the contralateral eye. It has variously been speculated that bacteria, fungi, and viruses (depending upon the current vogue) might provide the etiologic triggers for these events. When, after the turn of the century, anaphylaxis captured the attention of medical researchers, it was quickly called upon to help to explain sympathetic ophthalmia as well.

It was the Italian Santucci who in 1907 first drew attention to the fact that sympathetic ophthalmia might be due to the formation of cytotoxins (autoantibodies) following resorption of damaged ocular tissue in the first eye, which could then attack and cause disease in the hitherto normal fellow eye.[49] This conjecture was accompanied by experiments showing that injection of emulsified ocular tissues into rabbits and guinea pigs would produce endophthalmitis. Scarcely had this thesis begun to attract attention when a counterclaim for priority was submitted from Russia under the title "Hypothesis of the Autocytotoxic Origin of Eye Diseases."[50] In this, S. Golowin complained that no one in the West appeared to be aware of the fact that as early as 1904 he had published his theory in a Russian journal.[51] Golowin writes, "Soon after the appearance of the work of Bordet and others, it occurred to me to propose a new hypothesis for the still enigmatic pathogenesis of sympathetic ophthalmia, with the help of the doctrines of cytotoxins." He suggested that lesions of the ciliary body lead to the release of antigens, and to the formation of autocytotoxins which circulate and act specifically on the iris and ciliary epithelium of the fellow eye to cause inflammatory disease. Golowin named these autoantibodies *cyclotoxins*.

There now enters the story of the ophthalmologist Elschnig from Prague (later to become perhaps the foremost academic ophthalmologist of his day). Elschnig soon became the leading exponent of an autoimmune pathogenesis of sympathetic ophthalmia, and published a series of papers on this subject.[52] In the first of these, while acknowledging Golowin and Santucci, he implied that it was rather "the idea of Professor Weil on the origin of sympathetic ophthalmia that calls absolutely for further studies." Weil, one of the most active workers in immunology during that period, apparently served as Elschnig's immunologic mentor in these studies, and together they formulated the hypothesis that due to the resorption of antigen in the damaged uveal tissue, there develops a hypersensitivity (one of the earliest uses of this term) in the organism, and especially in the homologous organ, the second eye. This leads to a heightened ability to react, so that the slightest disturbance in the sensitized second eye would lead to inflammation with serious consequences. Elschnig performed numerous experiments in animals to test this theory, and eventually identified the pigment so abundantly present in the pigment epithelial cells of the iris and ciliary body as the antigenic culprit.

General observations on this period

The examples listed above, while they might soon fade from the view of mainstream immunology, permit some interesting conclusions to be drawn about the immunologic practices and the immunologic beliefs extant in the decades preceding World War I. First, it was clearly a "Golden Age" of immunologic and immunopathologic research, during which time the groundwork was laid for many later immunologic subspecialty areas. Secondly, it is clear that Ehrlich's theories held great sway (especially outside France), and that the concept of *horror autotoxicus* was not misunderstood then, as it would be later. Finally, the concepts of anaphylaxis and cytotoxins were extremely attractive to experimental pathologists, and were in the forefront of the candidates nominated to explain the pathogenesis of almost any disease then poorly understood.

The Dark Ages of autoimmunity research

We have seen how the fifteen years immediately preceding World War I witnessed an expansion of the young field of immunology into many interesting and fruitful areas of research. There was, at the same time, a flurry of interest in anaphylaxis and related mechanisms of disease, and a no less intense interest in the production and possible functions of autoantibodies. "Basic" scientists were interested, then as now, in what autoantibodies might tell them about the processes underlying and modulating the immune response, while clinicians were interested in their implications for disease pathogenesis. However, interest in autoantibodies and autoimmune diseases very rapidly slowed and then ceased within the mainstream of immunology, not to be resumed for another forty years or so. Of course, not *all* activity in these areas was terminated everywhere, and all previous knowledge was not lost. Just as occurred during the Dark Ages in Europe, some institutions continued their scholarly pursuits, and here and there isolated individuals appeared to revive and to extend past knowledge. The study of anaphylaxis and related phenomena passed, in the main, into the hands of clinical allergists; clinical ophthalmologists maintained an interest in the pathogenesis of sympathetic ophthalmia and of lens-induced ocular inflammation; and the occasional experimental pathologist or immunochemist might publish a sporadic study on autoimmune encephalitis, on the meaning of the Wassermann antibody, or on the autoantigens of the thyroid. It was evident, however, that the earlier continuity and interconnections of immunologic and biomedical thought had substantially waned between the two world wars. What was accomplished in the immunologic study of disease during this period received little consistent attention from immunologic leaders interested in other problems.

How can we account for this forty-year hiatus? The disruptions that accompanied and followed the 1914–1918 war surely contributed substantially to the lapse. Defeated Germany, formerly the leader in the field, went into eclipse in the biomedical sciences. Paul Ehrlich had died in 1915, and no one took his place to maintain the tradition. In Austria, conditions were just as bad; Karl Landsteiner lost his position, and was forced to emigrate to the United States to carry on his work. Even in victorious France, immunology went into decline. With the death of Metchnikoff in 1916, the Pasteur Institute too seemed to give up its long and glorious tradition of leadership in theoretical and experimental immunology. The war had caused the center of gravity of scientific research to shift from Europe to America, although even there little attention was paid to fundamental biomedical studies in immunology.

It is of interest that the slowdown in immunologic activity between the two world wars was not part of a more general phenomenon suffered by *all* biomedical research fields. Significant advances continued to be made in endocrinology and other physiologic pursuits, in genetics, and in virology. In biochemistry, the 1920s and 1930s were the halcyon years of nutrition and vitamin research. Even within immunology, immunochemistry continued its productive course, in the hands of Landsteiner, Heidelberger, Marrack, Pauling, Boyd, and many others. Here perhaps is one of the important clues to the decline of interest in autoimmunity. During its first thirty years, immunology had primarily been the domain of biologically- and medically-oriented individuals, interested in its implications for disease prevention and disease causation. With the exhaustion of the search for vaccines against the most important pathogens, and especially with the decline of the phagocytic theory of immunity at the hand of the more readily available and manipulable circulating antibody, biologists were replaced by chemists at the leading edge of immunologic research (see Chapter 17). These investigators focused their attention on the molecule rather than on the whole organism. They were more interested in the size, shape, and structure of the antibody than in its possible role in the pathogenesis of disease. Thus, the conceptual foundations of the new *Denkkollectiv* were markedly different from those of the old one, and the guidelines for research and for conceptual advance which accompanied this change were also notably different. This is well illustrated not only by the types of study deemed worthy of pursuit and of publication in journals of immunology, but also by how immunologic phenomena were now interpreted. This was the era of instructionist theories of antibody formation – theories which Macfarlane Burnet would later criticize as being too chemically oriented and too neglectful of biologic phenomenon and biologic realities.

It is no wonder, then, that interest in autoimmune diseases waned during the period between the wars. This may be best appreciated by an examination of Table 8.1, in which are listed for each organ or disease entity the date of the last significant contribution during the "classical" period, and of the first significant contribution during the "modern" era. In those systems for which we have both starting and ending dates, the average interlude is forty-four years! This is a long period in a field whose total lifespan up to the time when interest in autoimmunity was rejuvenated was scarcely eighty years.

As might be expected, interest in autoimmune disease was not completely extinguished during the interim. The ophthalmologists Verhoeff and Lemoine examined clinical cases of lens-induced ocular inflammation, and coined the term *phacoanaphylactic endophthalmitis* in 1922;[53] other ophthalmologists extended the study of retinal pigment as the autoantigen presumed to be responsible for sympathetic ophthalmia.[54] In 1933, the experimental pathologist Rivers created the model of experimental allergic encephalomyelitis[55] that would later be exploited so productively by other workers. Also in the 1920s, the immunologist Ludwig Hektoen and coworkers did careful studies on the autoantigenicity of thyroid antigens.[56] These latter investigators, however, were not interested in disease; rather, they utilized thyroid proteins to study species interrelationships by means of antigenic cross-reactions, in the tradition of Nuttal. One of the more distinctive curiosities of the time was the attempt to utilize autoimmunity for beneficial purposes. A number of efforts were made to utilize sperm antigens in antifertility vaccines[57] – a subject that was later revitalized.[58]

Such sporadic investigations as occurred during this interbellum period were, as we have mentioned, out of touch with most contemporary immunologic activity. They went substantially unremarked, if they were even seen, by the immunologic leaders of the day. The same situation appears to have been true of the early stirrings of activity in experimental immunopathology, by such investigators as Louis Dienes, Jones and Mote, Simon and Rackemann, and Arnold Rich. It required the biological sea-change in concept that followed World War II to attract interest once again in the biological and medical aspects of

Table 8.1 The "Dark Ages" of autoimmunity

Disease/system	Last "classical" contribution	First "modern" contribution
Hemolytic disease	1909	1945 (Coombs et al.)[a]
Sperm and testicular	1900	1951 (Voisin)[b]
Encephalomyelitis	1905	1947 (Kabat et al.)[c]
Sympathetic ophthalmia	1912	1953 (Collins)[d]
Phacoanaphylaxis	1911	1963 (Halbert and Manski)[e]
Thyroid	1910	1955 (Witebsky and Rose; Roitt et al.)[f]
Platelet disease	—	1949 (Ackroyd)[g]

[a]Coombs, R.R.A., Mourant, A.E., and Race, R.R., *Br. J. Exp. Pathol.* **26**:255, 1945. For an early history of the antiglobulin test, see Coombs, R.R.A., *Am. J. Clin. Pathol.* **53**:131, 1970. It is interesting that Moreschi in 1908 had described the same phenomenon as the Coombs antiglobulin test (Moreschi, C., *Zentralbl. Bakt.* **46**:51, 1908), a finding forgotten in the interval. We might have listed here the work of Dameshek and Schwartz in 1938 (*New Engl. J. Med.* **218**:75, 1938; *Am. J. Med. Sci.* **196**:769, 1938), although they only hinted at *auto*hemolysins. They were heard at the time only by fellow hematologists. The antiglobulin "Coombs test" is perhaps a better landmark, even though it involved at first only the detection of Rh *iso*antibodies in erythroblastosis fetalis. Its use very quickly showed autoantibodies in acquired hemolytic anemias, an observation of which the immunologic community was now fully aware.
[b]Voisin, G., Delaunay, A., and Barber, M., *Ann. Inst. Pasteur* **81**: 48, 1951. Important contributions to this field were also made by Freund, J., Lipton, M.M., and Thompson, G.E., *J. Exp. Med.* **97**:711, 1953.
[c]Kabat, E.A., Wolfe, A., and Bezer, A.E., *J. Exp. Med.* **85**:117, 1947; **89**:395, 1949.
[d]Collins, R.C., *Am. J. Ophthalmol.* **36**(Part II):150, 1953.
[e]Halbert, S.P., and Manski, W., *Prog. Allergy* **7**:107, 1963.
[f]Witebsky, E., and Rose, N.R., *J. Immunol.* **76**:408, 1956; Roitt *et al.*, *Lancet* **2**:820, 1956.
[g]Ackroyd, J.F., *Clin. Sci.* **8**:269, 1949.

immunology, and thus in the autoimmune diseases – i.e., the establishment of a new *Denkkollektiv* and a new paradigm.

The modern period of autoimmunity research

Conceptual progress

In the years immediately following World War II, biological phenomena and biological reasoning penetrated the field of immunology with ever-increasing effect. This was stimulated not only by Medawar's work on the immunology of skin-graft rejection,[59] by reports of immunologic deficiency diseases,[60] and by new sources of funding for biomedical research, but also, and equally importantly, by the entry into immunology of a new generation of young scientists with few ties to the old dogma. Here was a biomedical renaissance, in which autoimmunity research participated. One of the key observations that stimulated thought in the latter field was that of Ray Owen on chimerism in cattle twins.[61] This author showed that dizygotic calves whose circulatory systems were connected *in utero* would show, after birth, not only mixtures of their erythrocyte blood types, but also an inability to respond immunologically to one another's antigens. This new biological fact about immunology would demand consideration in any future concept of antibody formation, and focus attention on the ability *and inability* of these mechanisms to react to self. In addition, three other events made it easier for investigators to work with and to think about autoantibodies and autoimmunity. These were:

1. The Coombs test (see Table 8.1), which helped to detect such antibodies
2. The introduction of Freund's adjuvants,[62] which substantially simplified their production
3. Byron Waksman's review of 1959,[63] which helped to give focus and direction to these studies.

Waksman's call to arms was especially significant at the time in that it focused attention on the role of delayed hypersensitivity mechanisms in these autoimmune phenomena, and on the importance of a careful interpretation of accompanying cytologic and histopathologic changes.

Immunologic tolerance

Whereas Burnet's 1941 book[64] on *The Production of Antibodies* made no mention at all of autoimmunity or antoantibodies, this gap was rectified in his 1949 revision of the book with Frank Fenner.[65] They took note of Owen's observations and, even in the context of an instructionist theory of antibody formation, proposed an explanation for how the immunologic apparatus might distinguish between "self" and "not self." (Medawar's earlier work on graft rejection, and the blood group story, had already drawn attention to the antigenic differences among individuals of the same species.) According to Burnet and Fenner, there is established during fetal or neonatal life a set of "self markers" on every cell, subsequent recognition of which would inhibit an active immune response. Any antigen present during this maturational process (such as the foreign erythrocytes of Owen's calves) would be marked "self," and thus be exempt from future autoantigenicity. These events occur in other situations, with implications for congenital infections, as Burnet and Fenner pointed out. They cited the observations of Traub during the 1930s on lymphocytic choriomeningitis (LCM) virus infection of mice,[66] wherein fetal infection from the mother appeared to render the offspring incapable of mounting an immune response to the viral antigens after birth, resulting from "the development of a tolerance to the foreign microorganism during embryonic life"[67] – possibly the first use of this term in an immunologic context. (Perhaps the LCM story itself should have provided the trigger for speculations on immunologic tolerance during the 1930s, but the observation was apparently premature and even if known to immunologists would have been difficult to assimilate within the chemically-oriented paradigm of the times.)

Once Burnet had called attention to them, the implications of Owen's observations were considered so important that they figured significantly in every subsequent theory of antibody formation. However, it was Burnet himself who fully developed the concept of self and nonself and of immunologic tolerance in his clonal selection theory of antibody formation.[68] If the potential for antibody formation is preformed in clonal precursor cells (especially if generated by somatic mutations[69]), then Burnet insisted that this must be a random process. A mechanism should therefore exist to delete those anti-self clones that would threaten the integrity of the body. This could be accomplished by a mechanism of "clonal abortion," in which antigen present during embryonic life would somehow cause the destruction of such dangerous self-reactive clones. Here was a thesis with obvious experimental implications, and it was quickly put to the test and validated by Billingham, Brent, and Medawar[70] (for this, Burnet and Medawar shared the Nobel Prize in 1960).

It may be well to recall at this point that somewhat analogous observations had been made earlier on the immune response to polysaccharide antigens. In this case, *adult* animals had been rendered immunologically unresponsive by the administration of large doses of these antigens,[71] while modest dosages would lead to satisfactory levels of antibody formation. The explanation proposed was that the excess antigen had somehow "clogged" the apparatus, preventing its function. The phenomenon was termed *immunologic paralysis*, and appears to be related to the difficulty with which certain polysaccharides are metabolized (and thus to their long-term persistence).

The phenomenon of acquired immunologic tolerance attracted much attention and experimental confirmation during the 1950s, not only at the hands of Billingham, Brent, and Medawar working with mice, but also from Milan Hašek's group in Prague working with parabiotic chick embryos,[72] and later by numerous investigators who studied tolerance induced in neonatal rabbits.[73] The ability to induce tolerance with fairly large doses of antigen *in utero* and even during the neonatal period was abundantly confirmed, although tolerance was found in general not to be absolute, but rather to depend upon the persistence of antigen. This implied that the attainment of tolerance is not a single and irreversible event, but rather that the state of unresponsiveness has to be actively maintained. Another indication that immunologic tolerance is not absolute and qualitative, but rather a quantitatively variable condition, came from studies (especially in the transplantation field) showing that varying degrees of *partial* tolerance might exist.[74] Indeed, partial or "incomplete" tolerance to self-antigens may be the rule rather than the exception, since low levels of circulating antibody to a variety of tissue autoantigens is a fairly common finding in otherwise normal individuals.

Yet another observation that modified the view of how tolerance is induced came with the realization that the mammalian fetus may show a wide range of immunologic competencies even fairly early in gestation in some species,[75] and that the induction of tolerance may require *prior* immunologic competence to respond to the antigen in question, rather than taking place during an immunologic "null" state. All of these new facts suggested that tolerance is not a negative state, but instead a positive and even dynamically equilibrated regulatory mechanism.

Our understanding of the mechanism of induction of acquired immunologic tolerance was further complicated by the finding that extremely *low* doses of antigen administered repeatedly to an experimental animal might also induce the unresponsiveness, even in the adult animal.[76] Here was yet another clue that tolerance may not be based upon some form of antigenic cytotoxicity directed against clonal precursors, to induce a "gap" in the immunologic repertoire. (Such gaps have been experimentally produced, however, in the "immunological suicide" experiments of Ada and Byrt[77] and of Humphrey.[78] The injection of highly radioactive antigen into naive animals causes a radiation-induced death of specific clonal precursors, so that the animal is incapable thereafter of mounting an antibody response against the antigenic determinants involved.) Still another observation that argued against clonal deletion was the finding that tolerance to a given antigenic determinant might be "broken" by the administration of related, cross-reactive antigens.[79]

A new era in the interpretation of the basis of immunologic tolerance was ushered in by the finding that the cellular contributions to the immune response were divided among a variety of lymphocyte subsets, each with well-defined and highly specialized functions. B cells, originating in the bone marrow (and, in the avian, regulated by the bursa of Fabricius), are responsible for active antibody formation, and provide the plasma cells whose function Astrid Fagraeus had originally pointed out[80] and Albert Coons' fluorescent antibody immunohistochemical techniques had confirmed.[81] These cells, however, cannot act alone. They require the active intervention of macrophages to process and efficiently present the antigen to the immunocyte,[82] and of T cells (which undergo functional maturation of the thymus). Such "helper" T cells apparently interact first with antigen, to somehow provide the trigger for B cell activation.[83] The importance of the T cell receptor in recognizing self antigens was further emphasized by the finding that these cells, so important in defense against viral infections, act by responding not to virus alone but rather to viral antigen only when presented in the context of a self antigen (a Class II component of the major histocompatibility complex).[84]

All of these observations on the induction *and breakage* of immunologic tolerance, coupled with numerous reports on clinical and experimental examples of autoimmunity (to be detailed below), forced even the doubters to concede the reality of autoimmunity. But it did more; it forced acceptance of the fact that all antibody formation, all immunologic tolerance, and the presence or absence of pathological autoimmunity are the result of a complicated system of *immunoregulatory mechanisms*. Thus, Ehrlich's conjecture that *horror autotoxicus* means regulatory control of unwanted reactions against self was now apparently validated. While the microeconomics of the regulation of the cells active in the immune response would soon be assigned to a congeries of chemical signals (lymphokines, monokines, etc.), two new theories would compete to provide an explanation for the macroeconomics of immunoregulation. In an interesting replay of the old cellularist–humoralist debate in immunology which we described in Chapter 2, one of these theories would be predominantly cellular in its interpretation and the other predominantly molecular. Both would attempt to explain why autoimmune disease exists at all, and why it is not more common.

The basis of immunoregulation

The finding that B lymphocytes require the assistance of T lymphocytes in their antibody response to antigenic stimulus opened up a new avenue of research. Different subsets of T lymphocyte lineage would soon be described, each with its own distinctive set of surface membrane markers,[85] including helper cells, cytotoxic cells, and others. Many of these cell types appeared to function in the up-regulation of the immune response, but Gershon and Kondo[86] soon showed that some lymphocytes might contribute to down-regulation. These they called suppressor T cells which, like helper T cells, can be passively transferred to perform their functions in naive recipients. Indeed, these investigators showed that the information for down-regulation of the immune response might even pass from cell to cell, and they spoke of an "infectious immunosuppression."[87] Since most of these immunoregulatory cells appeared to be antigen-specific, here was an elegant hypothesis to explain how the immune response might be modulated. Depending upon their numbers and specificities, the intercommunication of these regulatory cell types among themselves and with primary immunocyte responders would decide whether the response to a given stimulus would be high or low. In terms of autoimmunity, a "normal" balance of helpers and suppressors would hold in check the ever-present threat of embarrassing responses to self, whereas an imbalance among the cells of the regulatory system might lead to serious autoimmune disease.

An alternative theory of immunoregulation arose at about the same time, from a different direction. It was discovered that the binding site on an antibody possesses a highly distinctive three-dimensional structure (the idiotype) that might itself act as an antigenic determinant to stimulate the formation not only of heteroantibodies in another species, but even of autoantibodies within the same host. With the realization that the anti-idiotype would possess a structure similar to that of the antigenic determinant (since both could interact with the same antibody combining site), and that nothing prevented the development of a cascade of anti-antibodies, anti-anti-antibodies, etc., Niels Jerne put forth a theory of immunoregulation based upon purely molecular considerations. This was his idiotype–anti-idiotype network theory,[88] in which the various levels of autoanti-idiotypes could interact with and inhibit the previous levels, thus establishing an immunoregulatory balance that would determine the extent of an immune response. (The details of this network theory will be examined more fully in Chapter 10.)

Phenomenological and technical progress

It is not our purpose here to provide a detailed description of each of the many diseases and syndromes that have been added to the ever-lengthening list of proved or probable autoimmune conditions. The bibliography of such findings has become too massive for this, and many useful summaries are available.[89] We would hope rather, in what follows, to provide the reader with a feel for the explosion of interest and activity that has taken place in this area since World War II, to indicate some of the more important observations that helped to stimulate this interest, and to touch upon some of the newer directions taken by laboratory and clinical research in the autoimmune diseases.

Single-organ autoimmune disease

When the conceptual dam that had blocked acceptance of the fact of auto-immune disease was broken in the late 1940s and early 1950s by clinical data from the hemolytical anemias and by laboratory data from tolerance studies, there followed a flood of new findings and new experimental models. Auto-immune orchitis with aspermatogenesis was shown to be a reality,[90] as were "allergic" encephalomyelitis,[91] sympathetic ophthalmia,[92] and phacoanaphy-laxis.[93] To these were added autoimmune thyroiditis,[94] adrenalitis,[95] pemphigus vulgaris,[96] bullous pemphigoid,[97] and numerous others. Particularly worthy of note is the fact that the pathogenesis of some of these diseases, such as the hemolytic anemias, thrombocytopenias, and pemphigoid, involves uniquely the participation of circulating antibodies and presumably of complement. On the other hand, there is a group of autoimmune diseases which, while they may be accompanied by the formation of circulating anti-bodies, seem to require cell-mediated immune mechanisms to effect the tissue destruction seen. Among these are such diseases as allergic encephalomyelitis and autoimmune thyroid disease, in which passive transfer of the disease state to naive recipients is possible only with sensitized lymphoid cells, and not with specific antibodies.

There is another group of antibody-mediated autoimmune diseases whose elucidation promises to lend an added dimension to the concept of autoimmu-nity. These involve the formation of autoantibodies directed at certain of the surface receptors of cells so important to their proper physiologic function.[98] The role of such receptors in cell nutrition and toxicity reactions had been stressed by Paul Ehrlich, and even before him Newport Langley had suggested, in 1878,[99] that the opposing actions of atropine and pilocarpine or of nicotine and curare involved the competition of the two drugs for the same "receptive substance." It is now known that all biological signals mediated by hormones, neurotransmitters, and other small molecules operate by attaching to specific cell receptors. Should an autoantibody be formed against the receptor itself to compete for the site with the active molecule, then normal function may be inhibited, with resulting disease. Recent evidence suggests that this is in fact what may underlie Graves disease, involving autoantibody to the receptor for thyroid-simulating hormone (TSH);[100] myasthenia gravis, in which autoantibody to the acetylcholine receptor at the neuromuscular endplate interferes with the electrical transmissions governing muscular responses;[101] and insulin-resistant diabetes, where autoantibodies against the insulin receptors in various tissues interfere with glucose metabolism.[102] These interesting studies open new pathways to the diagnosis and therapy of a number of important human diseases.

Multiple-system autoimmune disease

There are a number of diseases of probable autoimmune pathogenesis that target not a single organ, but rather multiple organs or organ systems throughout the body. Among these we may include systemic lupus erythematosus (SLE), rheumatoid arthritis, and Sjögren's syndrome. While each of these may be not so much a single disease as a group of related processes, each is characterized by a fairly well-defined immunopathology. In the case of SLE, most of the symptoms can be ascribed to the presence of antinuclear autoantibodies, which account not only for the LE cell phenomenon but also for the immune complexes that cause the damage in susceptible target organs (e.g., at the dermal–epidermal junction, yielding erythematous skin rashes, or in the glomeruli to cause lupus nephritis).[103] In rheumatoid arthritis, the autoantibodies are anti-type II collagen and anti-immunoglobulins, which form immune complexes whose presence in synovial linings activates complement to induce the effusion characteristic of rheumatoid synovitis.[104] Sjögren's syndrome differs from the previous two conditions in that, while autoantibodies and hypergammaglobulinemia may be present, the lesions of the lacrimal, salivary, and other exocrine glands appear rather to be due to the effects of immunocyte (and macrophage) activation.[105]

One of the more interesting consequences of the study of these autoimmune diseases was the growing realization of the importance of genetic constitution – a finding confirmed by the strong predilection of certain inbred strains of laboratory animals (e.g., the NZB, MRL, and RCS strains of mice and the BUF and BB/W strains of rats) to develop a variety of autoimmune diseases.[106] Among these genetic influences are certain predispositions for disease, located within the major histocompatibility gene complex (MHC) at loci that code for the formation of the principal human leukocyte antigens.[107] Yet another group of immunologic dysregulations appears to depend upon a set of immune-response genes at another location within the MHC.[108]

Technological advances

To the extent that most autoimmune diseases represent undesirable active responses to self antigens, it follows that they should be amenable to immunosuppressive therapy with the variety of chemotherapeutic agents that have emerged, primarily from developments in cancer chemotherapy. But other more elegant and more specific approaches to both diagnosis and therapy are in the process of development. The newer techniques of molecular biology have made available monoclonal antibodies and RNA–DNA molecular probes to identify and even to reproduce the antigenic epitopes that are the targets of the auto-immune response. This offers the possibility that genetically engineered antigens may be employed therapeutically to down-regulate these dangerous responses. Alternatively, where a receptor-specific autoantibody can be identified as the cause of a disease, its (temporary) alleviation may be obtained (as in myasthenia gravis) by removing the antibody using plasmapheresis[109] or perhaps, eventually, specific immunosorbents.

Conclusions

Both Metchnikoff with his phagocytic theory and Ehrlich with his side-chain theory understood that the mechanisms he proposed were part of a larger biological system whose evolution brought with it improvements in the organism's ability to nourish itself and, incidentally, to protect itself from infection. But Metchnikoff was always willing to concede that the advantages of such an evolution might be accompanied by certain disadvantages. Thus, he recognized that cellular inflammation might produce local tissue damage as well as overall benefit, and he accorded a role to the phagocyte in such deleterious aging processes as the greying of the hair, the wrinkling of the skin, and the deterioration with age of the brain and other organs.

Ehrlich, for his part, seemed to have been unwilling to concede a down-side to the Darwinian evolution of the receptor antibodies that he had proposed. His concept of *horror autotoxicus* was in fact his denial that some biological price might be exacted for the benefits that antibodies endow upon the individual organism. It was exactly this denial, strongly reinforced by the triumph of Ehrlich's humoralist views over the cellularist notions of Metchnikoff, that made it so difficult for immunologic theoreticians to accept the reality of autoimmunity for over fifty years.

In spite of the Donath–Landsteiner finding of an autoantibody in paroxysmal cold hemoglobinuria, and in spite of increasing evidence from clinical subspecialties of the existence of autoimmune diseases, the teleologic appeal of *horror autotoxicus* (= no autoantibody) made the acceptance of the reality of auto-immune disease difficult, if not impossible. To force such an acceptance would require a conjunction of events that many different scientific disciplines eventually experience: the accumulation of a large number of observations not explicable in terms of the current dogma, and a major change in the direction of thought in the field that would allow the previously unthinkable now to be thought. In the immunology of the 1950s and 1960s, it was the many new clinical and experimental models of autoimmune disease (and of immunopath-ologic responses in general) then being reported that drove this conceptual transition.

Notes and references

1. Among the few discussions of the history of autoimmunity in the literature are: Cruse, J.M., Whitcomb, D., and Lewis, R.E. Jr, *Concepts Immunopathol.* 1:32, 1985; Goltz, D., "*Horror autotoxicus*. Ein Beitrag zur Geschichte und Theorie der Autoimmunpathologie im Spiegel eines vielzitierten Begriffes," Thesis, Münster, 1980; Moulin, A.-M., *Le Dernier Langage de la Médicine: Histoire de l'Immunologie de Pasteur au SIDA*, Paris, Presses Universitaires, 1991; Eyquem, A., *Arch. Inst. Pasteur Tunis*, 58:281, 1981.

2. Ehrlich, P., *Verh. 73 Ges. dtsch. Naturf. Aerzte*, 1901; reprinted in *The Collected Papers of Paul Ehrlich*, Vol. 2, New York, Pergamon Press, 1957, p. 298.

3. Sigerist, H.E., *A History of Medicine*, Vol. II, New York, Oxford University Press, 1961.

4. See Temkin, O., "Health and Disease", in Wiener, P.P., ed., *Dictionary of the History of Ideas*, Vol. 2, New York, Charles Scribner, 1973, pp. 395–407. See also Canguilhem, G., *On the Normal and the Pathological*, Boston, D. Reidel, 1978.

5. Bernard, C., *Introduction to the Study of Experimental Medicine* (translation by H.C. Greene), New York, Macmillan, 1927.

6. Compte, I.A.M.F.X., *Cours de Philosophie Positive*, Vol. 3, Paris, 1830–1842.

7. Bouchard, C.J., *Union Méd.*, Apr. 10, 1886, p. 577; *idem, Lectures on Autointoxication* (translation by T. Oliver), Philadelphia, F.A. Davis, 1894.

8. See, for example, Paczkowski, T., *Die Autointoxication als Grundlage zu Erkrankung*, Leipzig, E. Demme, 1900; Jahn, V., "Die gastrointestinalen Autointoxikationspsychosen des späten 19 Jahrhunderts," *Zürcher medizingeschichtliche Abhandlungen*, 111, 1975.

9. Smith, J.L., *Ann. Intern. Med.* 96:365, 1982.

10. See, for example, Metchnikoff, E., *The Prolongation of Life* (English translation of *Essais Optimistes*, by P.C. Mitchell), New York, Putnam, 1908.

11. See especially Volume 14 (1900) of *Ann. Inst. Pasteur* for individual reports and a broad review by Metchnikoff on "cytotoxines."

12. Ehrlich, *Collected Papers*, note 2, Vol. 2, p. 315.

13. Bordet, J., *Ann. Inst. Pasteur* 12:688, 1899. Perhaps Ehrlich was aware that Rudolf Virchow, the greatest pathologist of his day and a believer in pathology as altered physiology, speculated that such alterations resulted from "insufficiency of the regulatory apparatuses." Virchow, R., "Über die heutige Stellung der Pathologie," *Naturforscherversammlung in Innsbruck*, 1869, reprinted in Sudhoff, K., *Rudolf Virchow und die deutschen Naturforscherversammlungen*, Leipzig, Akademische Verlags, 1922, p. 932.

14. Ehrlich, P., and Morgenroth, J., *Berlin klin. Wochenschr.* 38:251, 1901 (reprinted in English translation in Ehrlich, *Collected Papers*, note 2, Vol. 2, p. 253).

15. Goltz, note 1.

16. Rose, N.R., personal communication, 1986. As late as 1954, even while sitting on his own thyroid autoantibody data, Witebsky could say (at the Ehrlich centennial celebration) "The validity of the law [*sic*!] of *horror autotoxicus* certainly should be evident to everyone interested in the field of blood transfusion and blood disease. Autoantibodies – namely, antibodies directed against the receptors of the same individual – are not formed." (Witebsky, E., "Ehrlich's Side-Chain Theory in the Light of Present Immunology," *Ann. NY Acad. Sci.* 59:168, 1954). Even in the face

of positive Coombs tests in cases of acquired hemolytic anemias, Witebsky suggested in 1952 (*Proc. 4th Intl Cong., Intl Soc. Hematol., Mar del Plata, Argentina, 1952,* New York, Grune & Stratton, 19;54, p. 295) that the immunologist "should keep an open mind...toward the theory of autosensitization...it might be well to seek additional causes..." for acquired hemolytic anemia of the adult.

17. By the time that Witebsky accepted the implications of these data and they were finally published (Witebsky, E., and Rose, N.R., *J. Immunol.* 76:408, 1956; Rose, N.R., and Witebsky, E., *ibid.* 76:417, 1956), Roitt et al. (*Lancet* 2:820, 1956) had made similar findings in human Hashimoto's disease, and priority of discovery is now adjudged to be shared equally.

18. Metalnikoff, S., *Ann. Inst. Pasteur* 14:577, 1900.

19. Ehrlich and Morgenroth, note 14, p. 255 footnote.

20. Metchnikoff, E., "Sur les Cytotoxines," *Ann. Inst. Pasteur* 14:369–377, 1900; Sachs, H., in *Handbuch der Technik und Methodik der Immunitätsforschung*, Vol. 2, Jena, Fischer, 1909, p. 186.

21. Portier, P., and Richet, C., *C. R. Soc. Biol.* 54:170, 1902.

22. Arthus, M., *C. R. Soc. Biol.* 55:817, 1903.

23. von Pirquet, C., and Schick, B., *Die Serumkrankheit*, Vienna, Deuticke, 1906; English edition, *Serum Sickness*, Baltimore, Williams & Wilkins, 1951.

24. See, for example, von Szily, A., *Die Anaphylaxie in der Augenheilkunde*, Stuttgart, Ferdinand Enke, 1914.

25. Donath, J., and Landsteiner, K., *Münch. med. Wochenschr.* 51:1590, 1904. Other instances of autoimmune hemolytic anemias were reported a few years later by Chauffard A., and Troisier, J., *Sem. Méd. Paris* 28:345, 1908; 29:601, 1909.

26. Bordet, J., and Gengou, O., *Ann. Inst. Pasteur* 15:289, 1901.

27. See especially the extensive studies by Nuttal on comparative analyses of antigens from different species (Nuttall, G.H.F., Blood *Immunity and Blood Relationships*, Cambridge, The University Press, 1904). It is of interest that von Wassermann's original intent in his serologic studies was the detection of antigen, and not of antibody.

28. von Wassermann, A., and Bruck, C., *Deutsch. med. Wochenschr.* 32:449, 1906.

29. Citron, J., *Centralbl. Bakt.* 41:230, 1906.

30. Schaudinn, F., and Hoffmann, E., *Arb. ReichsgesundhAmt.* 22:527, 1905.

31. von Wassermann, A., Neisser, A., and Bruck, C., *Deutsch. med. Wochenschr.* 32:745, 1906.

32. von Wassermann, A., Neisser, A., Bruck, C., and Schucht, A., *Zschr. Hyg.* 55:451, 1906.

33. Detré, L., *Wien. klin. Wochenschr.* 19:619, 1906.

34. Fleck, Ludwik, *Entstehung und Entwicklung einer wissenschaftlichen Tatsache. Einführung in die Lehre vom Denkstil und Denkkollektiv*, Basel, Benno Schwabe, 1935; English translation, *Genesis and Development of a Scientific Fact*, Chicago, University of Chicago Press, 1979.

35. See, for example, Thomas Kuhn's acknowledgement of Fleck's influence in the introduction to his book *The Structure of Scientific Revolutions*, Chicago, University of Chicago Press, Chicago, 1980; and Cohen, R.S., and Schnelle, T., eds, *Cognition and Fact: Materials on Ludwik Fleck*, Boston, D. Reidel, 1986.

36. See the numerous broadsides fired by Wassermann, Bruck, and Weil in *Berl. klin. Wochenschr.*, 58, 1921.

37. This observation was reported almost simultaneously from many laboratories, by, among others, Marie, A., and Levaditi, C., *Ann. Inst. Pasteur* 21:138,1907;

Plaut, F., *Münch. med Wochenschr.* **44**:1458, 1907; Michaelis, L., *Berl. klin. Wochenschr.* **44**:1103, 1907; and Weil, E., and Braun, H., *Wien. klin. Wochenschr.* **20**:527, 1907.

38. Weil, E., and Braun, H., *Wien. klin. Wochenschr.* **22**:372, 1909.
39. Weill and Braun, note 38, p. 374.
40. Uhlenhuth, P., in *Festschrift zum 60 Geburtstag von Robert Koch*, Jena, Fischer, 1903.
41. Kraus, R., Doerr, R., and Sohma, M., *Wien. klin. Wochenschr.* **21**:1084, 1908.
42. Andrejew, P., *Arb. Kais. Gesundheitsamt* **30**:450, 1908.
43. Uhlenhuth, P., and Haendel, Z. *Immunitätsf.* **4**:761, 1910.
44. Römer, P., and Gebb, H., *v. Graefes Arch. Ophthalmol.* **60**:175, 1905.
45. Krusius, F.F., *Arch. Augenheilk.* **67**:6, 1910.
46. Römer, P., and Gebb, H., *v. Graefes Arch. Ophthalmol.* **81**:367, 387, 1912.
47. Römer and Gebb, note 46, p. 393.
48. Not only does Römer's use of Ehrlich's concepts and even phrases testify to their popularity at that time, but it was this same Römer who in 1904 wrote a 403-page treatise on *The Ehrlich Side-chain Theory and its Meaning for Medical Science*, Vienna, Hölder, 1904.
49. Santucci, S., *Riv. Ital. Ottal. Roma* **2**:213, 1906, abstracted in *Z. Augenheilk.* **17**:297, 1907.
50. Golowin, S., *Klin. Monatsbl. Augenheilk.* **47**:150, 1909.
51. Golowin, S., *Russky Vratch.* **22**, 29 May, 1904.
52. Elschnig, A., *v. Graefes Arch. Ophthalmol.* **75**:459, 1910; **76**:509, 1910.
53. Verhoeff, F.H., and Lemoine, A.N., *Proc. Intl Congr. Ophthalmol., Washington, 1922*, p. 234.
54. Woods, A.C., *Allergy and Immunity in Ophthalmology*, Baltimore, Johns Hopkins Univ. Press, 1933.
55. Rivers, T.M., Schwentker, F.F., and Berry, G.P., *J. Exp. Med.* **58**:39, 1933; Rivers, T.M., and Schwentker, F.F., *J. Exp. Med.* **61**:689, 1935.
56. Hektoen, L., and Schulhof, K., *Proc. Natl Acad. Sci. USA* **11**:481, 1925; Hektoen, L., Fox, H., and Schulhof, K., *J. Infect. Dis.* **40**:641, 1927.
57. Reviewed by Katsh, S., *Am. J. Obstet. Gynecol.* **77**:946, 1959.
58. Anderson, D.J., and Alexander, N.J., *Fertil. Steril.* **40**:557, 1983.
59. Medawar, P.B., *J. Anat.* **78**:176, 1944; **79**:157, 1945.
60. Bruton, O.C., *Pediatrics* **9**:722, 1952.
61. Owen, R.D., *Science* **102**:400, 1945.
62. Freund, J., and McDermott, K., *Proc. Soc. Exp. Biol. Med.* **49**:548, 1942; Freund, J., *Am. J. Clin. Pathol.* **21**:645, 1951.
63. Waksman, B.H., *Experimental Allergic Encephalomyelitis and the "Autoallergic" Diseases*, Suppl. to *Intl Arch. Allergy Appl. Immunol.* **14**, 1959. See also Waksman, B.H., *Medicine* **41**:93, 1962.
64. Burnet, F.M., *The Production of Antibodies*, Melbourne, Macmillan, 1941.
65. Burnet, F.M., and Fenner, F., *The Production of Antibodies*, 2nd edn, Melbourne, Macmillan, 1949.
66. Traub, E., *J. Exp. Med.* **64**:183, 1936; **68**:229, 1938; **69**:801, 1939.
67. Burnet and Fenner, note 65, p. 104.
68. Burnet, F.M., *The Clonal Selection Theory of Antibody Formation*, Cambridge, Cambridge University Press, 1959. An earlier version appeared in *Austral. J. Sci.* **20**:67, 1957.

69. Lederberg, J., *Science* **129**:1649, 1959.
70. Billingham, R.E., Brent, L., and Medawar, P.B., *Nature Lond.* **172**:603, 1953.
71. Felton, L.D., and Ottinger, B., *J. Bacteriol.* **43**:94, 1942. For later work on this problem, see Smith, R.T., *Adv. Immunol.* **1**:67, 1961; see also Coutinho, A., and Möller, G., *Adv. Immunol.* **21**:114, 1975.
72. Hašek, M., Lengerová, A., and Hraba, T., *Adv. Immunol.* **1**:1, 1961.
73. Weigle, W.O., *Adv. Immunol.* **16**:61, 1968. For earlier work in the field, see Hanan, R.Q., and Oyama, J., *J. Immunol.* **73**:49, 1954; Dixon, F.J., and Maurer, P.H., *J. Exp. Med.* **101**:245, 1955; and Cinader, B., and Dubert, J.M., *Br. J. Exp. Pathol.* **26**:515, 1955.
74. For a discussion of the theoretical and practical implications of partial tolerance, see Nossal, G.J.V., *Ann. Rev. Immunol.* **1**:33, 1983.
75. Šterzl, J., and Silverstein, A.M., *Adv. Immunol.* **6**:337, 1967; Solomon, J.B., *Foetal and Neonatal Immunology*, Amsterdam, North Holland, 1971.
76. Mitchison, N.A., *Proc. R. Soc. Med.* **161**:275, 1964.
77. Ada, G.L., and Byrt, P., *Nature Lond.* **222**:1291, 1969; Ada, G.L., et al., in Šterzl, J. and Říha, I., eds, *Developmental Aspects of Antibody Formation and Structure*, Vol. 2, New York, Academic Press, 1970, p. 503.
78. Humphrey, J.H., and Keller, H.U., in Šterzl, J., and Říha, I., eds, *Developmental Aspects of Antibody Formation and Structure*, Vol. 2, New York, Academic Press, 1970, p. 485.
79. Weigle, W.O., *J. Exp. Med.* **114**:111, 1961.
80. Fagraeus, A., *Acta Med. Scand.* (Suppl.) 204, 1948.
81. Coons, A.H., Leduc, E.H., and Connally, J.M., *J. Exp. Med.* **102**:49, 1955.
82. Unanue, E.R., *Adv. Immunol.* **15**:95, 1972. See also *Macrophage Regulation of Immunity*, Unanue, E.R., and Rosenthal, A.S., eds, New York, Academic Press, 1980.
83. Claman, H.N., Chaperon, E.A., and Triplett, R.R., *Proc. Soc. Exp. Biol. Med.* **122**:1167, 1966; Mitchell, G.F., and Miller, J.F.A.P., *J. Exp. Med.* **128**:801, 821, 1968; Mitchison, N.A., *Eur. J. Immunol.* **1**:18, 1971.
84. Zinkernagel, R.M., and Doherty, P.C., *Adv. Immunol.* **27**:52, 1979.
85. McKenzie, I.F.C., and Potter, T., *Adv. Immunol.* **27**:179, 1979.
86. Gershon R.K., and Kondo, K., *Immunology* **18**:723, 1970.
87. Gershon, R.K., and Kondo, K., *Immunology* **21**:903, 1972.
88. Jerne, N.K., *Ann. Immunol. (Paris)* **125C**:373, 1974; *Harvey Lect.* **70**:93, 1974.
89. Mackay, I.R., and Burnet, F.M., *Autoimmune Diseases*, Springfield, Charles C. Thomas, 1963; Schwartz, R.S., and Rose, N.R., *Ann. NY Acad. Sci.* **475**, 1986; Rose, N.R., and Mackay, I.R., eds, *The Autoimmune Diseases*, New York, Academic Press, 1985; a 4th edition of the Rose & Mackay book appeared in 2006.
90. See Voisin, Table 8.1, and Tung, K.S.K., and Menge, A.C. in Rose and Mackay's *The Autoimmune Diseases*, note 89, p. 537.
91. See Kabat et al., Table 8.1, and Arnason, B.G.W., in Rose and Mackay's *The Autoimmune Diseases*, note 89, p. 400.
92. See Collins, Table 8.1, and Faure, J.-P., "Autoimmunity and the Retina", in *Current Topics in Eye Research*, Vol. 2, New York, Academic Press, 1980, p. 215.
93. Marrack, G.E., Font, R.L., and Alepa, F.P., *Ophthalmic Res.* **8**:117, 1976; **9**:162, 1977.

94. See note 17 and Bigazzi, P.E., and Rose, N.R., in Rose and Mackay's *The Auto-immune Diseases*, note 98, p. 161.
95. Anderson, J., Goudie, R., Gray, K., and Timbury, G., *Lancet* 1:1123, 1957; Blizzard, R., and Kyle, M., *J. Clin. Invest.* 42:1653, 1963.
96. Beutner, E.H., and Jordon, R.E., *Proc. Soc. Exp. Biol. Med.* 117:505, 1965; Diaz, L.A., et al., in Rose and Mackay's *The Autoimmune Diseases*, note 98, p. 443.
97. Jordon, R.E., et al., *J. Am. Med. Assoc.* 200:751, 1967.
98. Harrison, L.C., in Rose and Mackay's *The Autoimmune Diseases*, note 89, p. 617.
99. Langley, J.N., *J. Physiol. Lond.* 1:339, 1878.
100. Manley, S.W., Knight, A., and Adams, D.D., *Springer Semin. Immunopathol.* 5:413, 1982; Davies, T. F., and De Bernardo, E., in T.F. Davies, ed., *Autoimmune Endocrine Disease*, New York, Wiley, 1983, p. 127.
101. Simpson, J.A., *Scot. Med. J.* 5:419, 1960; Nastuk, W.L., Plescia, O.J., and Osserman, K.E., *Proc. Soc. Exp. Biol. Med.* 105:177, 1960; Patrick, J., and Lindstrom, J., *Science* 180:871, 1973.
102. Flier, J. S., Kahn, C. R., Roth, J., and Bar, R. S., *Science* 190:63, 1975; Kahn, C.R., Baird, K.L., Jarrett, D.B., and Flier, J.S., *Proc. Natl Acad. Sci. USA* 75:4209, 1978.
103. McCluskey, R.T., *Arthritis Rheum.* 25:867, 1982; Tan, E.M., *Adv. Immunol.* 33:167, 1982.
104. Ziff, M., in Rose and Mackay's *The Autoimmune Diseases*, note 89, p. 59.
105. Miyasaka, N., et al., *Arthritis Rheum.* 26:954, 1983.
106. Howie, J.B., and Helyer, B.J., *Adv. Immunol.* 9:215, 1968.
107. Gorer, P.A., Lyman, S., and Snell, G.D., *Proc. R. Soc. Lond.*, 135:499, 1948; Zinkernagel, R.M., and Doherty, P.C., *Adv. Immunol.* 27:52, 1979; Kano, K., Abeyounis, C.J., and Zaleski, M.B., eds, *Immunobiology of the Major Histocompatibility Complex*, New York, Karger, 1981; Tiwari, J.L., and Teresaki, P.I., *HLA and Disease Associations*, New York, Springer, 1985.
108. McDevitt, H.O., and Benacerraf, B., *Adv. Immunol.* 11:31, 1969; Gonwa, T.A., Peterlin, B.M., and Stobo, J.D., *Adv. Immunol.* 34:71, 1983; Schwartz, R.H., *Adv. Immunol.* 38:31, 1986.
109. Vincent, A., *Physiol. Rev.* 60:756, 1980.

9 Allergy and immunopathology: the "price" of immunity

The conception that antibodies, which should protect against disease, are also responsible for disease, sounds at first absurd.

Clemens von Pirquet

The quotation that opens this chapter carries the same implication as that of Paul Ehrlich which introduced Chapter 8, on autoimmunity. It reflects the widespread contemporary view that the same mechanisms that provide for defense against infectious disease ought not to function also to embarrass the host. In the dawning years of the twentieth century, those investigators active in the young field of immunology had been brought up, with Metchnikoff and Ehrlich, to view the immune response as an eminently useful Darwinian adaptation. It had evolved, presumably, to defend the organism against an outside world heavily populated by highly pathogenic organisms and virulent toxins.

So deeply ingrained was this view of a benevolent immunity that the earliest observations that might have contradicted it were quickly attributed to other causes and mechanisms. Thus, Robert Koch's observations on the hyper-reactivity of tuberculous animals to new inoculations of tubercle bacilli (the Koch phenomenon) or to tuberculin (the inflammatory skin reaction) were attributed by him to the direct effect of local excesses of bacterial toxins.[1] Again, when Emil Behring reported in 1893 a "hypersensitivity" to diphtheria toxin in previously immunized guinea pigs, he called it a "paradoxical reaction," and followed Koch's lead in assigning it to the direct cumulative action of the toxin itself, rather than to any component of the acquired immune response.[2] Even the many workers who studied the formation and activity of a variety of antitissue iso- and xeno-antibodies (e.g., anti-erythrocyte, anti-spermatozoa, anti-liver, etc.) made little or no connection between these phenomena and human disease. They appeared to be more interested in what their results might tell them about antibody formation and antibody function.

It is not surprising, therefore, that the investigators who first reported on the phenomena that would open up the field of allergy and immunopathology, and many who first dared to speculate that these reactions might be an integral part of the "immune" response, did not come from within the classical tradition of bacteriologic immunology. Paul Portier and Charles Richet, who described anaphylaxis, were physiologists, as was Maurice Arthus, who discovered the phenomenon of local anaphylaxis ("the Arthus reaction"). The discoverers of the third of that famous triad, serum sickness, were Clemens von Pirquet and Bela

A History of Immunology, Second Edition
ISBN: 978-0-12-370586-0

Copyright © 2009, Elsevier Inc.
All rights reserved

Schick, both pediatricians. It was not long after these initial discoveries, with such obvious implications for human disease, that (for other reasons) immunology "shifted gears." It became a predominantly chemical science, so that it was left primarily to clinicians and later to experimental pathologists to expand upon these initial findings.

But throughout the course of the slow conceptual development of the field of allergy, one can detect a continuing and pervasive schizophrenic approach to the relationship between allergy and immunity, shared by both immunologists and allergists. Just as Ehrlich's maxim of horror autotoxicus inhibited free speculation and progress toward the understanding of autoimmune diseases, so did the general Darwinian teleologic view of a benign immune apparatus inhibit acceptance of allergic disease as another facet of the same response. The continuing desire to keep allergy separate from immunity fostered early suggestions that substances other than antibodies (such as toxic byproducts of the protein stimulant) were the immediate causes of these reactions. Even after full acceptance of the role of antibodies, this contemporary tendency was made evident by the ascription of these conditions to special classes of antibody ("atopic reagins"), or to those with special characteristics ("sessile," or "cell-bound" antibodies; see Chapter 15). Now, with the identification of IgE antibodies as the agents responsible for so many allergic diseases, but also as full-fledged members of the immunoglobulin family, this same teleologic drive may be an important contributor to the continuing search for some protective role for this class of immunoglobulins.[3]

Early observations[4]

Knowledge of the vexing problems of asthma and hayfever is almost as old as recorded history. The clinical signs and symptoms of these conditions were well described by the ancient Greeks, and appear also in the Talmud. In the Greek humoralist tradition, these conditions were lumped together with other reactions apparently unique to the individual, under the generic term "idiosyncrasies" (Gr. *idios*, self, and *syncrasis*, a mixture [of the humors]). From the time of Galen onward, the term was increasingly applied to abnormal reactions to drugs and to such conditions as poison sumac dermatitis, and were usually included in discussions of individual *sympathies* and *antipathies*. As the prefix *idio-* implies, these conditions were long felt to arise in the unique constitution of the individual (a conclusion later to be borne out by modern knowledge of the genetic predisposition to many of these diseases).

It is of interest that Edward Jenner provided a very good description *and illustration* of the wheal and erythema reaction in his 1798 report that introduced anti-smallpox vaccination to the world.[5] In 1839, the French physiologist Magendie described anaphylactic shock and death in dogs repeatedly injected with foreign proteins.[6] Again, in 1894 Simon Flexner provided a clear statement of the basic phenomenon of anaphylaxis in rabbits, reporting that "animals that had withstood one dose of dog serum would succumb to a second dose given after the lapse of some days or weeks..."[7]

Two other observations made during this period are of interest. Behring, working with diphtheria toxin in 1893, and Richet and Héricourt, working with eel toxin in 1898,[8] reported that animals would suffer enhanced responses and even death following a second dose of toxin too small to injure normal untreated animals. In each case, the phenomenon was interpreted as an increased susceptibility to the direct effects of the toxin, and indeed Behring coined the term hypersensitivity (*Überempfindlichkeit*) to describe these exaggerated reactions.

Little attention was paid to these early reports, or to their implications, until the studies of Portier and Richet[9] caught the attention of the immunologic world. In this oft-told story, these physiologists set sail on the yacht of the Prince of Monaco in order to study the mode of action of marine invertebrate poisons in mammals. They furnished careful descriptions of the clinical shock syndrome encountered in dogs given otherwise innocuous doses of the toxin, after previous experience with the same substance. Employing a somewhat questionable etymology, they named this new phenomenon *anaphylaxis* (to express its antithesis to the more familiar term for protection, *prophylaxis*). It is not widely appreciated that credit for this discovery ought to be shared also by Theobald Smith, who independently in 1902 studied analogous anaphylactic shock reactions in the guinea pig. Smith, however, failed to publish his results, and only communicated them to Paul Ehrlich several years later.[10] Ehrlich assigned the task of following up these studies to his colleague, Richard Otto, who published studies on "*das Theobald Smith'sche Phänomen*" in the years that followed.[11]

Now that investigators had been alerted to the hyper-reactivity that might accompany the injection of foreign proteins, there rapidly followed a series of new phenomenologic observations on analogous responses, and re-evaluations and reinterpretations of earlier observations. Thus, in 1903 Maurice Arthus described the heightened local hemorrhagic and necrotic response to repeated intradermal injections of protein antigens,[12] soon to be named the "Arthus reaction." In 1906, von Pirquet and Schick reanalyzed the now well-established observation that certain patients receiving diphtheria or tetanus antitoxic serum might suffer strange systemic and local symptoms, and they named it serum sickness.[13] For the first time, they identified this disease as the product of immunologic mechanisms. In order to describe these and related phenomena, they coined the term "allergy" (Gr. *allos, ergos*, altered reactivity), to set these responses apart from the customary minimal reactions expected of such otherwise innocuous substances.

Given the impetus provided by these widely publicized observations, many other investigators undertook the study of these interesting reactions, and made important contributions to their phenomenologic description and to the discussion of their causes. Foremost among these, in addition to Otto, were Rosenau and Anderson, who published an extensive series of papers on the quantitative and qualitative analysis of anaphylactic reactions.[14] In addition, significant contributions were made by such investigators as Gay and South-ard,[15] Auer and Lewis,[16] Biedl and Kraus,[17] Friedberger,[18] and Vaughan.[19]

Finally, the human conditions of hayfever and asthma were brought into this newly expanding immunologic fold, and joined conceptually to the new knowledge of anaphylaxis and allergy. In 1906, Alfred Wolff-Eisner made the connection of hayfever as a hypersensitivity state or reaction in the immunologic sense,[20] and in 1910 Samuel J. Meltzer did the same for asthma.[21]

The debate on the mechanism of allergy

Direct toxicity

One of the earliest views of anaphylaxis was that it results from the action of a potent toxin, either present intact in the injected material or split from its components by enzymatic action. Since many of his original observations were obtained using marine invertebrate toxins, Richet initially postulated that the material actually contained two active substances: *thalassin*, of only modest toxicity, which would induce immunity; and *congestin*, which, he suggested, far surpasses the original poison in toxicity and leads to "hypersensitiveness" by cumulative action.[22] Once it became known that even normal serum might serve to sensitize for and induce anaphylactic shock, Gay and Southard suggested that all sera capable of eliciting anaphylaxis contain such a toxic substance, which they called *anaphylactin*.[23] Vaughan, however, maintained that the active toxin could not be present in a free state, but rather was a toxic cleavage product of the injected protein.[24] He suggested that the cleavage process is initially slow, so that a first injection would generally not lead to a systemic response, but "the cells learn from this lesson" and a second injection results in the rapid liberation of large amounts of toxin, resulting in the typical shock syndrome.

As further knowledge of the specificity of anaphylactic reactions was gained, and especially after the demonstration that anaphylactic sensitivity, like protective immunity, might be passively transferred using the serum of sensitized animals,[25] the involvement of antigen–antibody interactions became more likely and a direct toxin theory less likely. However, as late as 1921 Maurice Arthus could claim a clear separation between anaphylaxis and immunity, and conclude "Thus, we may absolutely separate these two states and deny that they may be two different manifestations of a single and same state."[26]

Special antibodies – "misdirected" immunity

We noted earlier that the French school of immunologists – the followers of Jules Bordet – were given to a freer and more exuberant speculation than were their German counterparts who adhered to the doctrines of Paul Ehrlich. Workers at the Pasteur Institute in Paris felt unfettered by the tight doctrinaire strictures imposed by Ehrlich's side-chain theory, and it was predominantly they who led the way in proposing that antibodies might play the significant role in mediating anaphylaxis – a view that no firm adherent of Ehrlich's theory would share until years later. This is well illustrated in an extensive review on anaphylaxis by Ehrlich's student, Richard Otto, in 1909. After summarizing the increasingly strong evidence implicating circulating antibody in the pathogenesis of anaphylaxis, Otto finally credits the theory with "a certain likelihood," especially in view of the passive transfer experiments, but finally ends up on the fence, saying that "one must on this basis be cautious in using the term antibody."[27]

The most elaborate theory implicating an antibody in the development of anaphylaxis was that of Alexandre Besredka. In the French vernacular, the antibody was called a *sensibilisine*. The offending serum was held to contain a *sensibilisinogène* (antigen) which would stimulate the production of its corresponding antibody. Then, in a curious reprise of his anti-antibody immunoregulation theory of a few years earlier (see Chapter 10), Besredka postulated that the offending serum also contained an "anti-sensibilisin" (apparently not an antibody in this case) whose interaction with the sensibilisin antibody (purportedly attached to cells of the central nervous system) would result in the shock syndrome. This theory, supported initially by Richet and by Robert Doerr,[28] represents an interesting transition between the dualistic theories of Gay and Southard and of Vaughan (in which antigen and toxin coexist in the injected material, but specific antibody plays no role in anaphylaxis) and the unitarian theories to be described below, in which the antigen–antibody interaction is held to account for all aspects of the response.

In all of the considerations of the mediation of allergic reactions by "special" types of antibodies, perhaps none set the tone for the next fifty years quite so well as that of J. R. Currie in 1907. He employed the term "supersensitization" to denote the state of preparedness for anaphylactic shock and assumed, with others, that specific antibodies (precipitins) are the active factors. But, suggested Currie, two different antibodies may be formed against the same antigen, one protective and one destructive, "because these [sensitizing substances] are not normal noxious agents introduced through normal channels." He goes on to say:[29]

> But, if the active principle is introduced into the system neither through the customary channels nor under the form of a microorganism, whose power for mischief depends upon its liberty to grow and multiply, the procedure is out of accord with the course of nature, and the defensive powers of the animal, adapted to cope with natural infections, are somewhat at fault in their method of dealing with the artificial invasion. ...Extraneous sera appear to belong to an order of substances which effect immunization, not by inducing insusceptibility of tissue cells, but by means of an accelerated reaction [allergy], which may thus be regarded as the expression of a misdirected defense, a formal but useless immunity.

Here, in the expression "a formal but useless immunity" was a view that several generations of immunologists and allergists would fall back upon in trying to defend the notion that protective immunity and destructive allergy might both depend upon the same central mechanisms. The concept of a specific although somewhat special type of antibody was extended also to the Arthus reaction by Maurice Nicolle in his study of this phenomenon in 1907.[30] This attempt to set apart the antibody responsible for these deleterious reactions was repeated often, by the assignment of special names to the "allergic antibody" such as cytotropic or cytophilic antibody or atopic antibody. The antibodies responsible for allergy in man were given the special name *reagins* (not to be confused with the so-called reaginic antibody in the Wassermann test for syphilis), presumably in yet another attempt to set them apart from the more usual antibodies associated with defensive immune responses. Even IgE, when first discovered, was implied to be a special type of antibody unrelated to protective immune responses.

The unitarian approach

As might be expected from any young field in conceptual ferment, adherents could be found for each plausible theory, and even for many implausible ones. One of the earliest and strongest voices to be raised on behalf of the role of "ordinary" antibodies in allergic reactions was that of Clemens von Pirquet. In his book with Schick on serum sickness, von Pirquet assumed automatically that "precipitins" are the causative agents. It is the clinician and not the classically trained bacteriologist–immunologist who is able to say:[31]

> The conception that antibodies, which should protect against disease, are also responsible for the disease, sounds at first absurd. This has as its basis the fact that we are accustomed to see in disease only the harm done to the [host] and to see in the antibodies solely antitoxic [protective] substances. One forgets too easily that the disease represents only a stage in the development of immunity, and that the organism often attains the advantage of immunity only by means of disease. Thus, a mild disease leads to immunity in the normal way, and since the entry of non-multiplying agents (serum) into the body seldom takes place in nature, serum sickness represents, so to speak, an unnatural (artificial) form of disease.

In his 1911 book *Allergy*, von Pirquet expanded upon this thesis. He suggested that the *immune precipitate* of antigen and antibody is the pathogenic factor: "This explanation involves also a new conception of the antibody. ...A disease might be due indirectly to an antibody, an idea to which at that time [1906] adherents of the school of Ehrlich, like Kraus, took strong exception."[32] von Pirquet, in Vienna, was not bothered by the possibility that antibodies may be toxic as well as protective. Indeed, he pointed out that the symptoms of infectious diseases in general are not entirely due to the action of the microorganisms *per se*, but that the host takes an active part in the production of most of the symptoms through the interaction of *its* products with those derived from the infecting agent. This is a broad view of disease pathogenesis that echoes the theories of Metchnikoff, and assigns to the phenomenon of allergy a respectable position in the immunologic schema; taking the bad with the good, it is at once a harmful byproduct of the immune response and a potential contributor to the development of protective immunity.

This latter view of allergy as a step on the road to immunity was taken up during those early years by several other investigators. In one of their papers, Rosenau and Anderson suggest that "resistance to disease may be largely gained through a process of hypersusceptibility."[33] They expand further upon this view in their monograph on anaphylaxis written in 1906. They freely grant a role to antibody in the pathogenesis of anaphylaxis, and declare that "whether this increased susceptibility is an essential element or only one stage in the process of resistance to disease, must now engage our attention." They eventually conclude that "we cannot escape the conviction that this phenomenon of hypersuscepti-bility has an important bearing on the prevention and cure of certain infectious processes."[34] Finally, even Charles Richet reached the same conclusion, despite his apparent support of Besredka's ideas. As he pointed out, anaphylaxis can be stimulated by far smaller doses and much more rapidly than can protective immunity, and thus may enhance the production of protective antitoxins. He concludes that "anaphylaxis appears to us then, in the final analysis, to be a process of *rapid defense* and above all of *defense against small doses.* ...Put in another way, *immunity can be established because anaphylaxis has taken place* [his italics]."[35]

The name of Clemens von Pirquet is associated by most historians and immunologists only with the naming of the disease serum sickness and with the coinage of the term *allergy.* What is not generally appreciated is the remarkable quality of his early clinical and experimental observations, and the full significance of his interpretations of the data collected. This is nowhere better illustrated than in his interpretation of the pathogenesis of serum sick-ness in man, illustrated in Figure 9.1. This is taken from his book *Allergy*

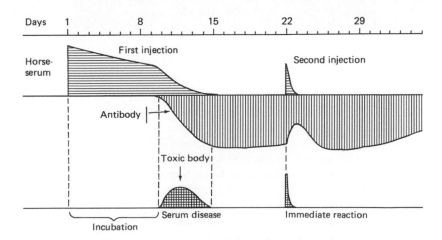

Figure 9.1 von Pirquet's concept of the steps in the development of serum sickness in man.
From von Pirquet, C., *Allergie*, Springer, Berlin, 1910; English translation, *Allergy*, Chicago, American Medical Association, 1911.

Plate 9 Clemens von Pirquet (1874–1929)

published during his brief tenure as Professor of Pediatrics at the Johns Hopkins Medical School.

From the very outset, von Pirquet had no difficulty in assigning to precipitating antibody the key role in the development of serum sickness, and indeed correctly identified immune complexes (which he called "toxic bodies") as the active agent. Almost in anticipation of a later generation's interest in immune complex disease, he also described many of the clinical accompaniments of such conditions, including glomerulonephritis, arthropathy, and such systemic changes as a drop in serum complement levels. But his diagram is even more revealing. In this description of the time-course of the response to intravenous horse-serum, he correctly attributed the initial incubation period to the time necessary to activate the immune system for the production of antibody.

He recorded the initial slow (metabolic) disappearance of antigen during this period, followed by a much more rapid *immune elimination* phase with the onset of antibody formation. It is precisely at this time that immune complexes are formed, said von Pirquet, accompanied by an active disease process. There is then a remission, after antigen has been completely eliminated and new immune complexes are no longer being formed. Then a second injection of antigen leads to an abrupt lowering of serum antibody levels, with rapid disappearance of free antigen, and the formation of significant amounts of immune complex, and an immediate exacerbation of the disease process.

Here, in a nutshell, is summarized the type of phenomenologic observation that would occupy so many investigators in the decades immediately following World War II – a prescient analysis of causes and effects for which von Pirquet has received little credit. von Pirquet's interest in the pathological effects of antigen–antibody complexes was pursued by many investigators – most notably by Fred Germuth[36] and Frank Dixon[37] – and their role in the pathogenesis of many different disease processes has been amply demonstrated.

The cellular theory – cytotropic antibody

Once it became generally accepted that specific antibody was somehow involved in the pathogenesis of various allergic reactions, the debate on mechanism became interestingly reminiscent of the earlier humoralist–cellularist debate on the basis of acquired immunity (see Chapter 2). For their part, those who favored the cellularist view pointed to the fact that anaphylactic shock could often be induced in animals in the absence of detectable circulating antibody, and proposed that the small amounts of antibody required were in fact affixed to the surface of appropriate target cells. There, any subsequent interaction with specific antigen would result in cell damage or death and a consequent shock-like syndrome. Additional support for this view was adduced from the observation that in passive anaphylaxis, a certain minimal time was required before which shock could not be induced by antigen administration. During this time, the passively administered antibody disappeared almost completely from the circulation, and it was assumed that the refractory period was that required by antibody to take up residence on target cells.

Perhaps the strongest support for this view was found in the experiments of Schultz,[38] who in 1910 excised portions of intestine from sensitized guinea pigs, suspended them in a bath of Ringer's solution, and showed the vigorous contraction of the isolated muscle upon exposure to specific antigen. These observations were confirmed and extended by Dale,[39] who substituted strips of uterus from sensitized guinea pigs for the intestinal preparation (whence the name Schultz–Dale phenomenon). With the demonstration that an anaphylactic shock-like syndrome can be induced by intravenous administration of histamine,[40] the humoral approach to anaphylaxis appeared to find strong support. However, the subsequent demonstration that histamine is a normal constituent of many tissues[41] tended to neutralize somewhat the implications of this

observation, since the cellular release of histamine might be one of the consequences of the interaction of antigen with cell-bound antibody.[42]

The humoralist view – anaphylatoxin

The observation that initially stimulated an interest in a humoral effector
mechanism in anaphylaxis was that of Friedemann in 1909.[43] He showed that
the characteristic symptoms of acute anaphylactic shock could be obtained by
injecting into the guinea pig the mixture of an antigen and its homologous
antibody, following a brief period of incubation *in vitro*. This view was
championed in a series of subsequent publications by Friedberger.[44] The interpretation given to this phenomenon was very much in accord with Erhlich's
then-popular theories. It was supposed that the reduction in circulating
complement that accompanies anaphylactic shock was due to its fixation onto
antigen–antibody complexes. The complement thus activated would exercise its
putative enzymatic activity and engage in proteolysis, the breakdown products
of which would constitute the toxic substances (*anaphylatoxins*). These would
account for the various local and systemic symptoms that accompany the shock
syndrome. This view was tempered somewhat by later observations that sera
could be rendered toxic in the same sense by a variety of other procedures in
which antigen and antibody played no part. Thus, anaphylatoxins may be
produced by incubating normal sera with kaolin, barium sulfate, talc, starch, or
agar, among others. In addition, similar shock-like syndromes could be induced
with a variety of other substances, including the heterophile Forssman antibody,
peptone, and even the simple chemical histamine, as indicated above.[45] Because
of the heterogeneity of these experimental models and their likely lack of relationship to the "true" anaphylaxis mediated by antibody, these reactions were
grouped together under the rubric of *anaphylactoid* reactions.

Progress in allergy – the clinical discipline[46]

We have noted earlier, in several contexts, that the period after World War I saw
a shift from biological to more chemical approaches to the study of immunity.
The disease-related aspects of antigen–antibody interactions were more and
more left to clinicians, while most laboratory-oriented immunologists chose to
follow the lead of Landsteiner, Heidelberger, Marrack, and Wells in studying
antibody formation and the chemistry of antigen–antibody interactions. This
development was not without its advantages, for it helped to foster the independent development of a new clinical specialty, allergy, with its own agenda
and its own distinctive avenues of research. When a Division of Immunology
was established in 1919 at Cornell Medical College (the first such unit in the
United States), its leadership was entrusted to Robert A. Cooke, the prototypical
clinical allergist, who organized a combination of laboratory and clinical studies
that would provide the academic model for the new field of allergy. One of

Cooke's first actions at Cornell was the appointment of Arthur F. Coca (the founding editor of the *American Journal of Immunology*) to his staff. Cooke and Coca contributed significantly to the development of allergy as a scientific discipline. It was Cooke who introduced to the allergy clinic the intradermal (actually intracutaneous) skin test for the etiologic diagnosis of allergic conditions, and Coca who pioneered in the purification of allergenic extracts for use in such tests. Together, Coca and Cooke attempted to classify the various hypersensitivity states and to distinguish among such conditions as hayfever, contact dermatitis, serum sickness, and experimental anaphylaxis in animals.[47] In recognition of the fact that hayfever and asthma in man might be genetically controlled, they coined the term *atopy* to set these conditions apart from other types of allergic conditions.

Coca and Cooke were also instrumental in the founding, in 1924, of the Society for the Study of Asthma and Allied Conditions (called the "Eastern" Society, to distinguish it from the "Western" Society for the Study of Hayfever, Asthma, and Allergic Diseases).[48] It is interesting that when Arthur Coca founded the Allergy Roundtable Discussion Group in New York,[49] all of its members were clinicians. Not until 1949 was a PhD "basic scientist" invited into the group. This was Merrill W. Chase, in apparent recognition of his pioneering work with Karl Landsteiner on the passive transfer of tuberculin hypersensitivity and of poison ivy-type contact dermatitis.

One of the more important contributions to the practical study and theoretical understanding of human allergies came in 1921 from Carl Prausnitz and Heinz Küstner.[50] Küstner was exquisitely sensitive to the cooked flesh of certain fish, but fish extracts failed to demonstrate the presence of precipitating antibodies in his serum. However, when a little of this serum was injected into Prausnitz' skin, a typical wheal and erythema hypersensitivity reaction could be elicited twenty-four hours later by local administration of the appropriate allergen. Here, finally, was the demonstration of the ability to transfer passively this human allergic condition, which strongly implicated antibody by analogy with other passive transfer reactions. Moreover, the persistence of this local hypersensitivity for more than four weeks after transfer implied a tight binding of the antibody involved to neighboring cells, thus reinforcing earlier speculations on so-called cytophilic antibodies.

With the discovery of a new class of immunoglobulins, IgA, in the late 1950s,[51] it was thought for a time that these might be the elusive "reaginic" antibodies responsible for human atopic allergies.[52] However, the report by Loveless in 1964 of allergy in an IgA-deficient patient,[53] and subsequent failure to transfer wheal and flare activity with purified IgA preparations, led to the demise of this theory. Then, in the mid-1960s, the husband-and-wife team of Kimishige and Teruko Ishizaka prepared an antiserum to a reagin-rich fraction from the serum of a person showing extreme hypersensitivity to ragweed, and demonstrated that this antibody would neutralize the Prausnitz–Küstner transferability of allergy with the patient's serum. Upon purification of the antibody, it was found that it would not react with any other known immunoglobulin class, and it was given the name -E (for Erythema) globulin,[54] since renamed IgE. Independently of the Ishizakas, Johanssen and Bennich isolated an atypical immunoglobulin from a myeloma patient, which they called IgND (after the patient's initials).[55] They went on to demonstrate that the serum of patients suffering from asthma or hayfever exhibited elevated levels of IgND, and it was soon concluded that this new class of human serum immunoglobulins was identical to IgE, and was the true mediator of the biologic and immunologic features formerly ascribed to reaginic antibodies.

With the discovery of IgE antibodies, and with improved methods for the isolation and purification of various allergens,[56] it was possible to work out many of the mechanisms and pharmacologic pathways involved in human allergic conditions.[57] Thus, IgE antibodies have the specialized ability to bind tightly to basophils and mast cells, and their interaction with specific allergen on those cell surfaces has been shown to result in degranulation of these cells, with the release of histamine and other agents that cause directly the symptoms of disease.

Desensitization

Not long after the discovery of anaphylactic shock, it was observed that those sensitized animals that escape death following the administration of large doses of specific antigen are unable for some time thereafter to respond to newly administered antigen with the development of a shock syndrome. Moreover, sensitized animals given repeated doses of antigen too low to provoke clinical symptoms would also develop this refractory state.[58] This was termed "immunity to anaphylaxis" by some investigators, and "anti-anaphylaxis" by Besredka and Steinhardt.[59] The interpretation of the phenomenon was similar in all cases, however. Besredka, whose theory of the mechanism of anaphylaxis involved the interaction of antigen with antibody (sensibilisin) bound to the surface of cells of the nervous system, postulated that the antibody became "saturated" with antigen, and thus could not react further.[60] Those who claimed that the primary stimulus for anaphylaxis resulted from an antigen–antibody interaction within the circulation suggested similarly that these precipitins were saturated with antigen and unable to participate further in the elaboration of shock-inducing substances. The relatively short duration of the refractory state was generally attributed to renewal of the supply of free antibody over the following days or weeks, so that conditions for the induction of anaphylaxis would be re-established.

While some such explanation seemed to be at least partially valid in the case of anaphylactic shock in the guinea pig, for example, it did not appear to offer an acceptable explanation for the occasional success obtained by clinicians in their efforts to desensitize patients suffering from hayfever or susceptible to allergic reactions to insect stings. This approach involves a regimen of subcutaneous injections of purified pollen allergens or insect venom, beginning with doses too feeble to support a clinical response. The dose is then slowly increased until "desensitization" is achieved. The recent understanding of the existence of different classes of immunoglobulin, with different biologic functions, has provided a more compelling explanation for the process of desensitization. It is no longer believed that the antibodies responsible for atopic allergy in man (IgE) are neutralized and thus prevented from acting, but rather that the production of specific antibodies of *other* immunoglobulin classes is stimulated during the "desensitization" series of injections.[61] Such "blocking" antibodies compete with IgE for the allergen, thus inhibiting the type of interaction that would result in the release of those pharmacologic mediators responsible for the allergic disease being treated. Desensitization is, therefore, something of a misnomer; the end result is the neutralization of allergen before it can embarrass the host, rather than neutralization of the culprit antibody.

The concept of "allergy of infection"

Tuberculosis, the "white plague," was one of the leading scourges of the nineteenth-century industrialized western world. Thus, when Robert Koch identified the tubercle bacillus as the responsible etiologic agent in 1882,[62] at a time when he and Louis Pasteur were demonstrating that many important infectious diseases might be controlled by preventive vaccination, the world looked forward to the conquest of this deadly disease. Koch's announcement[63] at an international congress in 1890 of a cure for tuberculosis was received, understandably, with a thrill of anticipation. The material which he proposed to employ was an extract of the broth used to culture tubercle bacilli, which he called tuberculin, and it was hoped that this bacterial product might serve both as a therapeutic agent to cure those already infected and as a vaccine to induce immunity. Unfortunately, neither of these aspirations was realized; the material was incapable of inducing acquired immunity to later infection by the organism, and its use in tuberculous patients proved to be extremely harmful in many cases. Intravenous injection of tuberculin in such patients often led to reactivation of old tubercles (the focal reaction), and to severe systemic reactions and, occasionally, death. But accompanying all of these disappointing results was an observation that would prove extremely valuable in the future – the discovery that small amounts of tuberculin administered into the skin of tuberculous patients results in a local inflammatory dermal reaction that allows the positive diagnosis of this infectious process. This finding was quickly seized upon by other workers, and a variety of other diagnostic applications of tuberculin were advanced, including the cutaneous reaction of von Pirquet,[64] the percutaneous reaction of Moro,[65] the intradermal reaction of Mantoux,[66] and the conjunctival ophthalmoreaction of Calmette.[67]

In the absence of any knowledge in the early 1890s of a relationship between such hypersensitivity reactions and the mechanisms of immunity, it is understandable that Koch should attribute such responses to tuberculin (as he attributed the heightened response of tuberculous animals to the subcutaneous administration of tubercula bacilli – the Koch phenomenon) to the incremental toxic effect of these inocula on tissues already saturated with the toxins thought to accompany tubercular infection. However, with the description, in the opening years of the twentieth century, of anaphylaxis, the Arthus reaction, and serum sickness, the mechanism of the tuberculin skin test was reinterpreted as an antibody-mediated "anaphylaxis," most notably by von Pirquet.[68] Given the ubiquity and importance of tuberculosis in contemporary society, and the clear-cut nature of the results obtained with the tuberculin test, it is not surprising that this disease should become the prototype for studies of the relationship between allergy and immunity in infectious processes, or that the tuberculin test should provide the focus for the future distinction to be made among different allergic mechanisms. Indeed, the tuberculin reaction was viewed as so archetypal that the term "tuberculin-type hypersensitivity" was long used to characterize all reactions of this type, only later giving way to such terms as "delayed-type hypersensitivity" and finally "cellular immunity."

Early speculations

We have already seen that as early as 1906 von Pirquet and Schick attempted a grand conceptual unification of all of the phenomena of allergy, including anaphylaxis, serum sickness, the tuberculin and other cutaneous reactions, and even the exanthems that accompany certain infectious diseases. These authors held that all of these responses were antibody-mediated, and that allergy in general was but a step on the road to immunity. This view was questioned only a few years later by Edward Baldwin, in the specific context of the tuberculin test and of hypersensitivity in tuberculosis.[69] Baldwin pointed to the fact that reactions to tuberculin were accompanied by fever, whereas *hypo*thermia generally accompanies anaphylactic shock. More telling, however, was Baldwin's argument about the oft-noticed inability to transfer typical tuberculin reactivity by passive transfer of serum, unlike the passive transfer of anaphylactic sensitivity, which is usually accomplished with ease. Baldwin also called attention, apparently for the first time, to differences in the passive transfer of hypersensitivity from mother to fetus *in utero*. As he says:[70]

> When we consider our results upon the progeny of tuberculous and [anaphylactically] sensitized females [animals] respectively there is an argument for a difference, because of the absence of any appreciable sensitiveness in the young of the former and the very great susceptibility of the latter. Clinical experience on the newly born from tuberculous mothers also indicates a lack of cutaneous and fever sensitiveness transmitted to the child, which is hard to understand if acute anaphylaxis and tuberculin hypersusceptibility have a common origin and mode of development.

While admitting that true anaphylactic sensitivity may be induced with bacterial products, Baldwin concluded that "enough is shown in these experiments to indicate a difference between the infected animals and those simply anaphylactic, in relation to Koch's old tuberculin applied in this way."[71]

The implications of Baldwin's work and the significance of his conclusions appear to have made little impression at the time. This is perhaps best illustrated by the absence of any mention of Baldwin's work in Hans Zinsser's comprehensive book *Infection and Resistance*,[72] published in 1914. Not only was Zinsser actively interested in this field; he is also usually dependably encyclopedic in his reviews. While he did write a chapter in this book on "bacterial anaphylaxis," his emphasis was on the relationship of hypersensitivity to pathogenesis and to immunity in infectious diseases, rather than suggesting that the tuberculin and analogous reactions might be set apart from "other forms of anaphylaxis."

Zinsser was not long in coming around to the view that basic differences do exist between the skin tests exhibited by tuberculous animals and those sensitized with protein for anaphylaxis; indeed, he soon became one of the foremost champions of the view that the hypersensitivity that accompanies infection is unique. After Calmette had insisted on the separation of tuberculin reactivity from anaphylaxis, in a widely read 1920 book on *Bacillary Infection and Tuberculosis*,[73] Zinsser published a series of papers[74] pointedly addressing these differences, in which he argued forcefully that different mechanisms must be at work. The basis for the conceptual separation of tuberculin sensitivity and anaphylactic sensitivity advanced by Zinsser may be summarized as follows:

1. *Conditions of induction.* Anaphylactic sensitivity may be induced by administration of almost any protein substance by almost any route, whereas active infection with live organisms is required to induce typical skin reactivity to tuberculin in tuberculosis, to typhoid in typhoid fever, to mallein in glanders, and to abortin in brucellosis; dead organisms or their extracts generally fail to induce this reactivity.
2. *The timing of the skin reaction.* Perhaps for the first time, in 1921 Zinsser applied the (still currently used) terms "immediate" skin reactions to anaphylaxis and "delayed" to the tuberculin-type of skin reaction. The former starts in a few minutes and fades after a few hours, whereas the latter does not commence until four to five or more hours after testing, and may not reach its highest development until about forty-eight hours.
3. *Clinical signs.* Anaphylactic reactions in the skin are characterized by a wheal of edema and a flare of hyperemia without residual local tissue damage, whereas tuberculin-type reactions are more often indurated and, where intense, may be accompanied by hemorrhage and central necrosis.
4. *Temperature changes.* As a rule, systemic reactions to tuberculin and similar substances are accompanied by a rise in the temperature of the host, while systemic anaphylactic reactions lead to a depression in body temperature.
5. *Passive transfer of hypersensitivity.* We have already cited above, in a discussion of the studies of Baldwin, the most important components of this argument. Indeed, it is only in 1921 that the now-converted Zinsser can say that "Baldwin's work is fundamental."[75]

From this point onward, the notion that some special mechanism of hypersensitivity was associated with infectious processes in general, and with tuberculin-type reactions in particular, was central to the formulation of experiments and to the interpretation of results in this area. There developed, for a time, almost a mystique about the nature of the infectious process that could engender so unique a form of allergic response, reflected in part in the oft-used term "allergy of infection."

The relationship of allergy to immunity

With the development of the notion that infectious diseases might be accompanied by a peculiar form of allergic response that not only appeared to exacerbate the disease but had also other deleterious consequences, an old and vexing dilemma was raised anew. How could these hypersensitivities, now so clearly a component of the general immune response, be conceptually integrated into a biological function so obviously evolved for the benefit and protection of the individual? As we saw above when simple anaphylaxis was considered, the easiest and most teleologically pleasing answer was to conclude with von Pirquet that hypersensitivity is merely a step (and even an important one) in the development of immunity.

The old debate was renewed in the context of tuberculin-type hypersensitivity and of the pathogenesis of infectious diseases. On one side, the foremost advocates of the position that hypersensitivity may be a protective component of the immune response were Dienes[76] and Topley and Wilson.[77] These authors pointed to the fact that the local allergic reaction, and especially the granulomatous responses that often accompany it (most notably the tubercle in tuberculosis) may serve an important function in walling off the infection and restricting the spread of the pathogenic organism. Further, in view of the lack of relationship between tuberculin sensitivity and circulating antibody, they assumed that the "antibodies" responsible for tuberculin hypersensitivity are cell-bound. In line with Ehrlich's theory, these antibodies (receptors) merely await the appearance of antigen to stimulate their exuberant release to provide for added humoral immunity. If this be accompanied by a local hypersensitivity reaction, then it is implied that this is a small price to pay for the protection that it affords. As Topley and Wilson put it: [78]

> There seems to be no valid reason for excluding resistance to tuberculosis from this general picture. Our present knowledge is compatible with the view that allergy represents a stage in the development of immunity when the antibodies are concentrated mainly on the surface of the cells, whereas so-called immunity is a stage further on, when there is a considerable amount of free antibody in the circulation, and the local disturbances caused by the meeting of antigen and antibody in the tissues are therefore less severe.

This view of an essentially benign role for tuberculin-type hypersensitivity was challenged by one of the world's foremost authorities on tuberculosis, Arnold Rich. In a long series of studies from his laboratory, summarized in his encyclopedic book *The Pathogenesis of Tuberculosis*,[79] Rich argued against the view that hypersensitivity and immunity are closely related. He cited the lack of obvious relationship between the level of tuberculin skin reactivity and the resistance to infection or the spread of disease in both man and experimental animals. He further pointed out: that simple hypersensitive inflammation is incapable of preventing the early spread of bacteria in the absence of specific acquired resistance; that immunity can be established without concomitant hypersensitivity; that immunity can be passively transferred without transfer of hypersensitivity; and that acquired resistance remains intact both in man and in lower animals after the abolition of hypersensitivity by desensitization.[80] He felt able to conclude with the dogmatic statement that "up to the present, *hypersensitive inflammation has never been satisfactorily shown to be necessary for the successful operation of acquired immunity at any stage of any infection under any condition whatsoever*" [his italics].[81]

The continuing desire to integrate allergy and immunity into a teleological pleasing single mechanism rose yet a third time, during the rebirth of interest in delayed-type hypersensitivity in the 1950s and 1960s. With the demonstration that "pure" delayed hypersensitivity to protein antigens[82] is followed by active antibody formation[83] and that delayed hypersensitive guinea pigs, though not yet forming antibody, are primed to yield an anamnestic antibody response,[84] it was once again speculated by Pappenheimer and colleagues[85] that delayed hypersensitivity might constitute an "early" or "immature" stage in the cellular mechanism of antibody formation. Final resolution of this relationship, however, had necessarily to await newer findings on cell–cell interactions and on the differentiation pathways of these cells – findings that were not long in coming.

Progress on delayed (tuberculin)-type hypersensitivity

The interest in tuberculin-type hypersensitivity reactions and in the mechanisms responsible for their peculiar features was couched initially in the terms and in the context of infectious disease processes, as we saw above. Soon, however, new information would appear that would significantly broaden the interest in and implications of this phenomenon.

Hypersensitivity to bland proteins

It was in 1929 that Louis Dienes first showed that tuberculin-type hypersensitivity was not restricted to substances of bacterial origin.[86] He injected egg albumin directly into the tubercles of tubercular animals, and demonstrated that they would then develop typical "delayed" hypersensitivity skin reactions to the bland protein itself. This was quickly followed up by similar studies at the hands of Jones and Mote,[87] and of Simon and Rackeman.[88] With the introduction of Freund's adjuvant,[89] containing emulsions of antigen with dead mycobacteria, Uhr, Salvin, and Pappenheimer were able to show that delayed hypersensitivity could also be obtained by immunizing with minute amounts of simple proteins, or with antigen–antibody complexes.[90] Skin reactions in such sensitized animals, elicited with specific antigen, were characterized by a similar time course of development, by similar histopathologic changes (save for the necrotic component), and by the same temperature rise in the systemic reaction so characteristic of the reaction to tuberculin. During the same period, Benacerraf and Gell showed that similar delayed hypersensitivity responses could be elicited by hapten–protein conjugates.[91] These studies were accompanied by two important new findings. First, it was discovered that the delayed hypersensitivity state induced by such bland proteins could easily be desensitized without untoward systemic reactions, by simple intravenous administration of specific antigen. So different was this from previous experience with the difficulty of desensitizing tuberculous individuals with tuberculin that Raffel and Newell argued that this was not typical *tuberculin*-type hypersensitivity, and should rather be placed in a separate category called "Jones–Mote hypersensitivity."[92]

The second curious finding that emerged from these studies was the demonstration by Benacerraf and Gell[93] of the "carrier effect." While the delayed skin reaction appeared to be specific for the simple chemical hapten employed, this reaction could be elicited only when that hapten was attached to the carrier protein employed for sensitization. In contrast to anti-hapten antibodies, which interact with the hapten regardless of the protein to which it is attached, the delayed hypersensitivity mechanism seemed to "see" the hapten and its neighboring carrier protein as a single entity. This led to the speculation that the combining site responsible for delayed hypersensitivity reactions might be larger than that on the surface of the normal antibody molecule. The demonstration of a carrier effect in delayed-type hypersensitivity would constitute one of the more important stimuli to the discovery of the role of cell–cell cooperation in the production of antibodies.

Histopathologic studies

In addition to the more obvious differences in the clinical symptoms that accompany immediate and delayed hypersensitivity reactions, significant differences were also found on the cytologic level. It was Dienes and Mallory[94] who first pointed out that the most prominent cytologic feature of the tuberculin reaction was the intense infiltration with mononuclear cells, and that polymorphonuclear infiltration was probably only secondary and in proportion to the degree of epithelial necrosis. In contrast, the picture of passive anaphylactic skin reactions in animals was a rapidly developing edema and hyperemia, quickly followed by intense polymorphonuclear infiltration. These studies were taken up and extended by Rich and coworkers,[95] who emphasized the presence of large numbers of lymphocytes in these delayed inflammatory infiltrates. (It was undoubtedly the significant presence of these cells of unknown function that prompted Rich's plaint in 1950, "The lack of more adequate information regarding the function of the lymphocyte is one of the most lamentable gaps in medical knowledge."[96])

The next significant step was taken by Gell and Hinde,[97] who not only confirmed the predominantly mononuclear nature of the delayed skin test, but also pointed out two additional features: first, that the temporal progression of Arthus skin reactivity shows a transition from an initial immediate-type skin response to a later cytologic picture more typical of the delayed-type response; and second, that significant local plasmacytosis and antibody formation would follow on the heels of the delayed hypersensitivity skin test. The significance of such cytologic studies both to distinguish among different forms of hypersensitivity as well as to provide leads for the study of mechanism was then emphasized in a series of detailed histopathologic studies by Waksman,[98] whose work so importantly pointed up the new directions of study of what would soon be called *cellular* immunity. The distinctive cytology of this delayed-type hypersensitivity reaction was amply confirmed by the extensive studies of Turk and colleagues,[99] and the similar cytologic characteristics of contact dermatitis reactions were made clear by the studies of Flax and Caulfield[100] and of Turk and colleagues.[101]

Passive transfer of delayed hypersensitivity

We have already mentioned that one of the characteristics that differentiated immediate- from delayed-type hypersensitivities was the ease with which the former could be passively transferred by serum – a characteristic not shared by the latter phenomenon. It was not until 1942 that Landsteiner and Chase were able to demonstrate the passive transfer of contact hypersensitivity to picryl chloride from a sensitized guinea pig into a naive recipient.[102] This was accomplished utilizing live peritoneal exudate cells from the donor, injected intraperitoneally into the recipient. Twenty-four hours later, a positive skin test could be elicited by application of picryl chloride to the skin of the recipient. No such transfer could be obtained using the fluid phase from the exudate, or with peritoneal exudate cells that had been killed prior to transfer. These results were confirmed by Stavitsky[103] and by Haxthausen,[104] who showed that transfer could also be effected using peripheral blood leukocytes. The generality of this system of passive transfer of delayed-type hypersensitivity using cells was made clear a few years later when Chase showed that tuberculin sensitivity could be transferred passively using a similar method.[105] Mitchison[106] demonstrated analogous passive transfer of transplantation immunity with cells, and similar demonstrations were made subsequently in other model systems.[107] A number of very early studies had suggested that passive transfer of tuberculin hypersensitivity could be attained using defibrinated blood or ground-up lymph nodes and spleen of tuberculous guinea pigs,[108] but in most of these instances it is not clear that the sensitization observed was truly passively acquired; active primary sensitization may have been produced by means of tubercle bacilli in the mixture transferred.

One of the important consequences of these passive transfer studies came with the use of donor cells marked with radioactive labels. It was shown from a number of different laboratories[109] that the proportion of specifically sensitized cells at the site of a passively-induced delayed skin test or allograft rejection reaction is minimal. These studies implied that only a small specific immunologic trigger might be required to initiate a predominantly nonspecific train of inflammatory events.

Stimulated by the rise of interest in the passive transfer of delayed-type hypersensitivity, H. S. Lawrence, in 1954, claimed that tuberculin hypersensitivity could be passively transferred to tuberculin-negative recipients using an *extract* of sensitized donor peripheral blood leukocytes.[110] Subsequent reports showed a similar passive transfer of delayed hypersensitivity to diphtheria toxoid, to coccidioidin, and to other stimulants of delayed-type reactions,[111] with a specificity that implied that these "transfer factors" were indeed informational molecules. They have since been shown to be of relatively low molecular weight, resistant to DNAse and RNAse, but little progress has been made in elucidating the mechanism of information carriage, or information transfer to the recipient, in this unique passive transfer system. It has been suggested, without proof, that it might be sensitizing antigen that is being transferred in these extracts.

Relationship to other phenomena

We have already seen that, based predominantly upon histopathologic criteria, in 1959 Waksman stressed the importance of delayed hypersensitivity in the pathogenesis of a variety of autoimmune diseases – a view that has since been amply confirmed (see Chapter 8). Even before this, the role of such cellular mechanisms had been extended to include a variety of other important biological systems. Thus, contact dermatitis had been brought within the fold of typical delayed-type hypersensitivities, in which sensitization is accomplished by the coupling of the active chemical or its metabolic intermediaries to proteins of the skin.[112]

Delayed-type hypersensitivity has also been implicated as the distinctive characteristic of a number of viral diseases, and even as an important contributor to their pathogenesis. Typical delayed hypersensitivity skin reactions can be demonstrated to vaccinia, herpes simplex, mumps, and measles viruses.[113] Indeed, it is now apparent that the measles rash is not so much the primary disease itself as the delayed hypersensitive response to dermal virus, and accompanies its clearance from the body. The primary disease, as seen in immunodeficient individuals, is rather a serious giant cell pneumonia, so that the symptoms seen in normal individuals may be considered more a part of the cure than of the disease itself. Perhaps the best-studied example of the role of delayed hypersensitivity mechanisms in the pathogenesis of a viral disease may be seen in the case of lymphocytic choriomeningitis infection in mice.[114] This infection is characterized in normal animals by a severe inflammatory infiltrate of the meninges and choroid plexus, whereas the immunosuppressed or immunologically tolerant animal shows no such pathologic changes. The brains of such animals, while harboring large amounts of virus, show no disease, but passive transfer of specifically sensitized T cells to these infected animals leads to typical choriomeningitis.

Early in his studies of allograft rejection, Peter Medawar was impressed by the predominance of mononuclear cells rather than polymorphonuclear leukocytes in skin allografts in the process of rejection.[115] While Medawar initially believed in an Arthus-type mechanism of rejection (see Chapter 11), this cytologic evidence was later taken as an indication that the process is related to delayed-type hypersensitivity – a suggestion confirmed by the demonstration that transplantation immunity may be transferred with cells rather than serum. Further evidence for the relationship was produced when Brent, Brown, and Medawar showed[116] that guinea pigs which had rejected a skin graft would show a delayed hypersensitivity reaction to subsequent intradermal injection of an extract of lymph node or spleen cells from the same donor, with all of the temporal and histologic features of a typical tuberculin reaction. Subsequent studies on the *in vivo* graft-versus-host reaction[117] and the *in vitro* mixed lymphocyte reaction[118] showed that these were indeed caused, like allograft rejection itself, by the response of lymphoid cells to the histocompatibility antigens of the other partner to the reaction.

Sorting out the mechanisms

Lymphocyte subset functions

It was the ability experimentally to induce defects in the immune response that first pointed the way to the assignment of different immunologic functions to different subsets of lymphocytes. The demonstration that the Bursa of Fabricius exercises (at least in birds) an important supervisory role over antibody formation,[119] and that the thymus in mammals appears to control delayed hypersensitivity responses,[120] led to the first functional division of lymphoid cells into T (thymus-dependent) cells and B (bursa- or bone marrow-dependent) cells.[121] The former includes those effector cells responsible for tuberculin-type skin reactions, allograft rejection, etc.; the latter comprises that developmental line of lymphoid cells responsible for antibody formation, the ultimate differentiated form of which is the plasma cell. Since then, as additional functions have been delineated, new subsets of T lymphocytes have been defined, usually characterized not only by function but by distinctive cell surface markers as well.[122] In their expanding role as regulators of immune responses, helper and suppressor T cells have been identified. (It is the requirement that T cells interact with the identical hapten-coupled polypeptide chain used for sensitization that resolved the paradox of the "carrier effect" mentioned above.) To perform the several helper and effector functions of T lymphocytes, other distinctive subsets have been identified.

Cell–cell intercommunications

During the course of studies on the mechanisms responsible for tuberculin-type hypersensitivity, in 1933 Rich and Lewis first demonstrated a curious response to antigen *in vitro*, on the part of mononuclear cells from tuberculin-sensitive animals.[123] Whereas cells normally migrate out of explanted bits of spleen in culture, such migration from the spleen of sensitized animals can be inhibited by the addition of specific antigen – in this case tuberculin. These studies were extended by George and Vaughan,[124] who studied the inhibition of migration by tuberculin of sensitized cells in capillary tubes. Further studies by David[125] and by Bloom and Bennett[126] helped to clarify the nature of this response and to open up a new dimension in the study of cellular-immune reactions. These investigators showed that:

- the phenomenon was immunologically specific;
- it was the migration of macrophages that was inhibited;
- it was, however, an antigen–lymphocyte interaction that initiated the inhibition;
- only very small numbers of sensitized lymphocytes were required to affect the activity of large numbers of macrophages; and
- the response was mediated by a small, nonspecific molecule released by the activated lymphocyte (called migration inhibition factor – MIF).

Here was a mechanism of intercellular communication and of amplification that helped to explain why so few specific cells are required in a tuberculin skin test site, or in the rejection of a tissue allograft. Here also was the cue to begin the search for other such intercellular signals, which resulted in the finding of a number of different monocyte-derived signal substances (monokines) and lymphocyte-derived substances (lymphokines and interleukins).[127] Each of these physiologically active substances has a more or less well defined role in the evolution of immunogenic inflammatory reactions, and in the protective and deleterious consequences of such reactions.

Other immunopathologic processes

We have noted, in the foregoing pages, many examples of the way in which immunology was reoriented along more biological and medical lines in the years following World War II. This shift was especially marked during the late 1950s by the surge of interest in diseases *caused* by the immune response, as is well attested to by the many reviews, symposia, and books devoted for the first time to this subject.[128] One of the landmarks of the period was Gell and Coombs' *Clinical Aspects of Immunology*, in which the authors proposed a new breakdown of immunopathologic processes into four distinct categories, which quickly proved extremely popular.[129] Their type I reaction includes all of the phenomena of anaphylaxis as well as human atopic allergies; type II includes antibody- and complement-mediated membrane-destructive reactions such as immune hemolysis and bacteriolysis; type III reactions include those attributable to the effects of immune complexes; and type IV those reactions variously termed tuberculin-type, delayed-type, or cellular-immune reactions. Since we have already in these pages addressed many of the disease processes and models covered by this classification, it may only be necessary here to fill in briefly some of the blank spaces in the picture, with special emphasis on those immuno-pathologic processes that do not fit into the simple categories proposed.

Lymphoproliferative diseases

Just as the organized lymphoid tissues of the body (the spleen and regional lymph nodes) respond to antigenic stimulus with an intense lymphoid cell proliferation and germinal center formation, so may the same response occur in other tissues, following chronic stimulation by antigen. This picture is a prominent component of such diseases as Hashimoto's thyroiditis, where a significant portion of the normal thyroid may be replaced by widespread lymphoid proliferation, plasmacytosis, and massive germinal center formation. Perhaps the best example of this process is seen in the blinding disease trachoma,[130] wherein the essentially noncytopathogenic organism *Chlamydia trachomatis* grows almost benignly in the conjunctival epithelium, but induces in the subepithelial tissues a chronic immunogenic inflammatory response with typical germinal center formation and surrounding mantles of T and B cells. While reactions of this type must necessarily be placed in the category of immunopathologic disease, they appear to be little more than the usual immune response expected of an organized lymphoid tissue subjected to chronic antigenic stimulus but which, in an ectopic and sensitive location, may present as serious disease.

Local organ hypersensitivity

It was Alexandre Besredka[131] who first called attention to the possibility that active immune responses might be localized to specific organs or special regions of the body, with possible important implications not only for host immunity to infection, but for immunopathologic disease as well. This view was taken up in the 1930s by Beatrice Seegal and coworkers,[132] who showed that the injection of antigen into an isolated organ like the eye could lead to the development of a local hypersensitivity such that each subsequent systemic administration of antigen induces an exacerbation of a potentially blinding ocular inflammatory disease. A similar mechanism occurring in the joints is thought to contribute to certain forms of arthritis. It has been suggested[133] that such processes may be nothing more than the establishment in the affected tissues of specific immu-nologic memory of the antigen involved, so that just as lymphadenitis accom-panies the booster antibody response in the lymph node, so might local inflammation (and clinical disease) accompany a booster antibody response in these ectopic locations.[134]

Immunologic deficiency diseases

We have thus far limited our discussion of immunopathologic processes to those induced *by* the immune system. The discussion would not be complete, however, without mention of pathologic processes *of* the immune system. Having already mentioned the important role of plasmacytomas in helping to define the nature of the antibody molecule, we will pass over other tumors of lymphoid cell lines and restrict this discussion to those immunologic *deficiency* diseases that have helped to sort out the complex mechanisms of the immune response. For the purposes of the historical record, only a brief account of origins will be given for each area discussed; more recent developments may be obtained from the numerous reviews that address these issues.[135]

Selective immunoglobulin deficiencies

The first defect of production of a single immunoglobulin class was that of IgA, reported in 1964 by Rockey and colleagues.[136] Despite its importance as a component of the secretory immune system, IgA-deficient patients are often clinically normal. The condition may be inherited through either a dominant or an autosomal recessive trait, may be due to a defect in chromosome 18, or may be acquired secondary to certain drug therapies, to viral infection, or to lymphoid malignancies.

Isolated IgM deficiencies are rare, perhaps due to the inability of further Ig class maturation and therefore to complete lack of all protective antibodies in such individuals. The first cases of this condition were reported by Hobbs and colleagues,[137] in two brothers whose father also showed low serum IgM.

Selective IgG deficiencies have also been reported.[138] The several subclasses of IgG are variously affected: one patient lacked IgG1 and 2; another IgG1, 2, and 4; another IgG1, 2, and 3; and a fourth lacked IgG2 and 4.

B cell defects

Sex-linked hypogammaglobulinemia was first described by Bruton in 1952.[139] There is a dearth of B lymphocytes, while most T cell functions appear to be normal. Patients with this condition have repeated severe bacterial infections, but handle most viral infections normally. The generally accepted cause of this disease is an arrest in the normal maturation of pre-B cells,[140] although others suggest that the pre-B cell may produce an abnormal and functionally useless chain in those afflicted.[141]

Hypogammaglobulinemia may be acquired later in life,[142] and while patients present with many of the same symptoms as the X-linked form, they often show in addition such complications as polyarthritis, autoimmune disease, and gastrointestinal complications. The causes of this condition are probably multiple, some apparently genetic in nature and others arising secondary to lymphoproliferative tumors.

T cell defects

The prototypical example of a defect in T cell function was reported by DiGeorge in 1965,[143] although Good and Varco had previously described a similar situation associated with thymoma in 1955.[144] B cell function and immunoglobulin levels are usually normal in such cases, while all typical T cell reactions are reduced or absent. A related condition was described by Neze-lof,[145] apparently due to an autosomal recessive defect. Other T cell defects have been described, due variously to deficiencies in such enzymes as nucleoside phosphorylase and adenosine deaminase, to the absence of HLA-A and -B antigens (the bare lymphocyte syndrome), or to diverse other causes.

Severe combined immunodeficiency

The first cases involving a defect in both B and T cell function were reported by Glanzmann and Riniker in 1950.[146] They termed it *essential lymphocytoph-thisis*, but it is now known as Swiss-type agammaglobulinemia, or severe combined immunodeficiency disease (SCID). Views of the pathogenesis of this disease complex are threefold: it may be due to a primary defect in thymic function, to a biochemical defect that prevents normal maturation of both T and B cells, or to an inability of stem cells to differentiate appropriately into T and B cell precursors. Recent experience with acquired immunodeficiency syndrome (AIDS) has shown that a virus may, by infecting and destroying specific subsets of lymphoid cells at an early stage in their differentiation, cause a somewhat similar clinical condition.[147]

Complement

Since complement has played such an important role in the advancement of our understanding of the functions of the immunologic apparatus (*viz.*, immune hemolysis, opsonization, anaphylaxis, and immune complex pathology), it may be well to pause here to review briefly the history of this complicated set of physiological processes.

The complement system[148]

It was Nuttall who in 1888 first pointed out the existence in normal serum of a protective substance,[149] soon called alexin by Buchner and complement by Ehrlich, who presciently assigned to it an enzymatic function. Its true significance was only shown a decade later by Bordet,[150] who demonstrated its crucial role in immune hemolysis (and, by implication, in bacteriolysis). In 1907, Ferrata[151] showed that hypotonic solution would separate complement into two inactive fractions, called "midpiece" and "endpiece" (later the basis for complement components one and two). The third component of complement (C3) was discovered with the finding that complement might be inactivated by cobra venom[152] or yeast.[153] A fourth component was found through the ability of ammonia to inactivate the hemolytic activity of fresh serum.[154] Then, beginning in the 1960s, "classical" C3 was shown to comprise a congeries of individual components and conversion products,[155] whose end effect is the result of a cascade of combinations and enzymatic alterations of the different components.[156] The existence of an alternative pathway for the activation of complement was heralded by the studies of Pillemer and Ecker on the effect of yeast on complement.[157] Pillemer described a new substance, *properdin*, which was held to be a significant contributor to natural (nonantibody-mediated) immunity.[158] It has since been shown that various bacterial polysaccharides can activate C3 directly, and thus initiate the alternative pathway to the complement cascade.[159]

A new facet of the complement story (and a justification of the old view that complement plays an important role in anaphylaxis) came with the reports that byproducts of the activation of C3 and C5 are significant pharmacologic contributors to inflammation.[160] Here at last was the long-elusive anaphylatoxin that had so fascinated an earlier generation of workers.

Complement deficiencies

A strain of complement-deficient guinea pigs was described first in the 1920s by Hyde,[161] but unfortunately the colony was lost, and the precise nature of the defect remains unknown. The first case of complement deficiency in a human was reported in 1960 by Silverstein.[162] This was an adult (in fact, an immunologist!) who, despite a severe deficiency in the second component of complement, was clinically normal. Numerous other cases have since been reported, usually showing autosomal codominant inheritance.[163]

Deficiencies in most of the components of the complicated pathway of complement activation have been reported.[164] As might be expected, defects in the components up to C5 are often (*but not invariably*) accompanied by systemic disease and increased susceptibility to infection, while deficiencies in the components which function later in the pathway are generally without significant consequence. A possible defect in the alternative pathway has also been described.[165]

Conclusions

New scientific concepts have often found it difficult to win acceptance, especially when they appear to conflict with teleologically pleasing arguments in favor of an older view. Thus, Ptolemaists (and churchmen) found Copernicus' theory of the solar system unacceptable, and atomists found it difficult to believe that their "ultimate indestructible particle" was composed of subunits, and could be fissioned. Similarly, early immunologists brought up to believe in an immune response evolved for the protection of the host found it hard to acknowledge that disease might result from its workings. Instead, they sought other explanations for allergic and immunopathologic processes, other mechanisms, or other "abnormal" antibodies or modes of antibody participation. Even now, when modern developments have shown how all of these factors are intimately tied together, one can still detect in modern writings on the subject a certain unease about the pathological aspects of the "immune" response, that harks back to the views of an earlier time. The immunopathologist, given his training, sees no problem here, but many of the rest retain at least a trace of that old schizophrenic feeling when contemplating the problems of that almost oxymoronic expression "immunologic disease." It is worth recalling, however, that every complicated physiological process that has evolved for our benefit carries with it the possibility of a harmful outcome – think only of the cost/benefit calculation involving our multi-component blood clotting system.

Notes and references

1. Koch, R., *Deutsch. med. Wochenschr.* 17:101, 1891; see also Koch, R., *Deutsch. med. Wochenschr.* 16:756, 1029, 1890; 17:1189, 1891.
2. Behring, E., *Deutsch. med. Wochenschr.* 19:389, 415, 543, 1893.
3. Byron Waksman recounts (personal communication, 1987) that Louis Dienes pointed out to him many years earlier that the sequence of events leading to the elucidation of the role of special antibodies (reagins and later IgE) in allergic and parasitic diseases depended very much on the development of immunology in industrial societies. In the tropics and among primitive cultures, parasitic diseases were so prominent that immunologic progress there would likely have taken a far different course – the discovery of IgE and its relationship to protective immunity against parasites would probably have occurred much sooner.
4. A highly detailed compendium of developments in all aspects of allergy will be found in Schadewaldt, H., *Geschichte der Allergie*, 4 vols., Düsseldorf, Dustri-Verlag, 1979.
5. Jenner, E., *An Inquiry into the Causes and Effects of the Variolae Vaccinae*, London, Sampson Low, 1798.
6. Magendie, *Vorlesungen über das Blut*, Krüpp, Leipzig, 1839, cited by Morgenroth, J., in Ehrlich's *Collected Studies in Immunity*, New York, Wiley, 1906, p. 332.
7. Flexner, S., *Med. News* 65:116, 1894.
8. Héricourt, J., and Richet, C., *C. R. Soc. Biol.* 50:137, 1898.
9. Portier, P., and Richet, C., *C. R. Soc. Biol.* 54:170, 1902.
10. The story of these events is told by R. Otto in his initial paper on the subject in *von Leuthold-Gedenkenschrift*, Berlin, 1906, p. 153; see also Otto, R., "Über Anaphylaxie und Serumkrankheit," in Kolle and Wassermann's *Handbuch der pathogenen Mikroorganismen*, Fischer, Jena, 1909.
11. See note 10, and especially Otto's summary in his chapter for Kolle and Wassermann's *Handbuch*, p. 239.
12. Arthus, M., *C. R. Soc. Biol.* 55:817, 1903.
13. von Pirquet, C., and Schick, B., *Die Serumkrankheit*, Vienna, Deuticke, 1906; English edition, *Serum Sickness*, Baltimore, Williams & Wilkins, 1951.
14. Rosenau, M.J., and Anderson, J.F., *Bull. Hygienic Lab., Washington DC*, 26, 1906; *J. Am. Med. Assoc.* 42:1007, 1906.

15. Gay, F P., and Southard, E.E., *J. Med. Res.* **16**:143, 1907; **18**:407, 1908; **19**:1, 1908.
16. Auer, J., and Lewis, P. A., *J. Am. Med. Assoc.* **53**:458, 1909; *J. Exp. Med.* **12**:151, 1910.
17. Biedl, A., and Kraus, R., *Z. Immunitätsf.* **7**:205, 408, 1910.
18. Friedberger, E., *Münch. med. Wochenschr.* **57**:2628, 1910; Friedberger, E., and Mita, S., *Z. Immunitätsf.* **9**:362, 453, 1911.
19. Vaughan, V.C., *J. Am. Med. Assoc.* **47**:1009, l906; Vaughan, V.C., and Wheeler, S.M., *J. Inf. Dis.* **4**:476, 1907.
20. Wolff-Eisner, A., *Das Heufieber*, Munich, 1906.
21. Meltzer, S. J., *Trans. Assoc. Am. Phys.* **25**:66, 1910; *J. Am. Med. Assoc.* **55**:1021, 1910.
22. Richet, C., *Bull. Soc. Biol.* **55**:246, 1073, 1903; **56**:302, 1904; *Ann. Inst. Pasteur* **21**:497, 1907.
23. Gay and Southard, 1907, note 15.
24. Vaughan, V.C., note 19.
25. Nicolle, C., *Ann. Inst. Pasteur* **21**:128, 1907; Otto, R., *Münch. med. Wochenschr.* **54**:1665, 1907; Friedemann, U., *Münch. med. Wochenschr.* **54**:2414, 1907; Doerr, R., and Russ, V K., *Z. Immunitätsf.* **3**:181, 706, 1909. The passive transfer of anaphylactic sensitivity from mother to newborn was also demonstrated at this time by Rosenau and Anderson, note 14.
26. Arthus, M., *De l'Anaphylaxie à l'Immunité*, Paris, Masson, 1921, p. 285.
27. Otto, 1909, note 10.
28. Doerr, R., "Die Anaphylaxie," in Kraus and Levaditi's *Handbuch der Technik und Methodik der Immunitätsforschung*, Jena, Fischer, 1909, pp. 856–894.
29. Currie, J.R., *J. Hygiene* **7**:35, 1907, p. 58.
30. Nicolle, note 25.
31. von Pirquet and Schick, note 13, p. 119.
32. von Pirquet, C., *Allergie*, Springer, Berlin, 1910; English translation, *Allergy*, Chicago, American Medical Association, 1911.
33. Rosenau and Anderson, *J. Am. Med. Assoc.*, note 14.
34. Rosenau and Anderson, *Bull. Hygienic Lab.*, note 14, p. 7.
35. Richet, C., *Ann. Inst. Pasteur* **21**:497, 1907, p. 524.
36. Germuth, F.G., *J. Exp. Med.* **97**:257, 1953; Germuth, F.G., and McKinnon, G.E., *Bull. Johns Hopkins Hosp.* **101**:13, 1957.
37. Weigle, W.O., and Dixon, F.J., *Proc. Soc. Exp. Biol. Med.* **99**:226, 1958; Dixon, F.J., *Harvey Lect.* **58**:21, 1963.
38. Schultz, W.H., *J. Pharmacol.* **2**:221, 1910.
39. Dale, H.H., *J. Pharm. Exp. Ther.* **4**:167, 1913.
40. Dale, H.H., *Bull. Johns Hopkins Hosp.* **31**:257, 310, 1920; Dale, H.H., and Laidlaw, P.P., *J. Physiol.* **52**:355, 1919.
41. Dale, H.H., *Lancet* **1**:1179, 1233, 1285, 1929.
42. The role of histamine in mediating many of the typical local symptoms of allergic reactions was especially well pointed out by Lewis, T., *The Blood Vessels of the Human Skin and their Responses*, London, Shaw, 1927. Lewis called the active factor "H-substance."
43. Friedemann, U., *Z. Immunitätsf.* **2**:591, 1909.
44. Friedberger, E., *Berlin klin. Wochenschr.* **47**:1490, 1922, 2303, 1910; *Z. Immunitätsf.* **3**:787, 1910.

45. A review of the many substances that might mimic anaphylactic shock is presented in a series of papers by Hanzlik, P.J., and Karsner, H.T., *J. Pharm. Exp. Ther.* **14**:379, 425, 449, 479, 1920; **23**:173, 1924.

46. The development in America of the clinical discipline of allergy is well summarized in the fiftieth anniversary issue of *J. Allergy Appl. Immunol.* (**64**:306–474, 1979), with an article by Merrill Chase on Cooke and Coca, one by Max Samter on future prospects for the field, and a lengthy history of the institutionalization of the discipline by Sheldon Cohen.

47. Coca, A.F., and Cooke, R.A., *J. Immunol.* **8**:163, 1923. See also Coca, A.F., Walzer, M., and Thommen, A.A., *Asthma and Hayfever in Theory and Practice*, Springfield, Thomas, 1931.

48. The story of the founding of and developments within and between the Eastern and Western Societies is recounted by Cohen, S.G., *NER Allergy Proc.* **5**:247, 342, 1984.

49. See Tuft, L., *NER Allergy Proc.* **6**:279, 1985.

50. Prausnitz, C., and Küstner, H., *Zentralbl. Bakt.* **86**:160, 1921.

51. Heremans, J.F., Heremans, M.T., and Schultze, H.W., *Clin. Chim. Acta* **4**:96, 1959.

52. Heremans, J.F., and Vaerman, J.P., *Nature* **193**:1091, 1962.

53. Loveless, M.H., *Fed. Proc.* **23**(2):403, 1964; see also Rockey, J.H., Hanson, L.A., Heremans, J.F., and Kunkel, H.G., *J. Lab. Clin. Med.* **63**:205, 1964.

54. Ishizaka, K., and Ishizaka, T., *J. Allergy* **37**:169, 1966; **38**:108, 1966.

55. Johanssen, S.G.O., and Bennich, H., *Immunology* **13**:381, 1967.

56. Berrens, L., *The Chemistry of Atopic Allergens*, Basel, Karger, 1971.

57. Kay, A.B., Austen, K.F., and Lichtenstein, L.M., eds, *Asthma: Physiology, Immunopharmacology, and Treatment*, Orlando, Academic Press, 1984.

58. See, e.g. Otto, note 10; Rosenau and Anderson, note 14; Besredka A., and Steinhardt E., *Ann. Inst. Pasteur* **21**:117, 384, 1907.

59. Besredka and Steinhardt, note 58.

60. Besredka, A., *Ann. Inst. Pasteur* **21**:384, 1907; **22**:496, 1908.

61. Cooke, R.A., Barnard, J.H., Hebald, S., and Stull, A., *J. Exp. Med.* **62**:733, 1935.

62. Koch, R., *Berl. klin. Wochenschr.* **19**:221, 1882.

63. Koch, note 1.

64. von Pirquet, C., *Berl. klin. Wochenschr.* **48**:644, 699, 1907.

65. Moro, E., and Doganoff, A., *Wien. klin. Wochenschr.* **20**:933, 1907; Moro, E., *Münch. med. Wochenschr.* **55**:216, 1908.

66. Mantoux, C., *Pr. Méd.* **18**:10, 1910.

67. Calmette, L.C.A., *C. R. Acad. Sci.* **144**:1324, 1907.

68. von Pirquet and Schick, note 13; Pirquet's ideas were further expanded and given additional substance in his book *Klinische Studien über Vakzination und Vakzinale Allergie*, Leipzig, Deuticke, 1907.

69. Baldwin, E.R., *J. Med. Res.* **22**:189, 1910.

70. Baldwin, note 69, p. 252.

71. Baldwin, note 69, p. 238.

72. Zinsser, H., *Infection and Resistance*, New York, Macmillan, 1914.

73. Calmette, A., *L'infection Bacillaire et la Tuberculose*, Paris, Masson, 1920.

74. Zinsser, H., *J. Exp. Med.* **34**:495, 1921; Zinsser, H., and Mueller, J.H., *J. Exp. Med.* **41**:159, 1925.

75. Zinsser, note 74, 1921, p. 499.

76. Dienes, L., *Arch. Pathol.* **21**:357, 1936.
77. Topley, W.W.C., and Wilson, G.S., *The Principles of Bacteriology and Immunity*, 2nd edn, Baltimore, William Wood, 1938, pp. 911 ff, 1044 ff.
78. Topley and Wilson, note 77, p. 1047.
79. Rich, A.R., *The Pathogenesis of Tuberculosis*, 2nd edn, Springfield, Thomas, 1951.
80. Rich, note 79, p. 565.
81. Rich, note 79, p. 568.
82. Uhr, J.W., Salvin, S.B., and Pappenheimer, A.M., Jr., *J. Exp. Med.* **105**:11, 1957; Salvin, S.B., *J. Exp. Med.* **107**:109, 1958.
83. Salvin, note 82; Salvin, S.B., and Smith, R F., *J. Exp. Med.* **109**:325, 1959; Benacerraf, B., and Gell, P.G.H., *Immunology* **2**:53, 1959.
84. Salvin and Smith, note 83; Sell, S., and Weigle, W.O., *J. Immunol.* **83**:257, 1959.
85. Pappenheimer, A.M., Jr., Scharff, M., and Uhr, J.W., in Shaffer, J.H., LoGrippo, G.A., and Chase, M.W., eds, *Mechanisms of Hypersensitivity*, London, Churchill, 1959, p. 417; see also Gell, P.G.H., and Benacerraf, B., *Adv. Immunol.* **1**:319, 1961.
86. Dienes, L., *J. Immunol.* **17**:531, 1929; Dienes, L., and Schoenheit, E.W., *Am. Rev. Tuberc.* **20**:92, 1929.
87. Jones T.D., and Mote, J.R., *New Engl. J. Med.* **210**:120, 1934.
88. Simon, F.A., and Rackeman, F.M., *J. Allergy* **5**:439, 1934.
89. Freund, J., and McDermott, K., *Proc. Soc. Exp. Biol Med.* **49**:548, 1942; Freund, J., *Am. J. Clin. Path.* **21**:645, 1951.
90. Uhr, Salvin, and Pappenheimer, note 82.
91. Benacerraf, B., and Gell, P.G.H., *Immunology* **2**:53, 1959.
92. Raffel, S., and Newell, J.M., *J. Exp. Med.* **108**:823, 1958.
93. Benacerraf and Gell, note 91. See also Salvin, S.B., and Smith, R.F., *Proc Soc. Exp. Biol Med.* **104**:584, 1960; Benacerraf, B., and Levine, B.B., *J. Exp. Med.* **115**:1023, 1962; Gell, P.G.H., and Silverstein, A.M., *J. Exp. Med.* **115**:1037, 1964.
94. Dienes, L., and Mallory, T.B., *Am. J. Path.* **8**:689, 1932.
95. These studies are well summarized by Rich in his *Pathogenesis of Tuberculosis*, note 79.
96. Rich, note 79, p. 600.
97. Gell, P.G.H., and Hinde, J.T., *Intl Arch. Allergy Appl. Immunol.* **5**:23, 1954. These local reactions were called "progressive immunization reactions" to reflect their maturation toward abundant local antibody production.
98. Waksman, B.H., *Experimental Allergic Encephalomyelitis and the "Autoallergic" Diseases*, Suppl. to *Intl Arch. Allergy Appl. Immunol.* **14**, 1959; Waksman, B.H., *Medicine* **41**:93, 1962.
99. Turk, J.L., and Heather, C.J., *Intl Arch. Allergy* **27**:199, 1965; Turk, J.L., Heather, C.J., and Diengdoh, J.V., *ibid.*, **29**:278, 1966; a fine summary of this and other aspects of delayed hypersensitivity will be found in Turk, J.L., *Delayed Hypersensitivity*, Amsterdam, North Holland, 1967.
100. Flax, M.H., and Caulfield, J.B., *Am. J. Path.* **43**:1031, 1963.
101. Turk, J.L., Rudner, E.J., and Heather, C.J., *Intl Arch. Allergy* **30**:248, 1966.
102. Landsteiner, K., and Chase, M.W., *Proc. Soc. Exp. Biol. Med.* **49**:688, 1942.
103. Stavitsky, A.B., *Proc. Soc. Exp. Biol. Med.* **67**:225, 1948.
104. Haxthausen, H., *Acta Derm.-Venereol.* **31**:659, 1951.
105. Chase, M.W., *Proc. Soc. Exp. Biol. Med.* **59**:134, 1945.
106. Mitchison, N.A., *Nature* **171**:267, 1953.

107. For example, the transfer of streptokinase sensitivity in the rabbit by Warwick, W.J., Archer, O., and Good, R.A., *Proc. Soc. Exp. Biol. Med.* **105**:459, 1960; to tuberculin in man by Lawrence, H.S., *Proc. Soc. Exp. Biol. Med.* **71**:516, 1949; to a variety of contact sensitizers in man by Epstein, W.L., and Kligman, A.M., *J. Invest. Dermatol.* **28**:291, 1957.

108. See, for example, Helmholtz, H.F., *Z. Immunitätsf.* **3**:370, 1909; Bail, O., *Z. Immunitätsf.* **4**:470, 1910.

109. Najarian, J.S., and Feldman, J.F., *J. Exp. Med.* **118**:341, 1963; Turk, J.L., and Oort, J., *Immunology* **6**:140, 1963; McCluskey, R.T., Benacerraf, B., and McCluskey, J.W., *J. Immunol.* **90**:466, 1964; Prendergast, R.A., *J. Exp. Med.* **119**:377, 1964.

110. Lawrence, H.S., *J. Clin. Invest.* **33**:951, 1954; see also Lawrence, H.S., *Adv. Immunol.* **11**:196, 1969.

111. Lawrence, H.S., and Pappenheimer, A.M., Jr., *J. Exp. Med.* **104**:321, 1956; Rapaport, F.T., et al., *J. Immunol.* **84**:358, 1960.

112. The early work is summarized by Landsteiner, K., *The Specificity of Serological Reactions*, New York, Dover, 1962, p. 197 ff; see also Sulzberger, M., and Baer, R.L., *J. Invest. Dermatol.* **1**:45, 1938.

113. Enders, J.F., Cohen, S., and Kane, L.W., *J. Exp. Med.* **81**:119, 1945; Rose, H.M., and Molloy, E., *Fed. Proc.* **6**:432, 1947; Urteaga, O., Wagner, H., and Chavez, A., *Arch. Peruan. Patol. Clin.* **16**:113, 1962; see also von Pirquet, notes 13 and 68.

114. Traub, E., *Science* **81**:298, 1935; Hotchin, J., *Cold Spring Harbor Symp.* **27**:479, 1962; Cole, G.A., Nathanson, N., and Prendergast, R.A., *Nature* **238**:335, 1972; Buchmaier, M.J., Welsh, R.M., Dutko, F.J., and Oldstone, M.B.A., *Adv. Immunol.* **30**:275, 1980.

115. Medawar, P.B., *J. Anat.* **78**:176, 1944; Billingham, R.E., Brent, L., and Medawar, P.B., *Proc. R. Soc. B.* **143**:58, 1954.

116. Brent, L., Brown, J.B., and Medawar, P.B., *Lancet* **2**:561, 1958; *Proc. R. Soc. B.* **156**:187, 1962.

117. Simonsen, M., *Acta Pathol. Microbiol. Scand.* **40**:480, 1957; Grebe, S.C., and Streilein, J.W., *Adv. Immunol.* **22**:120, 1976; *Immunol. Rev.* **88**, 1985.

118. The mixed lymphocyte reaction was first described by Bain, B., Vas, M., and Lowenstein, L., *Blood* **23**:108, 1964. The topic is reviewed by Dupont, B., Hansen, J.A., and Yunis, E.J., *Adv. Immunol.* **23**:107, 1976.

119. Glick, B., Chang, T.S., and Jaap, R.G., *Poultry Sci.* **35**:224, 1956; Warner, N.L., Szenberg, A., and Burnet, F.M., *Austral. J. Exp. Biol. Med.* **40**:373, 1956; Warner, N.R., and Szenberg, A., in Good, R.A., and Gabrielson, A.E., eds, *The Thymus in Immunobiology*, New York, Hoeber-Harper, 1964, p. 395.

120. Miller, J.F.A.P., *Lancet* **2**:748, 1961; Janković, B.D., Waksman, B.H., and Arnason, B.G., *J. Exp. Med.* **116**:159, 1962; Good, R.A., et al., *J. Exp. Med.* **116**:773, 1962.

121. Claman, H.N., Chaperon, E.A., and Triplett, R.F., *Proc. Soc. Exp. Biol. Med.* **122**:1167, 1966; Davies, A.J.S., et al., *Transplantation* **5**:222, 1967; Miller, J.F.A.P., and Mitchell, G.F., *J. Exp. Med.* **128**:801, 821, 1968.

122. It was N.A. Mitchison (in Landy, M., and Braun, W., eds, *Immunological Tolerance*, New York, Academic Press, 1969, p. 149) who first showed T and B cell cooperation in antibody formation. The first T lymphocyte differentiation marker was described by Reif, A.E., and Allen, J.M.V., *J. Exp. Med.* **120**:413, 1964; Raff, M.C., and Wortis, H.H., *Immunology* **18**:931, 1970. Subsequently many others

have been described, as reviewed by McKenzie, I.F.C., and Potter, T., *Adv. Immunol.* **27**:179, 1979 and in *Immunol. Rev.* **82**, 1984.

123. Rich, A.R., and Lewis, M.R., *Bull. Johns Hopkins Hosp.* **50**:115, 1933. Rich and Lewis believed that the antigen was directly cytotoxic for sensitized cells, but Waksman, B.H., and Matoltsy, M. (*J. Immunol.* **81**:220, 1958) later showed that the effect was one of cell stimulation.

124. George, M., and Vaughan, J.H., *Proc. Soc. Exp. Biol. Med.* **111**:514, 1962.

125. David, J.R., *Proc. Nat. Acad. Sci. USA* **56**:72, 1966.

126. Bloom, B.R., and Bennett, B., *Science* **153**:80, 1966.

127. Cohen, S., Pick, E., and Oppenheim, J.J., eds, *Biology of the Lymphokines*, New York, Academic Press, 1979; Rocklin, R.E., Bendtzen, K., and Greineder, D., *Adv. Immunol.* **29**:56, 1980. See also Chapter 19.

128. See, for example, the reviews by Waksman (note 98) and Turk's book on delayed hypersensitivity (note 99). Perhaps the most influential markers of this new interest were: Grabar, P., and Miescher, P., eds, *Immunopathology*, Basel, Benno Schwabe, 1959; Lawrence, H.S., ed., *Cellular and Humoral Aspects of the Hypersensitivity States*, New York, Hoeber-Harper, 1959; Shaffer, J.H., LoGrippo, G.A., and Chase, M.W., eds, *Mechanisms of Hypersensitivity*, Boston, Little Brown, 1959; and Wolstenholme, G.E.W., and O'Connor, M., eds, Ciba Symposium *Cellular Aspects of Immunity*, Boston, Little Brown, 1959.

129. Gell, P.G.H., and Coombs, R.R.A., *Clinical Aspects of Immunology*, 2nd edn, Oxford, Blackwell, 1968, p. 575.

130. Dhermy, P., Coscas, G., Nataf, R., and Levaditi, J., *Rev. Intl Trachome* **44**:295, 1967; Silverstein, A.M., and Prendergast, R.A., in Lindahl-Kiessling et al., eds, *Morphological and Fundamental Aspects of Immunity*, New York, Plenum Press, 1971, p. 583.

131. Besredka, A., *Ann. Inst. Pasteur* **33**:301, 557, 882, 1919; **34**:361, 1920; **35**:421, 1921. See also Besredka, A., *Immunisation Locale; Pansements Spécifiques*, Paris, Masson, 1925, and *Les Immunités Locales*, Paris, Masson, 1937.

132. Seegal, B.C., Seegal, D., and Khorazo, D., *J. Immunol.* **25**:207, 1933.

133. Silverstein, A.M., in Maumenee, A.E., and Silverstein, A.M., eds, *Immunopathology of Uveitis*, Baltimore, Williams & Wilkins, 1964.

134. Silverstein, A.M. and Rose, N.R., *Sem. Immunol.* **12**:173, 2000.

135. See, for example, Asherson, G.L., and Webster, A.D.P., *Diagnosis and Treatment of Immunodeficiency Diseases*, London, Blackwell, 1980; Aiuti, F., Rosen, F., and Cooper, M.D., eds, *Recent Advances in Primary and Acquired Immunodeficiencies*, New York, Raven Press, 1986.

136. Rockey, J.H., Hanson, L.A., Heremans, J.F., and Kunkel, H.G., *J. Lab. Clin. Med.* **63**:205, 1964.

137. Hobbs, J.R., Milner, R.D.G., and Watt, P.J., *Br. Med. J.* **4**:583, 1967.

138. Schur, P.H., et al., *New Engl. J. Med.* **283**:631, 1970.

139. Bruton, O.C., *Pediatrics* **9**:722, 1952.

140. Pearl, E.R., et al., *J. Immunol.* **120**:1169, 1978.

141. Schwaber, J., et al., *Nature* **304**:355, 1983.

142. Janeway, C.A., Apt, L., and Gitlin, D., *Trans. Assoc. Am. Phys.* **66**:200, 1953.

143. DiGeorge, A.M., *J. Pediat.* **67**:907, 1965; DiGeorge, A.M., in Bergsma, D., ed., *Immunologic Deficiency Diseases in Man*, New York, National Foundation, 1968.

144. Good, R.A., and Varco, R.L., *Lancet* **75**:245, 1955.

145. Nezelof, C., *Arch. Franç. Pédiat.* **21**:897, 1964.

146. Glanzmann, E., and Riniker, P., *Ann. Pediat. (Basel)* **175**:1, 1950.
147. See, for example, *Progr. Allergy*, **37**, 1986; Geraldo, G., et al., eds, *Recent Advances in AIDS and Kaposi's Sarcoma*, Basel, Karger, 1987.
148. A useful historical review of the complement system can be found in Mayer, M.M., *Complement* **1**:2, 1984.
149. Nuttall, G., *Z. Hyg.* **4**:353, 1888.
150. Bordet, J., *Ann. Inst. Pasteur* **12**:688, 1899.
151. Ferrata, A., *Berl. klin. Wochenschr.* **44**:368, 1907.
152. Ritz, H., *Z. Immunitätsf.* **13**:62, 1912.
153. Coca, A.F., *Z. Immunitätsf.* **21**:604, 1914.
154. Gordon, J., Whitehead, H.R., and Wormall, A., *Biochem. J.* **20**:1028, 1036, 1926.
155. Rapp, H.J., *Science* **127**:234, 1958; Linscott, W.D., and Nishioka, K., *J. Exp. Med.* **118**:795, 1963; Inoue, K., and Nelson, R.A., Jr., *J. Immunol.* **96**:386, 1966.
156. See note 148; see also Mayer, M.M., *Proc. Natl Acad. Sci. USA* **69**:2954, 1972.
157. Pillemer, L., and Ecker, E.E., *J. Biol. Chem.* **137**:139, 1941.
158. Pillemer, L., *Trans. NY Acad. Sci.* **17**:526, 1955. See also W.D. Ratnoff's interesting description of the history of the properdin controversy, *Perspect. Biol. Med.* **23**:638, 1979/1980.
159. Götze, O., and Müller-Eberhard, H.J., *Adv. Immunol.* **24**:1, 1976; Fearon, D.T., *Crit. Rev. Immunol.* **1**:1, 1979. Fearon, D.T., and Austin, K.F., *New Engl. J. Med.* **303**:259, 1980.
160. Shin, H.S., et al., *Science* **162**:361, 1968; Lichtenstein, L.M., et al., *Immunology* **16**:327, 1969. See also Hugli, T.E., and Müller-Eberhard, H.J., *Adv. Immunol.* **26**:1, 1978.
161. Hyde, R.R., *J. Immunol.* **8**:267, 1923; *Am. J. Hyg.* **15**:824, 1932.
162. Silverstein, A.M., *Blood* **16**:1338, 1960.
163. Lachmann, P.J., *Boll. 1st Sieroter. Milan* (Suppl.), **53**:195, 1974.
164. Lachmann, P.J., and Rosen, F.S., *Springer Sem. Immunopath.* **1**:339, 1978; Agnello, V., *Medicine* **57**:1, 1978.
165. Soothill, J.F., and Harvey, B.A.M., *Arch. Dis. Child.* **51**:91, 1976; *Clin. Exp. Immunol.* **27**:30, 1977.

10 Anti-antibodies and anti-idiotype immunoregulation 1899–1904

Everything of importance has been said before by someone who did not discover it.

Alfred North Whitehead

It happens occasionally in science that a discovery is made or a concept is advanced long before its full implications can be assessed or its utility exploited. It may then sink into oblivion, only to be "rediscovered" and put to use decades or even centuries later. It matters little that the original concept may have been based upon erroneous premises or upon misinterpretations of fact, so long as its heuristic value impelled its adherents to develop it further along some logically consistent and useful lines. The result of such an enterprise might even be classed, at a later date, as a milestone in the history of the science. Thus, in geography, the ancient Greeks considered the Earth to be a sphere based upon purely esthetic reasons, since the sphere was the most perfect of all solids. But the (to us) inadequate basis for this speculation did not prevent Eratosthenes from concluding, in the third century BC, that the logical consequence of a spherical earth should be that the sun's elevation at any moment would differ to two different observers standing in line with it. From this he was able to measure the circumference of the Earth with an accuracy not improved upon for almost two millenia.[1] This concept of a round Earth, and Eratosthenes' logical extension of its implications, faded from view, to be rediscovered by the demands of sixteenth-century transoceanic voyages.

Perhaps the best known instance of a "premature" theory in the field of immunology was Paul Ehrlich's side-chain theory of 1897.[2] This was, for all practical purposes, the first natural selection theory of antibody formation. It dominated the field for perhaps a decade or so, only to fall into disrepute and nearly to be forgotten for half a century until Niels Jerne[3] and Macfarlane Burnet[4] gave the selective theory of antibody formation its modern form and appeal. Indeed, it was the success of the clonal selection theory that refocused attention on Ehrlich's imaginative concept, and stood witness to his creativity and prescience.

But Ehrlich's side-chain theory was more than a concept of antibody formation; it speculated broadly also (within the obvious limitations of contemporary scientific knowledge) about the structure and function of antibody.[5] Implicit in the side-chain theory was an even more startling conceptual anticipation of the future – that the binding site of an antibody is a unique structure and might be immunogenic; that an anti-antibody might be formed against the specific site; that the shape of the anti-antibody combining site would be that of the corresponding

A History of Immunology, Second Edition
ISBN: 978-0-12-370586-0

Copyright © 2009, Elsevier Inc.
All rights reserved

antigen against which the original antibody was specific; and that this complex of antibodies and anti-antibodies might serve a very important immunoregulatory role in protecting against such undesirable events as autoimmunization.

All of these consequences of the side-chain theory, and more, were recognized and explicitly developed between 1899 and 1902 by the imaginative and implacably logical Paul Ehrlich, with occasional contributions also from Jules Bordet, Alexandre Besredka, and others. Here, save for the nomenclature, was a theory of idiotypes and anti-idiotypes, of mirror images, and of network immunoregulation that anticipated the modern development of this subject by almost seventy years, and that has hardly been improved upon on the theoretical level. Never mind that the earlier version was based upon experiments later shown to be flawed technically, that it was based upon fatal misinterpretations of the data, and that it died ignominiously within a very few years. It testifies nevertheless to the vitality of the early years of immunology, and to the imaginative approaches of its founders.

The full flavor and significance of the early work on anti-antibodies can best be appreciated by comparing it with modern developments in the theory and practice of the science of idiotypes. We shall, therefore, preface this look at the past with a brief review of more recent developments in this field.

Idiotypes and anti-idiotypes, 1963–1985

The first significant stirrings of modern interest in anti-antibodies date from the 1950s, and owe their origins to three different lines of attack. The first was purely theoretical, and stemmed from the fertile imagination of Victor Najjar.[6] Najjar assumed that the interaction of antibody with antigen leads to a change in molecular conformation which is immunogenic, leading to the development of a "cascade" of further anti-antibody reactants to the first and then to subsequent immune complexes, the entire process finally achieving a steady state. This appears to have been the first modern hint at the possible existence of an immunoregulatory network that might function by self-stimulation and *internal* controls.

The second approach to the study of anti-antibodies was sparked by the increasing interest in autoimmune diseases, and especially by the growing realization that rheumatoid factor might in fact be an anti-antibody. This proposal was made initially by Milgrom and Dubiski,[7] who suggested also that the agglutinins for erythrocytes sensitized with incomplete antibody might be "anti-antibodies." They further postulated that the immune globulins of the individual's own body may become antigenic, presumably because the antibody becomes denatured upon interaction with antigen, and that this might be the basis for the development of the putatively pathogenic rheumatoid factor. These studies were followed up by numerous other investigators,[8] all of whom were able to demonstrate that animals could mount an immune response to denatured autologous gamma globulin.

The third approach to anti-antibodies, and ultimately the most productive of all, originated in the earlier discovery that the immunoglobulins of humans and experimental animals possess genetically determined antigenic markers or *allotypes*,[9] based upon unique sequences of amino acids in the polypeptide chains of immunoglobulins whose composition and structure were even then being elucidated.[10] In the continuing search for new allotypes, three laboratories almost simultaneously discovered the phenomenon of idiotypy. From one direction, Jacques Oudin and Mauricette Michel in France,[11] and Philip Gell and Andrew Kelus in England[12] noted the development of antibodies with peculiar characteristics in animals isoimmunized with bacteria coated with specific antibody. These had none of the characteristics of anti-allotypes, which usually react with most immunoglobulins of certain other individual animals; rather, these new antibodies reacted only with those antibodies specific for the bacterial carrier employed. Thus, not only were they anti-antibodies, but indeed they appeared to react with the immunogenic combining site of the antibody employed for immunization. In the terminology of the increasingly popular clonal selection theory of Burnet, Gell and Kelus were the first to suggest that this was, in fact, an "anti-clone antibody," reactive with a unique antibody-combining site. They called the immunogenic site a "private" antigenic determinant, to distinguish it from the more public antigenic determinants due to allotypic markers. Oudin would soon coin the term *idiotype*[13] to characterize this distinction more precisely.

A somewhat different point of departure was employed by Henry Kunkel and his colleagues[14] in arriving at the same conclusions as those of Oudin and Michel and of Gell and Kelus. Myeloma proteins and those of certain macroglobuli-nemias had been shown to possess the characteristics of individual (monoclonal) antibodies, and to show unique antigenic specificities.[15] These workers therefore immunized rabbits with purified and fairly homogeneous preparations of human antidextran and antilevan antibodies. Several of the rabbits produced anti-antibodies specific only for the individual antibody employed for immunization. Clearly, each of these three groups had discovered anti-idiotypic antibodies, although the identity of the "new" antigenic determinant with the antibody combining site had yet to be firmly established.

In 1966, Gell and Kelus reviewed the then still limited literature on auto-antibodies and idiotypes, and speculated with impressive foresight on two aspects that would later loom large in this field. In a discussion of the possible biological significance of autoanti-idiotypes, they pointed out their relevance to our understanding of the mechanisms of autoimmunity and of immunological tolerance. As they said, "It is hard to believe that it [the body] can regularly react to 'private' determinants on its own antibodies, not so much because the process would be self-destructive as that it would lead to an infinite regress of anti-antibody production."[16] Here is yet another hint that each anti-idiotype is itself an idiotype, able to stimulate a further progression of immune responses at each stage. They suggested, however, that the immunogenicity of the idiotype may be quite low, thus cutting short the infinite regress at an early stage. Gell and Kelus further proposed in this review that self-tolerance to idiotypic determinants, in the sense of a Burnetian clonal deletion, does not provide an adequate expla-nation, since "it would entail the elimination of a number of clones equal to the number of possible antibodies."[17] This is the first intimation that the universe of potential anti-idiotypes may equal in size the universe of possible antibodies.

Despite the many fascinating problems posed by idiotypes and anti-idiotypes, this area did not yet grip the collective imagination of immunologists, and only modest progress was forthcoming for almost a decade.[18] It remained for Niels Jerne to rephrase the questions and the possibilities in another of his landmark speculative ventures, first hinted at in 1971[19] and then developed more fully in 1974.[20] This was his network concept of the immune system. Here was a theory in which idiotypes and anti-idiotypes play the central role not only in determining the ultimate size of the immune-response repertoire, but in furnishing as well a mechanism for the internal regulation of these responses.

Jerne's network theory, 1974

Plate 10 Niels Jerne (1911–1994)

In reviewing the pertinent data up to that point, Jerne called special attention to three aspects that he felt were especially significant:

1. A given antibody combining site structurally determines a set of immunogenic regions (idiotopes) in the variable region of the immunoglobulin molecule
2. The idiotope collective (the idiotype) forms an "internal image" of and interacts with the antigenic site (epitope) against which it is formed, but reacts also with a corresponding set of anti-antibodies (the anti-idiotopes) whose formation it may stimulate, thus forming a network of ever increasing size
3. The immune response is a measure of the balance that may exist at any time between active stimulus and active suppression – i.e., the result of an immunoregulatory network in which idiotypes and anti-idiotypes act upon both T and B lymphocytes.

(The reader will wish to compare Jerne's formulation with one by Jean Lindenmann,[21] apparently stimulated by informal communications from Jerne himself. Lindenmann noted that the antibody combining site contains a "negative image" of its respective antigen, and that the anti-idiotype, which he called a *homobody*, contains a "positive image" of the antigen. These form a network of interconnecting molecules which, Lindenmann suggested, "may have broad biological significance.")

It is difficult to overstate the extent of the interest and research activity generated by Jerne's concept. For the first time in the modern era, it was suggested that the production of autoantibodies against self-antigens might be the normal state of affairs, rather than the exception (see also Irun Cohen's thesis, Chapter 5). Again, the idiotype–anti-idiotype network offered an appealing molecular alternative to explain immunoregulation and the suppression of auto-immune phenomena, in place of the increasingly popular cell-dynamic model based upon the positive and negative contributions of T lymphocytes.[22] (Acceptance of the idiotype network theory of immunoregulation was not universal. Melvin Cohn[23] summarized the position of those who maintained that cellular interactions and intercellular signals contribute more to the regulation of the immune response than do the molecular interactions of idiotypes and anti-idiotypes.)

By analogy with earlier demonstrations that anti-allotypes can suppress the development of individual allotypes, it was quickly confirmed that anti-idiotypic antibodies may also suppress the production of specific idiotypes.[24] From this, it followed that passive administration of anti-idiotypic antibody might suppress specifically those immune responses to antigen that are characterized by a preponderance of certain idiotopes, although such responses may also be enhanced by anti-idiotypes under special conditions.[25] Further, idiotypes were found on both helper and suppressor T cells as well as on B cell membrane receptors,[26] suggesting that autoanti-idiotype antibody might contribute also to the regulatory functions of these lymphocyte subsets.

Autoanti-idiotypes were also found to be routinely present in association with such autoimmune disease states as systemic lupus erythematosus,[27] thyroid disease,[28] hemolytic anemias,[29] and autoimmune interstitial nephritis.[30] This implied, as Jerne had predicted, that autoanti-idiotypes might play a central role in the mechanism of self-tolerance and in the suppression of autoimmunity, and has been confirmed experimentally.[31] The speculation that the anti-idiotype should look like a positive image of the antigen and mimic it was also confirmed,[32] and as such could contribute to either positive or negative immunoregulation of the immune response, with possible applications also for use as vaccines. Thus, many believe that the implications of Jerne's network theory have been fully realized,[33] although others are unwilling to assign major importance to the phenomenon.

In order to appreciate fully the conceptual achievements of those engaged in anti-antibody research at the turn of the last century, to be recounted below, it may be well to keep in mind the salient aspects of the modern approach to this problem. These are as follows:

1. The antibody combining site (the idiotype) is unique and may itself constitute a new immunogenic site
2. The new antibody combining site may stimulate the formation of autoantibody against it (the autoanti-idiotype)
3. The idiotype presents an internal image of the antigenic site, and the anti-idiotype is structurally similar to and may act in place of antigen
4. The interaction of idiotypes and anti-idiotypes may constitute an effective molecular regulatory system, one of whose principal activities may be to inhibit the development of pathological autoimmune processes.

The background to anti-antibodies, 1890–1899

At the time that Emil von Behring and Shibasaburo Kitasato discovered antibodies in 1890,[34] infectious diseases were generally thought to be mediated by bacterial toxins, and protective antibodies were viewed solely as antitoxins. This view was strengthened by Ehrlich's demonstration that circulating antibodies could be induced even against such plant toxins as ricin and abrin.[35] The discovery of immune hemolysis by Bordet in 1899[36] appeared to confirm further the generalization that most destructive processes of interest to the bacteriologist-immunologist were toxic in nature, except that in this instance the antibody itself was the toxin. Indeed, Bordet and others repeatedly referred to hemolytic antibody as a "hemotoxin," and when other anti-cell or anti-organ antibodies were discovered, they were variously termed spermotoxins, neurotoxins, leucotoxins, and so on.[37]

The finding that "immunity" responses could be engendered by the cellular elements of the mammalian body as well as by bacteria caused little surprise in the late 1890s, for by this time it had generally become accepted that the immune response was part of a larger, normal physiologic process for the digestion and disposal of unwanted substances. As early as 1884, Metchnikoff had postulated that the immune response had evolved from primitive digestive functions, serving now to break down and to mitigate the effects of invading bacteria.[38] Of course, Metchnikoff assigned to the phagocyte the principal role in this process, and extended it also to the disposal of effete cells of the animal's own body.

With the rise of interest in antibodies, Metchnikoff's disciple Alexandre Besredka extended the concept to include the contribution of hemolytic anti-body toward the disposal of worn-out erythrocytes.[39] In a similar manner, Ehrlich utilized the same approach in his side-chain theory of antibody forma-tion, postulating that all antibodies are physiologic products which play an important part in the *normal* economy of the cell and of the body.[40]

Ehrlich's side-chain theory: 1897

The major aspects of Paul Ehrlich's side-chain theory of antibody formation have been described extensively elsewhere,[41] and need not be treated here in great detail. For the present purposes, it will suffice to recall the principal premises upon which Ehrlich based his concept:

1. Antibodies are naturally occurring substances that serve as receptors on the cell surface
2. The specificity of antibody for antigen is determined by a unique stereochemical configuration of atoms that permits the antibody to bind tightly and chemically to its appropriate antigen
3. The number of different combining site structures available is so great that each one differs from the others, with little or no cross-reactions among them
4. In order to induce active antibody formation, it is only necessary that appropriate receptors be present on the cells, and for antigen to interact with them and so stim-ulate their compensatory overproduction by the cell and liberation into the blood.

Thus the antibody, as first conceived by Ehrlich, appeared to be a rather nondescript blob of cytoplasm *with only one distinguishing feature* – a highly organized combining site (the "haptophore" group) that defines its specificity and thus its ability to bind to antigen.

With the subsequent discovery of immune hemolysis and of the participation of complement, Ehrlich expanded his theory to define hemolytic antibody as an "amboceptor," this time with two chemically defined combining sites – one for antigen and one for complement.[42] But if the combining site for antigen is a unique structure with a unique specificity, then so also should be the combining site for complement. This implied to Ehrlich that just as there exists a multi-plicity of different antigens, each defining and being defined by a distinctive combining site on antibody, so there should also be a multiplicity of different complements, one for the complementophile group on each different antibody.[43] Thus, in order for antibody to interact with antigen or with complement, its combining site must recognize and attach to a complementary, structurally-defined combining site on its partner. These conclusions were among the first of the many logical extensions which Ehrlich and his followers formulated to expand upon the rather simple premises initially embodied in his side-chain theory.

It was the further application of the side-chain theory and of Ehrlich's inex-orable logic to the question of anti-antibodies that invites comparison with modern theories of idiotypic interactions and immunoregulation.

The conceptual position of Jules Bordet

In contrast to Ehrlich, who was concerned primarily with the origin of anti-bodies and the chemical basis for their specificity, Bordet was more interested in their mode of action and biological implications. Here, he took exception to almost every aspect of Ehrlich's theories, and to the conclusions which Ehrlich drew from them.[44] Where Ehrlich argued for a stereochemical basis for specificity and a firm union between antigen and antibody, Bordet concluded that the antigen–antibody union was reversible and a physical one, based upon colloid-adsorptive processes.[45] It was especially in considering the mechanism of immune hemolysis that the two investigators differed most sharply. Ehrlich's theory demanded, in addition to a specific haptophore group for antigen, an equally specific complementophile group to mediate the firm interaction of antibody with complement. Bordet denied this, and suggested that the role of antibody was rather to "sensitize" the erythrocyte, whereupon complement would be fixed nonspecifically and in consequence permitted to effect the destruction of the erythrocyte.[46] Thus, while Ehrlich's logic seemed to require a multiplicity of different complements within a given serum, Bordet was content with but a single complement acting wherever required.[47]

Perhaps the most interesting difference between Ehrlich and Bordet lay in their scientific styles. One gets the impression from reading Ehrlich that he was by far the more doctrinaire of the two. Having advanced his side-chain theory, he (and his students) applied it relentlessly to each new observation, and at each step advanced a new modification or *ad hoc* hypothesis to help to fit the new information into the old theoretical mold. This tendency is best illustrated by Ehrlich's attempt to explain the shape of the diphtheria toxin–antitoxin neutralization curve, which required him to predict the existence not only of a toxoid, but also of a low affinity "toxon" and of alpha and beta modifications of proto-, deutero-, and trito-toxins of varying affinities.[48] Bordet, on the other hand, appears to have been more the pragmatist, and always claimed that he was less a theoretician than a describer of "the true state of affairs." Although he was firmly committed to his sensitization theory of immune hemolysis, contradictory data generally found him prepared to give up a previous conclusion quite readily, and to seek a new explanation for the phenomenon.

This difference in scientific styles between Ehrlich and Bordet is important to our story of anti-antibodies. Each investigator undertook to produce and to study anti-antibodies in 1899, to help bolster his own position and undermine that of his opponent. In their very first mention of anti-antibodies,[49] Ehrlich and Morgenroth suggested that these would provide further evidence for the multiplicity of amboceptors, and later utilized the same approach to help prove the existence of a multiplicity of complements. Bordet, for his part, felt that anti-antibodies would prove the unity of complement – a position which he also stated explicitly in his paper on this subject.[50] But it is this very difference in scientific styles which helps to explain the different roles played by Ehrlich and Bordet in the development of the anti-antibody story over the next five years. Ehrlich, the more rigid theorist, would be responsible for an impressive theoretical construct involving (to employ the terminology of a later era) idiotypes, anti-idiotypes, and internal and external images of antigen. Bordet would make little contribution to these theories; rather, it would be he who would pursue the problem experimentally, and finally provide the data to show that the entire theory was erroneous and based upon a misinterpretation of the experimental results.

Anti-antibodies, 1899–1904

Our story of the first discovery of anti-antibodies starts in 1898, with the almost simultaneous demonstration by Kossel in Germany[51] and Camus and Gley in France[52] that a protective antibody could be formed against the hemolytic toxin present in eel serum. Here was an antitoxin analogous to those formed against diphtheria or tetanus toxins, but one whose protective action could be demonstrated in the test tube in a system where the specific target of the toxic antibody (the erythrocyte) was established beyond question. This observation was immediately seized upon by Bordet and by Ehrlich and Morgenroth, who independently concluded that if an antibody could be prepared against hemolytic eel toxin, then it should surely be possible to prepare a similar neutralizing antibody against the "hemotoxic" antibody found in the serum of animals immunized with erythrocytes. Suiting action to speculation, both laboratories immunized animals of an unrelated species with whole anti-erythrocyte serum, and quickly reported the finding of an "anti-antibody" that would inhibit the destructive action of the hemolytic antibody employed for immunization.

The antibody-combining site as immunogen

In their "Second Communication on Hemolysis" in 1899, Ehrlich and Morgenroth record the theoretical basis for these experiments as follows:[53]

> This antibody, formed by an immunity reaction, thrusts itself into the
> hemotropic group [combining site] of the Zwischenkorper and thus deflects it
> from the erythrocyte. Our attempts, based on these premises, to produce an
> isolated [specific] antibody for some of the lysins have thus far been
> unsuccessful.

It was not that they were unable to demonstrate inhibition of hemolysis with their putative anti-antibody, but rather that the inhibition was not restricted to the species of erythrocyte employed to obtain the hemolytic antiserum. Since the hemolysis of many different species of erythrocytes was inhibited by the "anti-antibody," the demands of the side-chain theory forced Ehrlich and Morgenroth to conclude that the hemolytic serum must contain a multiplicity of antibodies against various erythrocytes, and therefore must stimulate a correspondingly large number of anti-antibodies.

The theoretical basis for the action of anti-antibodies was expanded upon during the following year in Ehrlich's Croonian Lecture before the Royal Society, in which he pointed out that:[54]

> the lysin, be it bacterial lysin or hemolysin (i.e., immune body plus complement), possesses altogether three haptophore groups, of which two belong to the immune body and one to the complement. Each one of these can be bound by an appropriate anti-group. Three anti-groups are thus conceivable any one of which, by uniting with one of the haptophore groups of the lysin, can frustrate the action of the lysin.

Jules Bordet interpreted his results on anti-antibodies somewhat differently. While concluding that the immune serum which he obtained by immunizing with a hemolytic antiserum might contain anti-antibodies specific for the *sensibilisateur* as well as anti-*alexine* antibodies (to use Bordet's terminology for hemolysin and complement), he argued that his results proved the unity of complement rather than its multiplicity. How else could one explain the ability of a rabbit anti-guinea pig hemolytic serum to neutralize all complement activity of guinea pigs, but not that of other species?[55]

(The modern reader will already have detected the basic flaw in these experiments. In 1899, hemolytic antisera were prepared by immunizing animals with defibrinated *whole* blood, so that antibodies would have been formed against many of the serum components as well as against the erythrocytes present in the inoculum. Similarly, "anti-antibodies" were prepared by immunizing with *whole immune serum*, and thus a variety of complement-fixing antiserum proteins would be formed. But in 1899, the only antibodies recognized were against bacterial toxins and such other cellular elements as erythrocytes, which were known [according to Ehrlich's side-chain theory] to contain specific receptors. Almost nothing was known about the composition of normal serum, so that an Ehrlich or Bordet would be justified *at that time* in expecting antibody formation only against those immunogenic receptors which they assumed to be present. In normal defibrinated blood, only the erythrocyte receptors were assumed to be immunogenic, whereas in an immune serum, immunogenicity could only be assumed for the combining sites on specific antibody or on complement.)

Autoanti-antibody immunoregulation

Bordet's observations on anti-antibodies were quickly extended by his colleague at the Institut Pasteur, Alexandre Besredka.[56] Arguing from Ilya Metchnikoff's idea that immunity is due to the normal processes of digestion, and that antibodies assist in the destruction of injected foreign cells, Besredka asked why antibodies are not formed against the cells of one's own body during their destruction. Why, for example, are autohemolytic antibodies not formed during the erythrophagocytosis of effete red blood cells in the spleen? Besredka's answer to this was that they are! Why, then, does this not cause continual hemolysis of all erythrocytes? Because, claimed Besredka, the body normally makes an anti-antibody against *every* potentially threatening autoantibody, thus interfering with its destructive activity.

Besredka justified his thesis by citing a series of experiments on hemolytic antibodies formed in species A against the erythrocytes of species B, utilizing complement from any other species. He showed that A anti-B serum lyses the washed erythrocytes of B, but that the addition of normal B serum inhibits this hemolytic action. However, sera from other species (C, D, etc.) exhibit no similar inhibitory effect. Further, B's serum will inhibit the hemolysis of an anti-B-erythrocyte serum formed in any other species as well. From these data, Besredka concluded that:

1. All normal sera contain anti-antibodies that protect their own erythrocytes from immune hemolysis
2. He had demonstrated the specificity of these anti-antibodies
3. The anti-antibody is not an anticomplement
4. Ehrlich was wrong about the multiplicity of amboceptors – all anti-B hemolysins are identical, since they are all inhibited by the anti-antibody normally present in every B serum.

When Besredka suggested in 1901 that autoantibody formation was the norm, with its pathogenicity controlled by the countervailing production of autoanti-antibodies, the existence of any type of autoantibody production had not yet been formally demonstrated. Indeed, Ehrlich had expressed the common view that while theoretically possible, the formation of destructive autoantibodies was "dysteleological" and unlikely. This point of view was epitomized in Ehrlich's famous dictum of *Horror autotoxocus*.[57] However, in 1904 Julius Donath and Karl Landsteiner published the first clear-cut description of a destructive autoantibody causing serious disease in man[58] – the autohemolysin responsible for paroxysmal cold hemoglobinuria (PKH). Ehrlich and his followers conceded at once that this phenomenon was a probable exception to the rule.[59] Besredka's adherents, however, offered up an alternative explanation based upon autoanti-antibody immunoregulation. They suggested[60] that Donath and Landsteiner were wrong in ascribing the disease to the presence of autohemolysins, since, according to Besredka's theory, everyone produced them. The defect in PKH, they claimed, was due to the *absence* in these patients of the regulatory autoanti-antibodies that normally inhibit spontaneous hemolysis!

Besredka's concept of the existence of autoanti-antibodies and of their immunoregulatory role appears to have been a purely speculative leap of the imagination; Paul Ehrlich, on the other hand, arrived at the same conclusions simply by pursuing the inner logic of his side-chain theory. In 1899, Morgenroth had shown that animals inoculated with the enzyme rennin would invariably produce antirennin antibodies.[61] But rennin is presumably one of the normal constituents of the animal's digestive tract, so that the formation of an "autoantibody" against a self-constituent could conceivably compromise the well-being of the host. Ehrlich and Morgenroth returned to this question the following year, and proposed a thought-experiment to explain the apparent paradox.[62] Here is the logical extension of the side-chain theory in its most elegant form.

Suppose that a hypothetical antigen is injected into an animal. Two consequences are then possible, according to Ehrlich and Morgenroth. If the animal lacks group α, then the specific site α on the injected antigen will seek out its corresponding receptor on the surface of the host's cells, react with the combining sites on these receptors, and thus stimulate the formation of anti-α antibodies. This is the usual course of the immune response. Suppose, however, that the immunized animal possesses antigenic group α within its body, as is the case with rennin. Anti-α antibodies will still be formed, but these will now appear as "autoantibodies." But these circulating antibodies with combining sites specific for antigenic group α will themselves find cells with α receptors on their surface (i.e., presumably those cells responsible for the original production of that antigen). Such cells will be stimulated to produce additional α molecules for release into the circulation. However, not only is α the original antigen; it is also functionally the autoanti-antibody able to combine specifically with the anti-α combining site to prevent its toxic action. Thus, an interactive network is established involving antigen, specific antibody, anti-antibody (= antigen), and so forth, all of which presumably reach a steady state self-regulated equilibrium to suppress autoimmune disease.

In this example of Ehrlich and Morgenroth's, we see epitomized both Ehrlich's inexorable logic and the full sweep of this turn-of-the-century theoretical construct. The antibody combining site is a unique structure, and may stimulate the formation of a specific autoanti-antibody; the antibody contains the negative image of antigen, and the anti-antibody contains the positive image of antigen; and, finally, a self-regulating network may be established to prevent self-intoxication.

The images of antigen

The foregoing discussion has already made it abundantly clear that Ehrlich viewed the interaction of antigen with antibody as the combination of two complementary stereochemical structures – the specific combining site on antibody and the immunogenic site on antigen. This required, in effect, that the structure of the antibody combining site present as a "negative image" of the antigen. Indeed, the diagrams published by Ehrlich and Morgenroth to represent these interactions make this conclusion explicit. In Figure 10.1a, we see a cartoon representation of different antibody combining sites, each with a unique shape into which the corresponding antigen will fit to mediate combination. Ehrlich extended this concept to include the individuality of the different combining sites for complements (Figure 10.1b), upon which he based his prediction of the existence of a multiplicity of different complements. It is

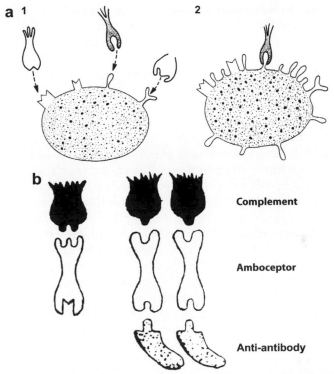

Figure 10.1 The Ehrlich conceptualization of antibody–antigen and antibody–complement binding sites in cartoon form. a, each antigen has a unique stereochemical structure that is matched by the specific combining site on antibody in the form of a negative image; b, similar unique structures define antibody–complement combining sites. Note that the anti-antibody will thus appear to be a "positive" image of the antigen. a, from Ehrlich, P., "Croonian Lecture – on Immunity…," *Proc. R. Soc. Lond.* **66**:424, 1900; b, from Ehrlich, P., *Klin. Jahrb.* **60**:299, 1897 (English translation in *The Collected Papers of Paul Ehrlich*, Vol. 2, Pergamon Press, New York, 1957, pp. 107–125).

also clear, from these diagrams, that the "anti-antibody" must contain a "positive image" of the antigen in question.

In their *Third Communication on Hemolysis*, Ehrlich and Morgenroth defined clearly the conditions necessary for antibody formation. Cell-surface receptors for the antigen must be present, for "if…an organism lack receptors…, the first essential for the production of an antibody will be wanting. In the development or non-development of antibodies, we shall have an indication of the presence or absence of receptors."[63] It is clear also, as these authors point out, that in order for a hemolytic antibody to function, antigenic receptors must be present on the target erythrocytes. But if an antibody is injected into an animal and finds appropriate receptors, then logic demands that anti-antibody be formed – and this is precisely what Ehrlich and Morgenroth report. It will be recalled here that for Ehrlich, the only identifiable structure on either antigen or antibody is the specific combining site. Thus, the anti-antibody is an anti-haptophore group and therefore, by definition, the positive image of the antigen combining site. The anti-antibody is in fact nothing less than the freed antigenic receptor itself! (While Ehrlich never says so explicitly, it is evident from this discussion that the two partners in the interaction of antigen and antibody cannot readily be distinguished. Either of them may be called the antibody, implying the active combining factor, and the other the antigen, implying the passive partner.)

The notion that the anti-combining site of antibody should be antigen itself was taken up by Pfeiffer and Friedberger[64] with extremely surprising results. These authors immunized a rabbit with goat anticholera serum, and obtained an antiserum which appeared to neutralize the anticholera antibodies employed for immunization. By studying the quantitative relationships between the anti-antibody and the original antiserum, they were able to show that the former was not an anticomplement, but rather an antibody against the original anticholera antibody. But such an anti-antibody would be, according to Ehrlich's theory, the antigenic determinant present originally on the cholera vibrio – a substance which surely should not be present on the cells of higher animals. While somewhat perplexed by these findings, Pfeiffer and Friedberger were forced to conclude that "from the present state of our knowledge, there is left no other remaining possibility than to conclude, in defiance of theory, the existence of anti-antibody in our serum."

August von Wassermann drew a similar conclusion from the logical imperatives of the Ehrlich theory, and attempted to apply this to clinical practice in a most interesting manner.[65] In his case, Wassermann had developed an anti-complement – i.e., an antibody of presumed specificity for the unique binding site on the complement "molecule." On the assumption that immunization with this anticomplement would result in its binding to complement receptors on the surface of appropriate cells, he sought to immunize animals with this anti-complement "...in the expectation of producing thereby, according to the applicable laws of immunity, an increase in the production of the respective complement, and thus a heightened resistance, or in such animals better thera-peutic results...." In the event, Wasserman's attempt did not succeed, and no increase in complement titer resulted, although the experiment appeared nevertheless to be firmly based upon the inner logic of Ehrlich's side-chain theory. Indeed, in discussing Wasserman's experiment, Ehrlich's student Hans Sachs pointed out that the anticomplement combining site should in fact have the character and structure of the complementophile group on amboceptor.[66] It should theoretically be possible, therefore, to employ anticomplement antibody for the production of specific anti-amboceptor. The theory is here carried one step further, and demonstrates clearly the understanding by these investigators of the reciprocal nature of specifically-interacting combining sites, where positive and negative images of a given structure would alternate at each succeeding stage of the immunization process.

The demise of anti-antibody theories, 1901–1905

The discovery of anti-antibodies and speculations about their significance stimulated a flurry of investigations,[67] of which we have mentioned here only the most prominent. But as new data appeared, it quickly became evident that the original interpretations required major modification. Somehow, the specificities of these "anti-antibodies" did not always obey the logical predictions of Ehrlich's theory. Thus, as early as 1900, Bordet showed that an antiserum against guinea-pig complement would neutralize all guinea-pig complement activity, but not that of other species.[68] He argued from his observation that Ehrlich was wrong in suggesting a multiplicity of different complement specificities, and that in fact there exists but a single type of complement within the given species. Similarly, Pfeiffer and Friedberger showed in 1902 that the inhibitory action of an "anti-amboceptor" extends to all of the amboceptors of the species, and suggested that therefore the anti-amboceptor (which they assumed to be directed against the complementophile receptor) had to be nonspecific.[69] Faced with this objection, Ehrlich was forced to retreat from his original stand, and to allow that:[70]

> we must assume that all the amboceptors of the same animal species are at least
> partly similar in structure so far as the complementophile apparatus is
> concerned. In a way, therefore, the amboceptor bears the stamp of the animal
> species from which it derives.

Here is the first concession by Ehrlich that not only might a haptophore group (combining site) not be a unique structure, but indeed that these active molecules might possess immunogenic species markers in addition to their specific combining sites.

In the end, it was Bordet who sounded the death knell of the anti-antibody theory, and this by means of two major experimental contributions. The first of these was the discovery in 1901 of the complement fixation test by Bordet and Gengou.[71] Hitherto, complement had been considered to function only in conjunction with anti-erythrocyte and antibacterial antibodies, to effect the lysis or death of these target cells. Now, numerous workers were able to show that any protein or bacterial antigen might fix complement in the presence of its specific antibody.[72] This led Neisser and Sachs to show that complement fixation might be employed for the determination of proteins for forensic purposes,[73] and Detré[74] and Wasserman and colleagues[75] to show that this approach might also be employed for the diagnosis of syphilis and other infectious diseases.

Complement fixation was therefore entirely nonspecific, as Bordet had originally claimed. The activity of many of the so-called anti-antibodies represented not specific anticomplements, but rather a nonspecific complement fixation due to the presence of unrelated antigen–antibody interactions in the test mixtures employed. These data forced Ehrlich almost to give up completely his belief in antibodies specific for complement or its combining site on amboceptor, and to confess that:[76]

> *in this case, therefore, the anticomplement action is brought about by the interaction of two components, one present in the serum of the immunized animal and the other in the serum of that animal species whose serum was used for immunization. It is clear, of course, that here the dissolved albuminous substances, not the complements, were the antigens. This being the case, the demonstration of anticomplements produced by immunization becomes extremely difficult...*

Thus was the possibility laid to rest of an anti-antibody specific for the combining site on complement, or for an anti-antibody specific for the com-plementophile site on amboceptor.

It now remained for Bordet to disprove the existence of an anti-antibody specific for the antigen-combining site, and this he did in a major publication in 1904.[77] In this paper, Bordet refuted the thesis of Ehrlich and Morgenroth that the anti-antibody might be directed against the specific combining site by showing that:

1. The putative anti-antibody appears to neutralize all of the antibodies (hemolysins, bacteriolysins, etc.) produced in the species employed to stimulate the anti-antibody
2. One can obtain an inhibitory "antihemolysin" by injecting into an animal of another species even normal serum, preabsorbed with erythrocytes to remove natural antibodies
3. The so-called antihemolysin will not neutralize hemolysins active against the same erythrocytes, but formed in a third animal species.

Bordet concluded from these results that the action of "anti-antibodies" is not directed against the specific site on the immunizing antibody; that the "anti-antibody" is not the positive image of the original antigen; and that it is in fact directed against something (not further specified) in the immunizing serum that is characteristic of the species. (The more subtle techniques that would permit Oudin and others to detect anti-idiotypes in the 1960s could not even be imagined by a Bordet sixty years earlier.)

This was the final nail in the coffin of anti-antibody theories. By 1905, such concepts were no longer seriously considered or experimentally pursued by most workers in the field. Those who held out the longest were the members of Ehrlich's school of immunology, but even Hans Sachs (who had assumed the role of spokesman for this group when Ehrlich's interests shifted to chemotherapy) was forced slowly to concede the earlier conceptual errors, in a series of major reviews on hemolysis and hemolysins.[78]

Conclusions

It must remain a tribute to the breadth of Ehrlich's side-chain theory of 1897 that within it lay the seeds of a concept of idiotypes, anti-idiotypes, and anti-idiotype immunoregulation that would foreshadow modern developments in this area by almost seventy years. It is, further, a reflection of the greatness of Ehrlich's intellect that he was able to extract from this theory all of its logical implications, and to construct a conceptual edifice that anticipated in almost every major respect those features of idiotype–anti-idiotype phenomenology that have been elucidated during the past few decades. What is most surprising about the earlier theoretical construct is not that it was subsequently disproved, but rather that it was so completely effaced from the collective memory of immunologists that no mention of it is to be found in modern writings in this field.[79]

Notes and references

1. Boorstin, D.J., *The Discoverers*, New York, Random House, 1983, p. 94 ff.
2. Ehrlich, P., *Klin. Jahrb.* **60**:299, 1897 (English translation in *The Collected Papers of Paul Ehrlich*, Vol. 2, New York, Pergamon Press, 1957, pp. 107–125).
3. Jerne, N.K., *Proc. Natl Acad. Sci. USA* **41**:849, 1955.
4. Burnet, F.M., *Austral. J. Sci.* **20**:67, 1957, and *The Clonal Selection Theory of Antibody Formation*, Cambridge, Cambridge University Press, 1959.
5. See Ehrlich's Croonian Lecture before the Royal Society; Ehrlich, P., "Croonian Lecture – on Immunity…," *Proc. R. Soc. Lond.* **66**:424, 1900, and Chapter 3.
6. Najjar, V.A., and Fisher, J., *Science* **122**:1272, 1955; Najjar, V.A., *Physiol. Rev.* **43**:243, 1963.
7. Milgrom, F., and Dubiski, S., *Nature* **179**:1351, 1957; Dubiski, S., *Folia Biol. (Cracow)* **6**:47, 1958. See also Milgrom, F., Dubiski, S., and Wozniczko, G., *Nature* **178**:539, 1956.
8. Abruzzo, J.L. and Christian, C.L., *J. Exp. Med.* **114**:791, 1961; McCluskey, R.T., Miller, F., and Benacerraf, B., *J. Exp. Med.* **115**:253, 1962; Williams, R.C., Jr., and Kunkel, H.G., *Ann. NY Acad. Sci.* **124**:860, 1965.
9. Oudin, J., *C. R. Acad. Sci.*, **242**:2606, 1966; Grubb, R., *Acta Path. Microbiol. Scand.* **39**:195, 1956.
10. Cohen, S., and Porter, R.R., *Adv. Immunol.* **4**:287, 1964; Kindt, T.J., *Adv. Immunol.* **21**:35, 1975.
11. Oudin, J., and Michel, M., *C. R. Acad. Sci.* **257**:805, 1963.
12. Gell, P.G.H., and Kelus, A., *Nature* **201**:687, 1964.
13. Oudin, J., *Proc. R. Soc. B.* **166**:207, 1966.
14. Kunkel, H.G., Mannik, M., and Williams, R.C., *Science* **140**:1218, 1963. See also Kunkel, H.G., *Harvey Lect.* **59**:219, 1965.
15. Slater, R.J., Ward, S.M., and Kunkel, H.G., *J. Exp. Med.* **101**:85, 1955.
16. Gell, P.G.H., and Kelus, A., *Adv. Immunol.* **6**:461, 1967, p. 476.
17. Gell and Kelus, note 16, p. 476.
18. Apart from the original laboratories of discovery, only those of Ramseier, H., and Lindenmann, J. (*Pathol. Microbiol.* **34**:379, 1969), of Cosenza, H., and Köhler, H. (*Proc. Natl Acad. Sci. USA*, **69**:2701, 1972), and of Hart, D.A., Wang, A.L., Pawlak, L.L., and Nisonoff, A. (*J. Exp. Med.* **135**:1293, 1972) made a significant early commitment to idiotypology, from which emerged important contributions to the field, extensively reviewed as early as 1971 by Hopper, J.E., and Nisonoff, A. (*Adv. Immunol.* **13**:57, 1971).

19. Jerne, N.K., in *Ontogeny of Acquired Immunity*, Ciba Foundation Symposium, London, 1971, pp. 1–15.
20. Jerne, N.K., *Ann. Immunol. (Paris)* **125C**:373, 1974; *Harvey Lect.* **70**:93, 1974.
21. Lindenmann, J., *Ann. Immunol. (Paris)* **124C**:171, 1973.
22. Mitchison, N.A., in Landy, M., and Braun, W., eds, *Immunological Tolerance*, New York, Academic Press, 1969, p. 149; see also Katz, D.H., and Benacerraf, B., *Adv. Immunol.* **15**:1, 1972.
23. Cohn, M., *Cell. Immunol.* **61**:425, 1981.
24. While Jerne's network concept appeared to provide a rational theoretical basis for immunoregulation, it was in fact preceded by isolated reports of anti-idiotype suppression of the immune response: see, for example, Lewis, M.G., Phillips, T.M., Cook, K.B., and Blake, J., *Nature* **232**:52, 1971; Cosenza and Köhler, note 18; Hart, Wang, Pawlak, and Nisonoff, note 18; Pawlak, L.L., Hart, D.A., and Nisonoff, A., *J. Exp. Med.* **137**:1442, 1973; see also Greene, M.I., Nelles, M.J., Sy, M.-S., and Nisonoff, A., *Adv. Immunol.* **32**:253, 1982.
25. Eichman, K., and Rajewsky, K., *Eur. J. Immunol.* **5**:661, 1975.
26. The pioneering work on the idiotypes of lymphocyte receptors was done by Ramseier and Lindenmann, note 18; *Transplant. Rev.* **10**:57, 1972; *Immunol. Rev.* **34**:50, 1977); see also Woodland, R., and Cantor, H., *Eur. J. Immunol.* **8**:600, 1977; Nisonoff, A., Shyr, T.J., and Owen, F.L., *Immunol. Rev.* **34**:89, 1977.
27. Adbou, N.I., Wall, H., Lindsley, H.B., Halsey, J.F., and Suzuki, T., *J. Clin. Invest.* **67**:1297, 1981.
28. Zanetti, M., and Bigazzi, P., *Eur. J. Immunol.* **11**:187, 1981.
29. Cohen, P.L., and Eisenberg, R.A., *J. Exp. Med.* **156**:173, 1982.
30. Neilson, E.G., and Phillips, S.N., *J. Exp. Med.* **155**:179, 1982.
31. Wigzell, H., Binz, H., Frischknecht, H., Peterson, P., and Sege, K., in Rose, N.R., Bigazzi, P., and Warner, N.L., eds, *Genetic Control of Autoimmune Disease*, New York, Elsevier, 1978, p. 327; Rajewsky, K., and Takemori, T., *Ann. Rev. Immunol.* **1**:569, 1983; Rajewsky, K., Takemori, T., and Müller, C.E., *Progress Immunol.* **5**:533, 1983; Thorbecke, G.J., Bhogal, B.S., and Siskind, G.W., *Immunol. Today* **5**:92, 1984.
32. Eichman, K., *Adv. Immunol.* **26**:195, 1978; Rajewsky, K., and Eichman, K., *Contemp. Topics Immunobiol.* **7**:67, 1977.
33. In addition to those listed above, see also: Urbain, J., Wuilmart, C., and Cazenave, P.A., *Contemp. Topics Molec. Immunol.* **8**:113, 1980; Janeway, C.A., in Sercarz, E., and Cunningham, A.J., eds, *Strategies of Immune Recognition*, New York, Academic Press, 1980, pp. 157–177; Bona, C., ed., *Lymphocyte Regulation by Antibodies*, New York, Wiley, 1981; Möller, G., ed., "Idiotypic networks," *Immunol. Rev.* **79**:1984; Janeway, C.A., Sercarz, E., and Wigzell, H., eds, *Immunoglobulin Idiotypes*, New York, Academic Press, 1981; Bona, C.A., and Kohler, H., eds, "Immune networks," *Ann. NY Acad. Sci.* **418**, 1983; Greene, M.I., and Nisonoff, A., eds, *The Biology of Idiotypes*, New York, Plenum, 1984; Kohler, H., Urbain, J., and Cazenave, P., eds, *Idiotypy in Biology and Medicine*, New York, Academic Press, 1984.
34. Behring, E., and Kitasato, S., *Deutsch. med. Wochenschr.* **16**:1113, 1890; Behring, E., and Wernicke, E., *Z. Hyg. Infektionskr.* **12**:10, 45, 1892.
35. Ehrlich, P., *Deutsch. med. Wochenschr.* **17**:976, 1218, 1891.
36. Bordet, J., *Ann. Inst. Pasteur* **12**:688, 1899.

37. A significant portion of Vol. 14 of *Ann. Inst. Pasteur* was devoted to immune "cytotoxins." The francophone use of these terms, and indeed of a completely different immunologic vocabulary from that of most of their German colleagues, has led to later misinterpretations, especially of the work of Karl Landsteiner, as discussed in Chapter 14.

38. Metchnikoff, E., *Virchows Arch.* **96**:177, 1884. Metchnikoff elaborated his theory more fully in *Lectures on the Comparative Pathology of Inflammation*, London, Keegan, Paul, Trench, Trübner, 1893, reprinted by Dover, New York, 1968.

39. Besredka, A., *Ann. Inst. Pasteur* **15**:785, 1901.

40. Ehrlich, note 2.

41. See, for example, Parascandola, J., and Jasensky, R., *Bull. Hist. Med.* **48**:199, 1974; Rubin, L.P., *J. Hist. Med.* **35**:397, 1980; Silverstein, A.M., *Paul Ehrlich's Receptor Immunology*, San Diego, Academic Press, 2002; a detailed summary of Ehrlich's side-chain theory has been presented in Chapter 3.

42. Ehrlich, P., note 5; Ehrlich, P., and Morgenroth, J., *Berl. klin. Wochenschr.* **36**:6, 1899 (English translation in *Collected Papers*, Vol. 2, p. 150).

43. Ehrlich, P., and Morgenroth, J., *Berl. klin. Wochenschr.* **36**:481, 1899 (English translation in *Collected Papers*, Vol. 2, p. 165); Ehrlich, P., and Sachs, H., *ibid*, **39**:297, 335, 1902.

44. The Ehrlich–Bordet debates are discussed more fully in Chapter 2.

45. Bordet's position is best summarized in his famous book *Traité de l'Immunité dans les Maladies Infectieuses*, Paris, Masson, 1920.

46. Bordet, J., *Ann. Inst. Pasteur* **14**:257, 1900.

47. Bordet, J., *Ann. Inst. Pasteur* **15**:303, 1901.

48. Ehrlich, P., *Berl. klin. Wochenschr.* **40**:793, 825, 848, 1903.

49. Ehrlich and Morgenroth, note 43.

50. Bordet, J., *Ann. Inst. Pasteur* **13**:273, 1899; *ibid.* **14**:257, 1900.

51. Kossel, H., *Berl. klin. Wochenschr.* **35**:152, 1898.

52. Camus, L., and Gley, E., *Ann. Inst. Pasteur* **13**:779, 1899.

53. Ehrlich and Morgenroth, note 43, p. 170.

54. Ehrlich, note 5; *Collected Papers*, Vol. 2, p. 94.

55. Bordet, note 50.

56. Besredka, note 39.

57. Ehrlich, P., and Morgenroth, J., *Berl. klin. Wochenschr.* **38**:251, 1901 (English translation in *Collected Papers*, Vol. 2, p. 253).

58. Donath, J., and Landsteiner, K., *Münch. med. Wochenschr.* **51**:1590, 1904.

59. Ehrlich, P., *Collected Papers*, Vol. 2, p. 444; Sachs, H., Lubarsch and Ostertag's *Ergeb. Allgem. Path.* **11**:515–644, 1907, p. 565.

60. Widal, G.F.I., and Rostaine, *C. R. Soc. Biol.* **58**:321, 372, 1905.

61. Morgenroth, J., *Zentralbl. Bakt.* **26**:349, 1899.

62. Ehrlich, P., and Morgenroth, J., *Berl. klin. Wochenschr.* **37**:453, 1900 (English translation in *Collected Papers*, Vol. 2, p. 205).

63. Ehrlich and Morgenroth, note 62, p. 208.

64. Pfeiffer, R., and Friedberger, E., *Berl. klin. Wochenschr.* **39**:4, 1902.

65. von Wassermann, A., *Z. Hyg. Infektionskr.* **37**:173, 1901.

66. Sachs, H., *Ergeb. Allgem. Path. Anat.* 7:714, 1901, p. 768.

67. See, for example, Müller, P.T., *Zentralbl. Bakt.* **29**:175, 860, 1901; Wechsberg, F., *Z. Hyg. Infektionskr.* **39**:171, 1902.

68. Bordet, note 46.

69. Pfeiffer and Friedberger, note 64.
70. Ehrlich, P., *Collected Papers*, Vol. 2, p. 445.
71. Bordet, J., and Gengou, O., *Ann. Inst. Pasteur* **15**:289, 1901.
72. Gengou, O., *Ann. Inst. Pasteur* **16**:734, 1902; Moreschi, C., *Berl. klin. Wochenschr.* **42**:1181, 1905, and **43**:100, 1906; Gay, F., *Zentralbl. Bakt.* **39**:603, 1905; Muir, R., and Martin, W.B.M., *J. Hyg.* **6**:265, 1906.
73. Neisser, M., and Sachs, H., *Berl. klin. Wochenschr.* **42**:1388, 1905.
74. Detré, L., *Wien. klin. Wochenschr.* **19**:619, 1906.
75. von Wassermann, A., and Bruck, C., *Deutsch. med. Wochenschr.* **32**:449, 1906; von Wassermann, A., Neisser, A., Bruck, C., and Schucht, A., *Z. Hyg.* **55**:451, 1906.
76. Ehrlich, P., *Collected Papers*, Vol. 2, p. 446.
77. Bordet, J., *Ann. Inst. Pasteur* **18**:593, 1904.
78. Sachs, note 59; Kraus and Levaditi's *Handbuch der Technik und Methodik der Immunitätsforschung*, Vol. II, Jena, Fischer, 1909, pp. 895–1075; Kolle and Wasserman's *Handbuch der Pathogenen Mikroorganismen*, Vol. II(2), Jena, Fischer, 1913, pp. 793–946.
79. To the best of our knowledge, R.S. Schwartz (*Progr. Allergy* **35**:1, 1984) is alone in having hinted that Ehrlich's work with "anti-immune body" resembled modern anti-idiotype interests.

11 Transplantation and immunogenetics

One of the distinguishing marks of modern science is the disappearance of sectarian loyalties. ...Isolationism is over; we all depend upon and sustain each other.

P.B. Medawar

In his essay "Two Conceptions of Science,"[1] Medawar suggests that biology before Darwin was almost all facts, and that the difficulty of dealing with an ever-increasing factual load caused the scientist to become ever narrower and more specialized (read "isolationist"). One of the characteristics of modern science is the willingness – indeed, the necessity – of the practitioner to cross disciplinary boundaries. This trend has surely accelerated in the biomedical sciences since the 1960s, and immunology has been one of the more important catalysts of this change. The molecular biologist studies the immunoglobulin gene superfamily; the oncologist studies T and B cell subsets; the internist and neurologist study autoimmune diseases and HLA predispositions; and everyone uses monoclonal antibodies and immunoassays. Indeed, it is becoming increasingly more difficult to know how to identify oneself in the now perhaps outmoded disciplinary terms.

Plate 11 Macfarlane Burnet (1899–1985) and Peter Medawar (1915–1987)

A History of Immunology, Second Edition
ISBN: 978-0-12-370586-0

Copyright © 2009, Elsevier Inc.
All rights reserved

   Just as one of the advantages of recent ecumenism is the rapid exchange of
information among disciplines, so one of the disadvantages of early isolationism
was the lack of recognition of significant conceptual advances in other disciplines. Perhaps the best example of this radical shift in the way that science
operates is to be found in the history of the transplantation of tissues and organs.
Five separate specialties are involved in this story, each with its own agenda.
The surgeons had for centuries been attempting various types of transplantation, and by World War I had more than hinted at an immunologic explanation
for their failures; it was they who would help to stimulate the renaissance in
transplantation following World War II. The tumor specialists wanted to
understand and to cure cancer, and to this end studied transplantable tumors; by
1916, they had described substantially all of the phenomenology and "rules" of
graft acceptance and rejection as we understand them today, but no one paid
attention. The Mendelian geneticists entered the picture in the 1920s and 1930s,
and began to characterize histocompatibility in scientific terms; it was their next
generation that would define the full significance of the HLA complex. After
World War II, a group of biologists (led by zoologist Peter Medawar) became
interested in transplantation; it was they who provided the scientific base upon
which modern transplantation biology is built. Finally, those who might have
called themselves immunologists paid little attention to this early work; they only
entered the fray in the late 1950s and early 1960s, in yet another reflection of the
major shift that occurred in immunology, from chemical to more biological
interests.
   We will, in this chapter, outline the early history of tissue transplantation. We
shall also attempt to explain, in terms of the shifting aspirations and disap-
pointments within the several disciplines, why the "laws of transplantation" had
to be discovered twice within the space of some forty years.

Transplantation biology

The surgeons begin

In 1597, the famous surgeon Gaspare Tagliacozzi of Bologna wrote that "the
singular character of the individual entirely dissuades us from attempting this
work [tissue transplantation] on another person..."[2] We do not know what
stimulated Tagliacozzi to write these prescient words a full three centuries before
the biological basis of individuality was firmly established. Since he himself
recorded the successful repair of a lost nose using a pedicle flap autograft from
the patient's own arm (a procedure still in use today), it is not unlikely that he
attempted similar procedures from allogeneic donors, with unhappy results. In
Tagliacozzi's century, medicine was still a curious mixture of primitive science
and of superstition, and the already long history of attempts at tissue transplantation reflected more the latter qualities than the former. Thus, instances of
miraculous transplants were recorded, such as that involving replacement of

a leg by Saints Cosmas and Damian, so well recorded by such famous artists as Fra Angelico and Ambrosius Franken. On the darker side, medieval bestiaries pictured monsters that represented chimerical mixtures of parts of different animal species, and it was suspected that some of them might have been put together by satanical surgeons.

As the centuries passed, medical science slowly replaced medical superstition, and the dream of replacing diseased or missing tissues with healthy ones was advanced with increasing frequency.[3] The ever-daring surgeons attempted skin transplantation for the most part, and, while success was not infrequent with autografts, the results were quite contradictory when allografts were employed and almost uniformly negative with xenografts. It was the ophthalmic surgeons, however, who led the way with successful transplantation of the cornea. In what may have been the first successful penetrating (full thickness) corneal allograft, the Irishman Samuel Bigger reported in 1837 the successful transplantation of an allogeneic cornea into the blind eye of a pet gazelle – an operation which he performed while a prisoner in Egypt.[4] Throughout the nineteenth century, continuing technical improvements and increasingly frequent trials brought a higher success rate among animals, and finally, in 1906, the first entirely successful case of a corneal allograft in the human was reported.[5] Therapeutic transplantation of the cornea thenceforth became a more-or-less standard procedure in the practice of ophthalmology, although no theoretical foundation existed that might explain why corneal transplantation should succeed and skin not, or why some corneal grafts were in fact rejected.

Meanwhile, attempts to transplant skin and other tissues continued in both animals and man. In 1902 Alexis Carrel perfected the technique of vascular anastomosis and, beginning in 1905, reported the experimental transplantation of limbs, kidneys, and other organs.[6] It quickly became apparent that while autografts generally succeed, allografts most often fail. He was forced to conclude that while the technical problems of organ grafting had been solved, "from a biological standpoint no conclusion has thus far been reached, because the interactions of the host and of the new organ are practically unknown." The increasing appreciation that the resistance to foreign grafts is systemic and somehow humoral in nature led to the repeated suggestion by surgeons that an immune response of the "anaphylactic type" was somehow responsible for graft rejection.[7] But despite the demonstration of second-set skin graft rejection in the human as early as 1924,[8] the successful exchange of skin between identical twins in 1927,[9] and the study of parabiotically united animal pairs as early as 1909,[10] no useful generalizations appear to have been drawn by the surgeons that might point the way to further elucidation of mechanisms involved. Thus, save for the ophthalmologists who continued with their moderately successful practice of corneal allotransplantation, plastic and other surgeons seemed to have concluded by the late 1920s that tissue and organ transplantation was impracticable – a fallen banner that would not be picked up and waved by them for another twenty years.

The tumor researchers continue

It is interesting to compare the relative contributions to the science of transplantation biology made by the surgeons on the one hand, and by the tumor workers on the other. Both had eminently practical ends in sight: the surgeons wished to repair defective tissues and malfunctioning organs by replacement, while the tumor specialists wanted to cure cancer by specific eradication. But surgeons are an admittedly practical group, and even those among them who engaged in transplantation research were at the same time practicing clinicians (almost the lone exception, Alexis Carrel, was a physiologist by training). On the other hand, those investigators interested in tumor biology were for the most part not clinical oncologists, but rather basic scientists attracted to the field from careers in anatomy, experimental pathology, physiology, etc. For them, success lay as much in understanding the basic mechanisms involved as in "solving" the problem. Indeed, their upbringing in the basic sciences had taught them that practical results most often emerge from strong theoretical underpinnings. It was perhaps this more basically scientific approach that enabled the tumor investigators not only to work out in detail the phenomenology and rules governing the transplantation biology of tumors, but indeed to contribute more to the understanding of skin and organ transplantation than had the surgeons.

The demonstrations during the last decades of the nineteenth century that immunization might protect against infectious diseases and that antibodies might be employed with sometimes startling therapeutic efficacy caught the attention of the medical world; it was only natural that those interested in cancer should ask whether similar approaches might not also help to solve their problems.[11] However, early efforts at preventive immunization or serum therapy of naturally-occurring tumors in man and animals met with almost uniform failure. It was only with the development of transplantable tumor lines in experimental animals[12] that a pathway was opened for the study not only of tumor pathophysiology, but of tumor immunology as well. From the very outset of these investigations with transplantable tumors, it was noted that there were very strict limitations on the ability of the tumor transplant to survive in the new host. The experimental work of barely a decade was summarized in 1912 in a remarkable book by Georg Schöne entitled *Heteroplastic and Homoplastic Transplantation*.[13] The importance attached to these studies is attested to by the fact that this summary lists almost 500 references. As Schöne made clear, tumor researchers had already established by 1912 the following general rules governing the acceptance or rejection of tumor grafts (and Schöne's coinage of the term "transplantation immunity" makes it clear what he believes to be the underlying basis for these rules):

1. Transplantation into a foreign species (heteroplastic = xenogeneic) invariably fails
2. Transplantation into unrelated members of the same species (homoplastic = allogeneic) usually fails
3. Autografts almost invariably succeed

4. There is a primary take and then delayed rejection of the first graft in the allogeneic recipient

5. There is an accelerated rejection of a second graft in a recipient that had previously rejected a graft from the same donor, or of a first graft in a recipient that had been preimmunized with material from the tumor donor

6. The closer the "blood relationship" between donor and recipient, the more likely is graft success.

Here, in 1912, are the "laws of transplantation" substantially as we understand them today. But Schöne's book does more than summarize a set of observations applicable only to the arcane world of tumor transplantation. He, and the other tumor specialists, went further and generalized these observations to encompass the transplantation of skin and organs! The reasons why the tumor investigators made such important contributions to the science of skin and organ transplantation (far beyond that accomplished by the surgeons) are interesting. It was not clear to them at the outset whether there was something unique about tumors that might differentiate them in this sense from other normal tissues; thus, the tumor workers felt obliged to employ skin and other tissues as controls. They quickly discovered that the phenomenology of graft rejection was similar in both cases. In fact, skin would presensitize a recipient for "second-set" rejection of a tumor, and vice versa. Further, they observed in the course of these studies of normal tissues that skin grafts fail more consistently than do grafts of other organs such as kidney. In a further review entitled "Tumor Immunity" in 1916 by E. E. Tyzzer,[14] the general findings reported by Schöne were confirmed and several further important advances were reported. Now the conclusions could be drawn in frank immunologic terms:[15]

> The degree of immunity which develops thus depends on the foreignness of the immunizing cell with respect to the organism into which it is introduced. The more foreign cells accordingly serve as the more effective and the more closely related cells as the less effective antigens.

Moreover, it was now noted that presensitization for "second-set rejection" requires *living* cells; that cytotoxic antibodies cannot be found; that "the delayed reaction of host tissue is difficult to explain except on the hypothesis that an immune body [of some sort] has been produced;" that lymphocytes predominate at the rejection site, i.e. "the reaction is not merely *exudative*, but is *proliferative* as well;" and that there is no tissue specificity, but rather "racial specificity with respect to the genetic origin...of the antigens."

With the availability of an inbred strain of mice, the "Japanese waltzing mouse," Tyzzer was able to report a further extension of these studies.[16] Breeding experiments showed that the F1 hybrid generation obtained by crossing two unrelated strains would accept parental tumors, that the backcross generation (F1 by tumor donor, i.e., parent–offspring mating) would also accept transplants, but that there was a decreasing incidence of acceptance in further sibling crosses. These results permitted Tyzzer to conclude that "it is quite apparent from these data that susceptibility is not inherited as a single Mendelizing factor. The only hypothesis… [nonsusceptibility] is dependent upon the presence of a complex of independently inherited unit factors."[17] Indeed, his data appeared to him to justify the existence of at least twelve to fourteen independently inherited factors to explain the ratios of tumor acceptance to tumor rejection.

It was during this period, and partly as a result of the tumor work described above, that the prominent biologist Leo Loeb undertook his long series of investigations into the basis of the individuality of tissues, as exemplified by the results of these transplantation studies. The results of this lifetime of work are well summarized in two major monographs, one on "Transplantation and Individuality" in 1930, and the other on *The Biological Basis of Individuality* in 1945.[18] Loeb recognized the genetic basis of individual differences and transplantation incompatibility, but would not ascribe the latter to specific immunologic mechanisms. Rather, he argued that there exists a specific capability for what we would today call self–nonself discrimination, but at the level of the somatic cell. Thus, a foreign tissue could not make the connections necessary to its physiologic survival in the new environment to which it had been transplanted.

Loeb's arguments were forceful, and probably exerted a degree of restraint upon truly immunologic speculation about transplantation, but evidence continued to mount in favor of an immunologic interpretation – especially from studies on the preimmunization of graft recipients. These studies are summarized in a massive review on "Immunity to Transplantable Tumors," written by William Woglom in 1929.[19] This review summarized the contributions of no fewer than 600 reports published since the appearance of Schöne's book in 1912. Broad confirmation was reported of the following observations: that *all* tissues would immunize for accelerated graft rejection;[20] that only living tissues would serve – dead cells were ineffective; that passive transfer of tumor immunity could not be achieved with serum (and that newborns of immune mothers were not themselves immune); and that tumor rejection was not accompanied by detectable cytotoxins or other "antitumor" antibodies. It was also shown that transplantation immunity is systemic and not local, but that certain sites such as the brain might be exempt from the systemic sensitization and thus able to support the grafted tissues (in current terminology, an "immunologically privileged site"). It had by this time also been shown that while washed erythrocytes would not immunize a recipient for graft rejection, whole blood would, and that the activity resided in the leukocyte moiety.[21]

One of the more fascinating sections of Woglom's review summarized a large number of studies of the cytologic changes that accompany tumor graft rejection. It had apparently been Da Fano[22] who first called attention to the fact that the bed of a rejecting tumor allograft characteristically contained large numbers of lymphocytes, rather than the polymorphonuclear leukocytes that might have been expected to surround dying cells. As we noted above, Tyzzer commented on the same observations in his 1916 review, and emphasized the fact that the lymphoid response was proliferative as well as infiltrative. Tyzzer's attention was undoubtedly drawn to this observation by the work of James Murphy, who contributed to the field a truly impressive series of reports starting in 1912. It was probably his work with Peyton Rous on the histopathology of fowl sarcoma rejection[23] that stimulated Murphy to pursue these studies, which he summarized in extensive detail in a monograph on *The Lymphocyte in Resistance to Tissue Grafting, Malignant Disease, and Tuberculous Infection.*[24] Not only did Murphy assign to the lymphocyte the predominant role in the rejection of tumor grafts, but he also tested and confirmed the speculation in an elegant series of experiments. First, he anticipated later studies on the ontogeny of the immune response by showing that a tumor transplanted into the chick embryo might enjoy uninhibited growth until the eighteenth day of embryonic life, when the tumor would undergo spontaneous rejection.[25] To prove that this maturational event involved the lymphocyte, Murphy demonstrated that tumor rejection could be induced in even younger embryos by the co-transplantation of bits of adult spleen or preparations of free "lymphocytes." In a similar manner, he showed that while a tumor might grow in the "privileged site" of the brain, rejection even in that site could be induced by local inoculation of lymphoid tissue preparations.

To further support his argument on the importance of lymphocytes in graft rejection, Murphy undertook to manipulate both the systemic lymphocyte levels in the grafted host as well as those available at the local site of graft implantation. Thus, he demonstrated that nonspecific stimulation of a lymphocytosis in the host would retard tumor growth and accelerate rejection. On the other hand, lymphopenia should inhibit the rejection of allogeneic tumors. To the latter end, Murphy was able to show that X-irradiation of the host, resulting in severe lymphopenia, would inhibit the development of immunity to the graft, and delay or even obviate the rejection process. Murphy and Sturm later showed that similar X-ray treatment would depress antibody formation[26] – an observation that had earlier been made by Ludwig Hektoen.[27]

Having implicated the lymphocyte in the rejection of tissue grafts, little more could be said on this subject at that time. The lymphocyte was then a cell of mysterious ancestry and function, and indeed even twenty-five years later Arnold Rich could say little about it. Writing in the context of the pathogenesis of tuberculosis, Rich said, as late as 1951:[28]

> *There are numerous reasons…for believing that the lymphocytes play a role of importance in acquired resistance, though the precise manner in which they act is still obscure, chiefly because so little is known about the function of these cells. The lack of more adequate information regarding the function of the lymphocyte is one of the most lamentable gaps in medical knowledge. Produced in enormous numbers…these cells undoubtedly must serve the body in a most important way; and yet little is known of their function.*

Enter the geneticists

It is a testimony to the hope that the study of transplantable tumors might lead to a more general and useful approach to the problems of cancer that for almost forty years geneticists examined transplantation almost entirely in the context of tumors. One of the first of the young geneticists to enter this field was Clarence C. Little, who very early became interested in the genetic differences that control the response of mice to transplantable tumors. In 1916 he published his first paper in this field with Tyzzer,[29] a follow-up study of Tyzzer's earlier work on the susceptibility to tumors of the Japanese waltzing mouse. Little maintained his interest in tumor genetics even after becoming, at age thirty-three, the President of the University of Maine. While limited in his ability to continue his experimental studies, he maintained an active interest in the field, as is witnessed by a review on the genetics of tissue transplantation written in 1924.[30] As he says:

> This paper has its justification in the fact that the subject of the genetics of tissue transplantation is likely to become in the not distant future of far greater general importance. ...There has not been brought to experimental biologists any considerable amount of evidence as regards the type of inheritance found in the case of tissue transplantation.

In fact, this paper was a strong criticism of Loeb's genetics, and of the basis for Loeb's conclusions on the basis of incompatibility between donor and host (a criticism that many subsequent workers shared in evaluating Loeb's work).

Throughout his career, Little was wedded to the idea that careful study of genetically homogenous animals would provide one of the more profitable approaches to the problems of cancer, and to this end he founded the Jackson Memorial Laboratory at Bar Harbor, Maine, in 1929. "The founding of the fledgling laboratory was to use the mice [inbred strains of which Little and others had developed] in research against that continuing scourge, cancer. In the differences of resistance and susceptibility between the inbred strains, they believed, might lie an answer to cancer's causes."[31] One of the more significant events during the early years of the Jackson Laboratory was the hiring in 1935 of George D. Snell, as committed as was Little to the use of inbred mice for the study of cancer. To this end, Snell "invented" congenic mouse strains, carefully selected from a parental inbred stock to differ at but a single locus from its congenic cousins. Very quickly, Snell discovered a locus intimately related to the rejection of tumor grafts that he labeled H (for histocompatibility). When Peter Gorer, in England, discovered a hemagglutinating antibody associated with the rejection of tumor grafts,[32] it appeared that the long-sought cytotoxic antibody responsible for tumor rejection might be associated with the blood group antigens. It was soon established that such an antibody response was not a general feature of graft rejection, but not before the antigen involved was named, by Gorer, antigen II, and it was established that the gene for the production of this antigen was located at Snell's H locus – hence the term H-2, which would eventually define an entire complex of murine histocompatibility genes.

Transplantation research in the 1930s

Having reviewed the impressive advances made in the previous quarter-century in the understanding of the immunologic basis for tissue graft rejection, and the "rules of transplantation" that were formulated, we may now inquire why so much of this work appeared to have been forgotten, and why Medawar's studies in the 1940s were received as "new" discoveries. In attempting an explanation of these events, we must consider separately the three principal disciplines that had been engaged in research on transplantation up to this time: the surgeons, the tumor specialists, and the geneticists.

Despite the technological improvements introduced by Alexis Carrel and others, surgeons (save for the ophthalmologists with corneal transplantation) had for decades seen all of their attempts at skin and organ transplantation fail in the face of the rejection process. The only method known to suppress the immune response to grafted tissue – that of whole body X-irradiation – had been shown to be at least partially successful in experimental animals, but was apparently deemed too radical an approach to be employed in man. Thus transplantation seemed doomed to disappointment as a clinically useful tool. The ever-practical surgeons therefore "gave up" on the procedure, as is witnessed by the decreasing reports in the literature on work along these lines, starting as early as the late 1920s.

The position of the tumor biologists was in many respects similar to that of the surgeons. Early on, approaches employing transplantable tumors had appeared to offer a fruitful avenue towards the solution of the problem of cancer, and indeed much valuable information was obtained about the immunology and genetics underlying the rejection of tumor grafts. But having worked out the basic phenomenology of the process by the mid-1920s, little further progress was made or appeared likely, so a "solution to the cancer problem" began to appear beyond reach – at least employing these approaches. We may therefore conclude that, like the surgeons, tumor researchers lost faith in their approach to tumor biology via transplantation, and moved on to other more promising areas.

The shift in the interest of the geneticists was somewhat more subtle than those described above. Since they too had entered the field of transplantation in the context of tumor biology, they were undoubtedly affected by the growing disenchantment of the cancer researchers. But something else occurred which merits attention, and this is well reflected in the work of George Snell. At the outset, Snell probably viewed (with Clarence Little) the inbred mouse as the perfect tool to study tumor graft rejection – i.e., as a fruitful approach not only to therapy, but to understanding initial susceptibility as well. If the early work to identify the histocompatibility locus in mice was aimed at understanding the immune response to tumor grafts, it quickly lost its oncologic and even immunologic motivations and became a study in pure genetics: in analyzing the size of the growing histocompatibility complex; in assessing the extent of the polymorphism at each locus; and in establishing the rules of segregation of these genes in backcrosses.[33] Only later would the studies return to the realm of transplantation biology, when their practical applicability to tissue typing and histocompatibility matching was appreciated.

We may therefore conclude that the science of tissue transplantation had, by the latter half of the 1930s, lost its appeal to those disciplines which had earlier maintained an active interest in and great hopes for this subject. We may also note in passing that despite the significant contributions made during the teens and 1920s to the immunology of tissue transplantation, no investigator who might have termed himself an immunologist had been involved in these efforts, and even the textbooks and reviews of immunology of that period failed to mention the studies of the nature of tumor or other tissue graft rejection.[34] New data, new technologies, and above all a new point of view would be required to rekindle the interest of research scientists in the value of transplantation studies.

The renaissance of transplantation biology

Among the many horrors that accompanied World War II, with its more mechanized methods and its incendiary bombings of cities, was a marked increase in the numbers of burn victims seen in both military and civilian hospitals. Such patients would previously have succumbed to such extensive burns, but with the advent of antibiotics to control infection and of the use of skin autografts to assist in the healing process, there was now hope for such individuals. Where the burn area was extensive, an adequate source of autograft skin might not be available; thus, while it was generally understood that skin homografts would invariably be rejected, the war provided the impetus to re-examine the question of homograft rejection in the hopes of developing techniques to circumvent it. It was in this context that zoologist Peter B. Medawar became interested in skin grafting, and was assigned by the War Wounds Committee of the British Medical Research Council to explore this question, first in a clinical setting with Thomas Gibson at the Burn Unit of the Glasgow Royal Infirmary and then using experimental animals at his home base at Oxford University.

Medawar's first paper, with Gibson in 1943,[35] revealed that they possessed a thorough knowledge of the earlier work in this field, including that of Schöne, Woglom, Loeb, and others who had used both tumors and normal tissues. They concluded, however, that the question of the mechanism of homograft rejection was still unsettled. They then reported their results in what would become the hallmark of Medawar's future work: an elegantly designed, carefully executed, and lucidly described report. In this paper, Gibson and Medawar described experiments on a single burn victim, and demonstrated with serial biopsies that:

1. Autografts succeed
2. Allografts fail, after an initial take
3. second-set homografts suffer an accelerated rejection
4. There is little evidence of a "local cellular reaction" (a term presumably employed in the original sense of Leo Loeb).

The authors then concluded that these data "suggest that the destruction of the foreign epidermis was brought about by a mechanism of active immunization."[36]

These results excited Medawar's scientific interests, and[37]

> It did not go unremarked that we were building rather a lot upon the study of the single case, and when I returned to Oxford I felt I should study the whole phenomenon of homograft rejection in laboratory animals to see if this renewed study gave results that would be fully compatible with our hypothesis that the rejection of homografts was an immunologic phenomenon.

The results of the extensive experiments in the rabbit that followed were compiled in two reports to the War Wounds Committee, and published in the *Journal of Anatomy* in 1944 and 1945.[38] In his studies of graft rejection in the rabbit, Medawar was able to confirm and to extend considerably his earlier work with Gibson. In the course of "the hardest stint of work I have ever undertaken in my life," Medawar carefully studied the pertinent parameters of timing, dosage, specificity, first- and second-set rejection, and the clinical and histologic changes that accompany the rejection process. If, in these studies, Medawar had "rediscovered" the laws of transplantation earlier summarized by Schöne in 1912, by Tyzzer in 1916, and by Woglom in 1929, it had been done now with a set of carefully devised and controlled experiments and a mass of supporting data that rendered the conclusions beyond any doubt. More than this, he had now convincingly demonstrated that the rejection process originates systemically and not locally, and his study of the mutual exchange of grafts between a large number of donor–recipient pairs allowed him to conclude that "the homograft reaction is governed by the operation of *at least* 7 antigens freely combined..."[39]

After acknowledging with reservation the immunologic nature of the homograft rejection reaction, Medawar went on to consider the likely mechanism involved. Given the prevailing view of the importance of circulating antibodies in *any* allergic (or hypersensitivity) event, and the absence in the contemporary literature of almost all reference to the notion of cellular immunity (in even its Metchnikovian sense), Medawar's conclusion is understandable. Even so, one must admire the acuity represented by the caveat that he inserted into his conclusion:[40]

> The inflammation which accompanies the homograft reaction in rabbits is very probably of the anaphylactic type. ...Yet, though all of the ingredients of the inflammatory process are present – vascular and lymphatic proliferation, edema, and the mobilization and deployment of mesenchyme cells of every type – the reaction is nevertheless atypical; for the lymphocyte takes the place of the polymorph in the "classical picture." ...It is as yet impossible to judge of the significance of this difference.

It is curious that in these first reports, and in those transplantation studies that followed during the next decade, Medawar did not review again any of the pertinent earlier transplantation literature, contenting himself with indicating that this had been covered earlier in his paper with Gibson, and in another paper published in 1943 in the transiently published *Bulletin of War Medicine*.[41] Not until his Harvey Lecture of 1957[42] did he provide a broad history of transplantation studies, and pay full attention to the earlier work on tumor transplantation and on skin and tissue grafting, and to the studies of Snell and Gorer.

Medawar continued his studies of skin transplantation in a further report in 1946 in the *British Journal of Experimental Pathology*.[43] In this paper, he demonstrated that skin-graft rejection is unaccompanied by the formation of isohemagglutinins for the donor's erythrocytes; that immunization with donor red cells confers no appreciable immunity to skin but that immunization with donor *leukocytes* does; and that the intradermal inoculation of leukocytes is eighteen times more effective than immunization via the intravenous route. As in his two earlier reports, Medawar is here repeating almost *seriatim* the decades-earlier work done with tumor grafts.

Despite the obvious immunologic orientation of these transplantation studies, it would appear from an examination of the indices of the several journals devoted to immunology, and of the programs of such meetings as those of the American Association of Immunologists, that the studies had not yet captured the attention of the immunologic world. There were, however, certain stirrings both within the field of transplantation as well as in other disciplines that would soon bring transplantation into the immunologic fold. Perhaps the most significant event for transplantation was provided by the work of Owen in 1945 on the immunologic consequences of natural vascular anastomoses established between nonidentical cattle twins *in utero*.[44] Owen showed by serologic tests that these animals are erythrocyte chimeras – i.e., they possess mixtures of their own and of their twin's red cells, but fail to produce isoantibodies against the foreign component. Burnet and Fenner called attention to this finding, and to its implication for an understanding of the immune response to antigenic stimulus, in their 1949 edition of *The Production of Antibodies*.[45] But it was the demonstration in Medawar's laboratory in 1951 and 1952 that such cattle are unable to reject one another's skin[46] that signaled that here was a finding not only of profound theoretical interest to immunology, but of possible practical interest to tissue transplanters as well.

It was these findings that focused the attention of Medawar's laboratory on this problem, which soon eventuated in the reports of Billingham, Brent, and Medawar on the experimental production in laboratory animals of the phenomenon of immunological tolerance,[47] akin to that seen by Owen in cattle twins. Here finally was at least the theoretical promise that the immune response to foreign tissue grafts might be overcome. This observation, together with subsequent demonstrations that the host response might be inhibited by the use of nitrogen mustard or corticosteroid therapies, fostered a renewal of interest by the surgical community in the possibilities of tissue and organ transplantation. For some years thenceforth, it was primarily the surgeons who paid attention to Medawar's work, and who organized the national and international symposia that helped to establish the foundations for future work in this field.

Meanwhile, other forces were at work that would help to incorporate transplantation into the mainstream of immunology – or rather, it might be more accurate to say that it was less a change in the nature of transplantation studies than a shift in the course of immunology itself that caused it to encompass the growing field of transplantation. As we saw in Chapter 8 (and will explore further in Chapter 17), thanks in part to Burnet's clonal selection theory; to the increasing realization that autoimmune diseases are real phenomena of great clinical importance; to the discovery of immunologic deficiency diseases; and now to the discovery of immunologic tolerance, the entire field of immunology found itself moving away from its former preoccupation with essentially chemical approaches to antibodies and the problems of specificity, and toward more biological questions of cellular mechanics and disease pathogenesis. A new generation of immunologists would quickly be attracted to the field, to sort out the biological basis of antibody formation, and the molecular and cellular mechanisms responsible for an increasing number of immunogenic inflammatory processes.

Thus, transplantation studies were at once a contributor to the stimulus for this phase-shift in immunologic interests, and one of the beneficiaries of the new movement. Now a concerted effort could be undertaken to relate transplantation immunity to other immunologic phenomena – an undertaking aided immeasurably by the renewal of interest in the implication of delayed-type hypersensitivity as one of the important contributors to immunopathologic reactions. Perhaps the key that enabled this doorway to be opened was provided by the demonstration by Mitchison and by Billingham, Brent, and Medawar in 1954 that transplantation immunity could be adoptively transferred with cells and not with serum from sensitized donors[48] – a finding that immediately put graft rejection into the same category as tuberculin hypersensitivity and contact dermatitis.[49] This view was strongly reinforced by the report of Algire and coworkers that grafts implanted within cell-impermeable chambers evaded rejection even in previously sensitized hosts,[50] but that inclusion in the chamber of immune spleen or lymph node cells would result in graft destruction.[51]

Two other observations contributed to the increasing attention drawn to transplantation studies, both from Medawar's laboratory. The first was that graft immunity could be elicited by *non*living cells, and even by cell extracts.[52] This discovery was quite important, for it removed from transplantation immunology the somewhat mystical quality that had been associated with the earlier belief in the requirement for *vital* cells, and made graft immunogenicity akin to all of the other more familiar antigen systems. The other significant observation was that of the graft-versus-host reaction.[53] The injection of genetically disparate but immunologically competent cells into a recipient would engender differing forms of host damage, depending upon the recipient. In the chick embryo, splenomegaly and death would result; in the neonatal mouse, "runt disease;" in the F1 hybrid given parental lymphoid cells, "F1 hybrid disease;" and in parabiotically attached animals, "parabiosis intoxication."

These clinical processes, and the mechanisms that underlay them, could not fail to excite interest.

As early as his Harvey Lecture of 1957, Medawar could give voice to the conceptual change that had taken place in the previous few years:[54]

> *The balance of evidence suggests that skin transplant immunity and*
> *hypersensitivity reactions of the delayed type are reactions which are*
> *fundamentally cellular as opposed to humoral, and which depend upon the*
> *activation, deployment, and peripheral engagement of the lymphoid cell –*
> *[but he was quick to add, wryly, that this was] a remark whose euphony will, I*
> *hope, distract attention from the fact that we are very ignorant of what these*
> *processes are.*

A further testimony to the fact that it was about this time that transplantation biology was integrated into the discipline of immunology comes from an analysis of how workers in a field perceive themselves, and are perceived by others. Prior to the early 1950s, Medawar and his colleagues directed their work primarily at surgeons and at biologists in general. From the mid-1950s onward, they would increasingly be invited to speak at symposia organized by immunologists and to contribute chapters to immunologic publications. Indeed, before this period, neither Medawar nor his colleagues considered themselves to be "immunologists" – an appellation to which thereafter they would feel entitled.[55]

The changing character of the developing science of tissue transplantation is perhaps best illustrated by the developments in the field recorded in the periodic symposia on transplantation organized by the New York Academy of Sciences.[56] What had started as an exchange predominantly among surgeons (with, admittedly, strong immunological overtones) gained momentum with the addition of geneticists interested in the basis of histocompatibility differences, serologists interested in histocompatibility testing, and immunologists and immunopathologists interested in underlying mechanisms. The marriage of these different disciplines was celebrated in the establishment in 1967 of the Transplantation Society, and the nuptials were recorded in the proceedings of the First International Congress of the Transplantation Society.[57] As the table of contents of this volume attests, not only were the surgeons, the geneticists, and the immunologists now joined together in a common cause, but even the tumor immunologists had finally been brought back into the fold. Indeed, one could see here the beginnings of a new discipline in its own right, that of the "transplantation biologist," with an organization, a set of aspirations, and even a language of its own.

Progress in transplantation research

The first decades following the founding of the Transplantation Society witnessed remarkable advances in the genetics and in the immunobiology of tissue grafting. The principal components of the major histocompatibility complex and the molecular biology of its constituency were well worked out, as were the major aspects of the host response to alloantigens and the underlying mechanisms of allograft rejection. Nevertheless, the *practice* of tissue transplantation even now remains a curiously pragmatic enterprise whose newer directions appear to be directed less by theory than by trial-and-error experiments to determine "what works." Indeed, modern transplantation studies have seemingly given up on the possibilities of success with the very tissue that prompted the renaissance in transplantation – the use of allogeneic skin transplants in burn victims – and are now devoted primarily to improvements in the transplantation of kidneys and other organs. This pragmatic approach to transplantation was recognized as long ago as 1977, when Leslie Brent, in his Presidential Address on the tenth anniversary of the establishment of the Transplantation Society,[58] discussed Roy Calne's view that progress in transplantation would come less from basic immunologic research than from the search for better immunosuppressive drugs.[59] Brent was, at the time, cautiously optimistic, but pointed out that the immunologic solution of graft rejection might involve "a time-scale of progress that is greater than self-interest and our natural urge for human advance demand." Little that has occurred during the thirty years since Brent made this statement suggests that the time-scale has been appreciably foreshortened.

The tissues employed in transplantation

It is now generally conceded that corneal transplantation succeeds in general because the avascular nature of the tissue endows the cornea with a degree of immunologic privilege, more or less isolating it from both the afferent and efferent arcs of the immune response. Skin, on the other hand, by virtue of its rich vascular bed and direct access to lymphatic channels, is highly immunogenic and equally highly susceptible to invasion by those effector cells that mediate the rejection process. Occupying an intermediate position in this hierarchy of immunologic susceptibilities are organs such as the kidney and the heart, which can be transplanted in their entirety by the anastomosis of only a few major blood vessels – a connection that serves at least in part to isolate them from the immune response of the host. The validity of this interpretation was elegantly demonstrated by Barker and Billingham,[60] who showed that even skin grafts might enjoy prolonged survival if transplanted onto a raised dermal pedicle connected to the host only by artery and vein, but not by lymphatic channels.

Yet another type of transplant that offers hope of success is exemplified by the transplantation of pancreatic islet cells for the production of insulin in diabetic recipients. While the basis for the success of this procedure is not fully understood, it may well be related to the small antigenic mass involved in the transplantation of only modest numbers of such cells. It will be recalled that Billingham showed many years ago that even epidermal melanocytes may be transplanted successfully onto the skin, if small enough numbers are employed;[61] presumably, the host does not even "see" (in an immunologic sense) the small antigenic mass, and fails to become sensitized.

Finally, there is yet another type of tissue graft that holds promise of success – that of bone marrow transplantation in the therapy of certain lymphatic leukemias. In this case, substantially the entire immunologic apparatus of the recipient is ablated chemotherapeutically and replaced with the immunocytes of the donor. This approach is currently accompanied by techniques which render it possible to induce immunologic tolerance of the transplanted cells on the part of any residual immunologic competence of the host. However, tolerance of the host is not shared by the donor lymphoid elements, so that systemic graft-versus-host reactions continue to constitute one of the more serious side-effects of this approach, although treatment to obviate this complication has been increasingly successful.

Tissue typing and donor–recipient matching

There exist within the major histocompatibility complex a number of loci which control the production of those antigens that play a major role in inciting the rejection process, and a larger number of loci coding for less important (but nonetheless contributory) histocompatibility antigens.[62] The polymorphism at each of these loci is impressively large, so that the number of different combinations that may exist within an individual makes it extremely difficult to match a recipient to an optimal donor. Two approaches to donor–recipient histocompatibility matching have been developed, each with its adherents. The first involves the assembling of a large library of antibodies specific for the different antigens that may exist at the important loci. Such tests provide an objective assessment of the antigenic differences between donor and recipient, and utilize prior experience to assess the practical importance of a mismatch at a given locus. The second approach is a more functional one, in which a one-way mixed lymphocyte reaction between donor and recipient cells is measured. Here, the extent of the response of recipient lymphocytes to the histocompatibility antigens of the donor provides a measure of the likely response to the grafted tissue.

It is clear that the better the histocompatibility match between donor and recipient, the more optimistic will be the prognosis for the grafted organ. Nevertheless, perfect matches (except between identical twins) are difficult to obtain, so that nationwide and even worldwide organizations have been established to direct available organs to the most promising (i.e., the closest matched) recipients.

Immunosuppressive therapy

Given the availability of donor tissues or organs not perfectly matched to the recipient, a greater or lesser degree of host sensitization with consequent attack on the graft would appear inevitable. The countervailing strategy employed in clinical transplantation studies has sought, therefore, to employ immunosuppressive agents in an attempt to minimize the effects of the rejection process. As we saw above, X-irradiation of the graft recipient was found to be moderately successful in preventing the rejection of tumor transplants in experimental animals and, despite deleterious side-effects, is still occasionally employed as an adjunct to immunosuppressive therapy in man. Advantage has also been taken of the anti-inflammatory action of corticosteroids in moderating the rejection process, a form of therapy especially useful when applied locally to interrupt the rejection of a corneal graft. It is not surprising that many of the chemical anti-metabolites applied to the problem of graft rejection have emerged from developments in cancer chemotherapy, since the aim in both fields is similar: to destroy or inhibit the tumor cell (or the threatening sensitized lymphocyte) without inflicting too much damage on other host cell types. Thus, such drugs as 6-mercaptopurine, methotrexate, 5-fluorouracil, cyclophosphamide, and many others have been applied to the inhibition of graft rejection with varying results. While a number of these drugs have been found to be effective, they are usually accompanied by undesirable and even serious side-effects, since efficacy is in general directly related to cytotoxicity. There has been much interest in the drug cyclosporine A, claimed to be highly effective in controlling graft rejection with but minimal side-effects.[63]

One of the more interesting approaches to immunosuppression in the field of transplantation is based upon an observation in 1937 by Chew and Lawrence that a serum with a powerful antilymphocytic effect *in vivo* can be prepared by immunizing an animal with suspensions of heterologous lymphocytes.[64] Interest in this preparation was revived in the 1960s,[65] and early reports suggested that antilymphocyte (or antithymocyte) serum might inhibit the rejection process without serious side-effects. With the advent of hybridoma techniques for the production of pure and specific monoclonal antibodies, it has become possible to ablate selectively one or another lymphocyte subset in the graft recipient, thus further limiting undesirable damage to the immune apparatus of the host.

The last approach to inhibition of the rejection process that we will mention is especially interesting, since it corroborates nicely the suggestion mentioned above that advances in transplantation biology are based more upon pragmatic observation than upon theoretical prediction. Immunologic theory would hold that great care should be taken that the graft recipient not be presensitized to the histocompatibility antigens of the donor. Since blood transfusion is a frequent accompaniment of organ transplantation, the possible consequences of pre-sensitization by transfusion (or that seen in multiparous women) seemed a thing to be feared. In fact, such presensitization appears now to favor graft survival, and indeed donor-specific transfusion appears to be more effective in this sense than that from unrelated donors.[66] It is not yet clear whether the beneficial effects of this treatment are due to the stimulation of suppressor T cells, or to the production of suppressive anti-idiotypic antibodies.

The promise of immunologic tolerance

Peter Medawar has suggested[67] that if the field of transplantation has borrowed most of its working concepts from orthodox immunology, it has in large measure

repaid that debt by contributing back to immunology the concept of immunologic tolerance. The possibilities inherent in tolerance were already implicit in
the work of Owen on cattle chimeras,[68] as pointed out by Burnet and Fenner,[69]
and its implications for transplantation research were made abundantly clear when
it was demonstrated that these animals were unable to reject one another's skin.[70]
The conditions required for the induction of transplantation tolerance were elaborated upon by Billingham and Brent, by Woodruff and Simpson, and most notably
by Hašek and his colleagues.[71] The generality of this phenomenon and its broad
implications for immunology were first made clear by Hannan and Oyama,[72] who
showed that tolerance may also by induced by nonliving antigens, which led to
extensive experimentation with a variety of simple proteins[73] – a subject that would
have important implications in the study of autoimmune diseases (see Chapter 8).

It was the initial hope and expectation that the induction of tolerance of donor
histocompatibility antigens in a graft recipient would offer the ultimate solution
to the problem of tissue and organ graft survival. However, while tolerance
might be induced readily in the immunologically immature fetus or neonate
(depending upon the species), it was found to be far more difficult to induce this
state in the immunologically mature adult who might require an organ graft.
Again, even if effective tolerance could be attained with simple protein antigens,
the difficulties attendant upon attempts to induce tolerance of the large complex
of histocompatibility antigen differences between donor and recipient render the
process more formidable. Thus far, immunologic tolerance has only proved of
practical value in the special case of bone marrow transplants, where some
practitioners utilize special preparatory regimens that favor the development of
tolerance. Nevertheless, tolerance is still viewed today as the Holy Grail of the
transplantationist, as it was fifty years ago.

There are some findings, however, that suggest that tolerance may participate
more subtly in the survival of organ transplants. A significant number of organ
graft recipients were found to have small numbers of donor leukocytes surviving
in their blood, suggesting the induction of what would be termed "micro-
chimerism," implying some degree of tolerance.[74] Such chimerism appears more
frequently in liver recipients than in those receiving kidneys or hearts, presumably
because of the greater supply of leukocytes in this organ. It has been suggested by
the chief proponent of this concept, Thomas Starzl, that a mild graft-versus-host
reaction may be a necessary prelude to this tolerance induction and its accompanying chimerism.[75]

Immunogenetics[76]

An International Society of Monists was founded in 1906 by Ernst Haeckel and
Wilhelm Ostwald, dedicated to the belief in the ultimate unity of all knowledge.
At a time when vitalism still infused much of biology, and when each discipline
followed its own guiding rules and methodology, the view that all sciences would
ultimately converge and even merge under the general laws of chemistry and

physics seemed somewhat questionable. Nevertheless, the history of much of biologic science in the first three-quarters of the twentieth century seems to point precisely at a convergence of disciplines. As Garland Allen has pointed out in *Life Science in the 20th Century*,[77] "embryology, biochemistry, cytology, and genetics began to come together [in the 1920s and 1930s] to form a unified, cellularly and physiologically oriented view of development." Similarly, as this chapter makes abundantly clear, immunology, genetics, biochemistry, and molecular and cellular biology have become so intimately intertwined that in recent years many investigators have been hard put to know how to identify themselves. In the study of blood groups, of histocompatibility relationships between graft donor and recipient, of the basis for the generation of immunologic diversity and the formation of the immunoglobulin molecule, and of the mechanisms for disease predisposition and resistance, the overlap between genetics and immunology has become increasingly evident, with each making major contributions to the other's progress.

Blood groups

The first report of the existence of naturally occurring isohemagglutinins in man was made by Karl Landsteiner,[78] apparently the result of a chance observation. Ehrlich and Morgenroth had described similar isoantibodies in goats that would lyse the erythrocytes of other goats, this time in animals immunized with red cells in the search for autoantibodies.[79] In a further examination of human sera Landsteiner was able to describe three erythrocyte groups, each of the first two containing its own red cell antigen, whereas the third group had erythrocytes that contained neither antigen.[80] Based upon these studies he was able to define "Landsteiner's rule," which held that the serum of any individual would contain hemagglutinins for those erythrocyte antigens *not* present on his own red cells. Subsequently, von Decastello and Stürli in Landsteiner's laboratory discovered a rarer fourth group of individuals, those whose erythrocytes contained both of the antigens present in Landsteiner's groups I and II.[81]

In 1910, von Dungern and Hirszfeld published two landmark papers[82] in which they showed that different blood groups also exist in the canine population. They went on to name the human blood groups A, B, AB, and O, reflecting the presence or absence of the two antigens A and B. Moreover, they showed that these human blood group antigens obey the normal Mendelian rules of inheritance; that A and B are dominant; and that it is possible to measure the frequency of these traits in the population; finally, they suggested that blood group identification might prove useful in forensics.[83] Very much in line with Ehrlich's chemical interpretation of the complementary structures of antigen and antibody, von Dungern and Hirszfeld implied that A and B antigens were inherited from one's parents as "chemical structures" – perhaps the first interpretation of Mendelian inheritance in terms of the formation of specific molecular entities.

These observations on the existence of different blood groups led, in the years that followed, to an impressive burst of research activity, in an effort to understand the origin and implications of this new phenomenon. These studies are well summarized in two massive reviews of the field, one in 1925 by Lattes[84] and the other in 1926 by Hirszfeld.[85] Of prime importance was the application of blood typing to the problem of blood transfusion, which was made immeasurably safer when donor and recipient could be matched. In addition, blood typing was applied to forensic medicine, and especially to the establishment of paternity. Throughout the world, anthropologists studied the incidence of different blood groups in various populations to establish racial interrelationships, or to confirm theories of mass migrations. ABO blood typing was also applied to the study of evolution and species relationships among primates,[86] and even to the confirmation of the existence and nature of interspecies hybrids.[87] Of special interest was the early realization that these genetic markers might be associated with an increased susceptibility or resistance to disease, and numerous investigators sought to correlate differing ABO frequency ratios with both infectious and noninfectious disease processes, and especially with different forms of cancer.

In 1927, Landsteiner and Philip Levine discovered the M, N, and P blood groups,[88] and in 1940 Landsteiner and Alexander Wiener discovered the RH system,[89] setting the stage for the eventual solution of the problem of hemolytic disease of the newborn (erythroblastosis fetalis). Since then numerous other blood group systems have been described (including Lutheran, Kell, Lewis, Duffy, Kidd, Diego, etc.), thus expanding the forensic value of blood group identification. In addition, some of these minor blood groups have proven especially important in anthropologic studies, as well as being extremely useful in linkage studies to establish chromosomal mapping.[90]

The genetics of atopic allergy

The establishment before World War I of a medical subspecialty devoted to allergic diseases permitted practitioners to see and to compare large numbers of cases, and the impression was rapidly gained that there was a high familial incidence of such diseases. The first careful study of this situation was published in 1916 by Cooke and Vander Veer,[91] who studied a large series of 621 cases (including identical twins). The high incidence of allergic disease in the children of allergic parents "warrants the conclusion that inheritance is a definite factor in human sensitization." By comparing the allergens to which their patients (and especially the twins) were sensitive, they were able to show that *specific* sensitization is not inherited:[92]

> To sum up, then, we must say that the results of a clinical study compel us to conclude that sensitized individuals transmit to their offspring not their own specific sensitization but an unusual capacity for developing bioplastic reactivities to any foreign proteins.

Their data permitted them further to conclude that the predisposition for sensitization is inherited as a dominant Mendelian characteristic. In a review of the problem some years later, J. Adkinson modified Cooke and Vander Veer's conclusion,[93] and suggested that while inheritance might be dominant in some cases, it appeared to be recessive in the majority. Adkinson also pointed out that "it is the tendency or power to develop asthma...which is transmitted, and not the condition itself." The data on the hereditary nature of these allergic diseases appeared so convincing that Coca and Cooke felt justified in suggesting in 1923 that the special term "atopy" be employed to designate those human hypersensitive conditions that are genetically inherited from one's parents.[94]

In a follow-up publication from Cooke's clinic, another series of families with allergic disease was reported.[95] In analyzing their results, Spain and Cooke questioned Adkinson's conclusion on the recessive nature of the transmission of allergic disease, and suggested rather that a multifactorial inheritance might explain why all children of atopes are not themselves allergic. Since hayfever and asthma are recognized as the two principal forms of atopic allergy, a comparison of their respective inheritance suggested that while there might be a hereditary connection between the two conditions, it was clear that there was a certain tendency to independent transmission.

The major histocompatibility complex (MHC)

We noted above that the H-2 region on chromosome 17 of the mouse (and the HLA region on chromosome 6 of the human) were discovered during the search for the basis of histoincompatibility between graft donor and recipient. Since then, these major histocompatibility complexes have been shown to comprise perhaps hundreds of different loci, only some of which encode for the cell surface glycoproteins responsible for allograft rejection; others code for gene products which play an important role in other facets of immunorecognition and immunoregulation, so that the alternative name "major immunogene complex (MIC)" has been suggested as the more appropriate.[96]

Among the many biological functions subserved by the different loci within the MHC, it has recently become apparent that the murine H-2D and H-2K regions (and their human counterparts) have not been conserved in evolution merely to confound the transplantation biologist. The products of these genes (called class I antigens) contribute also to an MHC-restricted recognition mechanism important for the function of cytotoxic T lymphocytes. The gene products of the multilocus I region (termed class II antigens) appear to be concerned principally with mediating positive or negative cooperative events, maturation signals, and sequential interactions among macrophages, T cells, and B cells. The S region of the murine H-2 complex codes for, among other products, the fourth component of complement, while the G region, which comprises almost half of the H-2 complex, is still relatively unexplored territory. In addition to the H-2 complex, there is another set of widely separated DNA sequences known collectively as T (for T cell) regions, which code for a large group of T cell membrane differentiation markers, of undoubted importance to the function of the various T cell subsets.

HLA and disease susceptibility

The 1960s and 1970s saw a remarkable burst of activity in the identification of certain HLA haplotypes that appear to predispose to an increased susceptibility to certain diseases. While some of these haplotype–disease associations involve the A, B, and C loci (e.g., A3 and B14 with idiopathic hemochromatosis; B27 with ankylosing spondylitis and Reiter's disease; B47 with congenital adrenal hypoplasia; and CW6 with psoriasis vulgaris), most associations appear to involve the D/DR region genes. Thus, D/DR3 is intimately associated with systemic lupus erythematosus, dermatitis herpetiformis, sicca syndrome, celiac disease, and others; D/DR4 is involved in insulin-dependent diabetes and pemphigus; and D/DR5 with Hashimoto's thyroiditis, pernicious anemia, and pauci-articular juvenile rheumatoid arthritis, etc. In general the mechanisms underlying these disease associations are not well understood, although there are some suggestions that the ABC associations in some way involve cytotoxic T cells, whereas the D/DR associations, so often characterized by autoimmune phenomena, may involve the participation of immune response (Ir) or immune suppression (Is) genes.

Immune response genes

One of the first demonstrations that genetic factors may play a role in the capacity to form specific antibodies was that of Scheibel in 1943.[97] Random-bred guinea pigs were separated into good and poor responders to diphtheria toxoid, and selective inbreeding of these two groups over several generations resulted in a positive selection for such high or low responders. The fact that so marked an effect could be had in a small number of generations indicated to Scheibel that there were relatively few genes segregating between the two groups. With the introduction by Sela of the use of antigens composed of polymers of L-amino acids of defined structure,[98] the search for genetic control of the immune response was substantially simplified. Levine et al. were able to show in 1963–1965[99] that random-bred guinea pigs varied in their response to dinitrophenyl poly-L lysine (DNP-PLL), and breeding experiments confirmed that a single gene controls the response to the PLL carrier, independent of the hapten to which most of the antibody is directed. The first systematic study of this phenomenon in mice was carried out by McDevitt and Sela,[100] using such branched copolymers as TGAL (tyrosine, glutamic acid, -alanine, -lysine) and similar compounds in which phenylalanine or histidine were used in place of tyrosine. The genes which control the level of response to these different antigens were termed immune response (Ir) genes, and it was shown by McDevitt and Chinitz[101] that the Ir loci are intimately linked to the murine major histocompatibility complex. In 1972, McDevitt and colleagues were able to map the Ir gene controlling responsiveness to TGAL to a new region of the H-2 complex called the I region, located between the K and S regions.

It is not yet clear how Ir genes function. While Ir genes seem to regulate T cell recognition of certain structures, and Ia molecules participate in this recognition process, it is not clear that the Ia molecule is itself the product of an Ir gene. Moreover, it must be recognized that Ir gene control is not absolute, but merely a regulatory factor controlling the *degree* of the immune response to the given hapten–carrier complex. This view is compatible with a second model of Ir gene function, which postulates that the gene products are primarily expressed on the surface of T cells, and are concerned with the production of specific helper and suppressor factors, coded for by loci in the I-A and I-J subregions respectively. Benacerraf and Germain have concluded that in all likelihood, both of these mechanisms may operate simultaneously as components in the general regulation of immune responses.[102]

The generator of immunological diversity

We must not conclude this brief overview of the interplay of genetics with immunology without recalling that most remarkable genetic contribution of all – the unique mechanism that has evolved for the generation of the myriad of specificities that constitute the immune response repertoire. As discussed in Chapter 4, evolution has devised a complex genetic mechanism based not on specific past experience (as is the general rule), but rather on a mechanism capable of anticipating all novel immunogens that might arise in the future!

Notes and references

1. Medawar, P.B., *Encounter* **32**(1), January, 1969; reprinted in Medawar, P.B., *Pluto's Republic*, New York, Oxford University Press, 1982, p. 30.
2. Tagliacozzi, G., *De Curorum per Insitionem*, Venice, 1597; English translation, Gnudi, M.T., and Webster, J.P., in *The Life and Times of Gaspare Tagliacozzi*, New York, Reichner, 1950, p. 185.
3. See, for example, Hamilton, D., "A History of Transplantation," in Morris, P.J., ed., *Tissue Transplantation*, Edinburgh, Churchill Livingstone, 1982, p. 1; Converse, J.M., and Casson, P.R., "The Historical Background of Trans-plantation," in Rapaport, F.T., and Dausset, J., eds, *Human Transplantation*, New York, Grune & Stratton, 1968, p. 3; Billingham, R.E., *J. Invest. Dermatol.* **41**:165, 1963; and the numerous references in Woodruff, M.F.A., *The Transplantation of Tissues and Organs*, Springfield, Charles C. Thomas, 1960. The best and most comprehensive history in this field is by Brent, L., *A History of Transplantation Immunology*, San Diego, Academic Press, 1997.
4. Bigger, S.L., "An inquiry into the possibility of transplanting the cornea...," *J. Med. Soc. Dublin* **11**:408, 1837.
5. Zirm, E., *von Graefes Arch. Ophthalmol.* **64**:580, 1906; earlier, A. von Hippel (*von Graefes Arch. Ophthalmol.* **34**:108, 1888) had devised the lamellar corneal graft (which uses only the superficial layers), and reported successes with allogeneic material.
6. Carrel, A., *Lyon Méd.* **98**:859, 1902; *J. Exp. Med.* **10**:98, 1908; *NY Med. J.* **99**:839, 1914.
7. See, for example, Underwood, H.M., *J. Am. Med. Assoc.* **63**:775, 1914; Holman, E., *Surg. Gynecol. Obstet.* **38**:100, 1924; Todd, C., *Proc. R. Soc. B* **106**:20, 1930.

8. Holman, note 7.

9. Bauer, K.H., *Beitr. klin. Chir.* **141**:442, 1927.

10. Sauerbruch, F., and Heyde, M., *Z. Exp. Pathol. Therapie* **6**:33, 1909; there is even the suggestion from these studies that something akin to the graft-versus-host reaction had occurred – see Schöne, G., *Die Heteroplastische und Homöoplastische Transplantation*, Berlin, Springer, 1912, pp. 73–74.

11. As early as 1895, there was a flurry of activity on the possibilities of serotherapy of cancer, which included such investigators as Richet, C., and Héricourt, J., *C. R. Acad. Sci.* **120**:948, 1895.

12. Morau, H., *C. R. Soc. Biol.* **3**:289, 1891.

13. Schöne, note 10. Much of the work on immunity to transplantable tumors came principally from E.F. Bashford, in the *Annual Reports of The Imperial Cancer Research Fund*, 1903–1910. A history of the early years and contributions of this organization will be found in Austoker, J., *The History of the Imperial Cancer Research Fund*, Oxford, Oxford University Press, 1990.

14. Tyzzer, E.E., *J. Cancer Res.* **1**:125, 1916.

15. Tyzzer, note 14, p. 131.

16. See also Little, C.C., and Tyzzer, E.E., *J. Med. Res.* **33**:393, 1916.

17. Tyzzer, note 14, p. 142.

18. Loeb, L., "Transplantation and individuality," *Physiol. Rev.* **10**:547, 1930; *The Biological Basis of Individuality*, Chicago, University of Chicago Press, 1945.

19. Woglom, W.H., *Cancer Rev.* **4**:129, 1929.

20. Bashford, E.F., *Trans. 17th Intl Congr. Med. London*, Sec. III, Part 2, 1913, p. 59.

21. Itami, S., *J. Cancer Res.* **10**:128, 1926.

22. Da Fano, C., *Z. Immunitätsf.* **5**:1, 1910; *Sci. Rep. Imp. Cancer Res. Fund*, **5**:57, 1912 (cited in Woodruff, *Transplantation*, note 3, p. 69).

23. Rous, P., and Murphy, J.B., *J. Exp. Med.* **15**:270, 1912.

24. Murphy, J.B., *The Lymphocyte in Resistance to Tissue Grafting, Malignant Disease, and Tuberculous Infection*. Monographs of the Rockefeller Institute Med. Res No. 21, 1926. The details and significance of Murphy's contributions are discussed in Silverstein, A.M., *Nature Immunol.* **2**:569, 2001.

25. Murphy, J.B., *J. Exp. Med.* **19**:181, 1914; **24**:1, 1916.

26. Murphy, J.B., and Sturm, E., *J. Exp. Med.* **41**:245, 1925.

27. Hektoen, L., *J. Infect. Dis.* **17**:415, 1915.

28. Rich, A.R., *The Pathogenesis of Tuberculosis*, 2nd edn, Springfield, Charles C. Thomas, 1951, p. 600.

29. Little and Tyzzer, note 16.

30. Little, C.C., *J. Cancer Res.* **8**:75, 1924.

31. Holstein, J., *The First 50 Years at the Jackson Laboratory*, Bar Harbor, The Jackson Laboratory, 1979, p. x.

32. Gorer, P.A., *J. Pathol. Bacteriol.* **44**:691, 1937; **47**:231, 1938.

33. See, for example, Snell, G.D., "Methods for the study of histocompatibility genes," *J. Genet.* **49**:87, 1948; Snell, G.D., Smith, P., and Gabrielson, F., *J. Natl Cancer Inst.* **14**:457, 1953.

34. For example, Hans Zinsser's 1914 book *Infection and Resistance* discusses tumors only in the context of the possibility of serologic diagnosis, while neither Karsner and Ecker's *Principles of Immunology* of 1921 nor Topley and Wilson's *Principles of Bacteriology and Immunology* of 1936 even list "tumors" or "transplantation" in their indices.

35. Gibson, T., and Medawar, P.B., *J. Anat.* **77**:299, 1943.
36. Gibson and Medawar, note 35, p. 309.
37. Medawar, P.B., *Memoirs of a Thinking Radish*, Oxford, Oxford University Press, 1986, p. 83.
38. Medawar, P.B., *J. Anat.* **78**:176, 1944; **79**:157, 1945.
39. Medawar, 1945, note 38, p. 174.
40. Medawar, 1944, note 38, p. 195.
41. Medawar, P.B., *Bull. War Med.* **4**:1, 1943.
42. Medawar, P.B., *Harvey Lect.* **52**:144, 1956–1957.
43. Medawar, P.B., *Br. J. Exp. Pathol.* **27**:15, 1946.
44. Owen, R.D., *Science* **102**:400, 1945.
45. Burnet, F.M., and Fenner, F., *The Production of Antibodies*, 2nd edn, Melbourne, Macmillan, 1949.
46. Anderson, D., Billingham, R.E., Lampkin, G.H., and Medawar, P.B., *Heredity* **5**:379, 1951; Billingham, R.E., Lampkin, G.H., Medawar, P.B., and Williams, H.L., *Heredity* **6**:201, 1952.
47. Billingham, R.E., Brent, L., and Medawar, P.B., *Nature (London)* **172**:603, 1953. (Peter Medawar always insisted that the authors of papers from his laboratory be listed in alphabetical order.)
48. Mitchison, N.A., *Proc. R. Soc.* **B142**:72, 1954; Billingham, R.E., Brent, L., and Medawar, P.B., *Proc. R. Soc.* **B143**:43, 1954. It was in this latter paper that the term "adoptive transfer" was coined.
49. Passive transfer of these sensitivity states had previously been demonstrated by Landsteiner, K., and Chase, M.W., *Proc. Soc. Exp. Biol. Med.* **49**:688, 1942; Chase, M.W., *Proc. Soc. Exp. Biol. Med.* **59**:134, 1945.
50. Algire, G.H., Weaver, J.M., and Prehn, R.T., *J. Natl Cancer Inst.* **15**:493, 1954.
51. Weaver, J.M., Algire, G.H., and Prehn, R.T., *J. Natl Cancer Inst.* **15**:1737, 1955.
52. Billingham, R.E., Brent, L., and Medawar, P.B., *Nature* **178**:514, 1956.
53. Billingham, R.E., Brent, L., and Medawar, P.B., *Ann. NY Acad. Sci.* **59**:409, 1955; Trentin, J.J., *Proc. Soc. Exp. Biol. Med.* **92**:688, 1956; **96**:139, 1957; Simonsen, M., *Acta Pathol. Microbiol. Scand.* **40**:480, 1957; Billingham, R.E., and Brent, L., *Transplant. Bull.* **4**:67, 1957. For a broad review of the early work in this field, see Grebe, S.C., and Streilein, J.W., *Adv. Immunol.* **22**:119, 1976.
54. Medawar, note 42, p. 165.
55. Billingham, R.E., personal communication, 1987; confirmed also by Brent, L., personal communication, 1987.
56. *Ann. NY Acad. Sci.* **59**:277, 1954; **64**:735, 1957; **73**:1, 1958; **87**:1, 1960; **94**:335, 1962; **120**:1, 1964.
57. Dausset, J., Hamburger, J., and Mathé, G., eds, *Advance in Transplantation*, Baltimore, Williams & Wilkins, 1968.
58. Brent, L., *Transplant. Proc.* **9**:1343, 1977.
59. Calne, R., in "Immunological tolerance," *Br. Med. Bull.* **32**:107, 1977.
60. Barker, C.F., and Billingham, R.E., in *Advance in Transplantation*, note 57, p. 25.
61. Billingham, R.E., and Sparrow, E., *J. Exp. Biol.* **31**:16, 1954.
62. M. Simonsen (*Transplant. Bull.* **19**:2765, 1987) called attention to the fact that the role of weak, non-MHC antigens was being neglected in recent work.
63. See, e.g., *Transplant. Proc.* **17**: 1985, *passim*.
64. Chew, W.B., and Lawrence, H.S., *J. Immunol.* **33**:301, 1937; see also Cruikshank, A.M., *Br. J. Exp. Pathol.* **22**:126, 1941.

65. Woodruff, M.F.A., and Anderson, N.A., *Nature (London)* **200**:702, 1963; Monaco, A.P., et al., *J. Immunol.* **96**:229, 1966; Levey, R.H., and Medawar, P.B., *Ann. NY Acad. Sci.* **129**:164, 1966.
66. See, for example, *Transplant. Proc.* **17**:2327–2375, 1985.
67. Medawar, P.B., note 42, p. 166.
68. Owen, note 44.
69. Burnet and Fenner, note 45.
70. Billingham et al., note 46.
71. Billingham, R.E., and Brent, L., *Transplant. Bull.* **4**:67, 1957; Woodruff, M.F.A., and Simpson, L.O., *Br. J. Exp. Pathol.* **36**:494, 1955; Hasek's many early contributions are summarized by Hašek, M., Lengerová, A., and Hraba, T., *Adv. Immunol.* **1**:1, 1961.
72. Hanan, R., and Oyama, J., *J. Immunol.* **73**:49, 1954. See also Billingham, R.E., Brent, L., and Medawar, P.B., *Phil. Trans. R. Soc.* **B239**:357, 1956; Simonsen, M., *Acta Pathol. Microbiol. Scand.* **39**:21, 1957.
73. Smith, R.T., *Adv. Immunol.* **1**:67, 1961.
74. Starzl, T.E., et al., *Surg. Forum* **20**:371, 1969; Starzl, T.E., et al., *Lancet* **339**:1579, 1992.
75. Starzl, T.E., et al., *Immunol. Today* **14**:326, 1993.
76. For general reviews of the entire field of immunogenetics, the reader is referred to Hildemann, W.H., Clark, E.A., and Raison, R.L., *Comprehensive Immunogenetics*, New York, Elsevier, 1980; Fudenberg, H.H., Pink, J.R.L., Wang, A.-C., and Ferrara, G.B., *Basic Immunogenetics*, New York, Oxford University Press, 1984.
77. Allen, G., *Life Science in the 20th Century*, New York, Wiley, 1975, p. 113.
78. Landsteiner, K., *Centralbl. Bakt. Orig.* **27**:357, 1900.
79. Ehrlich, P., and Morgenroth, J., *Berl. klin. Wochenschr.* **37**:453, 1900.
80. Landsteiner, K., *Wien. klin. Wochenschr.* **14**:1132, 1901.
81. von Decastello, A., and Stürli, A., *Münch. med. Wochenschr.* **49**:1090, 1902.
82. von Dungern, F., and Hirszfeld, L., *Z. Immunitätsf.* **4**:531; **6**:284, 1910.
83. In this latter connection, see also Moss, W.L., *Bull. Johns Hopkins Hosp.* **21**:63, 1910.
84. Lattes, L., *Die Individualität des Blutes*, Berlin, Springer, 1925.
85. Hirszfeld, L., *Ergeb. Hyg. Bakt.* **8**:367, 1926.
86. Landsteiner, K., and Miller, P., *J. Exp. Med.* **42**:863, 1925; *Science* **61**:492, 1925.
87. Landsteiner, K., and van der Scheer, J., *J. Immunol.* **9**:213, 221, 1924.
88. Landsteiner, K., and Levine, P., *Proc. Soc. Exp. Biol. Med.* **24**:600, 941, 1927.
89. Landsteiner, K., and Wiener, A.S., *Proc. Soc. Exp. Biol. Med.* **43**:223, 1940.
90. See, for example, Mourant, A.E., *The Distribution of Human Blood Groups*, Oxford, Blackwell, 1964; Race, R.R., and Sanger, R., *Blood Groups in Man*, 6th edn, Philadelphia, F.A. Davis, 1975.
91. Cooke, R.A., and Vander Veer, A., *J. Immunol.* **1**:201, 1916.
92. Cooke and Vander Veer, note 91, p. 215.
93. Adkinson, J., *Genetics* **5**:363, 1920.
94. Coca, A.F., and Cooke, R.A., *J. Immunol.* **8**:166, 1923.
95. Spain, W.C., and Cooke, R.A., *J. Immunol.* **9**:521, 1924.
96. Hildemann et al., note 76.
97. Scheibel, I.F., *Acta Pathol. Microbiol. Scand.* **20**:464, 1943.
98. Katchalski, E., and Sela, M., *Adv. Protein Chem.* **13**:243, 1958; Sela, M., *Adv. Immunol.* **5**:29, 1966.

99. Levine, B.B., Ojeda, A., and Benacerraf, B., *J. Exp. Med.* **118**:953, 1963; *Nature* **200**:544, 1963; Levine, B.B., and Benacerraf, B., *Science* **147**:517, 1965.
100. McDevitt, H.O., and Sela, M., *J. Exp. Med.* **122**:517, 1965.
101. McDevitt, H.O., and Chinitz, A., *Science* **163**:1207, 1969.
102. Benacerraf, B., and Germain, R.N., *Immunol. Rev.* **38**:70, 1978. See also Schwartz, R.H., *Advances Immunol.* **38**:31, 1986.

Part Two

Social History

12 The uses of antibody: magic bullets and magic markers

The immune substances, ...in the manner of magic bullets, seek out the enemy.
Paul Ehrlich[1]

Throughout its history, immunology has made consistent contributions to other sciences by virtue of a unique technological asset – the specific antibody. Since the development and application of many of these technics were often unrelated to the history of immunologic *ideas* described in the foregoing chapters, they may have been neglected. Nevertheless, each of them represented a forward step in the science of immunology, and many contributed significantly to other disciplines as well. This chapter will therefore be devoted to these disparate technologies, in order to examine briefly their roles in immunology itself as well as the applications to which they have been put in other fields.[2] In entitling this chapter "The uses of antibody," I employ the term antibody in its widest sense, to include all of the products, interactions, and ancillary factors that are properly associated with the immune system.

During the initial period when immunology developed as an offspring of bacteriology, its principal technological contributions lay in the areas of preventive immunizations and serum therapy. Somewhat later, the diagnosis of disease took advantage of the specific immune response of the individual to infection. However, a new dimension in the young field was opened in 1906 when Obermeyer and Pick[3] showed that the chemical treatment of proteins could confer upon them new antigenic specificities. Thus, immunization with diazotized proteins would induce the formation of antibodies specific for the attached chemical group – an approach that would later be applied to the study of antibody specificity by Karl Landsteiner and others. The use of labeled antigens for specificity studies proved to be only the tip of a technological iceberg; labeled antigens (and antibodies) would soon be employed as immunohistochemical reagents for localization purposes. Then, with the advent of monoclonal antibodies, these would be modified to furnish exquisite tools for the separation of cell mixtures and for immunodiagnosis and immunotherapy. The popularity and importance of these many uses of labeled proteins in immunology is apparent from the great proliferation of publications – literally tens of thousands of citations in the US National Library of Medicine search engine.[4]

The development of each of the many applications of labeled antigens and antibodies has its own inner logic, which we shall explore briefly in this historical account. Some technical advances involve the combination of two methods (often from outside disciplines) to form a useful third one; others

A History of Immunology, Second Edition
ISBN: 978-0-12-370586-0

Copyright © 2009, Elsevier Inc.
All rights reserved

represent the step-by-step accretion of complexity as methods are adapted to answer new questions or to better answer old ones. Equally interesting is the interplay of technique and theory; sometimes an idea would demand a new technique for its pursuit, while in other situations application of a new method might push concept and understanding in entirely new directions. As Keating and Cambrosio would say in their study of technology evolution, "either scientific instruments and techniques are derived from theory, or they are the ground upon which theory is based."[5]

Immunotherapy

Active prophylactic immunization

The eighteenth-century demonstrations of the efficacy of inoculation and Jennerian vaccination to protect against smallpox were based on pragmatic observation rather than on an understanding of the mechanisms involved. It was only with acceptance of the germ theory of disease, and with Louis Pasteur's demonstration of etiologic specificity and of the use of attenuated organisms, that preventive immunization was able to build upon a firm theoretical base. At the outset it was thought that only live organisms would suffice to furnish protective immunity, as in the case of Pasteur's demonstrations with fowl cholera, anthrax, and rabies (where the pathogen could not even be isolated and cultured at the time). Then, with the demonstration that dead organisms and even their components might serve, many new approaches to prophylactic immunization were revealed.

The discovery that diphtheria and tetanus are mediated by exotoxins permitted these substances to be used *prospectively* – an approach rendered immeasurably more efficient with the development of detoxified antigens, the toxoids.[6] It was observations of this type that paved the way for the development of effective vaccines not only for a wide variety of bacterial pathogens, but also for such viral diseases as measles, mumps, rubella, and poliomyelitis.[7] In an interesting variation on the preventive vaccine approach, it has been proposed to vaccinate young girls against the human papilloma virus to prevent the later development of its precancerous lesions that may lead to cervical cancer in future years.[8]

However, not all infectious diseases were amenable to these approaches. Syphilis, trachoma, and essentially all parasitic diseases have resisted efforts toward the development of preventive vaccines. The use of killed organisms or toxoid for cholera and of bacille Calmette–Guérin for tuberculosis are only partially effective, and the latter has not gained acceptance everywhere. In some diseases, such as influenza and trypanosomiasis, the pathogens elude the immunologists' efforts by changing their antigenic coat almost faster than specific vaccines can be developed.

Recent years have witnessed radical changes in the approach to vaccine development. Along with the concept of idiotype–anti-idiotype immunoregulation came the recognition that the anti-idiotype possesses the same three-dimensional combining site as the antigen initially employed to induce the formation of the specific idiotypic antibody. The use of such anti-idiotypes as immunogens, in place of antigens derived from the pathogen itself, appeared initially to yield promising results.[9] Yet another new approach to vaccine development has accompanied the current revolution in genetic engineering. Now, those portions of a pathogen's genome that encode for the desired antigen can be excised and reincorporated into the genome of an appropriate carrier (such as a vaccinia virus). Introduced into the vaccine recipient by this route, the antigen will be actively produced within the host during viral replication, rendering immunization much more efficient. Alternatively, those antigens that are difficult to isolate may be produced *in vitro* in large quantities, using cultures of bacteria or yeasts into whose genome the antigen-encoding plasmid has been inserted.

Passive serotherapy

The discovery of diphtheria and tetanus antitoxins by Behring and Kitasato in 1890[10] seemed at the time to guarantee victory in the war against infectious diseases. Not only would these antibodies protect prospectively against infection by their respective pathogens, but timely use might even arrest the disease once started. Unfortunately, this approach was limited in the main to those few diseases mediated by exotoxins, and so serotherapeutic approaches to most other infectious processes met with little success. Moreover, the readily available horse antitoxins employed to treat diphtheria and tetanus in humans often led to the development of systemic serum sickness, caused by the massive amounts of xenogeneic protein contained in the antitoxin preparation. With the availability of suitable toxoids for preventive immunization and boosting, this form of serotherapy has substantially disappeared in the economically developed world.

The passive transfer of protective antibody has found two other interesting applications in modern times. In such diseases as infectious hepatitis, where an effective vaccine has not yet been developed and where circulating antibody efficiently neutralizes the infecting virus, human immune globulins have been utilized to advantage. The second application of passive antibody depends upon the observations that specific antibody may actually interfere with host sensitization and active antibody formation.[11] This approach has been applied to the prevention of erythroblastosis fetalis, in which the red cells of a Rh-positive fetus are hemolyzed *in utero* by anti-Rh antibodies, produced by the Rh-negative mother stimulated by transplacental passage of fetal erythrocytes. Where such maternal–fetal incompatibility is anticipated, human gamma globulin containing antibody specific for the Rh antigen may be given to the mother in small amounts, to inhibit her active sensitization by the fetal cells.[12]

In addition to the suppression of specific antibody formation, passively administered antibodies have also been used to suppress the production of entire classes of immunoglobulins – an approach most often employed to study the fundamental mechanisms of the immune response. Thus, since any given B lymphocyte can only employ one of the two allelic genes that govern immuno-globulin formation (*allelic exclusion*), specific antibody against one or the other allelic gene product administered during the early ontogenetic development of the B cell repertoire will selectively inhibit those cells destined to produce that allotype, so that only the other is formed.[13] Again, the maturation of an anti-body response generally involves the sequential utilization of those genes which encode for the constant region of heavy chains (*isotype switching*, in the order IgM, IgD, IgG, IgE, and IgA). The administration of a heterologous antibody against an immunoglobulin isotype, especially during the initial immune responses of experimental neonates, results in suppression of the formation of that isotype and of the others further along in the switching sequence.[14] Along the same lines, entire classes of T lymphocytes may be suppressed by appropriate use of monoclonal antibodies directed against those surface marker antigens that characterize the particular subset.[15]

Immunotoxic agents

The term *immunotherapy* was employed as early as 1906 by Paul Ehrlich, not so much to illustrate a successful achievement as to describe the ideal model for his new ventures into chemotherapy. It was already clear by this time that many of the diseases that afflict mankind could not be prevented by immunization, or alleviated by serotherapeutic approaches. Specific antibody was to be the model for a new pharmacology, in which a drug would be so endowed with a combining site (haptophore group) that it would, like a magic bullet, unerr-ingly seek out its pathogenic target. Moreover, appropriate chemical manipu-lation would ideally confer upon the molecule a toxophore group that would, like a poisoned arrow, kill the target organism while sparing the neighboring tissues of the host. Ehrlich's grand goal for the new pharmacology was nothing less than a *therapia magna sterilisans* – the specific chemotherapy of infectious diseases. He recognized, however, that "magic substances like the antibodies, which affect exclusively the harmful agent, will not be so easily found."[16]

One of the major reasons for Ehrlich's shift from immunology to chemo-therapy was the disappointment experienced by his and other laboratories in their attempts to apply serotherapy to the problem of cancer. In the mid-1890s, such investigators as Héricourt and Richet in France[17] and Salvati and de Gaetano in Italy[18] had reported on attempts to treat various forms of cancer with "antitumor antibodies," along the lines of von Behring's brilliant success with diphtheria and tetanus. This approach excited a great deal of interest and activity, but the bright promise was not fulfilled. Five to ten years later, Ehrlich's Royal Institute for Experimental Therapy in Frankfurt saw the future in drugs rather than in antibodies. This disappointing assessment of the potential of immunotherapy in cancer found confirmation in a major summary of the field in 1908 by Bashford, Murray, and Haaland from the Imperial Cancer Research Fund in London. These authors concluded that "As regards the hope of a prac-tical outcome, we consider that it is not at present to be sought in the direction of a curative serum..."[19] Forty-five years later, in a lengthy review of the immunologic aspects of cancer, Hauschka could offer little more hope.[20] Later developments, however, would open up new therapeutic possibilities.

Radiolabeled antibodies

Throughout the fifty years' search for an immunotherapy of cancer, it had never been clear whether the failure of this approach was due to the inability to raise specific antibodies against *unique* tumor antigens or whether, conversely, such antibodies exist, but fail to exert a cytotoxic effect on the tumor targets. To the best of my knowledge, the first person to suggest that the latter alternative could be corrected by the preparation of immunotoxic antibodies was David Pressman. Pressman had trained in immunochemistry with Linus Pauling, and in 1946 established a laboratory at the Sloan-Kettering Institute for Cancer Research to investigate the localization of antitissue and antitumor antibodies, employing the now-popular techniques of radioactive tracers.[21] Here was not only the forerunner of the future use of radiolabeled antibodies in tumor diagnosis (see below), but an approach with important therapeutic potential as well.[22] As Pressman speculated:[23]

> it is not impossible that if antibodies can be found which go specifically to
> a certain tissue, they can be made to carry physiologically active amounts of
> radioactivity to the tissue. Thus, reliance need not be placed upon the capacity of
> the antibodies themselves to produce a physiologic effect.

Indeed, Pressman's studies led him to conclude that as much as 100 mCi of radioiodine might be carried to the target tissue to effect its local irradiation and destruction.[24]

It was the demonstration of the validity of this finding that stimulated the imaginative "suicide" experiment simultaneously in the laboratories of Gordon Ada and John Humphrey in the late 1960s.[25] By this time it was known that the B cell has immunoglobulin receptors on its surface specific for the antigens for which it is genetically programmed, as Paul Ehrlich had predicted in 1897. Thus, injection into an animal of a highly radioactive antigen might be expected lethally to irradiate only those clonal precursors to which it is specifically bound. This in fact occurred, and animals so treated were rendered incapable of responding thereafter against the antigens involved. Here was a true clonal deletion, in the sense that Macfarlane Burnet had originally hypothesized to explain immunological tolerance.[26] Specific deletion of anti-hapten B cells would later be similarly achieved using ricin-labeled antigen.[27]

Unfortunately, the efforts by Pressman and others to develop a therapeutic modality using radioactive antibodies were doomed to failure. Despite all efforts to absorb out all cross-reacting antibodies using appropriate tissue extracts, preferential localization of these crude antibody preparations (if indeed they were truly antitumor) was at best marginal. Too high a portion of the radioactive antibody was found to localize in such undesired targets as the lung, spleen, and kidney.

It was only with the advent of monoclonal antibody technics that the sharp specificity required of an immunotoxin approach to tumor therapy could be realized.[28] One might now hope to localize more of the immunotoxin in the desired target tissue. Thenceforth, monoclonals labeled with a variety of radioisotopes would be utilized for the treatment of any tumor that might be associated with a "tumor-specific" antigen.[29] Of course, the ultimate efficacy of this approach depends upon the finding of antigens unique to the tumor in question, which are not shared by normal tissues. Unfortunately, such unique antigens may not often exist. Tumor cells appear, in the main, to be normal cells which suffer from a dysregulation. Thus, most tumors do not produce qualitatively different neo-antigens; approaches to immunodiagnosis and immunotherapy most frequently depend on quantitative (or timing) differences in the formation of otherwise-normal cell products. These take the form either of differentiation markers posted on the cell membrane, or of more-or-less organ-specific proteins exported by the cell.[30] Included among the former are the Cluster of Differentiation (CD) markers of lymphocytes, and glycoprotein and other receptor molecules for the many substances that mediate the cell's specialized functions. The exportable proteins of interest to the oncologist include, among others, "oncofetal" antigens such as alpha-fetoprotein and carcinoembryonic antigen (CEA), prostate specific antigen (PSA) and prostatic acid phosphatase for prostate tumors, ferritin and lactic dehydrogenase for liver tumors, and beta-human gonadotropin for testicular tumors. There are, however, instances in which true "tumor-specific" antigens may be formed. These include idiotypes on non-Hodgkin's lymphomas and myelomas, products that appear to be altered-self peptides, viral antigens in tumors incited by these agents, and the putative products of oncogenes. Each of these targets is a candidate for therapy using radiolabeled (or toxin-labeled – see below) specific antibody.

It is also possible to irradiate a tumor by indirect means, employing non-radioactive elements that can be activated by exposure to neutrons.[31] It was Bale, as long ago as 1952, who first proposed the use of boron in the radio-therapy of cancer,[32] and this element appears to be the most commonly employed. Boron-10 undergoes fission when exposed to thermal neutrons with the release of alpha particles and gamma rays, and neutron-activated gadolinium-157 releases beta and gamma rays, resulting in the destruction of cells in the immediate surround. The tumors found most amenable to this approach have been brain tumors and malignant melanomas.

Immunotoxins (ITs)

In contrast to radiolabeled antibody therapy, where the radioactivity need only be brought near the tumor target, the use of toxin-labeled antibodies imposes a more stringent requirement. This is because the toxin must not only attach to the target cell, but must also penetrate into it to exert its toxic action. Thus all toxins have various subunits or domains, some devoted to binding to cells and/or to translocation across the cell membrane while other subunits act within the cell by interfering with protein synthesis.[33] These different domains may be separated or genetically manipulated to delete those that may be unwanted.

The first immunotoxin studied was by Moolten and Cooperband in 1970, using diphtheria toxin conjugated to polyclonal antibodies.[34] Since then, many different bacterial, fungal, and plant toxins have been tried, including ricin, abrin, pseudomonas exotoxin, single chain ribosome-inactivating proteins, saporin-S6, and momordin.[35] Most applications of immunotoxins have been for the treatment of various cancers, but ITs have also been used to inhibit the rejection of transplanted organs[36] and to suppress graft-versus-host reactions in bone marrow transplant recipients by *ex vivo* destruction of active lymphocytes in the donor marrow.[37]

Immunodrugs

Just as toxins may be coupled to antibodies to act against tumor cells, so also may conventional cytotoxic drugs be similarly employed. The first such experiment was attempted as far back as 1958, when Mathé and coworkers attached amethopterin to hamster anti-L1210 cells in an attempt to treat murine leukemia.[38] Since then, many different substances have been used, attached to various monoclonals.[39] Similar approaches have taken advantage of the activity of such stimulatory agents as lymphokines and growth factors.[40] One compound, calicheamycin coupled to a monoclonal anti-CD33, has been approved for the treatment of acute myeloid leukemia.[41]

One of the more imaginative variations on the immunodrug theme involves the attachment to the monoclonal antibody of a specialized enzyme capable of converting an inactive "prodrug" into its active form (termed ADEPT – antibody-directed prodrug therapy). After the enzyme-coupled antibody affixes to its tumor-associated antigen, the prodrug is introduced systemically, to be converted into an active cytotoxic form only at the targeted tumor site.[42]

Photoimmunotherapy

Another indirect approach involves the attachment to the antibody carrier of an inert photosensitizer whose activation by light leads to cytotoxic damage. The first such effort was by Mew and colleagues in the early 1980s, employing hematoporphyrin attached to an anti-myosarcoma monoclonal.[43] This approach using antibody-directed photolysis has found its principal application in the treatment of epithelial tumors, where exposure to light is more readily accomplished.[44]

Liposome vehicles

Having demonstrated the efficacy of the antibody-mediated delivery of small amounts of cytotoxic agents, a more wholesale approach was sought, and liposomes were adapted for this purpose. Liposomes are vesicles formed of phospholipid membranes, within which may be placed aqueous solutions of drugs, toxins, vaccines, or other active materials. They are endowed with specificity by the attachment of as many as 50–1,000 molecules of a specific monoclonal antibody to a 200-nanometer liposome. These tumor-directed antibodies may be coupled to the liposome by such heterobifunctional reagents as N-hydroxylsuccinimidyl 3-(2-pyridyldithio) propionate. The liposome can thus carry large, concentrated therapeutic doses directly to the tumor target.[45]

Technical problems

We have already referred to the problem of target specificity, wherein most "tumor" antigens are shared by normal tissues, so that toxicity may not be limited to the cancerous tissue. Related to this is the absence of a lethal bystander effect associated with either immunotoxin or immunodrug therapies; these ITs act only on the targeted cell. But since uniformity of antigen expression is rare among tumor cells, use of a cocktail of labeled monoclonals directed against more than one surface antigen should improve therapeutic efficiency.[46]

One of the major problems encountered in immunotoxin therapy is the associated damage to normal vascular endothelium, leading to a vascular leak syndrome. Attempts are being made to genetically engineer the toxins to delete the offending moieties.[47] Yet another problem involves the difficulty that ITs have in penetrating into large solid tumors; they are much more efficient in treating leukemias and lymphomas than bulky carcinomas.[48] In the latter case, they will most likely find their greatest application in the final "mopping up" of residual disease after other therapeutic modalities have been employed.

Another problem encountered in the use of monoclonals for therapeutic purposes involves their immunogenicity in the patient. The antibodies initially used in humans were xenogeneic, primarily murine. Their repeated use led to decreasing efficacy, due to the formation in the patient of neutralizing antimouse antibodies. Many solutions to this problem have been proposed, most involving genetic engineering to "humanize" the monoclonal. Initially, the mouse Fc portion was replaced by its human counterpart to reduce immunogenicity. Then increasingly larger murine segments were replaced, until finally only the mouse complementarity determining regions remained on the otherwise humanized immunoglobulin.[49] Another approach has been to attach the mouse antibody combining sites directly to the toxin.[50]

Each of these approaches ameliorated but did not completely solve the problem of the immunogenicity of the immunotoxin. Only a fully human monoclonal might take care of this aspect of the problem, and this was quick to come (although an anti-idiotype response might still occur). Mice may be induced to form human antibodies by "knocking out" the genes for endogenous heavy and light chains and replacing them with human H and L chain genes.[51] An even more efficient approach takes advantage of the ability to produce a recombinant library of Ig chain segments displayed on bacterial, yeast, or even mammalian cells. Fvs, single chain Fvs (scFvs, where VH and VL domains are linked in a single polypeptide chain), and even divalent scFvs can thus be produced.[52] The practical consequences of putting 10^9 single chain variable fragments on yeast cells will allow a combinational expression of the entire repertoire of specificities, the desired one being selectable with fluorescent antigen and flow cytometry.[53] (Equally interesting are the theoretical consequences; these antibody and T cell receptor products can be affinity matured[54] *in vitro* to affinity constants far exceeding the *in vivo* "affinity ceiling" thus far observed.[55])

Nonspecific immunotherapy

There is a two-fold rationale behind the numerous nonspecific approaches to the therapy of various infectious diseases and of cancer. On the one hand, where an organism or a tumor-specific antigen may be only poorly immunogenic, one attempts to enhance the protective immune response engendered in such a system with the use of nonspecific adjuvants, in order to stimulate further one or another of the components of the immune response. Alternatively, where the principal attack is mediated by essentially nonspecific factors (such as by macrophages acting against dermal melanomas, or by killer T cells acting against certain solid tumors), the approach is to render these components more plentiful or more active at the desired location. This is reminiscent of the way that Metchnikoff, over ninety years ago, was able to enhance the destruction of cholera organisms by activating macrophages in peritoneal exudates.

Other approaches to the immunologic control of tumors involve efforts aimed at increasing the immunogenicity of tumor cells, employing BCG, neuraminidase treatment, and muramyl dipeptides. Alternatively, attempts have been made to stimulate individual components of the immune response, including thymosin activation of T cells, levamisole activation of T cells and macrophages, interleukin-2 activation of cytotoxic T lymphocytes, or the use of hapten-conjugated tumor cells to enhance helper T cell function. In seeking to foster a greater participation of macrophages and natural killer cells in the destruction of tumors, systemic activation has been attempted employing BCG adjuvants and interleukin-2, respectively.[56] To accomplish the same end at the precise site of tumor growth, dermal tumors have been locally infiltrated with such bacterial products as BCG and PPD.[57] The immunogenic inflammation that results at the site brings in activated macrophages and other cells that may contribute incidentally to tumor destruction.

Immunodiagnosis

The active vertebrate immune response to antigenic stimulus is almost invariably accompanied by detectable changes in the blood of the host. Within days, there appear in the circulation new or increased titers of specific antibody and/or specifically sensitized T cells. In either case, the circulating products of the immune response can be utilized to identify the inciting antigen, and therefore provide many approaches to the etiologic diagnosis of a wide variety of infectious or autoimmune diseases.

Serologic tests

One of the most widely used of all immunodiagnostic tests was that involving complement fixation, most notably the Wassermann test for syphilis first introduced in 1906.[58] This procedure was based upon two fundamental demonstrations from the laboratory of Jules Bordet: that complement would mediate the hemolysis of antibody-sensitized erythrocytes,[59] and that any antigen–antibody interaction would result in the nonspecific fixation of complement.[60] The presence of antibody in a patient's serum can be detected and even quantitatively measured by mixing it with specific antigen and a measured amount of complement. A positive serology results when the antigen–antibody complex fixes complement, as measured by a reduction in the ability of the residual free complement to hemolyze the test erythrocytes.

It is interesting that while most complement fixation tests do in fact measure the interaction of bacterial or viral antigen with specific antibody, the prototypical serologic test – the Wassermann reaction for syphilis – actually utilizes a nontreponemal lipid antigen, cardiolipin, and measures an antibody that is incapable of interacting with the offending pathogen. But whatever its basis, the reaction proved reasonably satisfactory, and its offshoots were widely employed for diagnosis and screening. In addition to numerous modifications of the Wassermann complement fixation test for syphilis, a flocculation test was introduced by Reuben Kahn,[61] involving the formation of visible aggregates of cardiolipin–lecithin mixtures by the serum of syphilitic patients. Perhaps the most specific of the tests for syphilis was the *Treponema pallidum* immobilization test of Nelson and Mayer.[62] Although the treponeme can still not be grown *in vitro*, viable preparations of motile treponemes can be obtained from infected rabbits. Specific antibody in the serum of an infected patient is detected by its ability, with added complement, to inhibit the motility of these treponemes in test suspensions.

Another serologic approach to the etiologic diagnosis of infection takes advantage of the availability of standardized cultures of various bacterial pathogens. The mixture of a patient's serum with a suspension of the suspected organism results in a visible agglutination of these organisms in the test tube, permitting identification not only of the genus of the pathogen, but often of the species and even the strain as well. In the case of such pathogens as pneumococci, which are surrounded by large polysaccharide capsules, the various types can be differentiated by observing the swelling of the capsule induced by type-specific antibodies.[63]

The etiologic diagnosis of autoimmune diseases can similarly be established by examining the sera of affected patients for specific antibodies, whether or not those antibodies are intimately involved in the pathogenesis of the disease in question.[64] Given the availability of an appropriate antigen, any one of a number of tests for antibody may be employed, including complement fixation, precipitin reactions in aqueous solution or in gels, or one of the more sensitive radioimmunoassays currently available (see below). It was such approaches that established the autoimmune nature of Hashimoto's thyroiditis, Sjögren's syndrome, and the presence of anti-immunoglobulins in the serum of rheumatoid patients and of antinuclear antigens in the serum of patients with systemic lupus erythematosus. Similar tests have implicated an autoimmune pathogenesis in such diseases as insulin-dependent diabetes, Addison's disease, and a variety of autoimmune diseases which depend upon the development of autoantibodies directed at such physiologically important receptors as those for acetylcholine, insulin, and thyroid-stimulating hormone.

Dermal reactions

Ever since the observation by Maurice Arthus in 1903 that antibody-mediated inflammation could be elicited in the skin of sensitized animals upon local introduction of an appropriate antigen,[65] such reactions have been utilized extensively for both diagnostic and pathogenetic purposes. Although the Arthus phenomenon was originally described as "local anaphylaxis," it was later shown to be a mixture of antibody- and cell-mediated mechanisms. With the isolation of a variety of pollen and bacterial allergens, the intracutaneous skin test soon became the approach of choice for the etiologic diagnosis of hayfever, asthma, and other allergic disorders.[66] It was common practice in allergy clinics to subject the backs of patients to massive arrays of skin tests, utilizing entire libraries of allergenic extracts.

The demonstration by Prausnitz and Küstner in 1921 that skin reactivity could be transferred passively with the serum of sensitized individuals (*reverse passive anaphylaxis*)[67] firmly established this response as an antibody-mediated phenomenon. Now, with the modern knowledge that the antibody involved is a cytophilic IgE that binds to the patient's leukocytes, etiologic diagnosis can also be made by exposing these leukocytes *in vitro* to various allergens. A positive response is indicated by the release of histamine from these cells, indicative of a specific IgE-allergen interaction on their surface membranes.[68]

Delayed-type skin tests

The first such test described was that by Robert Koch in 1891.[69] Koch showed that tuberculin, a product of the culture of tubercle bacilli, would elicit a severe local inflammatory reaction in tuberculous patients upon intradermal application. The tuberculin test has proved to be invaluable in the diagnosis of past or present exposure to tubercle bacilli, and set the stage for the development of numerous other diagnostic skin tests. Thus, luetin, an extract of treponemes, has been used for the diagnosis of syphilis; mallein, an extract of *Pfeifferella mallei*, is used for the diagnosis of glanders; and lepromin, an extract of Hanson's bacillus, is used for the diagnosis of leprosy, etc.[70] Similar positive skin tests have been employed to demonstrate immunity to such viruses as measles and mumps.

It had originally been thought that such skin tests, like local anaphylactic reactions, were mediated by circulating antibodies, but differences in their temporal development, cytology, and the frequent inability to find circulating antibodies in such patients soon demanded their reclassification as "delayed" skin reactions.[71] Finally, the demonstrations that delayed skin reactivity to tuberculin and to poison ivy could only be transferred passively with cells[72] helped to rule out the participation of classical antibodies in these reactions, and ushered in the new field of cellular immunology.

Diagnostic antibody libraries

The ability of specific antibodies to distinguish among even closely related antigenic structures is limited only by the availability of the appropriate specific antisera. This has led to the development of entire libraries of immune sera, with a wide variety of applications in the clinical laboratory. For example, the availability of antibodies against the various blood group substances is widely used to type blood donors and recipients in order to avoid incompatible blood transfusion.[73] Again, rapid etiologic diagnosis of bacterial infection is made possible by the availability of specific antibodies capable of identifying the genus, species, and even strain of the microorganism. Finally, libraries of antibodies specific for the most important histocompatibility antigens have been widely employed to assure the best possible match of donor and recipient for tissue and organ transplantation.[74]

Radioimmunodiagnosis

The application of immunoconjugates to the diagnosis of cancer obviously requires use of a label that can be detected after localization, preferably by non-invasive methods. This might involve the use of radio-opaque substances, or metals whose paramagnetic properties can be detected by nuclear magnetic resonance analysis. But by far the most useful group of labels is that of those which employ radionuclides in one form or another, and that depend upon the detection of their radioactive emissions.[75]

It was David Pressman who first suggested in the 1940s that radiolabeled antibodies might be employed to determine the location of tumors and their metastases.[76] The requirements for diagnosis using radioactive labels are quite different from those demanded by therapy. In diagnosis, one wants small quantities of an emission (generally a gamma ray) energetic enough to escape the body and be registered, but without causing excessive damage to normal tissues on the way. For therapy, large quantities of a less penetrating emission are desired (such as strong alpha- or moderately strong beta-emitters), to irradiate the immediate target and surround, without reaching normal tissues.

There are many radioactive candidates for diagnostic imaging, including indium-111, technicium-99, gallium-67, rhenium-186, and iodine-132 and -135. There are also many methods to attach the radiolabels to the antibody:[77] these include direct attachment (as in the iodination of the tyrosine residues of the antibody); attachment to one or another protein that is then coupled to the antibody; and attachment via heavy metal chelating agents.[78] Once the radioactive antibody has been localized, it can be detected by computerized whole-body scan to detect the radioactive emissions. This approach is especially useful in detecting tumor metastases in ectopic sites where the target antigen is unlikely to be produced by normal tissues. The preponderance of clinical applications of this diagnostic method employ radioiodine isotopes due to their ease of use, the energy of their emissions, and their short half-life.

Another approach to the detection of tumors employs certain paramagnetic compounds which, after localization by virtue of their attachment to monoclonal antibodies, may be detected by magnetic resonance imaging. The most commonly employed substance is trivalent gadolinium, which is attached to the antibody following its binding by a chelating agent.[79] Iron oxide nanoparticles conjugated to antibody have also been used for the same purpose.[80]

Identification, assay, and localization

All of the techniques described below take advantage of that unique quality of immunologic interactions, the fine specificity of antibody for antigen. Some of these approaches yield only qualitative results, while others are capable of being quantified to an extraordinary degree. It will be apparent that in almost all instances, if a known antibody is available, it can be employed to test for its specific antigen; similarly, if the antigen is known, its antibody can be examined.

Immunoprecipitation

Ever since the precipitin reaction was described by Kraus in 1897,[81] the technic has been employed for the qualitative identification of antigens and antibodies, simply by observing the development of turbidity in an aqueous mixture of antiserum and antigen. A rapid form of this qualitative approach is known as the *ring test*. Here, a dilute solution of antigen is carefully layered over the denser immune serum in a test tube, and the diffusion of the specific reactants results in the formation of a visible immune precipitate at the interface. The precipitin reaction was adapted in the 1930s by Heidelberger and Kendall[82] to permit the quantitative measurement of antibody and antigen. This is accomplished by allowing the precipitate of optimal ratios of antibody to antigen to form under standard conditions, whereupon it is washed extensively and then subjected to nitrogen analysis.

Plate 12 Michael Heidelberger (1888–1991)

Gel diffusion analysis

In the mid-1940s, Jacque Oudin showed[83] that if an antigen is incorporated into an agar gel and antiserum layered over it, diffusion of specific antibody into the gel will result in the formation of discrete lines of precipitate down the length of the gel – in theory, one line per pair of specific interactants. The utility of this one-dimensional approach was enhanced immeasurably by Ouchterlony,[84] who adapted it for two-dimensional double diffusion. Ouchterlony showed that the addition of an antigen and an antibody into adjacent wells bored into a thin layer of agar would permit their diffusion toward one another through the medium. A thin line of immune precipitate forms where the reactants meet, its position depending upon the relative concentrations of antigen and antibody and upon their diffusion rates. Not only was this approach useful in determining the purity of an antigen solution, but it could also establish the partial cross-reaction of two antigens (containing both common and unique epitopes) by the development of precipitin spurs where the respective immune precipitate lines partially fuse in the agar plate.

Immunoelectrophoresis

The utility of gel double diffusion was further enhanced by Grabar and Williams,[85] in first separating one of the components (most often the antigen) in one dimension by electrophoresis, and then permitting the antiserum to diffuse from the other direction from a long trough cut in the agar. The antigen could now be characterized and mixtures separated by electrical charge as well as by specific antibody – an approach especially useful for the study of the complex mixture of proteins in serum. In addition, the characteristics of different antibodies could now be studied by subjecting an immune serum to electrophoresis and placing the antigen in the lateral well.

Radioimmunoassay

This technique, of exquisite sensitivity, was devised by Yalow and Berson[86] primarily for the quantitative assay of peptide hormones. This involves measurement of the competition between the test sample and a standardized iodine-labeled antigen for binding to an immobilized antibody. In this manner, nanogram to picogram amounts of certain substances could be measured (such as the hormones that interested Yalow). This valuable technic won Yalow the Nobel Prize in Physiology or Medicine for 1977.

ELISA assays

Except in endocrinologic laboratories, the enzyme-linked immunosorbent assay (ELISA) has largely replaced radioimmunoassays for the quantitative determination of antigens, and of antibody as well.[87] The approach is based on the ability to immobilize one of the immunologic reagents (antigen or antibody) on the surface of an insoluble carrier, without compromising its activity. The reciprocal reagent is then linked to an enzyme so that, once fixed to the carrier and washed, the amount of enzyme-coupled reagent can be estimated by permitting it to react with a substrate that yields a colored product. The most common approach employs a "sandwich" technique, in which excess antibody (or antigen) is absorbed to the plastic microplate, washed, and then treated with the test sample. The amount of antigen (or antibody) fixed from that sample is then measured by using enzyme-linked antibody (or antigen). The intensity of the resulting substrate color then provides a precise measure of the amount of reagent in the middle layer of the sandwich. In an indirect adaptation of this method, used in the assay of antibodies, the final step can employ an enzyme-linked antiglobulin specific for the appropriate species, or even labeled anti-immunoglobulin isotypes to measure the isotype composition of the unknown antibody sample.

Fluorescence-activated flow cytometry and sorting

From a variety of different sources came evidence that lymphocytes possess distinctive surface membrane antigens, the so-called CD (cluster of differentiation) nomenclature.[88] It soon became apparent that these surface markers might be used to characterize functionally distinct lymphocyte subsets and their lineages. Adapting the cell sorter previously used in hematology, the Herzenbergs and their associates perfected the fluorescence activated cell sorter (FACS). This takes advantage of the ability to separate out single cells with a given surface marker by treating them with specific fluorescein-labeled monoclonal anti-receptor antibody.[89] The FACS approach was then applied as a powerful tool for the analysis and isolation of lymphocyte subsets.[90]

Immunohistochemistry

Dye labels

The first use of a distinctive colored label on an antibody was in 1933 by Heidelberger and Kendall,[91] who attached the colored compound "R salt" to antigen for the quantitative analysis of antigen–antibody precipitates. Marrack utilized the same approach[92] to show that antibodies coupled to R salt would color bacteria that had undergone specific agglutination. Then Florence Sabin, in 1939, showed that labeled antigens might be used to study their distribution *in vivo*.[93] Wherever the labeled antigen localized in sufficient quantity, the red color of the dye would show up on histologic section.

Fluorescent labels

Perhaps the most significant development in the field of immunohistochemistry came from the laboratory of Albert Coons. In 1941, Coons, Kreech, and Jones[94] reported the use of a fluorescent-labeled antitype III pneumococcus (coupled with β-anthrylcarbamido groups), and were able to detect the distribution of the specific carbohydrate by examination of histologic sections in a fluorescence microscope. The fluorescent antibody technique was substantially improved in 1950 with the shift to a fluorescein label,[95] and thenceforth it found broad application to the *in vivo* localization of a wide variety of antigens, both autologous and exogenous. Employing an indirect method, even the site of antibody formation or antibody deposition could be determined. This involves treatment of the histologic section first with an antigen and then with specific labeled antibody for that antigen. In addition, a more generally applicable "sandwich technic" was also developed, in which an unlabeled antibody is localized on a section and then the fluorescence picture developed using a fluorescein-labeled anti-immunoglobulin.[96] Eventually, fluorescent labels with other colors were introduced, most notably the orange-red derivatives of rhodamine;[97] these were often employed together with fluorescein in double-labeling experiments.

The uses to which fluorescence immunohistochemistry were put were legion. To cite but a few examples, it was possible to locate antigens (both native and foreign) with great precision in tissues and cells. The plasma cell was confirmed as the one involved in antibody formation,[98] and the Russell body plasma cell inclusion was identified as immunoglobulin.[99] Finally, the pathogenesis of many infectious diseases could be clarified by tracking the antigens involved.[100]

Enzyme labels

A new approach to immunohistochemistry was furnished in the mid-1960s by Nakane and Pierce and by Avrameas and Lespinats.[101] This involves the coupling of an enzyme such as horse-radish peroxidase to the desired antibody. The antibody moiety affixes to its specific antigen in tissue section, and then the immunostain is developed by treating the enzyme portion with a suitable substrate to yield an insoluble colored product.[102] Just as double fluorescent labels of contrasting colors may be used to compare two antigens, so also may double enzyme labels be similarly employed. Thus, alkaline phosphatase may be coupled to one antibody, and horse-radish peroxidase to the second.[103] The enzyme-linked antibody technique has also been adapted for luminescence detection. In this case, the localized enzyme acts upon a substrate to produce a luminescent product, which may provide a more sensitive probe for immuno-assay.[104] More modern applications of immunohistochemistry have entailed the development of two additional procedures. In the one, by analogy with the ELISA test described above, an avidin–biotin immunoperoxidase method is employed. Biotin-linked antibody is localized on a tissue section which is then treated with peroxidase-linked biotin. The avidin–biotin–peroxidase complex then acts upon a suitable substrate to form an insoluble colored deposit wherever the enzyme has been localized.[105]

Electron-opaque labels

It did not take long, after the demonstration of the uses of labeled antibodies to answer structural questions, for the possibility of their use for ultrastructural studies to be raised.[106] The ability to attach electron-opaque labels (or labels from which electron-opaque products can be developed) to antibodies opened new vistas in immunohistochemistry.[107] As early as 1959, Singer demonstrated that the iron-containing protein ferritin could be coupled to antibody.[108] The high content of iron (up to 2,000 atoms per molecule) and the unique appearance of this molecule renders it highly recognizable in the electron photo-micrograph. (A variety of labels to carry other heavy metals, such as uranium and mercury, were introduced,[109] but have not received widespread general application.)

One of the very few approaches that does not require a chemical linkage of the label to the antibody probe involves the use of colloidal gold.[110] This takes advantage of the fact that gold is extremely electron-opaque, and in very fine colloidal suspension will firmly adsorb proteins onto its surface. When these are antibodies, it is found that their specific binding sites are left free to interact with antigen.[111] Another approach that does not require the direct chemical conjugation of label to antibody uses a different sandwich technique. Here, advantage is taken of the ability of staphylococcal protein A to attach firmly to the Fc portion of the immunoglobulins of many species. After treating the section with an antibody specific for the antigen sought, colloidal gold-treated protein A will develop an electron-opaque stain.[112]

As with labeling for light microscopy, contrasting labels may be used for the electron microscope.[113] Because of their different appearances in the electron microscope, ferritin, the several enzyme labels, and colloidal gold may be utilized together in pairs. Indeed, since the size of colloidal gold preparations can be controlled so well, double labels may be employed using colloidal gold of two different sizes.

Radioactive labels

The localization of antigen using radioactive tracers was first reported by Libby and Madison in 1947, using radiophosphorus-labeled tobacco mosaic virus.[114] There are several approaches available for the use of radioactive tracers to localize antigens on tissue sections.[115] The most direct method involves the labeling of the desired antibody, usually with radioiodine. Alternatively, use may be made of the biotin–avidin bridge. In one approach, the antibody is coupled to avidin, localized, and then treated with tritiated biotin; the location of the label is revealed by radioautography. Alternatively, the antibody may be coupled with unlabeled biotin, localized, and then treated with radiolabeled biotin–avidin complex (^{3}ABC), where one of the four avidin sites for biotin is free to bind to the biotin on the antibody. The avidin–biotin systems may also be used to carry fluorescent and enzyme labels as well as radioactivity,[116] as described above.

Immunoblots

One of the more imaginative and practically useful applications of labeled antibodies emerged with the introduction of the immunoblot (Western blot) technique. The protein components of a complex antigen are separated by polyacrylamide gel electrophoresis (PAGE) and then transferred onto a nitrocellulose membrane, where they are identified by treatment first with specific antibody and then with a radiolabeled or enzyme-coupled antiglobulin.[117] This system has found broad application in such areas as the diagnosis of infection, the screening of blood to avoid HIV transmission, and the testing of the efficacy of vaccines.

Plaquing technics

The idea of a technic for detecting and enumerating antibody-forming cells was developed independently by Jerne and Nordin and by Ingraham and Bussard in 1963.[118] This involves the distribution of a single-cell suspension of anti-erythrocyte-forming cells in agar, together with the species of erythrocytes employed for immunization. The red cells immediately surrounding a single antibody-forming cell become sensitized by the antibody diffusing out, and are hemolyzed with added complement to yield a clear circular "plaque" in the otherwise cloudy suspension. Since the technic is most effective for the detection of IgM antibody-producing cells (because that isotype most efficiently fixes complement), a modification was introduced that facilitates hemolysis by specific IgG antibodies. This involves the intermediary treatment of the agar plate with anti-IgG antibodies and then complement, to enhance the hemolysis of the IgG-sensitized erythrocytes. The method has also been extended to the study of antibody formation against other antigens, where these can be attached to the surface of erythrocytes to mediate immune hemolysis by their respective antibodies. These approaches have been especially useful to study the distribution and cellular kinetics of the antibody response.

Taxonomy and anthropology

The demonstration in the 1890s of the specificity of antibodies, and that this specificity could be utilized easily and with visible results in bacterial agglutination, the precipitin reaction, and immune hemolysis, led rapidly to widespread application of these procedures. It was clear from the early work of Ehrlich and Morgenroth not only that different species can be identified by their response to hemolytic antibodies, but also that even closely related species would show cross-reactions among their erythrocyte antigens. Even closely related plants could be identified with the precipitin reaction, utilizing appropriate antisera to test for the degree of cross-reaction shown by their protein antigens.[119] It thus became evident that serologic approaches might offer a powerful tool to study the taxonomic relationships among various species, and this was undertaken in a major way by G. H. F. Nuttall.

Nuttall's aim was nothing less than to establish a complete evolutionary tree based upon serologic studies, but the subtitle of his 1904 book on *Blood Immunity and Blood Relationships*[120] was the more modest "A demonstration of certain blood-relationships amongst animals by means of the precipitin test for blood." This massive report described 16,000 tests employing a large variety of antisera to compare the blood of many species, ranging from lobsters to primates. It was not possible with this technic, and with the limited number of tests made, to establish a clear sequence of evolutionary development, but certain close relationships could be confirmed, such as that between man and the higher primates, and between these and lower primates. Many of these relationships found later confirmation with the study of blood group antigens, the best example being the sharing of the Rh antigen between man and other primates.

However, the serologic approach to animal (and plant) taxonomy was found to be somewhat flawed, for at least two reasons. First, with the crude antisera available, directed against a multiplicity of antigens, the apparent closeness of the relationship between two species would depend upon the relative titers for the many different antigens contained in a given antiserum. Depending upon the specificities involved, one antiserum might show a close relationship while another might suggest a greater disparity between the species tested. Even with the availability of more specific sera, however, the relationship between species might be misinterpreted, depending upon whether the antibody specificity was directed against a more highly evolutionarily conserved protein, or one more subject to the mutational hazards of time. The same variability might result, if the specificity were directed against a more highly conserved or less highly conserved region of the same protein. Given these shortcomings, the use of serologic approaches to the study of taxonomy has given way to the estimation of amino acid sequence homologies for various proteins. Perhaps the best example of this is the evolutionary tree for vertebrates established by the comparative study of the amino acid sequences of immunoglobulin light and heavy chains.[121]

One of the most useful applications of serology has been to the study of anthropologic relationships among the different races of man. This is based upon the identification of the numerous blood group systems possessed by the human. Having shown that the inheritance of these blood groups follows fairly simple Mendelian genetic rules,[122] Ludwig Hirszfeld went on to illustrate the broad applicability of blood typing for anthropologic study in his 1926 review "Serologic Predispositions and Blood Group Research."[123] It has since become well established that the gene frequencies for one or another blood group vary from one race or subpopulation to another,[124] so that comparison of the incidence of the several blood groups among different populations may reveal significant relationships between peoples, and provide important information on their mass migrations and intermixing. In a similar fashion, the variable incidence of many of the human lymphocyte antigens among different populations has contributed importantly not only to such anthropologic studies, but also to the epidemiologic distribution of certain HLA-related diseases.[125]

Forensic pathology

Just as the antisera prepared against the serum proteins of different species may be used to determine similarities, so may they be employed also to identify the species of origin of a blood by measuring antigenic differences. The possibilities inherent in this approach were recognized and adapted early by forensic laboratories, to distinguish between human and animal bloods in criminal cases, or to test for the substitution of an inferior species in meat and meat products.

A more common medico-legal application of serologic technics involves the analysis of the red cell group of blood specimens. As early as 1903, Landsteiner and Richter[126] noted the possible application of blood group differences to forensic practice. Relying only upon the major ABO blood group system, Lattes pointed out in 1915[127] that even if the certain identification of a blood stain with a given individual could not be made, definite exclusion of that individual was possible if his blood group differed from that of the specimen. In 1923, the same author pointed out that similar considerations apply to questions of paternity.[128] Comparison of the blood group of an infant with that of its putative father might yield results only suggestive of paternity if they are compatible, but would definitely rule out that possibility if incompatible. Since spermatozoa also manifest the major blood group antigens on their surface, blood typing has also been applied forensically in cases of rape, with an interpretation of results similar to that employed in the analysis of blood specimens.

If all of the minor blood groups are included in these analyses, then the number of combinations is so large as to provide a red cell identification almost as unique as that of a fingerprint. However, recent advances in molecular biology have demonstrated the uniqueness of an individual's DNA sequence, and this approach is rapidly gaining judicial acceptance for identification purposes. It may soon supplant the forensic applications of serologic approaches.

Comment

The study of the history of technologies shows how difficult it is to predict at the outset how important any one of them will be to the field, and how the modifications of a technique may extend its applications far beyond its original use. Thus, for example, the first chemical treatment of an antigen was a simple exercise to see how its antigenicity would be affected. But Karl Landsteiner would show that such treatment might be adapted to the study of immunological specificity. This approach eventually led to an understanding of the heterogeneity of the response, the size of the antibody combining site, its valence, and the thermodynamics of its interactions. Then such labels were adapted to serve as tracers for both the light and electron microscopes, leading to new concepts regarding the structure and function of the immune system. The shift from dye and radioactive to fluorescent tracers and from the labeling of antigens to that of antibodies not only increased the many applications of the technique, but also led to the development of fluorescence-activated cell sorting. Without this technical advance, our knowledge of lymphoid cell lineages and of the markers that define functional subsets would not only have been incomplete, but would probably also have taken a quite different and unpredictable form. Again, few technical innovations would produce more significant (though unintended) consequences than did the hybridoma and its monoclonal antibody product.[129] Similarly, the discovery of immune hemolysis mediated by antibody and complement led to a shower of unpredictable technological innovations, as described in Chapter 20.

We may note also how a concept and the terminology with which it is expressed may change, as techniques are adapted to the changing nature of the research program. A "label" in immunology originally meant a small molecule dye or hapten, or even a radioactive atom attached to a larger protein carrier molecule. Then it became possible to "label" one protein with another, one peptide chain with another, and even a peptide chain with multiple small portions of another chain, as the methods of molecular biology were increasingly applied. The result was that it is sometimes difficult to decide which is the label and which the carrier.

At each step in the increasingly complex technical development, we see an expansion of the applications to answer new questions. A productive cycle emerges; data from one experiment stimulates a new idea, leading to another question, perhaps a new modification of the technique and a new application, and then new data – and the science advances. This is no less true of the practical consequences of a technological approach than of its theoretical implications. Indeed, the technical method may even assume a life of its own, resulting in what Rheinberger has termed an "autonomously driven research trajectory."[130] In this capacity, the technique becomes a sort of experimental system with its own intrinsic and expanding heuristic value, where the investigator is prompted to ask new questions and to develop new theories unimaginable in the absence of these technological advances.

A technique may not only extend its hegemony far beyond the purpose for which it was originally developed; it may even serve as the basis for a new discipline. Thus, there came a time during the expansion of fluorescent antibody studies when some individuals began to study and perfect the method itself (as happened with electron microscopy), rather than continuing to use it as a tool to explore some other question. The ultimate transition could be seen following the development in 1906 of the Wassermann complement fixation test for syphilis. Originating in the basic immunology laboratory, this serological test was then applied so broadly that soon not only did "serologists" appear, but indeed serology became a discipline in its own right. In the present context, we may note that already the field of oncology includes "cancer immunotherapists."

From a certain point of view, therefore, it may not be unreasonable to say about immunology what Lord Adrian said about electrophysiology – that its history has been decided by the history of its technologies.[131]

Notes and references

1. Ehrlich, P., *Collected Papers of Paul Ehrlich*, Vol. 3, p. 59, London, Pergamon, 1960.

2. In addition to the specific references detailed below, descriptions of the many tech-nics described, with their applications, may be found in: Weir, D.M., ed., *Handbook of Experimental Immunology*, 4 vols, Cambridge, Blackwell, 1996; Rose, N.R., Friedman, H., and Fahey, J.L., eds, *Manual of Clinical Laboratory Immunology*, 3rd edn, Washington, DC, American Society of Microbiology, 1986.

3. Obermeyer, F., and Pick, E.P., *Wien. klin. Wochenschr.* **19**:327, 1906; Pick, E.P., "Biochimie der Antigene," *Handbuch der pathogenen Mikroorganismen*, 2nd edn, Part I, Jena, Fischer, 1912, pp. 685–868.

4. Further testimony to the rapid expansion of interest in these techniques is provided by Polak, J.M., and Van Noorden, S., eds, *Immunocytochemistry: Practical Applications in Pathology and Biology*, Bristol, J. Wright, 1983. In just three years between the first and second editions (1983–1986), the size increased from 396 to 703 pages.

5. Keating, P., and Cambrosio, A., *J. Hist. Biol.* **27**:449, 1994.

6. It was Ehrlich in the 1890s who first described toxoids, and Glenny and Ramon thirty years later who showed how they might be prepared efficiently.

7. See, for example, one of the early editions of Topley and Wilson's *Principles of Bacteriology and Immunity*, 2nd edn, Baltimore, Williams & Wilkins, 1938; Parish, H.J., *A History of Immunization*, London, Livingstone, 1968.

8. Gissmann, L., et al., *Ann. NY Acad. Sci.* **690**:80, 1993; Paavonen, J., et al., *Lancet* **369**:2161, 2007.

9. See, for example, Greene, M.I., and Nisonoff, A., eds, *The Biology of Idiotypes*, New York, Plenum, 1984; Köhler, H., Urbain, J., and Cazenave, P., eds, *Idiotypy in Biology and Medicine*, New York, Academic Press, 1984; *Immunol. Rev.* **90**, 1986.

10. Behring, E., and Kitasato, S., *Deut. med. Wochenschr.* **16**:1113, 1890.

11. Uhr, J.W., and Baumann, J.B., *J. Exp. Med.* **113**:935, 1961; see also Uhr, J.W., and Möller, G., *Adv. Immunol.* **8**:81, 1968.

12. Stern, K., Davidsohn, I., and Masaitis, L., *Am. J. Clin. Pathol.* **26**:833, 1956.

13. Dray, S., *Nature Lond.* **195**:677, 1962; see also Kindt, T.J., *Adv. Immunol.* **21**:35, 1975.

14. Kincade, P.W., Lawton, A.R., Bockman, D.E., and Cooper, M.D., *Proc. Natl Acad. Sci. USA* **67**:1918, 1970.

15. See *Immunol. Rev.* **74**, 1973.

16. This declaration was made by Ehrlich in 1906 at the dedication ceremony marking the opening of the Georg-Speyer Haus at his Royal Institute for Experimental Therapy in Frankfurt; in Ehrlich, *Collected Papers*, note 1, Vol. III, p. 60.

17. Héricourt, J., and Richet, C., *C. R. Acad. Sci.* **120**:948, **121**:567, 1895; Richet, C., and Héricourt, J., *Sem. Méd. Paris* **3**:510, 1895.

18. Salvati, V., and de Gaetano, L., *Riforma Med.* **11**:495, 507, 1895.

19. Bashford, E.F., Murray, J.A., and Haaland, M., *Sci. Report Imper. Cancer Res. Fund* **3**:396, 1908.

20. Hauschka, T.S., *Cancer Res.* **12**:615, 1953.

21. Pressman, D.P., and Keighley, G., *J. Immunol.* **59**:141, 1948; a long series of papers on this subject by Pressman and coworkers is summarized in Pressman, D.P., *Ann. NY Acad. Sci.* **69**:644, 1957.

22. Pressman's first published suggestion on the therapeutic possibilities of radiolabeled antibodies is in *Conference on Biological Applications of Nuclear Physics*, Upton, Brookhaven National Labs, 1948, p. 66.

23. Pressman, D.P., *J. Allergy* **22**:387, 1951.
24. Pressman, note 23. The same suggestion was made by Bale, W.F., Spar, R.L., Goodland, R.L., and Wolfe, D.E. (*Proc. Soc. Exp. Biol. Med.* **89**:564, 1955) and by Lawrence, J.H., and Tobias, C.A. (*Cancer Res.* **16**:185, 1956) with respect to the carriage of radioactive isotopes to a tumor.
25. Ada, G.L., and Byrt, P., *Nature*, **222**:1291, 1969; Humphrey, J.H., and Keller, H.U., in Šterzl, J., and Řiha, I., eds, *Developmental Aspects of Antibody Formation and Structure*, Vol. 2, pp. 485–502, New York, Academic Press, 1970; Ada, G.L., et al., *ibid*, vol. 2, pp. 503–519.
26. Burnet, F.M., *The Clonal Selection Theory of Acquired Immunity*, London, Cambridge University Press, 1959.
27. Volkman, D.J., et al., *J. Exp. Med.* **156**:634, 1982; Vitetta, E.S., et al., *Science* **219**:644, 1983.
28. Kennett, R.H., Bechtol, K.B., and McKearn, T.J., eds., *Monoclonal Antibodies and Functional Cell Lines: Progress and Applications*, New York, Plenum, 1984.
29. Goldenberg, D.M., ed., *Cancer Therapy with Radiolabeled Antibodies*, Boca Raton, CRC Press, 1995; Wahl, R.L., in Sandler, M.P., et al., eds, *Diagnostic Nuclear Medicine*, 4th edn, New York, Lippincott Williams & Wilkins, 2003, pp. 969–985.
30. Sell, S., ed., *Serological Cancer Markers*, Totowa, Humana Press, 1992; Garrett, C.T., and Sell, S., eds, *Cellular Cancer Markers*, Totowa, Humana Press, 1995.
31. Mishima, Y., ed., *Cancer Neutron Capture Therapy*, Boston, Kluwer, 1996.
32. Bale, W.F., *Proc. Natl Cancer Conf.* **2**:967, 1952. Bale, W.F., and Spar, R.L. (*J. Immunol.* **73**:125, 134, 1954) suggested that even nonradioactive isotopes such as boron-10, with a high absorption cross-section for slow neutrons, might be attached to antibody and localized at a tumor, so that exposure to a high neutron flux would lead to local irradiation of the desired target.
33. Hall W.A., ed., *Immunotoxin Methods and Protocols*, Totowa, Humana Press, 2000.
34. Moolten, F.L., and Cooperband, S.R., *Science* **169**:68, 1970; see also Thorpe, P.E., et al., *Nature* **271**:52, 1978.
35. Houston, L.L., and Ramakrishnan, S., in Vogel, C.-W., ed., *Immunoconjugates: Antibody Conjugates in Radioimaging and Therapy of Cancer*, New York, Oxford University Press, 1987, pp. 71–96. A list of the various immunotoxins and their applications is given by Thrush, G.R., et al., *Annu. Rev. Immunol.* **14**:49, 1996.
36. Kahan, B.D., Rajagopalan, P.R., and Hall, M.L., *Transplantation* **67**:276, 1999; Vincent, F., et al., *N. Engl. J. Med.* **338**:161, 1998.
37. Vallera, D.A., in Frankel, A.E., ed., *Immunotoxins*, Boston, Kluwer, 1988, pp. 515–535; Bachier, C.R., and LeMaistre, C.F., in Grossbard, M.L., ed., *Monoclonal Antibody-Based Therapy of Cancer*, New York, Marcel Dekker, 1998, pp. 211–227.
38. Mathé, G., Loc, T.B., and Bernard, J., *C .R. Acad, Sci.* **246**:1626, 1958. Ghose et al. later used a chlorambucil conjugate, *Br. Med. J.* **3**:495, 1972.
39. These include aminopterin, methotrexate, idarubicin, doxorubicin, vindesine, cisplatin, mitomycin C, adriamycin, daunamycin and cytosine arabinoside; Blair, A.H., and Ghose, T.I., *J. Immunol. Methods* **59**:129, 1983; Sela, M., and Hurwitz, E., in Vogel, C.-W., ed., *Immunoconjugates: Antibody Conjugates in Radioimaging and Therapy of Cancer*, New York, Oxford University Press, 1987, pp. 189–216; Pietersz, G.A., et al., *Antibody Immunoconj. Radiopharmaceut.* **3**:27, 1990.

40. Vogel, C.-W., in Vogel, *Immunoconjugates*, note 39, pp. 170–188.
41. Sievers, E.L., et al., *J. Clin. Oncol.* **19**:3244, 2001.
42. Bagshawe, K.D., *Br. J. Cancer* **56**:531, 1987; Senter, P.D., et al., *Proc. Natl Acad. Sci. USA* **85**:4842, 1988; Melton, R.G., and Knox, R.J., *Enzyme-Prodrug Strategies for Cancer Therapy*, New York, Kluwer, 1999.
43. Mew, D., et al., *J. Immunol.* **130**:1473 1983; Yarmush, M.L., et al., *Crit. Rev. Ther. Drug Carrier Syst.* **10**:197, 1993.
44. Wat, C.-K., et al., *Progr. Clin. Biol. Res.* **170**:351, 1984; Oseroff, A.R., et al., *Proc. Natl Acad. Sci. USA* **83**:8744, 1986.
45. Torchilin, V.P., *Immunomethods* **4**:244, 1994. The entire issue is devoted to liposome technology and applications.
46. Ghetie, M.A., et al., *Blood* **80**:2315, 1992; Flavell, D.J., et al., *Intl J. Cancer* **62**:1, 1995.
47. Smallshaw, J.E., *Nature Biotechnol.* **21**:387, 2003; Kreitman, R.J., *Nature Biotechnol.* **21**:372, 2003.
48. Bjorn, M.J., and Villemez, C.L., in Frankel, A.E., ed., *Immunotoxins*, Boston, Kluwer, 1988, pp. 255–277; Fidias, P., in Grossbard, M.L., ed., *Monoclonal Antibody-Based Therapy of Cancer*, New York, Marcel Dekker, 1998, pp. 281–307.
49. Morrison, S.L., et al., *Proc. Natl Acad. Sci. USA* **81**:6851, 1984; Boulianne, G.L., Hozumi, N., and Shulman, M.J., *Nature* **312**:643, 1984; Reichmann, L., et al., *Nature* **332**:323, 1988; Winter, G., and Harris, W.J., *Immunol. Today* **14**:243, 1993.
50. A recombinant IT, by fusing two antibody variable domains to pseudomonas toxin, was described by Chaudhary, V.K., et al., *Nature* **339**:394, 1989; Brinkmann, U., et al., *Proc. Natl Acad. Sci. USA* **90**:7538, 1993; Pastan, I., and Kreitman, R.J., *Investigational Drugs* **3**:1089, 2002.
51. Green, L.L., et al., *Nature Genet.* **7**:13, 1994; Little, M., et al., *Immunol. Today* **21**:364, 2000.
52. For Fvs, Owens, R.J., and Young, R.J., *J. Immunol. Methods* **168**:149, 1994; for TCRs, Shusta, E.V., *Nature Biotechnol.* **7**:754, 2000.
53. Boder, E.T., and Wittrup, K.D., *Nat. Biotechnol.* **15**:553, 1997; Feldhaus, M.J., et al., *Natl Biotechnol.* **21**:163, 2003.
54. Boder, E.T., Midelfort, K.S., and Wittrup, K.D., *Proc. Natl Acad. Sci. USA* **97**:10701, 2000; Holler, P.D., et al., *Proc. Natl Acad. Sci. USA* **97**:5387, 2000; van den Beuken, T., et al., *FEBS Letts* **546**:288, 2003.
55. Foote, J., and Eisen, H.E., *Proc. Natl Acad. Sci. USA* **97**:10697, 2000.
56. See, for example, Chirigos, M.A., ed., *Control of Neoplasia by Modulation of the Immune System*, New York, Raven, 1977; Mihich, E., ed., *Immunologic Approaches to Cancer Therapeutics*, New York, Wiley, 1982; Salmon, S.E., ed., *Adjuvant Therapy of Cancer*, New York, Grune & Stratton, 1987.
57. Mastrangelo, M.J., and Berd, D., in *Immunologic Approaches to Cancer Therapeutics*, note 56.
58. von Wassermann, A., Neisser, A., Bruck, C., and Schucht, A., *Z. Hyg.* **55**:451, 1906.
59. Bordet, J., *Ann. Inst. Pasteur* **12**:688, 1899.
60. Bordet, J., and Gengou, O., *Ann. Inst. Pasteur* **15**:289, 1901.
61. Kahn, R.L., *Serum Diagnosis of Syphilis by Precipitation*, Baltimore, Williams & Wilkins, 1925.

62. Nelson, R.A., and Mayer, M.M., *J. Exp. Med.* **89**:369, 1949.
63. This "Quellung Reaction" was first described by Neufeld (*Z. Hyg. Infektionskr.* **40**:54, 1902).
64. These diagnostic procedures are reviewed in Rose, N.R., and Mackay, I.R., eds, *The Autoimmune Diseases*, Orlando, Academic Press, 1985. The evolution of these technics can be appreciated by comparing with the fourth edition of this book, 2006.
65. Arthus, M., *C. R. Soc. Biol.* **55**:817, 1903.
66. Coca, A.F., Walzer, M., and Thommen, A.A., *Asthma and Hay Fever in Theory and Practice*, Springfield, Thomas, 1931.
67. Prausnitz, K., and Küstner, H., *Zbl. Bakt.* **86**:160, 1921.
68. Kay, A.B., Austen, K.F., and Lichtenstein, L.M., eds, *Asthma, Physiology, Pharmacology, and Treatment*, Orlando, Academic Press, 1984.
69. Koch, R., *Deut. med. Wochenschr.* **16**:756; **17**:101, 1189, 1891.
70. These and other such tests are described in Topley and Wilson's *Principles of Bacteriology and Immunity*, note 7.
71. See discussion in Chapter 9.
72. Landsteiner, K., and Chase, M.W., *Proc. Soc. Exp. Biol. Med.* **49**:688, 1942; Chase, M.W., *Proc. Soc. Exp. Biol. Med.* **59**:134, 1945.
73. Sanger, R., *Blood Groups in Man*, 6th edn, Philadelphia, F.A. Davis, 1975.
74. See section on "Immunogenetics and Transplantation Immunology" in Rose, Friedman, and Fahey, *Manual*, note 2, pp. 822–920.
75. Vogel, C.-W., *Immunoconjugates: Antibody Conjugates in Radioimaging and Therapy of Cancer*, New York, Oxford University Press, 1987; *Antibody Immunoconj. Radiopharmaceut.* **2**, 1989, special issue; Chatal, J.-F., ed., *Monoclonal Antibodies in Immunoscintigraphy*, Boca Raton, CRC Press, 1989; Goldenberg, D.M., ed., *Cancer Imaging with Radiolabeled Antibodies*, Boston, Kluwer, 1990; Perkins, A.C., and Pimm, M.V., *Immunoscintigraphy: Practical Aspects and Clinical Applications*, New York, Wiley-Liss, 1991.
76. Pressman, D.P., and Keighly, G., *J. Immunol.* **59**:141, 1948. This approach was then utilized by Pressman, D.P., and Korngold, L., *Cancer* **6**:619, 1953, and by Pressman, D.P., and Day, E.D., *Cancer Res.* **17**:845, 1957.
77. Saccavini J.C., Bohy J., and Bruneau J., in Chatal, *Monoclonal Antibodies*, note 75, pp. 61–73.
78. Gansow, O.A., et al., in Goldenberg, *Cancer Imaging*, note 75, pp. 153–171.
79. Gohr-Rosenthal, S., et al., *Invest. Radiol.* **28**:789, 1993.
80. Remsen, L.G., et al., *Am. J. Neuroradiol.* **17**:411, 1996.
81. Kraus, R., *Wien. klin. Wochenschr.* **10**:736, 1897.
82. Heidelberger, M., and Kendall, F.E., *J. Exp. Med.* **50**:809, 1929.
83. Oudin, J., *Ann. Inst. Pasteur* **75**:30, 1948.
84. Ouchterlony, Ö, "Antigen–antibody reactions in gels," *Acta Pathol. Microbiol. Scand.* **26**:507, 1949.
85. Grabar, P., and Williams, C.A., *Biochem. Biophys. Acta* **10**:193, 1953; see also Grabar, P., and Burtin, P., *Immunoelectrophoretic Analysis*, Amsterdam, Elsevier, 1964.
86. Yalow, R.S., and Berson, S.A., *J. Clin. Invest.* **40**:2190, 1961.
87. Avrameas, S.P., Druet, P., Masseyeff, R., and Feldmann, G., *Immunoenzymatic Techniques*, Amsterdam, Elsevier, 1983.

88. "Nomenclature for clusters of differentiation (CD) of antigens defined on human leukocyte populations," *Bull. World Health Org.* **62**:809, 1984; Shaw, S., *Immunol. Today* **8**:1, 1987.

89. Loken, M.R., and Herzenberg, L.A., *Ann. NY Acad. Sci.* **254**:163, 1975; Herzenberg, L.A., De Rosa, S.C., and Herzenberg, L.A., *Immunol. Today* **21**:383, 2000. For the application of FACS to cell surface markers, see Herzenberg, L.A., ed., *Weir's Handbook of Experimental Immunology*, 5th edn, Vol. II. *Cell Surface and Messenger Molecules of the Immune System*, Cambridge, Blackwell Scientific, 1996. For an interesting ethnographic account of the FACS story, see Cambrosio, A., and Keating, P., *Med. Anthropol. Q.* **6**:362, 1992. See also Cambrosio and Keating's *Biomedical Platforms*, Cambridge, MIT Press, 2003.

90. Chess, L., and Schlossman, S.F., *Advances Immunol.* **25**:213, 1977.

91. Heidelberger, M., and Kendall, F.E., *J. Exp. Med.* **58**:137, 1933.

92. Marrack, J.R., *Nature* **133**:292, 1934.

93. Sabin, F.R., *J. Exp. Med.* **70**:67, 1939. These studies were extended by Smetana, H., *Am. J. Pathol.* **23**:255, 1947, and by Kruse, H., and McMaster, P.D., *J. Exp. Med.* **90**:425, 1949.

94. Coons, A.H., Creech, H.J., and Jones, R.N., *Proc. Soc. Exp. Biol. Med.* **47**:200, 1941; *J. Immunol.* **45**:159, 1942. Coons described the early history of his involvement with fluorescent labels in his 1961 Presidential address to the American Association of Immunologists, *J. Immunol.* **87**:499, 1961.

95. Coons, A.H., and Kaplan, M.H., *J. Exp. Med.* **91**:1, 1950; see also Nairn, R.C., *Fluorescent Protein Tracing*, 4th edn, Edinburgh, Churchill Livingston, 1976.

96. Coons, A.H., Leduc, E.H., and Connolly, J.M., *J. Exp. Med.* **102**:49, 1955. Another "sandwich" method involves the conjugation of a hapten to the antibody, with development of color using fluorescein-labeled anti-hapten antibody, Lamm, M.E., et al., *Proc. Natl Acad. Sci.* **69**:3732, 1972; Wofsy, L., et al., *J. Exp. Med.* **140**:523, 1974.

97. Silverstein, A.M., *J. Histochem. Cytochem.* **5**:94, 1957; Hiramoto, R., Engel, K., and Pressman, D.P., *Proc. Soc. Exp. Biol. Med.* **97**:611, 1958.

98. Coons, Leduc, and Connolly, note 96. This conclusion had earlier been reached by Astrid Fagreaus, "Antibody production in relation to the development of plasma cells," *Acta Med. Scand.* **Suppl. 204**, 1948.

99. Ortega, L.G., and Mellors, R.C., *J. Exp. Med.* **106**:627, 1957.

100. Chu, A.C., "Immunocytochemistry in dermatology," in Polak, J.M., and Van Noorden, S., *Immunocytochemistry*, 2nd edn, Bristol, Wright, 1986, pp. 618–637; Unanue, E.R., and Dixon, F.J., *Advances Immunol.* **6**:1, 1967; Elias, J.M., *Immunohistopathology: A Practical Approach to Diagnosis*, Chicago, ASCP Press, 1990.

101. Nakane, P.K., and Pierce, G.B., *J. Histochem. Cytochem.* **14**:929, 1966; Avrameas, S., and Lespinats, G., *C. R. Acad. Sci. Paris* **265**:1149, 1967; Avrameas, S., *Immunochemistry* **6**:43, 1969.

102. The many uses of this approach are detailed in Chapters 3–7 of Cuello, A.C., ed., *Immunohistochemistry*, New York, John Wiley, 1983. The technique has largely been superseded by an immunoenzyme sandwich method in which the enzyme binds to an antibody combining site rather than being attached chemically (Sternberger, L.A., and Cuculis, J.J., *J. Histochem. Cytochem.* **17**:190, 1969; Sternberger, L.A., *Immunocytochemistry*, 4th edn, New York, John Wiley, 1989).

103. Mason, D.Y., and Sammons, R.E., *J. Clin. Pathol.* **31**:454, 1978. For applications of this technique using monoclonal antibodies, seee Boorsma, D.M., *Histochemistry.* **80**:103, 1984.

104. Edwards, J.C., and Moon, C.R., in Herberman, R.B., and Mercer, D.W., eds, *Immunodiagnosis of Cancer*, 2nd edn, New York, Dekker, 1990, pp. 95–106.

105. See, for example, Avrameas and Lespinats, note 101; Sternberger, L.A., *Immunocytochemistry*, 2nd edn, New York, Wiley, 1979.

106. Polak, J.M., and Varndell, I.M., eds, *Immunolabeling for Electron Microscopy*, Amsterdam, Elsevier, 1984.

107. Mason, D.Y., et al., in McMichael, A.J., and Fabre, J., eds, *Monoclonal Antibodies in Clinical Medicine*, London, Academic Press, 1982, pp. 585–635.

108. Singer, S.J., *Nature* **183**:1523, 1959.

109. Uranium: Sternberger, L.A., et al., *Exp. Mol. Pathol.* **4**:112, 1965; iron: Yamamoto, N., *Acta Histochem. Cytochem.* **10**:246, 1977; mercury: Kendall, P.A., *Biochim. Biophys. Acta* **97**:174, 1965.

110. Faulk, W.P., and Taylor, G., *Immunochemistry*, **8**:1081, 1971; De May, J.R., in Cuello, *Immunohistochemistry*, note 102, New York, John Wiley, 1983, pp. 347–372.

111. Roth, J., *J. Histochem. Cytochem.* **31**:987, 1983.

112. Romano, E.L., and Romano, M., *Immunochemistry* **14**:711, 1977.

113. Varndell, I.M., and Polak, J.M., in Polak, J.M., and Varndell, I.M., eds, *Immunolabeling for Electron Microscopy*, Amsterdam, Elsevier, 1964, pp. 155–177.

114. Libby, R.L., and Madison, C.R., *J. Immunol.* **55**:15, 1947.

115. See, for example, Hunt, S.P., Allanson, J., and Mantyh, P.W., "Radioimmunocytochemistry," in Polak and Van Noorden, *Immunocytochemistry*, note 100, pp. 99–114.

116. Coggi, G., Dell'Orto, P., and Viale, G., in Polak and Van Noorden, *Immunocytochemistry*, note 100, pp. 54–70.

117. Towbin, H., Staehelin, T., and Gordon, J., *Proc. Natl Acad. Sci. USA* **76**:4350, 1979.

118. Jerne, N.K., and Nordin, A.A., *Science* **140**:405, 1963; Ingraham, J.S., *C. R. Acad. Sci.* **256**:5005, 1963; Ingraham, J.S., and Bussard, A., *J. Exp. Med.* **119**:667, 1964; see also Mishell, R.I., and Dutton, R.W., *Science* **153**:1004, 1966.

119. Kowarski, A., *Deut. med. Wochenschr.* **27**:442, 1901.

120. Nuttall, G.H.F., *Blood Immunity and Blood Relationships*, Cambridge, University Press, 1904.

121. Fitch, W.M., and Margoliash, E., *Science* **155**:279, 1967; Grey, H.M., *Adv. Immunol* **10**:51, 1969; Hood, L., and Prahl, J., *Adv. Immunol.* **14**:291, 1971.

122. von Dungern, F., and Hirszfeld, L., *Z. Immunitätsf.* **4**:531; **6**:284, 1910.

123. Hirszfeld, L., *Ergeb. Hyg. Bakt.* **8**:367, 1926.

124. Mourant, A.E., *The Distribution of Human Blood Groups*, Blackwell, Oxford, 1964.

125. Dawkins, R.L., et al., eds, "HLA and disease susceptibility," *Immunol. Rev.* **70**, 1983; Tiwar, J.L., and Terasaki, P.I., eds, *HLA and Disease Association*, New York, Springer, 1985.

126. Landsteiner, K., and Richter, M., *Z. Medizinalbeamte* **16**:85, 1903.

127. Lattes, L., *Arch. Antropol. Crim. Med. Leg.* **36**:4, 1915; *Arch. Ital. Biol.* **64**:3, 1915; Lattes, L., *Individuality of the Blood*; English translation, Oxford, Oxford University Press, 1932.

128. Lattes, L., *Riforma Med.* **39**:169, 1923; the history of the numerous applications of blood grouping is discussed by Schneider, W.H., *Bull. Hist. Med.* **57**:545, 1983.
129. The technological revolution that would accompany monoclonal antibodies was not immediately apparent at the time, even to its authors. Thus, the editors of *Nature* might be forgiven for having rejected the Köhler–Milstein manuscript as not worthy to appear as a scientific report. They consigned it to the middle of a very long list of letters!
130. Rheinberger, H.-J., *Stud. Hist. Phil. Sci.* **23**:305, 1992.
131. Adrian, E.D., *The Mechanism of Nervous Action. Electrical Studies of the Neurone*, Philadelphia, University of Pennsylvania Press, 1932, p. 4.

13 The royal experiment: 1721–1722

The small-pox…is here [in Turkey] entirely harmless.
Lady Mary Wortley Montagu, 1717

It is not unusual for events that occur outside the realm of science to exert a profound effect on the directions that science takes and on the rate of its forward progress. One need only look at the role played in the seventeenth century by governments in the founding of the Accademia dei Linci in Italy (1603), the Royal Society in England (1660), or the Académie des Sciences in France (1665). In more modern times, the rise of German science (and especially medicine) in the late nineteenth century was profoundly affected by the support of Minister Friedrich Althoff in Bismark's Germany; he found posts and Institutes for Koch, Behring, and Ehrlich, among many others.[1] Similarly, in America the Rockefeller Foundation did much for medical research in the early years of the twentieth century,[2] as did the expanded funding for the National Institutes of Health and the National Science Foundation following World War II.

Each time that the ruler of a nation suffered a disease – be it Louis XIV's anal fistula, George III's porphyria, Franklin Roosevelt's poliomyelitis, or Dwight Eisenhower's heart attack – interest in that disease was heightened. In this chapter, we shall see how the use of a prophylactic method for the prevention of smallpox by the Royal Family of Great Britain stimulated the medical profession to investigate the practice further, and the public to submit to it more readily.

Smallpox and its prevention

By the middle of the seventeenth century, smallpox (along with typhus) had replaced the plague as the leading infectious cause of death in the adult population of Europe.[3] Epidemics of smallpox appeared with increasing frequency,[4] and were all the more noticed because, unlike many other contemporary diseases, they afflicted the rich and powerful as well as the poor.[5] Many feared the disease as much for the disfigurement suffered by its survivors as for its impressive mortality, which averaged some 15 to 20 percent of those infected.[6]

It is customary to credit Edward Jenner with the development of the first effective immunization procedure to protect against an infectious disease. However, when Jenner first published his findings on the use of cowpox vaccination in 1798[7] there already existed an equally effective and (for the times) reasonably safe immunization procedure. This was smallpox inoculation, involving the (usually) dermal infection of the subject with the wild virus, which

A History of Immunology, Second Edition
ISBN: 978-0-12-370586-0

Copyright © 2009, Elsevier Inc.
All rights reserved

most often resulted in a mild, transient illness that thenceforth protected the individual against more severe forms of the disease. Prior to Jenner's publication, this procedure had been practiced with generally favorable results for some three-quarters of a century in "polite" society in Europe, and long before that was employed in many countries as a standard practice in the folk-medicine of "more primitive" peoples. It was, of course, common knowledge in the eighteenth century (and earlier) that a case of smallpox confers lifelong immunity.

The manner in which the practice of smallpox inoculation was introduced into England, where it first attained broad recognition and application,[8] is of great interest in several respects. First, it illustrates the role of the medical "establishment" in conferring dignity upon, and gaining acceptance for, a new procedure. Second, it provides an interesting contrast between the eighteenth and later centuries, with respect to the prestige of the physician and scientist in society. Whereas a Pasteur, a Behring, or a Koch could feel free in the 1880s to introduce novel procedures to the practice of medicine with no appeal to other than his own authority, the physician of a century and a half earlier hesitated, and felt impelled to appeal elsewhere for more powerful patronage and support. Third, the introduction of inoculation to England involved what was probably the first recorded clinical trial in the history of immunity. And, finally, it involved the use of human guinea pigs chosen from among the underprivileged and "inferior" peoples – a common practice that persisted well into the twentieth century, and has only recently begun to be questioned.

The introduction of inoculation into England

In Boswell's *Life of Johnson*, the learned doctor is recorded as suggesting that it was foolish to send Radcliffe traveling fellows to the Continent in order to add to medical knowledge; they should rather go "out of Christendom" and visit "barbarous nations."[9] This was a common sentiment of his day, when the accounts of travelers to distant parts were eagerly read by an interested public, and when it was often stated that every single effective new remedy, such as cinchona or ipecacuanha, stemmed from its use among primitive peoples. It was, however, often easier to learn about such foreign arts and inventions than to introduce them and gain their acceptance by a basically conservative society. This was especially true of medical innovation in the early eighteenth century, since not only was the general view of disease and of therapy tradition-bound to 2,000-year-old humoralist concepts, but its practitioners had also not yet become the independent social innovators that the following century would witness.

The Royal Society of London was established for the promotion of learning, and by 1700 had already achieved a prominent position as a clearinghouse for communications of scientific and medical interest from all parts of the world. It was common for Fellows of the Royal Society to receive letters from individuals at some distant outpost, relaying new information or sending interesting specimens for study. One such communication, dated January 5, 1700, came to Dr Martin Lister, a prominent London physician and Fellow of the Royal Society, from an East India Company trader stationed in China.[10] This reported "a Method of Communicating the Small Pox," involving "opening the pustules of one who has the Small Pox ripe upon them and drying up the Matter with a little Cotton,...and afterwards put it up the nostrils of those they would infect." This procedure was preferable to natural infection, since the patient could be prepared for the illness and it could be done at an appropriate age and season for an optimal outcome. Apparently Dr Lister did not relay this information further, but by a curious coincidence this Chinese practice was reported to the Royal Society at its February 14, 1700 meeting by Dr Clopton Havers,[11] even before the letter to Lister could have reached its destination.

There is no evidence that this initial notice excited any attention within the medical establishment, and nothing more was heard of it in London for thirteen years. Then, on May 27, 1714, Dr John Woodward reported to the Royal Society extracts of a letter dated Constantinople, December 1713, "An Account, or History, of the Procuring the SMALL POX by Incision, or Inoculation; as it has for some time been practiced at Constantinople."[12] The writer of the letter was a Dr Emanuele Timoni, born in Greece of Italian parents, who had taken his medical degree at Padua, had taken another degree at Oxford, and had been elected Fellow of the Royal Society in 1703. At the time that he wrote on smallpox inoculation, Timoni had been practicing medicine in Constantinople for some years and was in fact family physician to the British Ambassador to the Porte, Sir Robert Sutton, and to his successor, Lord Edward Wortley Montagu.

In the letter, Timoni described inoculation as a familiar practice "for about the space of 40 years among the Turks and others..." He gave a careful description of the choice of an appropriate donor, of the manner in which the patient was to be inoculated, and of the clinical course of the resulting mild disease. He observed:

> that altho' at first the more prudent were very cautious in the use of this Practice; yet the happy Success it has been found to have in thousands of Subjects for these eight Years past, has now put it out of all suspicion and doubt; since the Operation having been perform'd on Persons of all Ages, Sexes, and different Temperaments, and even in the worst Constitution of the Air, yet none have been found to die of the Small-Pox; when at the same time it was very mortal when it seized the Patient in the common way, of which half the affected dy'd.

The Timoni report provoked several discussions of inoculation at the Royal Society meetings, resulting in a motion to instruct the Secretary to obtain further information on it from the British Consul at Smyrna. Two years later, the Royal Society reprinted in full a more extensive description of smallpox inoculation,[13] prepared by Dr Jacobo Pylarini, then serving in Smyrna as Venetian Consul. Pylarini also testified to the efficacy and relative safety of the inoculation procedure. However, despite the widespread publicity given to the practice of inoculation in the *Philosophical Transactions of the Royal Society* and in other leading scientific periodicals of the day, there seemed to be little inclination on the part of a cautious medical profession to adopt the practice. Physicians were apparently afraid to risk their reputations by testing this novel procedure.

However, two individuals (neither, significantly, a physician) did independently undertake to popularize inoculation. One was Cotton Mather of Boston, who sent a letter in July of 1716 to Dr John Woodward of the Royal Society, asking why the practice of inoculation had not been tried in England, and stating that he intended to persuade the Boston physicians to employ it the next time that smallpox entered the city.[14] He was as good as his word; during the smallpox epidemic of 1721 in Boston he persuaded Dr Zabdiel Boylston to undertake inoculation, and later speculated on the nature of acquired immunity in smallpox infection.[15]

The second individual who espoused the doctrine of smallpox inoculation was Lady Mary Wortley Montagu, wife of the British Ambassador to Constantinople. It is not clear that Lady Mary was aware of the Timoni and Pylarini reports, but she soon learned of the local custom of smallpox inoculation, or ingrafting, and it made a great impression upon her. As she wrote in a letter to her friend Sarah Chiswell in April 1717:[16]

> *I am going to tell you a thing that I am sure will make you wish yourself here. The small-pox, so fatal, and so general amongst us, is here entirely harmless by the invention of* ingrafting...*I am patriot enough to take pains to bring this useful invention into fashion in England; and I should not fail to write to some of our doctors very particularly about it, if I knew any one of them that I thought had virtue enough to destroy such a considerable branch of their revenue for the good of mankind...Perhaps, if I live to return, I may, however, have courage to war with them.*

Lady Mary had obviously not much respect for the medical profession. She was, however, a firm convert to inoculation, and in March 1718 had her six-year-old son inoculated by Charles Maitland, surgeon to the Embassy.[17]

Lady Mary returned to London in 1719 but apparently did little to advance the cause of inoculation over the next two years, until the deadly smallpox epidemic of 1721 stimulated her to action. She thereupon sent for Maitland (who by this time had retired to a small town near London) and requested him to see to the inoculation of her three-year-old daughter, Mary. Maitland was somewhat hesitant to perform this Eastern practice in London, and insisted that outside physicians be called in as witnesses – he appeared to be reluctant to accept sole responsibility for the procedure. But finally, in late April of 1721, the young Montagu child was successfully inoculated. When the pocks appeared, three members of the College of Physicians examined her separately, and one of them was so convinced by what he saw that he had Maitland inoculate his own son – the only one of his children who had not yet already contracted the disease.

The importance of Lady Mary's demonstration of inoculation in the events that followed is not entirely clear. Undoubtedly it aroused considerable professional interest, although it was not reported in the newspapers of that time. History has usually credited Lady Mary with the dominant role in the introduction and acceptance of inoculation in England, based primarily on Voltaire's effulgent praise, and his report that she enjoyed the close friendship of the Princess of Wales,[18] but this explanation has been open to question.[19] Be that as it may, the groundwork had been laid for further experimentation with the inoculation procedure, and the currently raging smallpox epidemic, which was exacting a heavy toll among the upper classes, provided excuse enough to exploit any hopeful approach.

The royal experiment

One of the remarkable figures of eighteenth-century England then emerged as a champion of smallpox inoculation. This was Sir Hans Sloane, Bart, President of the Royal College of Physicians of London, one of the two Secretaries to the Royal Society (to whose Presidency he would succeed Isaac Newton in 1727) and court physician first to Queen Anne and then to the current ruler, George I. Sloane had an inquiring mind, and was interested in everything. (His personal library of over 50,000 books, and his vast collection of botanical specimens, coins, and antiquities would later be sold to the nation to form the core of the British Museum.[20]) Sloane was thus in an excellent position to advance the cause of smallpox inoculation, both directly and indirectly.

During the height of the smallpox epidemic in early May, 1721, the youngest child of the Prince and Princess of Wales fell ill with what was at first thought to be smallpox but later provided to be a milder ailment. It is not clear that this was the spark that stimulated an interest in inoculation by the intellectually and scientifically oriented Princess of Wales, Caroline of Ansbach – but every mother must have feared for her children in the midst of a deadly epidemic.

It is not known exactly who was responsible for the royal interest in inoculation. Voltaire's candidate was Lady Mary Wortley Montagu, as we have seen. A contemporary account published by a German visitor maintained that it was Dr Maitland who appealed to the Princess for permission to experiment further with the procedure,[21] while Sloane's own "Account of Inoculation," written some fifteen years after the event (and, unaccountably, not published until 1756), held that Princess Caroline "to secure her other children, and for the common good, begged the lives of six condemned criminals,…in order to try the experiment of inoculation upon them."[22] But Sloane's retrospective account contains other inaccuracies, and perhaps does not do justice to his own role in the affair as one of the most powerful promoters of the new scientific movement in England.

In any event, within a month of the illness of the royal Princess, the newspapers recorded that "some physicians" had made a representation to the King to obtain permission to carry out experiments on condemned criminals in

Newgate Prison, on condition that the prisoners receive subsequent pardon.[23] The King apparently acquiesced, for on June 14 the Secretary of State, Lord Townsend, addressed a letter to the Attorney and Solicitor Generals, asking them to advise the Crown "Whether His Majesty may by Law Grant his Gracious Pardon to two Malefactors under Sentence of Death upon Condition that they will suffer to be try'd upon them the Experiment of Inoculating the Small pox."[24] Three days later, an opinion was returned that:

> the Lives of the persons being in the power of His Majesty, he may Grant a Pardon to them upon such lawful Condition as he shall think fit; and as to this particular Condition We have no objection in point of Law, the rather because the carrying on this practice to perfection may tend to the General Benefit of Mankind.

This interest in science on the part of George I was not surprising, and the new scientific movement in England, exemplified by the Royal Society, was fortunate in finding itself under Hanoverian rule. Before becoming George I of England, the Elector of Hanover had patronized learning by supporting men like Leibnitz, and once in England both he and the Prince of Wales (later George II) had maintained their interest in science. They attended lectures on experimental philosophy, studied Newtonianism, and sponsored experiments in mechanics at Hampton Court. Indeed, the Princess of Wales' intellectual curiosity led Voltaire to refer to her, after her husband ascended the throne, as the *"Philosophe aimable sur le Trône."*

Whether Sir Hans Sloane was instrumental in prompting the initial appeal to the King is not known, but it was he who asked Maitland to perform the experiments upon the condemned criminals, and it was he who sought additional confirmation of the efficacy of the inoculation practice. Under the supervision of the two Royal Physicians, Sir Hans Sloane and Dr John George Steigherthal, three male and three female prisoners were inoculated by Maitland at Newgate Prison on the morning of August 9, 1721. The importance of the proceedings and the significance of the royal patronage is attested to by the extensive coverage that the experiment received in the press, and by the fact that it was witnessed by no less than twenty-five physicians, surgeons, and apothecaries, including eminent members of the College of Physicians and of the Royal Society.[25]

Incisions for the insertion of the smallpox pus were made on the arms and right legs of the convicts. Because, on August 12, the incisions were not as inflamed as expected, Maitland obtained sufficient fresh donor pock material to repeat the operation on five of the patients. Symptoms appeared on the next day in all but one man, and after a brief illness of varying severity, each of the patients recovered. (The one prisoner who never showed any reaction was found to have had smallpox the previous year.) Interest in these experiments was maintained, and the patients were visited almost daily by physicians and other interested persons. In addition, the newspapers continued their coverage of the experiments, and even reprinted the original Timoni description of five years earlier.

Shortly after the initial experiments, another prominent London physician, Dr Richard Mead, obtained permission to try the Chinese method of inoculation on a young woman prisoner. This involved intranasal administration of matter from a favorable case of smallpox. In this case the usual symptoms of smallpox appeared promptly, and although the woman was more seriously ill than any of the others, she recovered fully.[26] As agreed, all of the convicts who had been involved in these experiments were pardoned by the King and his Council, and were released from Newgate on September 6, 1721.

While the Newgate experiment convinced most people that smallpox inoculation was safe, there was still the question of its efficacy in preventing subsequent attacks of the natural disease. Some claimed that inoculation did not give true smallpox, but rather chicken- or swine-pox, and that it would not confer immunity. To test this, Sloane and Steigherthal arranged, at their own expense, to send one of the pardoned women to a small town near London where Maitland lived in retirement, and where a very severe smallpox epidemic was then raging. Under Maitland's supervision, the nineteen-year-old was ordered "to lie every Night in the Same Bed [with a 10-year-old smallpox victim], and to attend him constantly from the first Beginning of the Distemper to the very End." For a period of some six weeks, she was exposed to the most serious form of the disease without contracting it herself – a fact to which numerous witnesses testified.[27]

To further confirm the original experiment on the safety of smallpox inoculation, Maitland successfully inoculated an additional six persons the following February, again under royal sponsorship. In this instance the experiment was announced officially from Whitehall, and it was stated that:[28]

> the Curious may be further satisfied by a Sight of those Persons at Mr. Forster's House in Marlborough Court at the Upper-End of Poland-street and Berwick-street in Soho, where Attendance is given every Day from Ten till Twelve before Noon, and from Two till Four in the Afternoon.

Not satisfied with the inoculation experiments on adults, the newspapers announced in the middle of November that instructions had been given by the Princess of Wales to draw up a list of all orphan children in St James Parish, Westminster, who had not yet had smallpox, so that they might be inoculated at her expense. Although by this time Maitland had successfully inoculated several children in private, it was apparently felt that a public demonstration of the safety of the procedure in children was also necessary. Here again, the role of the royal advisors is well illustrated by a letter from the Royal Surgeon, Claude Amyand, to the Royal Physician, Sir Hans Sloane, in which, commenting on the availability of orphan children, Amyand writes:[29]

> What I thought proper to urge was, that these fresh instances might reconcile those that were yet diffident about the success of the inoculation. ...The princiss will be glad to know wuther you think these wanting, and Therefore came to waite on you on this account.

Five orphan children were inoculated successfully in March, and once again the newspapers published the details, indicating where interested persons might view the patients.

Primarily as a consequence of these inoculation experiments, confidence in the new method had mounted, by April of 1722, to the point where several prominent political figures called upon Maitland to inoculate their children. Finally, on April 17, 1722, the Prince and Princess of Wales, after consultation with Sir Hans Sloane, had the operation performed on two of their daughters – eleven-year-old Princess Amelia and nine-year-old Princess Caroline. While Sloane was hesitant to urge inoculation of the royal Princesses, neither would he attempt to dissuade the Princess of Wales "in a matter so likely to be of such advantage." Caroline resolved to have the inoculation performed, and sent Sloane to the King to obtain his permission. Sloane advised the King that "it was impossible to be certain but that raising such a commotion in the blood, there might happen dangerous accidents not foreseen." The King replied that such accidents might happen from being bled or taking *any* medicine, no matter how much care was taken, and gave the Royal consent to proceed. The Princesses were thereupon inoculated by Amyand, with the assistance of Charles Maitland, and under the supervision of the Royal Physicians, Sloane and Steigherthal. So important an event was naturally mentioned in all of the newspapers, and helped to insure at least the temporary popularity of the inoculation procedure in polite society.

Discussion

Opposition to inoculation

Despite the success of the Royal Experiment, and the prominent example set by the inoculation of the royal children, smallpox inoculation did not escape severe criticism at the time, nor did it become widely practiced. Indeed, in the seven years following the Royal Experiment, only 897 inoculations were reported in the British Isles, America, and Hanover.[30] Of these, seventeen patients (2 percent) had died. This figure was substantially confirmed in a 1759 pamphlet by one B. Franklin of Philadelphia, who argued forcefully nevertheless for the advantages of inoculation.[31] But while the mortality that accompanied smallpox inoculation was appreciably better than the estimated one person in each six to ten who might expect to die of natural smallpox (not to mention those who would be disfigured), it did indicate that the inoculation procedure was not without its hazards. In addition to citing the dangers of inoculation, including further spread of the infection by the inoculated individual, those opposed to the practice claimed variously that it did not give true smallpox; that it did not protect against subsequent disease; and, above all, that it went against both God and Nature.

The anti-inoculation banner was raised as early as 1722 by the surgeon Legard Sparham in his *Reasons against the Practice of Inoculating the Small-Pox.*

Among other concerns, Sparham argued against the insertion of "poisons" into wounds, and held that "till now, [we] never dreamt that Mankind would industriously plot to their own Ruin, and barter Health for Diseases."[32] Only a little later, the Reverend Mr Massey preached from the pulpit of the Parish Church of St Andrew's, Holborn, on "the Dangerous and Sinful Practice of Inoculation." Mr Massey held that the power to inflict disease rests upon God alone, and it is He who gives the power to heal. The fact that one possesses the physical power or knowledge to perform an act does not imply that one has the moral right to do so:[33]

> I shall not scruple to call that a Diabolic Operation, which usurps an Authority founded neither in the Laws of Nature or Religion, which tends in this case to anticipate and banish Providence out of the World, and promotes the encrease of Vice and Immorality.

If inoculation had its difficulties in England, however, it met with even greater opposition in France, where word of the London inoculation experiments soon arrived. Here, the medical establishment itself rose in opposition to inoculation. The Faculté de Médicine of the University of Paris was not only a teaching body, but also supervised medical police measures, drug inspection, medicolegal questions, and various other aspects of public health. When in 1723 the Faculté sponsored a disputation on inoculation – and then voted in favor of the thesis that inoculation was a useless, uncertain, and dangerous practice, and should be condemned – it was clear that inoculation would face great problems of acceptance in France. In fact, the adoption of inoculation in France lagged some forty years behind its introduction into England.[34]

In general, interest in smallpox inoculation in England waxed and waned in parallel with the comings and goings of epidemics of the disease. Thus, the return of epidemic smallpox to London in 1746 led to the founding of a Smallpox and Inoculation Hospital, established for the provision of free care for smallpox patients, and to make inoculation more readily available. But the epidemic ran its course, and interest in inoculation decreased, so that after ten years it could be reported that only 1,252 inoculations had been given in the hospital.[35] Interest in smallpox inoculation persisted for many years, even surviving the introduction of Jennerian vaccination, until, in 1840, an Act of Parliament outlawed the practice, in final recognition of the additional safety offered by Jenner's innovation.

Human experimentation

Few persons would have dreamed in 1721 of raising any question of the ethics of employing human prisoners for the type of experiment we have described here – indeed, few voices were raised against similar practices even in twentieth-century America.[36] Two or three centuries ago, as in most of recorded history, it was implicitly accepted and often explicitly stated that a life was valued increasingly more as one ascended the social ladder, and numerous examples of these valuations may be found. Thus, the Wangensteens, in their book *The Rise of Surgery*, cite the experiments done in an attempt to find a cure for Louis XIV's anal fistula.[37] In 1686, the King's physician, Guy Fagon, recommended that the surgeon Charles-François Félix relieve Louis XIV of his distressing ailment by operation. Provision was made for Félix to obtain knowledge and deftness in the procedure at a Paris hospital, and patients afflicted with such fistulas were sent there for surgery. These experiments went on for some months, until Félix felt ready to work on the King and, as the Wangensteens point out, "tradition has it that a number of these did not survive Félix's operation, and that the bodies were disposed of at night and the deaths were attributed to poisoning." In the end, the experiments succeeded, and the King's discomfort was alleviated.

Another example of this approach may be found in a suggestion made by J. Bellers in 1714 that hospitals be established in Britain to improve the treatment of disease.[38] Bellers was obviously a humanitarian, but the times permitted him to suggest that "one Hospital should be more particularly under the Care and Direction of the QUEEN'S [Anne's] Physicians; that they may take into it such Patients whose Infirmities at any time our SOVEREIGN may be subject to…"

The royal experiment of 1721–1722 was not the only clinical trial of smallpox inoculation, and neither was it the only one to use social inferiors as human guinea pigs. During the spring of 1751, the Geneva Council gave its permission to l'Hôpital Générale et de la Bourse Française to experiment on inoculation, using "subjects entirely dependent upon the Directors, and principally upon bastards;" eighteen persons were successfully inoculated in these tests.[39]

On the validity of experimental evidence

The design of the royal experiment might surprise the modern investigator, and would certainly fail to qualify it for publication in a modern journal. No controls were employed, and the initial experiment tested the safety of the inoculation procedure in only six individuals, and its efficacy in preventing subsequent infection in only one. But these were simpler times, when a procedure or a phenomenon either worked or did not work in a clear-cut fashion; the concept of utilizing large numbers to establish statistical significance had not yet been introduced.

It is interesting, in the same connection, to examine the data in Edward Jenner's original publication which introduced smallpox vaccination to the world. In all, Jenner published twenty-three case reports, only seven of which involved inoculation of cowpox virus into normal human recipients, and only two individuals were subsequently challenged with the wild smallpox virus to assess efficacy. But it was not the limited numbers in Jenner's report that led members of the Royal Society to reject an earlier draft of the *Inquiry*, "lest he damage his reputation."[40] Apparently, they simply did not believe Jenner's claims and their implications, thus depriving him of the prestige of publication in the *Philosophical Transactions of the Royal Society*. Jenner was thereupon forced to have his *Inquiry* published privately, with consequences that are history.

Why was it that the practice of inoculation made such slow progress despite its royal patronage, whereas vaccination gained substantially worldwide acceptance and use within a very few years, though lacking any official backing at the outset? There was, as we have seen, no substantial difference in the quality of the initial data on safety or efficacy, and indeed vaccination had many early opponents also. The difference may well rest less on the force of scientific evidence than on a variety of unrelated social factors, as is often true of medical vogues even today. In 1722, such measures as smallpox inoculation were restricted to a relatively small class of the educated and landed gentry that composed "Society," and knowledge of or interest in scientific advance was rare in the rest of the population. But throughout the eighteenth century the concept of public health spread fairly widely, as was witnessed by the founding of many hospitals for the poor and of societies devoted to the health and welfare of the underprivileged. Indeed, inoculation itself helped pave the way for vaccination, since much of the population saw the latter technique as an improved version of one already in use. Thus, by the start of the nineteenth century, we may tentatively suggest that society as a whole was better prepared than it had been seventy-five years earlier to accept what appeared to be a significant medical advance against the dread smallpox.[41] The role and status of medicine and of the medical scientist had changed appreciably in the public mind.

Notes and references

1. Lischke, R.L., *Friedrich Althoff*, Berlin, Sigma, 1990; see also *Minerva* **291**:269, 1991.
2. Fosdick, R.B., *The Story of the Rockefeller Foundation*, New York, Harper, 1952.
3. Creighton, C., *A History of Epidemics in Britain II*, Cambridge, 1894. See also Kübler, P., *Geschichte der Pocken und der Impfung*, Berlin, 1901.
4. It has been suggested by Carmichael, A.E., and Silverstein, A.M. (*J. Hist. Med. Allied Sci.* **42**:147, 1987) that only a mild form of smallpox existed in Europe prior to the seventeenth century, and that virulent *Variola major* substantially replaced it by the 1660s.
5. La Condamine, C.M., *Mémoires de l'Académie Royale des Sciences*, 1754, p. 615. "It is especially in the towns and in the most brilliant Courts that one sees it [smallpox] exert its ravages."
6. Jurin, J., *An Account of the Success of Inoculating the Small-pox in Great Britain, for the Year 1724*, London, 1725. Analogous reports were published for the following two years.
7. Jenner, E., *An Inquiry into the Causes and Effects of the Variolae Vaccinae*, London, Sampson Low, 1798.
8. See, for example, Klebs, A.C., *Bull. Johns Hopkins Hosp.* **24**:69, 1913; Stearns, R.P., and Pasti, G., *Bull. Hist. Med.* **24**:103, 1950; Miller, G., *The Adoption of*

Inoculation for Smallpox in England and France, Philadelphia, University of Pennsylvania Press, 1957.

9. *Boswell's Life of Johnson*, Vol. II, London, Oxford University Press, London, 1946, p. 551.
10. Lister MSS. 37, fol. 15, Bodleian Library, Oxford, quoted in Stearns and Pasti, note 8, p. 107.
11. *Journal Book IX, 1700*, Royal Society Library, p. 194.
12. *Phil. Trans. R. Soc. London* 29(339):72, 1714.
13. *Phil. Trans. R. Soc. London* 29(347):393, 1716.
14. Blake, J.B., "The inoculation controversy in Boston: 1721–1722," *New Engl. Q* 25:490, 1952. See also Kittredge, G.L., Introduction to C. Mather's *Several Reasons Proving that Inoculating or Transplanting the Small Pox, Is a Lawful Practice, and That It Has Been Blessed by God for the Saving of Many a Life* (1721), Cleveland, 1921.
15. Mather, C., *The Angel of Bethesda* (1724), G.W. Jones, ed., Barre, American Antiquarian Society, 1972. Mather's theory of acquired immunity in smallpox has been discussed in Chapter 1.
16. Lady Mary Wortley Montagu, *Letters and Works*, Vol. I, pp. 308–309, quoted in Halsband, R., *J. Hist. Med.* 8:390, 1953.
17. Maitland, C., *Account of Inoculating the Small Pox*, London, 1722.
18. Voltaire, "Sur l'insertion de la petite vérole", 1727, in *Lettres Philosophiques*, 3rd edn, Paris, Gustave Lanson, 1924, pp. 130–151. "*Cette dame* [Lady Mary] *de retour à Londres fit part de son expérience à la Princesse de Galles…*"
19. Miller, G., "Putting Lady Mary in her place: a discussion of historical causation," *Bull. Hist. Med.* 55:2, 1981; see also Miller, note 8.
20. *Dictionary of Scientific Biography*, Vol. XII, p. 456 ff.
21. Boretius, M.E., "Special-Nachricht von der neuen Invention der Inoculationis variolarum, oder des Blatter-Peltzens, wie solche in London exerciret worden," *Sammlung von Natur und Medicin…Geschichten* (Leipzig and Bautzen) 17:206, 1723.
22. Sloane, H., "An Account of Inoculation by Sir Hans Sloane, Bart., given to Mr. Randy to be published, Anno 1736, Communicated by Thomas Birch, D.D. Secret R.S.," *Phil. Trans. R. Soc. London* 49:516, 1756.
23. *The Weekly Journal; or, British Gazetteer*, June 17, 1721, p. 1952; *Applebee's Original Weekly Journal*, June 17, 1721, p. 2087.
24. *Hardwicke Papers*, British Museum, Vol. 786, Add. MS. 36:134, f. 58.
25. *Applebee's Original Weekly Journal*, Aug. 12, 1721, p. 2134; *The Weekly Journal; or, British Gazetteer*, Aug. 12, 1721, p. 1999; *The Weekly Journal or Saturday's-Post*, Aug. 12, 1721, p. 844.
26. Mead, R., *A Discourse on the Small Pox and Measles*, London, John Brindley, 1748, pp. 88–89.
27. Maitland, *Account*, note 17, p. 20.
28. *The London Gazette*, No. 6040, Mar. 6–10, 1722.
29. Sloane MSS., British Museum, MS. 4076, f. 331, Mar. 14, 1722.
30. Scheuchzer, J.G., *Journal-Book of the Royal Society*, XIII, 319, March 27, 1729.
31. Franklin, B., *Some Account of the Success of Inoculation for the Small-Pox in England and America*, London, 1759. (Reprinted in Reddis, L.H., *Edward Jenner and the Discovery of Smallpox Vaccination*, Menasha, George Banta, 1930; also in *Military Surg.* 65:645, 1929.

32. Sparham, L., *Reasons against the Practice of Inoculating the Small-Pox As Also a Brief Account of the Operation of This Poison, Infused after This Manner into a Wound*, London, 1722, p. 26.
33. Massey, Edmund, *A Sermon against the Dangerous and Sinful Practice of Inoculation. Preach'd at St. Andrew's Holborn, on Sunday, July the 8th, 1722*, London, 1722.
34. Miller, *The Adoption*, note 8.
35. *Gentleman's Magazine* 28:41, 1758.
36. See articles by Brieger, G.H., Capron, A.M., Fried, C., and Frankel, M.S., in the Section on Human Experimentation, *Encyclopedia of Bioethics*, Vol. II, New York, The Free Press, 1978, pp. 683–710.
37. Wangensteen, O.H., and Wangensteen, S.D., *The Rise of Surgery*, Minneapolis, University of Minnesota Press, 1978, p. 13.
38. Bellers, J., "An Essay Toward the Improvement of Physick in Twelve Proposals," 1714, cited in Woodward, J., *To Do the Sick no Harm: A Study of the British Voluntary Hospital System to 1875*, London and Boston, Routledge and Kegan Paul, 1974.
39. Gautier, Léon, *La Médecine à Genève jusqu'à la fin du dix-huitième siècle*, Geneva, 1906, cited in Miller, note 6.
40. Crookshank, E.M., *History and Pathology of Vaccination*, Vol. I, London, H.K. Lewis, 1889, p. 250 ff. Crookshank was, ninety years after Jenner's publication, an ardent anti-vaccinationist(!), so that his account may be suspect – see anonymous contribution to the "Jenner Centenary Number," *Br. Med. J.* 1:1257, 1896.
41. The rapid spread of acceptance of vaccination throughout the world, and the resulting adulation of Jenner, can best be compared with the similar response to Pasteur's anti-rabies treatment of the 1880s. In both instances, the initial evidence on safety and efficacy was incomplete, but the world responded with alacrity and enthusiasm, and even their governments rewarded them. When Jenner interceded with the French on behalf of an English prisoner, Napoleon, then at war with England, remarked that he could not refuse anything to this great benefactor of mankind.

14 The languages of immunological dispute

Therefore is the name of it called Babel; because the Lord did there confound the languages of all the earth.

Genesis 11:9

A scientific discipline sometimes flourishes best when opposing theories compete with one another to furnish the intellectual foundation of the science. To the heuristic value of each concept is added the impetus provided by the conflict itself, as the protagonists vie to gather evidence to support their own ideas, or to contradict those of their opponents. The secondary literature in the history of medicine contains numerous reports of such conflicts, but little attention has been paid to the linguistic aspects of these disputes. Opposing theories are often associated with different terminologies, whose semantic implications may be unacceptable and even incomprehensible to the other side. The historian of a given dispute may not be able to describe it and its context adequately, unless the languages of debate are fully understood.

A particularly apt example of this situation occurred during the early years of immunology, involving a hard-fought conflict between opposing theories of how antibodies are formed and how they function, as we saw in Chapter 6. It will be seen that failure to appreciate the nature of the different languages employed by the opposing camps may lead to later misinterpretation of their views and of their contributions.

The Donath–Landsteiner discovery: 1904

In an impressive series of studies commencing in 1899,[1] Paul Ehrlich and Julius Morgenroth sought to identify the constituents and to define the mechanisms involved in the phenomenon of immune hemolysis, which Jules Bordet had only recently described.[2] Such studies involved the immunization of animals with foreign red blood cells, resulting in an immune serum whose thermostable antibody would collaborate with a thermolabile substance (variously termed complement, alexin, or cytase) to cause the specific destruction *in vitro* of the erythrocyte species employed for immunization. During the course of these studies, Ehrlich and Morgenroth attempted repeatedly to induce the animal to form hemolytic antibodies toxic for *its own* cells, using as the immunizing agent blood from members of the same species or even from the immunized animal itself. These attempts to elicit the formation of *auto*antibodies were uniformly unsuccessful, and at best only *iso*antibodies were detectable, able to agglutinate

A History of Immunology, Second Edition
ISBN: 978-0-12-370586-0

Copyright © 2009, Elsevier Inc.
All rights reserved

or to hemolyze the red cells of certain other members of the same species. (Indeed, it was the discovery of isohemagglutinins in the serum of *normal* humans that enabled Karl Landsteiner to describe the ABO blood group system,[3] so important for the later success of human blood transfusion and for forensic medicine. It was this discovery, along with that of the M, N, and P blood groups in man,[4] that earned Landsteiner the Nobel Prize for Physiology or Medicine in 1931.)

The failure to detect the formation of hemolytic autoantibodies did not disturb Ehrlich unduly. Although he had postulated, in his landmark paper of 1897,[5] that antibody formation was part of the normal physiologic process of cellular digestion, and so might theoretically be stimulated by autochthonous as well as by foreign substances, he pointed out that "It would be dysteleologic in the highest degree, if under these circumstances self-poisons of the parenchyma – autotoxins – were formed."[6] Thus, "we might be justified in speaking of a *horror autotoxicus* of the organism."[7] When Metalnikoff described the ability of the guinea pig to produce autoantibodies against its own spermatozoa,[8] Ehrlich pointed out that even this did not speak against the concept of *horror autotoxicus* because while this autoantibody would attach specifically to the sperm, it appeared incapable of utilizing complement to effect their destruction, and was thus not an *autotoxin* "within our meaning."[9]

It was in the context of a widespread acceptance of Ehrlich's dictum that Julius Donath and Karl Landsteiner published their famous study of the mechanism of hemolysis in patients with paroxysmal cold hemoglobinuria (PKH) in 1904.[10] This fascinating but relatively uncommon disease is characterized by acute episodes of intravascular destruction of erythrocytes and an accompanying hemoglobinuria upon exposure of the patient to the cold. Donath and Landsteiner were able to reproduce this phenomenon *in vitro*, demonstrating that exposure of the patient's (or other human) washed red cells to the patient's serum in the cold permitted the fixation of a thermostable substance, whereupon the cells would be hemolyzed on warming by a thermolabile agent available even in normal serum. Here, then, was the first report that appeared to contradict Ehrlich's generalization of *horror autotoxicus*. A naturally-occurring disease, paroxysmal hemoglobinuria, seemed to involve the same two agents shown to function in all previous studies of *immune* hemolysis – a thermostable antibody-like substance and a thermolabile complement-like substance. This discovery by Donath and Landsteiner, extended in several subsequent papers over the next twenty years,[11] has been widely acclaimed as the first description of an autoantibody and of an autoimmune human disease.

For the purposes of this analysis of the linguistic aspects of a scientific dispute, we will analyze a stimulating article on the Donath–Landsteiner discovery written by Dietlinde Goltz in 1982.[12] In a carefully reasoned reassessment of the Donath–Landsteiner publications, Goltz suggests that in fact the ascription to these authors of the first discovery of autoantibody is erroneous. Donath and Landsteiner, she claims, did not even *themselves* believe that they were dealing with an antibody, or even with an immunologic phenomenon; she pointed out that nowhere in the early articles do they use the accepted terms "antibody," "*Ambozeptor*," "antigen," or even "immune." Rather, they persist in employing such apparently nonspecific terms as "hemolysin," "lytic substance," "toxin," and "poison." From this, Goltz concludes that Donath and Landsteiner subscribed to a nonimmunologic toxin-theory for the pathogenesis of paroxysmal hemoglobinuria, and that the "myth of their discovery" is ascribable to the desire of later immunohematologists "to create pioneers and heroes that mark the starting point of their science."[13]

One of the incidental purposes of this chapter is to demonstrate that Donath and Landsteiner did indeed discover the first autoantibody, and that they understood full well what they were about. To this end, it will suffice merely to examine what they *and their contemporaries* thought that they had accomplished in their report, and to recall that throughout his career Landsteiner's scientific style was one of extreme conservatism and caution in his writing, rarely speculating beyond the limits of his data. But if this were merely a question of priority in scientific discovery, it would scarcely be of broad interest either to the immunologist or to the historian of medicine. Dr Goltz's thesis, however, raises a much more fundamental issue for the historian (and sociologist) of science – the need to recognize that at any given time, competing schools in a science may speak completely different scientific languages, which are often incommensurable and which may reflect totally different world views.

Linguistic aspects of the great immunological debate

When Paul Ehrlich advanced his famous side-chain theory of antibody formation and antibody function in 1897,[14] he found it necessary to coin a new vocabulary to describe the various participants in the several immunologic reactions under consideration. Each of the terms that he employed not only described a discrete physical entity, but also carried full semantic implications about how that entity was supposed by Ehrlich to originate and to act. Jules Bordet, on the other hand, disagreed strongly with almost every aspect of Ehrlich's theories, and elected to employ a completely different set of terms to describe the same immunologic reactants – terms that either reflected his opposing views of the nature of the phenomena or which, at minimum, were semantically noncommittal as to the mode of action of these substances.

Karl Landsteiner adhered to the concepts of Jules Bordet, and they both violently disagreed with the Ehrlich theories (see Chapter 6). They employed a completely different language, based upon diametrically opposite views of the origin, nature, and workings of the immunologic reactants. To seek, therefore, for the Ehrlich terms "antibody" or "amboceptor," or even for the more universal terms "immune" or "antigen" in a Landsteiner paper on paroxysmal hemoglobinuria in 1904 is fruitless, since Landsteiner would not and probably could not have brought himself to use these words at that time. While Landsteiner's vocabulary in 1904 may have been different from that of Ehrlich, it was no less broad, and it permitted him no less to express what he intended. An understanding of these different languages of early immunologic dispute is thus crucial for an understanding both of the protagonists and of their technical and philosophical positions. It is equally important for a clear understanding of the nature and value of their contributions.

As we shall see, the members of each school would be loath to accept and to employ the language of the other. But while they would not willingly speak one another's language, it is clear that they understood one another's terms and their implications quite well. Thus, in their writings, the Bordet school would often refer to the "so called *Ambozeptor*" or "...in the terminology of the Ehrlich theory," while the Ehrlich school might write of "the so-called *sensibilisatrice*," or "*Komplement* (= Bordet's *alexine*)." In addition, each school would, in their general reviews of the subject, provide full translation dictionaries for the use of the public at large.[15] It is important to note here that while these two languages might be semantically incommensurable, and might lead to certain difficulties of translation at a later time, contemporary participants had perforce to understand the nuances of *all* current languages, in order to remain successfully in the forefront of their discipline.

The conceptual basis of Ehrlich's terminology

With the discovery by von Behring and Kitasato[16] that antitoxic immunity in tetanus and diphtheria infections could be ascribed to circulating "anti-bodies," perhaps the most significant conceptual problem faced by the young field of immunology was to explain the origin and basis for specificity of these new substances. Paul Ehrlich met this challenge in 1897, by advancing a theory broad enough to encompass all aspects of the formation and mode of action of these newly discovered antibodies.[17] Based upon his extensive earlier work with dyes, he assumed that the interaction of antibody to neutralize a toxin involved a purely chemical union between two chemically-defined complementary structures, whose stereochemical "fit" would assure initial binding. Indeed, he named the specific binding site on the antibody molecule the "haptophore group" (Greek *aptein*, to bind to). To explain the provenance of these antibody molecules, Ehrlich advanced a simple explanation. Just as nutrients might be assumed to be ingested by cells by virtue of specific surface receptors (side-chains), so toxins should be able to damage only those susceptible cells that carry specific toxin-receptors. But the utilization of these receptors should lead to their regeneration by the cells concerned, and any over-regeneration of receptors would result in their being cast off into the blood, to appear as circulating antitoxin antibody. Here for the first time was a plausible suggestion that not only explained antibody formation and antibody specificity, but also appeared to integrate the immune response into the more general laws of biology and chemistry.

When Jules Bordet described the phenomenon of immune hemolysis in 1898,[18] Ehrlich was quick to undertake studies of this fascinating new area with his colleague Morgenroth, and quick also to integrate his findings into the conceptual framework of the side-chain theory. The thermostable hemolytic antibody was, he suggested, a specialized molecule with not one but two binding sites – a haptophore group specific for the erythrocyte antigen, and a separate site which bound the thermolabile substance responsible for lytic action. Although this latter substance had been termed alexin (Greek *aleksein*, to protect) by its discoverer, Buchner,[19] Ehrlich chose to call it *Komplement*, since its function appeared to complement that of the anti-erythrocyte antibody. For the antibody itself, Ehrlich chose the terms *Zwischenkörper* (intermediary body) or *Ambozeptor*, the former term reflecting its role in mediating the interaction of complement and erythrocyte and the latter term describing the presence of two combining sites. Both of these sites were understood by Ehrlich to mediate highly specific and purely chemical interactions.

It will be noted that for Ehrlich, the specific antibody found in the circulation of an animal immunized with the corresponding antigen was a natural *and pre-existing* substance, normally present on appropriate cells in the body of the host. Thus, it was easy for the Ehrlich theory to explain another of the current findings that had perplexed immunologists – the presence in hitherto un-immunized animals of small amounts of antitoxins, antibacterial substances, or even substances that would cause the hemagglutination or hemolysis of certain erythrocytes. According to the Ehrlich theory, these were simply cell receptors that had been prematurely cast off into the circulation of "normal" individuals. For Ehrlich, then, "normal" antibodies and "immune" antibodies (obtained in response to active immunization with antigen) were identical, and the presence of the former in individuals with no known previous exposure to a specific antigenic stimulus posed no particular conceptual problem.

The conceptual basis of Bordet's terminology

As he continued his exploration of the mechanisms involved in immune hemo-lysis, Jules Bordet put together a conceptual framework that differed markedly from that of Ehrlich.[20] Bordet was more interested in the functions of antibodies than in their origin, and he rejected Ehrlich's speculations in this latter area as contributing little to one's understanding of the basic process.[21] Bordet fought continuously against Ehrlich's theory of a firm chemical interaction between antigen and antibody, and suggested rather that the interaction was a physical one, resembling more the adsorption so characteristic of colloids.[22] Bordet viewed the interaction of antibody with erythrocyte as reversible, in opposition to Ehrlich's view of a tight chemical union. Moreover, Bordet suggested that the antibody did not *first* fix complement and then attach to the erythrocyte to effect its hemolysis, as Ehrlich had suggested, but rather that the antibody would interact with the erythrocyte to cause such a change in the surface pattern – a sensitization – that the altered configuration would then fix complement and result in hemolysis. For Bordet, anti-erythrocyte antibody was neither *Zwi-schenkörper* nor *Ambozeptor*, but rather *la substance sensibilisatrice*.[23] Again, he chose to employ Buchner's less committal name *alexine* for the final active factor in hemolysis, rather than use Ehrlich's more suggestive term *Komplement*.

The running battle between Bordet and Ehrlich over the nature and activity of hemolytic antibody carried over into an ancillary dispute about complement or alexin. For Ehrlich, the existence of a corresponding specific receptor for complement on the antibody molecule implied a specific site on the complement molecule, and thus Ehrlich was logically forced to postulate the presence of many different complements – one for each type of antibody.[24] Bordet, on the other hand, was content with but a single complement in each individual, which would affix to *any* antibody-sensitized erythrocyte in an almost nonspecific manner.[25]

Another aspect of the Bordet–Ehrlich dispute pertinent to our present consideration of their language differences relates to Bordet's concept of the significance of those antibody-like substances found in the serum of normal individuals. For Ehrlich, as we have seen, these were *identical* to those antibodies obtained by active immunization, and thus could logically be called "immune" substances or "antibodies." This Bordet would not concede. He (together with Landsteiner) viewed these naturally occurring substances as, at best, primitive precursors from which specific antibody might later be formed following upon their "adaptation" (*perfectionnement*) by interaction with injected antigen.[26] Thus, *at that time* (at least during the first decade or two of the twentieth century), Bordet would no more use the terms immune or antibody in referring to these naturally occurring substances than he would use the term amboceptor in referring to hemolytic antibody.[27]

A final aspect of the Bordet language then current at the Pasteur Institute and among its outside adherents is deserving of attention. It involves the generic terms that Bordet most generally employed in referring to antibodies or antisera capable of destroying cells. While he would occasionally utilize the terms *anticorps* or *hemolysine*, he more frequently in this era called them *cytotoxines*. He makes this abundantly clear in a broad review of the field published in 1900, in which he says:[28]

> We shall employ very frequently the terms "hemolytic sera" or preferably "hemotoxins" to refer to anti-erythrocyte sera. If we adopt this latter word, then the antitoxin against a hemolytic serum will be called an "antihemotoxin." These terms are highly convenient. They were suggested by M. Metchnikoff who, as we know, calls the serum active against spermatozoa a spermotoxin, and that which destroys leukocytes a leukotoxin. The general term "cytolytic" or "cytotoxic serum" will designate the various immune sera capable of destroying energetically either microbes or other cells (erythrocytes, etc.).

This same terminology was employed again and again in the same volume by other Pasteurians. Thus, Metchnikoff reviewed the general field of anti-cell antibodies in a paper entitled "Sur les Cytotoxines;"[29] Besredka wrote on "leukotoxines";[30] Metchnikoff and Besredka collaborated on an article on "hémotoxines;"[31] and Metalnikoff described "spermotoxines."[32] In each of these reports, it is abundantly clear that the authors meant specific antibody when using the term toxin, and did not imply some sort of nonimmunologic toxic action. It is this peculiarly Francophone convention and lexicon that Landsteiner chose to employ in his report with Donath on paroxysmal cold hemoglobinuria – language which Goltz interpreted to imply that Landsteiner actually believed in a toxic (in the narrow sense) pathogenesis for the disease. In fact, Landsteiner was only writing in the immunologic language of Bordet.

Landsteiner's allegiances and his language

Karl Landsteiner's first exposure to immunology came at the hands of Max von Gruber in Vienna, in whose Institute of Hygiene he served as second assistant from January 1896 to October 1897.[33] Gruber was noted for his significant contributions to the new field of serology, but perhaps even more as one of the foremost, and often the most outspoken and vitriolic, critics of Paul Ehrlich's theories.[34] Thus, the young Landsteiner's formative period in immunology took place in an environment that was decidedly hostile to Ehrlich's views. It is no wonder, therefore, that the majority of Landsteiner's studies over the next dozen years or so were devoted to experiments often seemingly devised to contradict the basic elements of Ehrlich's theory. He followed the lead of Gruber[35] in claiming that Ehrlich's side-chain theory was too complicated, and that the universe of potential antigens and therefore of specific antibodies was too large to be accounted for by *naturally occurring* products.[36] More importantly, however, Landsteiner developed very early a high regard for the work and the concepts of Jules Bordet, which Bordet reciprocated,[37] and he followed Bordet in arguing variously against Ehrlich: that immunologic interactions are physical and colloidal rather than chemical;[38] that hemolytic antibody is not a *Zwischenkörper* or *Ambozeptor* with a uniquely determined specificity, but rather a Bordet-type sensitizer;[39] and that the active substances in the serum of normal individuals are not the same as immune antibodies, but at best antibody precursors.[40]

As Mazumdar has pointed out,[41] while some of Landsteiner's immunologic beliefs (such as that on the colloidal nature of the antigen–antibody interaction) were perhaps more casual and transient, others were deeply rooted in a philosophical world-view that he retained throughout his life. Thus, Landsteiner argued repeatedly against Ehrlich's notion of discrete and uniquely separable specificities for antibody, suggesting instead that a multiplicity of only slightly differing molecules provides a continuous spectrum of cross-reactions more dependent upon quantitative than upon qualitative variations.[42] In this, as Mazumdar shows, Landsteiner's argument with Ehrlich was only the latest engagement in a longstanding philosophical debate between those who joined Leibnitz in arguing for an underlying continuity in a seamless physical world and those who held, with Kant, that Nature is marked by sharp discontinuities in most of its aspects. Ehrlich, in the tradition of the botanist Ferdinand Cohn and the bacteriologist Robert Koch, argued for sharp discontinuities and major qualitative jumps. Landsteiner, in the tradition of the botanist Karl von Nägeli and the bacteriologist Max von Gruber, favored smooth transitions and minor quantitative variation.[43] In the event, it is not surprising that Landsteiner chose to employ, during the early years of this century, the language of Bordet – of continuity, of colloids, and of physical chemistry – rather than Ehrlich's terminology of discontinuity, of discrete molecules, and of structural organic chemistry.

However, no language is static. New terms are continually added, and old terms change their meaning in any developing science. Only an analysis based upon comparative linguistics can reveal what a scientist meant by the words that he employed *at that time*. Thus, close attention must be paid to subtle changes in Landsteiner's vocabulary as the years passed. In his very first publications on hemolysis, Landsteiner referred repeatedly to the work of Gruber and Bordet, and denoted the interactants as *sensibilisierende Substanz* and *Alexin*, while studiously avoiding the then-current Ehrlich terms *Ambozeptor*, *Zwischenkörper*, and *Komplement*.[44] Just a few years later,[45] though, he gave up the usage of Bordet's *alexine* in favor of Ehrlich's *Komplement* (apparently feeling that this term, used uniquely by German-language writers, no longer carried with it an unwanted semantic burden). But never thenceforth, in any of his writing, did he utilize the terms *Zwischenkörper* or *Ambozeptor*, unless it was preceded by a modifier such as "*sogenannte* [so-called]," or followed by "in the terminology of the Ehrlich school." It was some time before Landsteiner used even the term "*Antikörper*," preferring instead, in line with French practice, to utilize such less committal terms as "*Antistoffe*," "agglutinating [or hemolytic] substances," or, in the case of cell-destructive reactions, "cytotoxic sera" or more simply "poisons."

Landsteiner's most notable early contribution to immunology was the demonstration that the serum of normal humans contains substances (isoagglutinins) capable of clumping the erythrocytes of certain other individuals.[46] He returned again and again to the study of such naturally occurring substances, making it clear from his reports that he felt that their characteristics provided one of the strongest arguments against Ehrlich's ideas. In a series of papers between 1905 and 1907,[47] Landsteiner sought to show that the side-chain concept was improbable, in that these normally occurring substances and "immune antibodies" were not identical, as the theory required. Rather, they differed markedly in their specificity and in their susceptibility to various treatments. Throughout, he was careful to call them "normal agglutinins," "normal hemolysins," or simply *Normalstoffe*, in contrast to "immune agglutinins" and "*Immunstoffe*." Landsteiner never offered an explanation for the origin of these normal substances, but suggested that Ehrlich was wrong, and that immune antibodies result from the effect of antigen on these normal substances to "adapt" them for greater specificity.[48] Only in his lengthy review of 1909, while still emphasizing the differences between normal and immune agglutinins, does he permit himself to refer to "*die Antistoffe der normalen und*

Immunsera," but goes on to say that "The specific immune bodies...are newly formed during the immunization process, and different from the physiologic antibodies."[49]

This is the very first time in his writings (1909) that Landsteiner, while still protesting their difference, permits himself to refer to these active substances in normal serum as "antibodies." The use of such a term five years earlier, to refer to the presumably naturally occurring agent responsible for hemolysis in paroxysmal cold hemoglobinuria, would have been unlikely and probably even impossible. His extensive publications in this and closely related fields *at that time* show that the word in this special context was not yet in his vocabulary.

Landsteiner's views on the nature of the active agents present in normal sera have other implications for the language that he employed to describe them, *and for the words that he did not employ*. Landsteiner fought Ehrlich, in part, by demonstrating the differences between these normal agents and "immune" antibodies. Landsteiner considered them, in fact, to be the relatively nonspecific stuff of which true specific antibodies are formed. Their presence could therefore not be ascribed to an "immune response," and their activity could not be classified as an "immune reaction." Thus, while admitting the antibody-like function of these normal substances, Landsteiner consistently reserves the term "immune" for the *active* response to defined antigens, introduced to the host either naturally through infection or artificially by immunization. Similarly, since antigen (i.e., the *gen*erator of *anti*body) was assumed by Landsteiner to act *on* these natural substances, it would be illogical for him to postulate that antigen was also responsible for their presence.

It is evident, then, that the immunologic language employed by Karl Landsteiner during the first decade of the century differed markedly from that used by Paul Ehrlich. Each of the key terms used by one school embodied within it a distinctive viewpoint about origin or function, and would be avoided by adherents of the opposing school. During the ensuing decades it was the language of Bordet and Landsteiner that substantially disappeared, and the language of Ehrlich that survived in large measure and which is currently used in modern immunology. Without knowing, therefore, what a given word meant to its user at that time, within the context of his *then-current* conceptual position, the modern reader might easily misconstrue his meaning.

Karl Landsteiner's scientific style

Throughout his long career, Karl Landsteiner had the reputation among all who knew him for the careful execution of his laboratory studies, and for the precision and conservatism with which he wrote up his results for publication. Perhaps he was merely following the lead of his conceptual mentor, Jules Bordet, who insisted repeatedly that he himself was not a theorist and that any ideas that he had advanced were not even worthy to be called theories, but merely represented "a description of the true state of affairs." Landsteiner would rarely argue beyond the strict confines of his data. This was true even in his earliest publications, with one notable series of exceptions! Only when he was attempting to refute Paul Ehrlich would Landsteiner not only employ his data to counter Ehrlich's ideas, but also use them to advance alternative concepts sometimes only weakly supported by the data. Thus, in arguing against Ehrlich's notion of a firm chemical union between antigen and antibody, he speculated about the colloidal nature of the reactants, and their physicochemical (adsorptive) combination.[50] Again, in arguing against Ehrlich's side-chain theory of the origin of antibody, he theorized that antibody was formed through the adaptation of "natural antibodies" under the influence of antigen.[51] Only when Ehrlich and Ehrlich's concepts were not at issue was Landsteiner content to let his facts speak for themselves, and to curb any impulse toward speculation.

Landsteiner's usual scientific style is perhaps best illustrated in a statement by one of his former students, Dr John L. Jacobs: [52]

> *Dr. Landsteiner had a gift for building patiently, step by step...the rigid limitation of his experiments to the exploration of facts (avoiding theories) – advancing by one limited hypothesis at a time, kept his work close to objective reality. ...In writing papers, Dr. Landsteiner was never ready to put pen to paper until he had definitely established a new fact. ...He limited himself severely to pointing out the highly probable implications and relationships of the facts observed, almost completely omitting opinion and theory. Thus, discussion in Dr. Landsteiner's papers consisted of relating the new fact or facts observed...in the manner that held hypothesis in check, to the point that such hypotheses as were advanced, represented only one short step forward with obviously a high probability of accuracy. ...a large element of his genius consisted in the humility with which he would forego the opportunity to draw broad theoretical conclusions in the interest of maintaining a high degree of accuracy and objective reality.*

A long-term collaborator of Landsteiner's, Dr Merrill W. Chase, finds Jacob's assessment of Landsteiner's style highly accurate. Indeed, Chase has suggested that Landsteiner's care in performing and reporting experiments, and his general disinclination to theorize, were rooted in a basic dread of being proved wrong in anything by his colleagues in the scientific community.[53]

Since the 1904 report by Donath and Landsteiner on paroxysmal cold hemoglobinuria did not directly concern any of Paul Ehrlich's fundamental precepts, it may be reasonable to assume that Landsteiner brought to this paper, and to the others on the same subject, the conservatism of approach that typified all of his other publications which were not "anti-Ehrlich" in nature. Thus, we would not expect to find in the report a broad theory on the possible immunologic basis for the pathogenesis of this disease, since this had not yet rigorously been proved. Given the nature of his theoretical base and the restraints on language that this imposed, Landsteiner would also not have employed in this paper the immunologic key words that an Ehrlich might have used, or that a later generation of immunologists might expect.

But did Landsteiner himself actually write the Donath–Landsteiner report of 1904, and does it really well reflect Landsteiner's views on the subject? In answer, we must conclude that he either wrote the paper, or (since it is not as tight and crisp as his other writings at the time) that he carefully revised a draft that Donath might have written. This conclusion appears warranted, based upon a comparison of the Donath–Landsteiner report with one written earlier in that same year by Donath alone.[54] Donath reported his own study of three cases of paroxysmal cold hemoglobinuria, and reviewed at length the various possible mechanisms that may cause the disease. Ruling out various physical mechanisms, Donath then spent seven pages discussing the possibility of the participation of a *hemolysin*, and there is no question that he here meant antibody. He suggested that the attack is elicited by a hemolysin which, "like Ehrlich's normal [immune] serum hemolysins, is composed of two components (complement and amboceptor)."[55] Throughout the paper, Donath refers to Ehrlich's theory, and uses the Ehrlich terminology *Komplement* and *Ambozeptor*. These are terms that do not appear only a few months later in the paper with Landsteiner. This would imply not that an antibody is ruled out in the latter paper, but more probably that it is Landsteiner who is calling the tune on nomenclature in this joint publication, and will permit neither the use of Ehrlich's language nor a theoretical overcommitment not yet fully warranted by the facts.

Contemporary views of the Donath–Landsteiner report

In contesting the priority of the Donath–Landsteiner discovery of the first autoantibody, Dietlinde Goltz suggested not only that these authors did not believe that an antibody was involved, but also that the attribution was not made until many years later, most notably in the 1940s to 1960s by that most famous of immunohematologists, William Dameshek.[56] But in fact Landsteiner repeatedly claimed priority for this discovery, *and his contemporaries readily conceded this claim*.

Despite the limitations imposed upon Landsteiner by his conservative style and by his arcane vocabulary, he could, when the situation demanded, bring himself to employ more explicit language in describing the agent responsible for paroxysmal cold hemoglobinuria. As early as the year following the 1904 report, Landsteiner and Leiner wrote a paper on the isoagglutinins and isohemolysins in normal and diseased patients. They pointed out that while none of these has been shown to be pathologic, "in fact, diseases and even disease symptoms have been shown reliably to be caused directly by auto- and isolysins, in a special series of experiments on cases of paroxysmal hemoglobinuria."[57] Again, in a broad review of immunology written in 1910, Landsteiner says that "Donath and Landsteiner found a strongly active hemolysin (autolysin) in the serum of people…with paroxysmal cold hemoglobinuria." And then, after outlining the phenomenon itself, he concludes that, "The entire event occurs in two separate phases. In the first, the hemolytic '*Immunkörper*' [Landsteiner's quotation marks] is bound to the blood cells...in the second phase...only the presence of complement is necessary."[58]

The initial report on paroxysmal cold hemoglobinuria by Donath and Landsteiner attracted much attention. Repeated reference was made thereafter to the phenomenon *and to its interpretation* by numerous authors in both scientific reports and literature reviews. All of these make it quite evident that even if Landsteiner's language might be misinterpreted at a later period, his contemporaries surely understood him. Thus, in 1905, Widal and Rostaine from Paris published on PKH.[59] These authors credited Donath and Landsteiner with the description of an autohemolytic substance in patients' serum, but, following the lead of Besredka[60] and the language of Bordet, they claimed that such substances are normally present in *everyone*. They suggested rather that the proximate cause of the disease is the *absence* of an anti-autohemolytic substance (in modern terms, an anti-antibody). This, according to Landsteiner, was impermissible, and in the Donath–Landsteiner paper of 1908 the Widal–Rostaine thesis was criticized as follows:

> it may also be said of the Widal and Rostaine hypothesis that both
> substances...the autohemolytic and the anti-autolytic substance have not until
> now been experimentally observed. ...such a hypothesis is, however, manifestly
> superfluous, since one can simply omit the supposed combination (i.e., the anti-
> autolytic substance) without altering the way of thinking about the
> phenomenon, and we must accordingly give preference to our interpretation
> that assumes only the actually observed hemolysin.

It is clear from this that Donath and Landsteiner understood that their own explanation of the pathogenesis of paroxysmal cold hemoglobinuria involved a hemolytic autoantibody. Indeed, they even define the putative "antilysin" of Widal and Rostaine in a footnote, as "i.e., *Antiambozeptor, antisensibilisierende Substanz.*"[61]

In the same paper, Donath and Landsteiner also contest the priority for their discovery with the British physician John Eason. Eason had published two papers in 1906,[62] claiming to have discovered the pathogenesis of paroxysmal hemoglobinuria in work purportedly done prior to that of Donath and Landsteiner. Using the Ehrlich language then popular in England, Eason acknowledges that it is his view, *as well as that of Donath and Landsteiner*, that an "intermediary body (Ehrlich's *Zwischenkörper*) anchors to the red blood cell, requires low temperature, and then a rise in temperature sufficient to allow complement to participate in the process."[63] There is, he says, a potential toxin composed of two bodies, one of which possesses the characteristics of amboceptor and the other those of complement. He concludes "that paroxysmal hemoglobinuria is attributable to the activities of an intermediary body (which is, in fact, an immune body to corpuscles of the affected individual)..."[64] Here, in unmistakable (i.e., Ehrlich's) language, is a purely immunologic explanation of the disease. Do Donath and Landsteiner take exception to Eason's proposal?

On the contrary, they merely state that "Eason joined [himself] to our inter-pretation of the mechanism of hemolysis,"[65] and contest not the theory but Eason's claim to its priority.[66] Indeed, in as explicit a statement as they have permitted themselves thus far, Donath and Landsteiner conclude that:

> *Since the development of the hemolysin is connected to the course of certain infections [most notably syphilis], so does our earlier-mentioned concept become more apt, that the development of autotoxic substances, which are bound to the organism's own cells, can be related to the process of antibody formation, a possibility which, so far as we know, has not previously been discussed.*[67]

The recognition that Donath and Landsteiner had described an autoantibody in their 1904 paper was not restricted to Britain and France, but was acknowledged even within the "enemy camp" itself. In his review of recent advances written in 1906 expressly for the English edition of his collected works, Ehrlich already referred to Donath and Landsteiner as observing "hemolytic autoambo-ceptors."[68] Again, Ehrlich's principal disciple, Hans Sachs, published an extensive review of "Hemolysins and Cytotoxic Sera" in 1906, conceding that "Donath and Landsteiner have produced information of the highest interest, that in the serum of this disease [PKH] an amboceptor is present that acts upon its own red cells..."[69] Not only did Sachs concede the concept to Donath and Landsteiner; he also went so far as to dispute their priority! Apparently unwilling to yield too much to an acknowledged opponent, Sachs claims that the fact that "the serum of a hemoglobinuric patient dissolves its own blood cells *in vitro*, i.e., contains an *Autoambozeptor*, has already been reported from other quarters."[70] Sachs repeated this concession to Donath and Landsteiner, and the accompanying counter-claim, two years later in another extensive immunologic review.[71]

Other adherents of the Ehrlich school also conceded the autoantibody discovery to Donath and Landsteiner. In an extensive review of cytotoxins, Rössle discussed the general evidence for the existence of autoantibodies, and stated that:[72]

> *There are also cases, however, in which direct evidence for the presence of autoamboceptor is splendid. The best known instance concerns paroxysmal hemoglobinuria. ...Already in their first paper, Donath and Landsteiner advanced the conjecture that in paroxysmal hemoglobinuria the production of autotoxic substances (hemolysins) was involved.*

Rössle concluded the discussion with: "Even in their first report, Donath and Landsteiner called our attention to the possibility that such a substance might be the result of a self-immunization."[73]

In 1909, Meyer and Emmerich published an extensive report on paroxysmal hemoglobinuria,[74] and it is they who are credited by Goltz with advancing the first clear hypothesis of the autoimmune character of paroxysmal cold hemoglobinuria.[75] Meyer and Emmerich worked in Munich, and spoke the language of Paul Ehrlich. They did indeed expand upon many of the immunologic aspects of the mechanism involved, but claimed no priority for themselves in discussing this concept. Indeed, they referred to "In very pretty and numerous investigations [of Donath and Landsteiner]...the hemolysin so observed proves to be of a complex nature, composed of a complement destructible at 56°, and a thermostable amboceptor."[76] They conclude their paper with the revealing statement that "In [our] four cases of typical paroxysmal cold hemoglobinuria, the autohemolysin found by Donath and Landsteiner was observed."[77]

It will be evident from the foregoing that Donath and Landsteiner did indeed understand from the outset that they were describing an autoantibody and an immunologic process, despite the curious terminology that they employed. Moreover, *all* of their contemporaries understood precisely what they meant, and the full significance of their report. When necessary, they were always quick to translate the crucial terms from the language of Jules Bordet (which Landsteiner employed in 1904 and for some time thereafter) into their own language (most generally that of Paul Ehrlich). In order to compete effectively in the immunologic science of the first decade of the twentieth century, it was absolutely necessary that a German understand French, and that a Francophone understand German. No less important in this science was that a follower of the Bordet school understand "Ehrlichese," and that an adherent of the Ehrlich school understand the language of the Pasteur Institute. The latter language has become substantially extinct, and thus may lead to modern difficulties of translation; however, fortunately, an appropriate Rosetta stone exists, and is available to us throughout the journals of that period.

The lexicon of scientific dispute

The most popular philosophical view of the scientific endeavor during the 1940s and 1950s, advanced most notably by Sir Karl Popper,[78] was that science is unique among intellectual pursuits in building in linear and cumulative fashion an ever-clearer picture of the physical world, and that scientific progress is characterized by a remarkable consensus of view about both fact and theory among its participants. The principal features of this point of view were adopted also by many early sociologists of science, led by Robert Merton.[79] The philosophers sought to explain the bases for agreement among scientists by examining their epistemological underpinnings, and suggested that all scientists adhere to the same set of logical principles of scientific inference. If these are followed rigorously, then it was inferred that scientific consensus would be inevitable. The sociologists, for their part, looked to the behavioral rules that govern individual scientists and the scientific community, and found that consensus is based upon a set of "shared social norms," the observance of which guides all reasonable individuals toward agreement.

This rosy picture of the workings of science was challenged by historian Thomas Kuhn,[80] by philosophers Imre Lakatos[81] and Paul Feyerabend,[82] and by sociologist Michael Mulkay,[83] among others. Pointing to the innumerable instances in science of conceptual debate, they suggest that it may be more important to seek explanations of scientific *disagreement* than of scientific *agreement*, and that indeed the former may be more productive of scientific progress than the latter. Kuhn points out that when anomalies are encountered in the workings of normal science, a crisis may develop leading to the proposal of a new "paradigm," with a resulting disagreement and conflict between proponents of the old theory and supporters of the new one.[84] Because the theories are usually incommensurable, the two schools of thought generally have little basis for a reasonable exchange of views. Even if the words employed are the same, they frequently mean fundamentally different things to the opposing parties, and thus translation is often impossible.

Although Kuhn's approach to the study of scientific dispute has been criticized severely,[85] it is now clear that *dis*sensus constitutes an important aspect of science. It may be instructive to examine Bordet and Landsteiner's dispute with Ehrlich within the context of this larger question. When scientific dissensus arises within a *single* discipline, language problems often arise – in part because old terms may be given new meanings, and in part because new concepts may demand new terminology. Such apparently was the case in the neurosciences in arguments between "brain" and "mind," in physics between wave and particle theories of light, in geology between gradualists and saltationists, in chemistry between Priestley and Lavoisier, etc. Sometimes, however, the new concept is so radically different that an entirely new lexicon must be devised, such as that which accompanied Einstein's relativity theory and quantum mechanics.

In each instance, theories and lexicons may have been incommensurable, but the historian must be cautious in joining to his conclusion that the theories were mutually incompatible the further conclusion that the respective languages were untranslatable, and thus quite incomprehensible to the opposition. Mutual and total incomprehensibility of language does occasionally occur in scientific dispute, most commonly when the same question is approached by representatives of two distinct scientific disciplines. Perhaps the best instance of this in biology was the decades-long conflict over the driving force for evolution and the basis of speciation, by geneticists on the one hand and field naturalists and paleontologists on the other. As Ernst Mayr has pointed out,[86] the geneticists studied the genotype, argued proximate causes, and evolved a concept of evolutionary speciation based upon saltationism. For their part, the naturalists confined their attention to the phenotype, argued ultimate causes, and arrived at a concept based upon gradualism. Each camp had as its point of departure a scientific training and tradition, and a world-view diametrically opposite to that of its opponents; thus both developed not only a set of incommensurable theories, but a set of languages that were incommensurable as well. Communication between the two schools was almost nonexistent for a long period, only in part because they could not understand one another's language. The major factor appears to have been that each thought so little of the other's approach that they felt little need even to attempt the translation.[87]

This was not the case in the immunologic dispute discussed above. Paul Ehrlich and Jules Bordet each had a theory to describe the origin, nature, and mode of action of the major components of the immune response based upon quite different philosophical viewpoints, and each disagreed violently with the other. In turn, each protagonist coined his own lexicon to describe the several substances – terms which carried full semantic implications about the governing theories, and were thus incommensurable. The opponents would no sooner accept the other's terms than they would their theoretical origins. And yet, while the languages were incommensurable, they were nevertheless fully understood by all parties to the dispute. How else, in an actively moving discipline, would one be able to plan the next experiment to advance one's own theory, or to refute the opponent, than by understanding precisely what he had done, and what he meant in his report? In this example of immunologic dissensus, as perhaps in many other scientific disagreements, incommensurability need not necessarily imply incomprehensibility.[88]

Notes and references

1. Ehrlich, P., and Morgenroth, J., The six landmark communications on hemolysis appeared in *Berl. klin. Wochenschr.* 36:6, 481, 1899; 37:453, 681, 1900; 38:251, 569, 1901. These also appear in *The Collected Papers of Paul Ehrlich*, Vol. II, New York, Pergamon Press, 1957, in both German and English translation, and in English alone in *Collected Studies on Immunity*, C. Bolduan translation, New York, John Wiley, 1906.
2. Bordet, J., *Ann. Inst. Pasteur* 12:688, 1898.
3. Landsteiner, K., *Centr. Bakt. Orig.* 27:357, 1900.
4. Landsteiner, K., and Levine, P., *Proc. Soc. Exp. Biol. Med.* 24:600, 1926/1927.
5. Ehrlich, P., *Klin. Jahrb.* 60:299, 1897 (English translation in Ehrlich, *Collected Papers*, Vol. II, p. 107). Ehrlich expanded on this concept in his Croonian Lecture to the Royal Society, *Proc. R. Soc. Lond.* 66:424, 1900.
6. Ehrlich, P., *Verh. 73 Ges. deutsch. Naturf. Aerzte*, 1901; reprinted in Ehrlich, *Collected Papers*, Vol. II, pp. 298–315.
7. Ehrlich and Morgenroth, fifth communication, note 1, p. 255; Ehrlich *Collected Papers*, Vol. II, p. 253. See Chapter 8 for a more extensive discussion of *horror autotoxicus* and of autoimmunity.
8. Metalnikoff, S., *Ann. Inst. Pasteur*, 14:577, 1900.
9. Ehrlich and Morgenroth, fifth communication, note 1, p. 255; Ehrlich *Collected Papers*, Vol. II, p. 253, footnote.
10. Donath, J., and Landsteiner, K., *Münch. med. Wochenschr.* 51:1590, 1904.
11. Donath, J., and Landsteiner, K., *Z. klin. Med.* 58:173, 1906; *Zentralbl. Bakt. Parasitenk.* 45:205, 1908a; *Wien. klin. Wochenschr.* 21:1565, 1908b; *Z. Immunitätsf.* 18:701, 1913; *Ergeb. Hyg. Bakt.* 7:184, 1925.
12. Goltz, D., "Das Donath–Landsteiner-Hämolysin. Die Entstehung eines Mythos in der Medizin des 20 Jahrhunderts," *Clio Medica* 16:193, 1982.

13. Goltz, note 12, p. 193, summary.
14. Ehrlich, note 5; see also Chapter 3.
15. Thus H. Sachs, a disciple of Ehrlich, opened the section on hemolysis in a broad review of cytotoxins (*Handbuch der Technik und Methodik der Immunitätsforschung*, Vol. 2, Jena, Fischer, 1909, p. 896) with definitions as follows: "Thermostable substance: Ambozeptor, Immunkörper, Zwischenkörper (Ehrlich and Morgenroth); Substance sensibiliratrice [sic!] (Bordet); Copula (P. Mueller); Desmon (London); Philocytase, Fixateur (Metchnikoff); Präparateur (Gruber); Hilfskörper (Buchner). Thermolabile substance: Komplement, Addiment (Ehrlich and Morgenroth); Alexin (Buchner, Bordet); Cytase (Metchnikoff)."
16. Behring, E., and Kitasato, S., *Deutsch. med. Wochenschr.* **16**:113, 1890. See also Behring, E., and Wernicke, E., *Z. Hyg. Infektionskr.* **12**:10, 45, 1892.
17. Ehrlich, note 5.
18. Bordet, note 2.
19. Buchner, H., *Zentralbl. Bakt. Parasitenk.* **6**:561, 1889.
20. The best summary of Bordet's general position is provided in his *Traité de l'Immunité dans les Maladies Infectieuses*, Paris, Masson, 1920. The Ehrlich–Bordet dispute is discussed more extensively in Chapter 6.
21. Bordet criticized Ehrlich's theories as being too complex, and said of them: "One knows with what luxuriance they have been developed on the fertile ground of immunology, where so much of the unknown still stimulates the imagination and invites audaciously synthetic concepts from the schools desirous of affirming their superiority...conceptions that are defended with all of the partisanship that amourpropre mixed with chauvinism so readily inspire" (Bordet, note 20, p. vi ff.). Bordet's use of the epithet "chauvinism" is carefully chosen. We have commented in Chapter 2 on the contributions of Franco-German enmity to immunologic disputes, and Bordet's book was written in Belgium during World War I.
22. Bordet, note 20, p. 546. See also Mazumdar, P.M.H., *Karl Landsteiner and the Problem of Species 1838–1968*, Thesis, Johns Hopkins University, 1976, Vol. II, p. 320 ff, for a discussion of the broader aspects of the debate about chemical vs. physical interactions around the turn of the nineteenth century. This debate is well summarized in Mazumdar's *Species and Specificity: An Interpretation of the History of Immunology*, Cambridge, Cambridge University Press, 1995.
23. Bordet, J., *Ann. Inst. Pasteur* **14**:257, 1900.
24. Ehrlich, P., and Sachs, H., *Berlin klin. Wochenschr.* **39**:297, 335, 1902.
25. Bordet, J., *Ann. Inst. Pasteur* **15**:303, 1901.
26. Bordet, J., *Ann. Inst. Pasteur* **13**:273, 1899, p. 288.
27. Much later (1920), Bordet would employ the term *anticorps* to denote these naturally-occurring substances (Bordet, note 20), but he would continue to protest their differences from "immune" antibodies.
28. Bordet, J., note 23, p. 257.
29. Metchnikoff, E., *Ann. Inst. Pasteur* **14**:369, 1900.
30. Besredka, A., *Ann. Inst. Pasteur* **14**:390, 1900.
31. Metchnikoff, E., and Besredka, A., *Ann. Inst. Pasteur* **14**:402, 1900.
32. Metalnikoff, S., note 8.
33. Speiser, P., and Smekal, F.G., *Karl Landsteiner*, English translation, R. Rickett, Vienna, Verlag Brüder Hollinek, 1975.
34. The Ehrlich–Gruber debates are discussed in Mazumdar, *Species and Specificity*, note 22, pp. 123–135, and in Chapter 6.

35. von Gruber, M., *Münch. med. Wochenschr.* **48**:1827, 1924, 1901. Gruber concludes (p. 1927) that "The entire Ehrlich nomenclature must be given up, since it is based upon false premises."

36. Landsteiner, K., *Wien. klin. Wochenschr.* **22**:1623, 1909. Landsteiner says of Ehrlich's theory: "In reality, this hypothesis does not assist our understanding, it offers no principle that is more simple than the phenomenon itself...the physiologic presence of countless substances whose utility to the organism...cannot be perceived."

37. Max von Gruber noted, in a letter written in 1908 (*Wien. med. Wochenschr.* **81**:309, 1931), that "His [Landsteiner's] recognition in Austria seems to be hindered by his not belonging to [the school of] Paltauf [a pro-Ehrlich institute in Vienna], and in Germany by his having set himself up against Ehrlich. But in France...his reputation is of the highest. Bordet once said to me that in his opinion Landsteiner is the only brilliant mind among Austrian bacteriologists."

38. Landsteiner, K., *Münch. med. Wochenschr.* **49**:1905, 1902; Landsteiner, L., and Jagić, N., *Münch. med. Wochenschr.* **50**:764, 1903; Landsteiner, L., and Jagić, N., *Wien. klin. Wochenschr.* **17**:63, 1904.

39. Landsteiner, K., in Oppenheimer, C., ed., *Handbuch der Biochemie*, Vol. I, Jena, Fischer, Jena, 1910; see also notes 3 and 36.

40. Landsteiner, K., and Reich, M., *Wien. klin. Rundschau.* **19**:568, 1905; also Landsteiner, note 39.

41. Mazumdar, *Species and Specificity*, note 22, pp. 214–256; see also Mazumdar, P.M.H., *J. Hist. Biol* **8**:115, 1975.

42. Landsteiner, K., notes 36 and 39. Landsteiner's continuing search for continuity in antibody interactions is well reflected in his lifelong study of the cross-reactions of related antigens, well summarized in his book *The Specificity of Serologic Reactions*, Boston, Harvard University Press, 1945. A reprint of the second edition (New York, Dover, 1962) contains Landsteiner's complete bibliography.

43. Mazumdar, *Species and Specificity*, note 22, pp. 154–155.

44. Landsteiner, note 3.

45. Donath, J., and Landsteiner, K., *Z. Hyg.* **43**:552, 1903.

46. Landsteiner, note 3; see also Landsteiner, K., *Wien. klin. Wochenschr.* **14**:1132, 1901.

47. Landsteiner and Reich, note 40; Landsteiner, K., and Raubitschek, H., *Zentralbl. Bakt.* **45**:660, 1907; Landsteiner, K., and Reich, M., *Z. Hyg.* **58**:213, 1907.

48. Landsteiner and Reich, 1907, note 47, pp. 230–231.

49. Landsteiner, note 36, p. 1626.

50. Landsteiner, note 38. Bordet would later credit Landsteiner with the leading role in this aspect of the dispute with Ehrlich, and would claim that "The affinity of adsorption is sufficiently delicate, graduated, and elective, so that the notion of its participation in antigen–antibody reactions is compatible with that of specificity" (Bordet, note 20, p. 546).

51. Landsteiner and Reich, note 40.

52. Jacobs worked with Landsteiner from 1931 to 1936. The quotation is from a letter to Dr Paul Speiser dated October 31, 1962, published in Speiser and Smekal, note 33, pp. 119–121.

53. Chase, M.W., personal communication, 1984. Dr Chase worked closely with Landsteiner at the Rockefeller Institute from 1932 until Landsteiner's death in 1943.

54. Donath, J., *Z. klin. Med.* **52**:1, 1904.

55. Donath, note 54, p. 27.
56. See, for example, Dameshek, W., Schwartz, R., and Oliner, H., *Blood* **17**:975, 1961; and Dameshek, W., "Theories of Autoimmunity," in *Conceptual Advances in Immunology and Oncology*, New York, Harper and Row, 1963, p. 37.
57. Landsteiner, K., and Leiner, K., *Zentralbl. Bakt.* **38**:548, 1905.
58. Landsteiner, note 39, pp. 492–494.
59. Widal, G.F.I., and Rostaine, S., *C. R. Soc. Biol.* **58**:321, 372, 1905.
60. Besredka, A., *Ann. Inst. Pasteur*, **15**:785, 1901.
61. Donath and Landsteiner 1908a, note 11, p. 211.
62. Eason, J., *J. Path. Bact.* **11**:167, 203, 1906.
63. Eason, note 62, p. 176.
64. Eason, note 62, p. 183.
65. Donath and Landsteiner, 1908a, note 11, p. 206.
66. In fact, Leonor Michaelis had published a case of paroxysmal cold hemoglobinuria in 1901 (*Deutsche med. Wochenschr.* **27**:51, 1901), in which he speculated on the possible involvement of an autolysin. Michaelis, much involved with immunology at the time, provided no experimental support for this insight, and thus received little credit from later workers.
67. Donath and Landsteiner, 1908a, note 11, p. 213.
68. Ehrlich, *Collected Studies*, p. 581; also in *Collected Works*, Vol II, p. 444.
69. Sachs, H., Lubarsch and Ostertag's *Ergeb. allgem. Path.* **11**:515–644, 1907, p. 565.
70. Sachs, note 69, p. 566. Sachs is referring here to the work of R. Kretz (*Wien. klin. Wochenschr.* **16**:528, 1903) and of G. Mattirolo and E. Tedeschi (*Gior. Acad. Med. Torino*, **9**:58, 1903), which fall far short of providing the evidence for their conjectures that Donath and Landsteiner did.
71. Sachs, H., note 15, pp. 902 and 927.
72. Rössle, R., Lubarsch and Ostertag's *Ergeb. allgem. Path.* **13**:124, 1909, p. 228.
73. Rössle, note 72, p. 250.
74. Meyer, E., and Emmerich, E., *Deutsch. Arch. klin. Med.* **96**:287, 1909. A brief report had earlier been presented by these authors in *Verh. Ges. Deutscher Naturf. Aerzte*, 80 Versammlung, Leipzig, Vogel, 1909, Part II, p. 66.
75. Goltz, note 12, p. 210.
76. Meyer and Emmerich, note 74, p. 289.
77. Meyer and Emmerich, note 74, p. 326.
78. Popper, K., *The Logic of Scientific Discovery*, New York, Basic Books, 1959.
79. Merton, R.K., 'The Normative Structure of Science,' in *The Sociology of Science*, Chicago, University of Chicago Press, 1973, p. 267 ff.
80. Kuhn, T., *The Structure of Scientific Revolutions*, 2nd edn, Chicago, University of Chicago Press, 1970.
81. Lakatos, I., "Falsification and the Methodology of Scientific Research Programs," in Lakatos, I., and Musgrave, A., eds, *Criticism and the Growth of Knowledge*, Cambridge, Cambridge University Press, 1970.
82. Feyerabend, P., *Against Method*, London, Verso, 1978.
83. Mulkay, M., "Sociology of the Scientific Research Community," in Spiegel-Rosing, H., and Price, D.J., eds, *Science, Technology, and Society*, Beverly Hills, Sage, 1977.
84. Kuhn, note 80, pp. 77–91.
85. See, for example, Lakatos and Musgrave, note 81, and Suppe, F., ed., *The Structure of Scientific Theories*, Urbana, University of Illinois, 1971. L. Laudan has

summarized the positions in "Two puzzles about science: reflections about some crises in the philosophy and sociology of science," *Minerva* 20:253, 1984.

86. Mayr, E., "Some Thoughts on the History of the Evolutionary Synthesis," in Mayr, E., and Provine, W.B., eds, *The Evolutionary Synthesis*, Cambridge, Harvard Univ. Press, 1980, p. 1.

87. It is interesting that they finally did come to an understanding of sorts, in what was called the modern synthesis by Julian Huxley (*Evolution, the Modern Synthesis*, London, Allen and Unwin, 1942). Some of the history of the famous meeting that helped to reconcile the parties is described in Mayr and Provine, note 86.

88. The dispute between Bordet and Ehrlich was actually never resolved, in the broadest sense. Both made many lasting technical contributions to the field, but their theories were already disbelieved by the 1920s. The fact that Ehrlich's concept would be recast in the now-prevailing paradigm of immunology, the clonal selection theory of antibody formation, was almost accidental; its proponents often did not acknowledge Ehrlich's influence sixty years later (see Chapter 3). No scientific revolution resulted from this debate – immunology *itself* was the revolution that changed the face of nineteenth-century bacteriology and pathology.

15 The search for cell-bound antibodies: on the influence of dogma

...in tuberculous individuals...these antibodies, however, remain to a preponderant extent sessile...

Hans Zinsser, 1914[1]

Even as early as 1914, when immunology was still in its infancy, Hans Zinsser could already summarize the general observation that the tuberculin skin test, though quite specific, appeared to function independent of circulating anti-bodies. But since it was clear to all that immunologic specificity depends upon the interaction of antibody, the logical conclusion must be that the antibody involved was *sessile*. Sessile literally means "seated" or "attached" – i.e., the antibody was most likely attached to the cells that one sees infiltrating the site of the skin test. Over the next fifty years a variety of other curious observations would be made about the tuberculin test and related reactions, but nothing could shake the belief that specific antibodies must be involved in their development. Thus, Russell Weiser could still conclude in 1963 that "Indirect evidence of the presence of immunologically active antibodies has received its greatest support from work on delayed hypersensitivity and especially tuberculin hypersensi-tivity."[2] (Note, however, the use of the term "indirect evidence"; all of the evidence that could be adduced was limited to the demonstration of specificity, not to the detection of actual antibody.)

Weiser actually used these words at a meeting entitled *Cell-bound Antibodies*[3] that was convened in May 1963 at the National Academy of Sciences in Washington, DC.[4] It sought to discuss and hopefully resolve a paradox that had challenged immunology for many years, and that had recently reached the level of an acute intellectual embarrassment. We may define the basic problem that confronted the immunological community as follows:

1. Much of the phenomenology of immunology and immunopathology – e.g., the precipitin, agglutination, and hemolytic reactions, the necrotizing (immune complex) Arthus reaction, anaphylaxis and the human allergies, serum sickness, a few autoimmune diseases (acquired hemolytic anemia, thrombocytopenic purpura, etc.) – were demonstrably associated with the action of *specific* circulating antibodies of one type or another;

2. However, such phenomena as "delayed" hypersensitivities (tuberculin-type, contact dermatitis), allograft rejection, numerous viral infections, and a large number of autoimmune diseases appeared to be unrelated to the presence of circulating anti-bodies, *even though each of these reactions exhibits the exquisite specificity expected*

A History of Immunology, Second Edition
ISBN: 978-0-12-370586-0

Copyright © 2009, Elsevier Inc.
All rights reserved

of acquired immunity. Indeed, by 1963 it had been shown that many of these reactions could be transferred passively only by immune lymphoid cells, and not by immune sera, whereas the "immediate" hypersensitivities were all transferable with the serum of sensitized individuals.[5]

From the initial discovery in 1890 of circulating antibodies to explain specific immunity,[6] and for the next three-quarter century, no other basis for immunologic specificity was known. Ilya Metchnikoff had proposed a central role for phagocytic cells,[7] but no specificity for their action could be identified,[8] and the cellularist idea soon succumbed to the avalanche of humoralist data.[9] Not only did all the early discoveries involve antibodies; Paul Ehrlich's popular side-chain theory of their origin and the chemical basis of their specificity also gave early conceptual support to all these observations.[10] Thus, every demonstration of immunological activity and every explanation of specificity would thenceforth be couched in terms of either protective or destructive antibody, and the analysis of any new observation would be made in these terms.

A review of the history of these developments, leading up to the separation of the immunological armamentarium into two distinct divisions – B cells and T cells – and the momentous discovery of the T cell receptor (TCR), points up an important lesson. Too firm a commitment to a contemporary dogma may inhibit objectivity and constrain speculation. When the investigator becomes a prisoner of this dogma, then it will surely begin to control the direction of his thoughts, the design of his experiments, and perhaps even color the interpretation of his data.

As Thomas Kuhn pointed out,[11] the commitment to what he termed *normative science* restricts both experiment and thought to the context of the "old" accepted paradigm, and inhibits the type of speculative flight into uncharted areas that so often moves science in an unanticipated leap forward. Only when forced by the overwhelming inability of the old ideas to explain new phenomena will the field change its direction. The older generation often refuses to acknowledge the change, and it may be the younger, conceptually uncommitted generation that leads the charge in new directions.

The historical background to the problem

The widening dichotomy

We have listed above the principal phenomena discovered during the first thirty years of research in immunology. With the single exception of the tuberculin skin test, an identifiable, passively transmissable antibody could be associated with each specific result seen. Then, as the field of immunology matured over the years, further inconsistencies appeared that could not readily be explained within the old framework. Whereas all of the phenomena associated with antibodies, including their skin reactions, appear within minutes or at most within a very few hours, the tuberculin skin reaction (and, later, the luetin test for syphilis and the lepromin test for leprosy, etc.) as well as the rash of contact dermatitis do not develop for some twenty-four to forty-eight hours. Based upon these timing differences, the former were called *immediate* hypersensitivities and the latter *delayed* hypersensitivities. This led to the suggestion that there is a unique response to microorganisms known as "immunity of infection" or "bacterial allergy," until Dienes and Schoenheit showed in the late 1920s[12] that this type of allergy could be elicited against almost any bland protein by injecting it directly into the tubercles of infected animals. The procedure was later generalized with the introduction of Freund's adjuvants containing dead mycobacteria.[13] Now delayed-type hypersensitivity (DTH) could be elicited by *any* protein, even by the hapten protein conjugates[14] that Karl Landsteiner had so fruitfully employed to study antibody specificity.[15]

Another phenomenon apparently related to these delayed hypersensitivities was poison ivy contact dermatitis and related responses, elicited in sensitized individuals by such chemically active substances as picryl chloride and dinitro-fluorobenzene. In all of these tuberculin/DTH reactions, there seemed to be no correlation between the degree of hypersensitivity and the titer of circulating antibodies.[16] Similarly, when Peter Medawar did his classical studies of trans-plantation immunology, he sought in vain the antibodies that must account for these specific destructive processes.[17]

Since it had by then been well established that immunity to many toxins, infectious agents, and allergens could be transferred passively using immune serum,[18] the quandary posed by the failure to find antibodies was only made starker by the demonstration by Landsteiner and Chase that contact dermatitis could only be transferred passively with cells and not with antisera;[19] by anal-ogous results with the passive transfer by cells of tuberculin hypersensitivity[20] and of delayed hypersensitivity to proteins;[21] and by similar results with the passive transfer with cells of the ability to reject allografts.[22] This latter finding received strong support from the study that showed that tissue grafts would survive in a chamber permeable to antibodies, but impermeable to cells.[23]

Other phenomena appeared to demonstrate further differences between immediate and delayed hypersensitivities. Among the earliest was the demon-stration that whereas immediate (e.g., Arthus-type immune complex) skin reac-tions are characterized by predominantly polymorphonuclear cell infiltrates, those of the delayed type (including allograft rejection) involve significant round cell (lymphocyte and histiocyte) infiltrates.[24] Again, the induction of anaphylactic reactions and attempts to desensitize them with antigen injections led generally to a lowering of the temperature of the host, while the injection of specific antigen into delayed-sensitive hosts resulted in an increase of body temperature.[25] Finally, the response to antigen injection into the avascular cornea of the sensitized host differs. A positive tuberculin reaction (corneal inflammation) can be elicited in the central avascular cornea of the sensitized anomal, but not an Arthus-type reac-tion, unless vascularization of the cornea is induced beforehand.[26]

In his customary precise way, Arnold Rich devoted some fifty pages in his *The Pathogenesis of Tuberculosis* to the many apparent differences in phenome-nology and mechanism between the immediate and delayed hypersensitivities.

One of the more curious differences was the demonstration in 1932 by Rich and Lewis[27] that tuberculin would inhibit the migration of macrophages from explants of spleen fragments (forerunner of the discovery of migration-inhibition factor by Bloom and Bennett and by David,[28] the first of a congeries of what would later be termed lymphokines). In the spirit of the times, Rich could do no better than suggest that "One of the most persuasive reasons for believing that antibody is concerned in tuberculin type hypersensitivity is the high degree of specificity of the phenomenon."[29]

Immunological deficiency diseases

In 1952, there appeared a report, by Ogden Bruton, of a congenital, sex-linked agammaglobulinemia in a child.[30] No antibodies could be formed by such patients, but what perplexed pediatricians was that while they suffered repeated infections by (mostly gram-positive) bacteria, they resisted most viral diseases normally and showed definite DTH reactions.[31] (It is interesting, however, as Robert Good later confessed,[32] that some of the initial reports concluded that these patients were *un*able to develop DTH and to reject allografts, but soon the picture changed. Can these investigators have found initially only what their mindset had prepared them for?)

An immunodeficiency associated with defects of the thymus then surfaced, but with the difference that now it was viral infections that could not be controlled, whereas bacterial infections were controlled by circulating immunoglobulins.[33] Clearly, defense against these viral diseases could not involve typical circulating antibodies. It then became apparent that an earlier report of what has become known as "Swiss-type" disease,[34] or severe combined immunodeficiency (SCID), involves both of these defense systems.

The underlying causes of these various immunodeficiencies was then clarified by a series of imaginative experiments. It was shown that excision of the Bursa of Fabricius in birds leads to a severe defect in the formation of antibodies,[35] and it became clear that somewhere there exists in mammals an equivalent – possibly in the bone marrow. Then it was shown that thymectomy depresses or abolishes the ability to mount delayed-type responses and to reject allografts.[36] From these starting points, and with the support of a number of imaginative *in vitro* experiments, would emerge the two-component concept of lymphoid development,[37] later to become known as the B cell (bursa/bone marrow) and T cell (thymus) systems.[38]

The hegemony of the antibody paradigm

DTH and antibody formation

I mentioned earlier the extensive search for a relationship between hypersensitivity and immunity. This was pursued most notably by Arnold Rich in the context of tuberculosis, and the results were generally disappointing.[39]

However, Gell and Hinde's demonstrations of plasma cells in resolving DTH lesions, and their observation that delayed sensitivity seemed to prepare the animal for a booster antibody response, were impressive.[40] These authors also described what they termed a "progressive immunization reaction," in which certain protocols led initially to DTH followed by Arthus sensitivity. Taken together, these results seemed to imply an extremely close connection between DTH and antibody formation. It was then a simple jump to the suggestion by Benacerraf and Gell that DTH is "an early, immature stage of immunity."[41]

In a lengthy review of the subject, Alwin Pappenheimer suggested that Ag interacts in two stages with "complementary structures."[42] He suggested that the first interaction is to "sensitize" cells to DTH, and the second to push them to differentiate to antibody producers. The structures (receptors) with which the antigen interacts in both steps were implicitly assumed to be identical.

Specificity of DTH

Even while a close interrelationship between DTH and antibody was being suggested, almost incompatible differences were reported in the specificities of the two reactions. One of the most telling of these was the observation by Gell and Benacerraf that there is a marked cross-reaction in the DTH system between native and denatured proteins.[43] Such cross-reactions are not normally seen in reactions involving serum antibodies.

Next was the oft-repeated finding that whereas anti-hapten antibodies are specific primarily for the hapten alone, DTH reactions are specific only for the entire hapten–protein conjugate.[44] This means that sensitization for DTH by a hapten on carrier protein A cannot be demonstrated by testing with the same hapten on protein B. This came to be known as the "carrier effect," and led to the assumption that the mediator of DTH specificity (implicitly some unusual type of antibody) sees both the hapten *and a portion of the surrounding carrier protein*. In other terms, this was taken to mean that the DTH determinant is larger than that controlling the reactions of "regular" antibodies. This view found support from a study using a Landsteiner-type homologous series of haptens.[45] DTH was shown to exhibit less specificity with respect to the hapten than does antibody, but a greater specificity with respect to the protein to which it is linked.

Perhaps the most telling demonstration of the difference between antibody and DTH specificity involved the attachment of a hapten to different sites on the same carrier protein.[46] If an animal is immunized with the nitrobenzene hapten attached to either the lysine or the tyrosine residues on albumin, the resulting antibodies will react no matter what the attachment site on the test homologous albumin. In the DTH system, however, there is absolutely no skin-test cross-reaction between the two attachment sites, indicating that not only is there carrier specificity in DTH, but site-attachment specificity as well.

High-affinity antibody

The preoccupation of many immunologists up to the 1960s was with the chemistry and thermodynamics of antibodies[47] – thus the resistance of many to accepting Burnet's ideas about the importance of cell dynamics in immunological responses. The notion that some non-antibody intrinsic property of lymphoid cells might account for delayed-type reactions was distasteful to many. An imaginative theory was advanced by Karush and Eisen in 1962,[48] in an effort to "save" immunochemistry from what seemed to be a most improbable biological explanation. Based upon the very small amounts of protein antigen required to elicit a delayed skin test, they proposed that as little as 10^{-10} molar antibody was required (far below the level detectable in serum), and that the cells involved in the process would continue to produce these small amounts even after passive transfer to a new host. The affinity constants of these putative antibodies was assumed to be some 100 to 10,000 times that of anaphylactic antibodies.

The theory of high-affinity antibodies carried with it several testable implications. The first was that in order to react at these extremely low concentrations, the antigen must be able to contribute significantly to the interaction energy – i.e., the specificity of the antibody must be for more than the simple hapten, and include at least the neighboring three amino acids of the protein carrier. This prediction seemed to accord well with the demonstration of the large determinant size found experimentally, as described in the previous section. However, it also predicted that a saccharide, by virtue of its chemical composition, could not possibly contribute sufficient energy to such an interaction, and this seemed in line with the observation that polysaccharides can neither induce nor elicit DTH. It was soon shown, however, that mono- and di-saccharide *haptens* attached to carrier proteins will serve this purpose quite well.[49]

Another prediction essential to the theory involved the possible concentration range in which antigen might interact with the putative high-affinity antibody. If an antibody association constant of between 10^{-10} and 10^{-12} molar is postulated, then this restricts the antigen concentration required to elicit a DTH reaction to approximately similar values. But it was shown that amounts of antigen at skin concentrations of the order of 10^{-14} molar can elicit delayed reactions[50] – an amount too low to support adequate interaction with the postulated antibodies. Moreover, it was found that a delayed skin test could be elicited in a guinea pig partially desensitized with specific antigen, such that the residual concentration of antigen in the animal's circulation approached 10^{-7} molar. This much antigen should have neutralized any high-affinity antibody present, and thus aborted the reaction.

The search for IgT

In their review of T cell receptors years later, Marrack and Kappler put their finger on the danger of preconceptions:[51]

Early attempts to isolate these proteins relied heavily on the idea that T cell
receptors [IgT] might be similar, if not identical, to immunoglobulin. In
retrospect, although this idea was not unreasonable, it certainly created a good
deal of confusion in the field.

Many laboratories sought the elusive T cell immunoglobulin receptor on the
surface of these active cells, and one group claimed success. In 1972 Marchalonis
and coworkers claimed to have isolated it,[52] but the claim was strongly contested.[53] In that same year, an entire issue of *Transplantation Reviews* was
devoted to the topic "Interaction between humoral antibodies and cell-mediated
immunity." Although much emphasis was placed on immunoregulation, some of
the discussion related to the role of the T cell receptor, and Feldmann and Nossal
could specify in a footnote that "The terms 'T cell IgM' and 'IgT' are used
synonymously."[54]

Resolution: T and B cells and the TCR

During the several decades under discussion here, the explanation of the
complicated process responsible for the generation of antibody diversity and
for the B cell Ig receptor was even then on the horizon. It represented an
evolutionary development apparently unique in biology. But was the independent evolution of *two* such remarkable systems readily imaginable at the
time? (One is reminded here of Felix Haurowitz's argument against the
bivalence of antibody. He claimed that the formation of one combining site on
an antibody is already a miracle; can we reasonably expect two such miracles
on the same molecule?[55]) Suffice it to say that Felix Haurowitz's double miracle
had indeed taken place, not once but twice! On the molecular level was the
bivalent antibody, and on the systems level was the evolution of two distinct
mechanisms employing two differently constituted receptors (Ig and TCR) to
mediate immunity.

It is not necessary to outline in detail here the remarkable series of
investigations that resolved the antibody–cell paradox, nor need we review
the many steps in the discovery of the nature of the T cell receptor and the
genetic mechanisms responsible for its production. These were briefly
reviewed in Chapter 7. In addition, the steps in the maturation of the several
lymphocyte lineages have been well established, as well as the differentiation
markers that characterize them and the variety of molecular signals that each
produces.

It may be well, however, to point out how the final solution of the T cell
receptor problem resolved so elegantly the perplexities and paradoxes posed
by the earlier observations. First was the demonstration that cellular and
humoral immunities are in fact separate processes, although the two do inter-
sect at various points – e.g., the role of helper and suppresser T cells in regulating antibody formation and other responses.[56] It is these functions that

account in part for the earlier suggestion that DTH might be only a stage in the production of antibody.

Next, the demonstration of the structure of the T cell receptor and its mode of function proved that a "cell-bound antibody," or IgT, does not exist. This resolved the problem of the different specificities of the two reactions; the antibody sees only a relatively small region (usually on the surface) of the protein antigen, whereas the TCR sees an entire amino acid sequence (from within the protein) attached to an MHC molecule.[57] This explained why denatured and native proteins are able to cross-react in DTH, and why a hapten attached to lysine does not cross-react with the same one attached to tyrosine on the same protein. It also clarified the reason why a "carrier effect" operates – the antibody combining site sees primarily the hapten itself, whereas the TCR sees in addition the polypeptide portion of the carrier to which the hapten is affixed.

The structural basis of the TCR also explained well the paradoxical observation that DTH could not be elicited against polysaccharide antigens, although it did function against saccharide haptens attached to protein. This is because the polysaccharide lacks the polypeptide structure demanded by the TCR, while the peptide carrying a saccharide hapten does fulfill this requirement.

Comment

Sociologists of science point out that the great advances of one generation may often retard progress in the next. This is because each advance may induce a mindset in the scientist that slants the interpretation of data and may inhibit new speculations. Thus, we saw above how the brilliant progress in antibody research early in the century led to later delays and false leads in the exploration of cellular-immune reactions. In the same way, Chapter 8 showed how Paul Ehrlich's concept of *horror autotoxicus* was so widely accepted that progress in understanding autoimmune diseases was delayed for decades. Similarly, the initial firm conviction that the immune system is protective (whence its very name) long prevented the development of an appreciation of the proper place of immunopathology in the overall functions of the immune system. Thus, the facile explanation was advanced that this was "immunity gone wrong;" there were "good antibodies" responsible for protective immunity and "bad antibodies" (so-called reagins) responsible for allergic reactions, autoimmunity, and other immunopathologic processes.

Every science can point to earlier, overly dogmatic commitments that interfered with progress; as noted above, the commitment to antibody as the sole basis for the exercise of immunologic specificity is but one of several that have involved immunology. Its study carries an important lesson; we must take care always to question our preconceptions, rather than taking the easier path of designing experiments to confirm them.

Notes and references

1. Zinsser, H., *Infection and Resistance*, New York, Macmillan, 1914, p. 443.
2. Weiser, R.S., in Amos, B., and Koprowski, H., eds, *Cell-bound Antibodies*, Philadelphia, Wistar Institute Press, 1963, p. 72.
3. At this meeting and in the present book, the use of the term "cell-bound" antibodies is not meant to include "cytophilic antibodies." The former could never be found in the circulation, whereas the latter are circulating antibodies, primarily IgE active in allergic reactions, that are readily shown to adhere to certain cells.
4. It is significant that this meeting was convened at the behest of the National Academy of Science's Committee on Tissue Transplantation, since by this time allograft rejection was understood to be specific, but independent of circulating antibodies.
5. Thus Behring, E., and Kitasato, S. (*Deutsch. med. Wochenschr.* **16**:1113, 1890) and Behring, E., and Wernike, E. (*Z. Hyg.* **12**:10, 45, 1892) had shown that immunity to diphtheria and tetanus toxins could be transferred using immune serum. Similarly, Prausnitz, C. and Küstner, H. demonstrated that serum could transfer a food allergy to a normal human recipient.
6. Behring and Kitasato, note 5.
7. Metchnikoff, I., *Virchows Archiv.* **96**:177, 1884; see also Metchnikoff, E., *Immunity in the Infectious Diseases*, New York, Macmillan, 1905, reprinted by Johnson Reprints, New York, 1968.
8. Macrophage specificity was suggested initially by the studies of Elberg, S.S., and Faunce, K., Jr., *J. Bacteriol.* **73**:211, 1957; Elberg, S.S., *Bacteriol. Rev.* **24**:67, 1960; Suter, E., *J. Exp. Med.* **97**:235, 1953. Soon, however, this view was successfully challenged by Mackaness, G.B., *Am. Rev. Tuberc.* **69**:495, 1954. For more general reviews, see Rowley, D., *Adv. Immunol.* **2**:241, 1962; Suter, E., and Ramseier, H., *Adv. Immunol.* **4**:117, 1964; and Nelson, D.S., *Macrophages and Immunity*, Amsterdam, North Holland, 1969.
9. For a discussion of the history of the cellularist–humoralist debate, see Chapter 2.
10. Ehrlich, P., *Klin. Jahrb.* **6**:299, 1897; see also Ehrlich, P., Croonian Lecture, *Proc. R. Soc. London*, **66**:424, 1900.
11. Kuhn, T., *The Structure of Scientific Revolutions*, 2nd edn, Chicago, University of Chicago Press, 1970.
12. Dienes, L., and Schoenheit, E.W., *Am. Rev. Tuberc.* **20**:92, 1929; Dienes, L., *J. Immunol.* **17**:531, 1929.
13. Freund, J., and McDermott, K., *Proc. Soc. Exp. Biol. Med.* **49**:548, 1942.
14. Gell, P.G.H., and Benacerraf, B., *Advances Immunol.* **1**:319, 1961.
15. Landsteiner, K., *The Specificity of Serological Reactions*, revised edn, New York, Dover, 1962.
16. Arnold Rich comments at length on the disparity between the degree of tuberculin hypersensitivity, immunity to tuberculosis, and the titer of circulating antibodies in infected individuals in *The Pathogenesis of Tuberculosis*, Springfield, Charles Thomas, 1944, pp. 509–613.
17. See Medawar's summary of his work in *Harvey Lect.*, **52**:144, 1956–1957. Medawar would soon abandon antibodies as the effectors in allograft rejection (Medawar, P.B., in Lawrence, H.S., ed., *Cellular and Humoral Aspects of Hypersensitivity*, New York, Hoeber-Harper, 1959, pp. 504–529, but Peter Gorer persisted

in this belief (see L. Brent, *A History of Transplantation*, New York, Academic Press, 1997, p. 74 ff).

18. See the studies in note 5.

19. Landsteiner, K., and Chase, M.W., *Proc. Soc. Exp. Biol. Med.* **49**:688, 1942.

20. Chase, M.W., *Proc. Soc. Exp. Biol. Med.* **59**:134, 1945.

21. Lawrence, H.S., *J. Clin. Invest.* **34**:219, 1955.

22. Mitchison, N.A., *Proc. R. Soc. London* **142**:72, 1954; Billingham R.E., Brent, L., and Medawar, P.B., *Proc. R. Soc. London* **143**:58, 1954.

23. Weaver, J.M., Algire, G.H., and Prehn, R.T., *J. Natl Cancer Inst.* **15**:1737, 1955.

24. The significance of these cytological differences was very early pointed up by Dienes, L., and Mallory, T., *Am. J. Pathol.* **8**:689, 1932, and more forcefully later by Gell, P.G.H., and Hinde, I.T., *Br. J. Exp. Path.* **32**:516, 1951 and *Intl Arch. Allergy* **5**:23–46, 1954; and by Waksman, B.H., *Intl Arch. Allergy Appl. Immunol.* **14**:Suppl., 1959.

25. See Zinsser, *Infection and Resistance*, note 1, p. 366.

26. Rich, A.R., and Follis, R.H., Jr., *Bull. Johns Hopkins Hosp.* **66**:106, 1940.

27. Rich, A.R., and Lewis, M.R., *Bull. Johns Hopkins Hosp.* **50**:115, 1932. They thought that the antigen killed the target macrophages specifically, but see Waksman, B.H., and Matoltsy, M., *J. Immunol.* **81**:220, 1958.

28. Bloom, B.R., and Bennett, B., *Science* **153**:80, 1966; David, J.R., *Proc. Natl Acad. Sci. USA* **56**:72, 1966.

29. Rich, *The Pathogenesis*, note 16, p. 404.

30. Bruton, O.C., *Pediatrics* **9**:722, 1952.

31. Good, R.A., and Zak, S.J., *Pediatrics* **18**:109, 1956; Janeway, C.A., and Gitlin, D., *Adv. Pediatrics* **9**:65, 1957.

32. See Good, R.A., et al., in Shaffer, J.H., et al., eds, *Mechanisms of Hypersensitivity*, Boston, Little Brown, 1959, pp. 467–476.

33. Nezelof, C., et al., *Arch. Franç. Pediat.* **21**:897, 1964; DiGeorge, A.M., *J. Pediat.* **67**:907, 1965.

34. Glanzmann, E., and Riniker, P., *Ann. Paediat. Basel*, **175**:1, 1950.

35. Glick, B., Chang, T.S., and Jaap, R.G., *Poultry Sci.* **35**:224, 1956.

36. Miller, J.F.A.P., *Lancet* **ii**:748, 1961. Similar results appeared independently by Janković, B.D., Waksman, B.H., and Arnason, B.G., *J. Exp. Med.* **115**:159, 1962 and by Good, R.A., et al., *J. Exp. Med.* **116**:773, 1962. See also Defendi, V., and Metcalf, D., eds, *The Thymus*, Philadelphia, Wistar Institute Press, 1964.

37. Cooper, M.D., et al., in Bergsma, D., and Good, R.A., eds, *Immunological Deficiency Diseases of Man*, New York, National Foundation, 1967, pp. 7–16. See also Mitchison, N.A., in Landy, M., and Braun, W., eds, *Immunological Tolerance*, New York, Academic Press, 1969, p. 115 ff.

38. Claman, H.N., Chaperon, E.A., and Triplett, R.F., *Proc. Soc. Exp. Biol. Med.* **122**:1167, 1966; see also Claman, H.N., and Chaperon, E.A., *Transplant. Rev.* **1**:92, 1969.

39. Rich, *The Pathogenesis of Tuberculosis*, note 16; Rich, A.R., *Harvey Lect.* **42**:106, 1946–1947.

40. Gell, P.G.H., and Hinde, I.T., *Intl Arch. Allergy Appl. Immunol.* **5**:23, 1954; see also Turk, J.L., *Delayed Hypersensitivity*, Amsterdam, North Holland, 1967.

41. Benacerraf, B., and Gell, P.G.H., *Immunology* **2**:53, 1959. See also Salvin, S.B., *J. Exp. Med.* **107**:109, 1958.

42. Pappenheimer, A.M., Jr., Scharff, M., and Uhr, J.W., in *Mechanisms of Hypersensitivity*, note 32, pp. 417–434. See also Gell, P.G.H., and Benacerraf, B., *Adv. Immunol.* **1**:319, 1961.
43. Gell, P.G.H., and Benacerraf, B., *Immunology* **2**:64, 1959.
44. Benacerraf and Gell, note 41; Salvin, S.F., and Smith, R.F., *J. Exp. Med.* **111**:465, 1960; Benacerraf, B., and Levine, B.B., *J. Exp. Med.* **115**:1023, 1962; Gell, P.G.H., and Silverstein, A.M., *J. Exp. Med.* **115**:1037, 1962.
45. Silverstein, A.M., and Gell, P.G.H., *J. Exp. Med.* **115**:1053, 1962.
46. Gell and Silverstein, note 44.
47. See, for example, Pressman, D.P., and Grossberg, A., *The Structural Basis of Antibody Specificity*, New York, Benjamin, 1968; and Kabat, E.A., *Structural Concepts in Immunology and Immunochemistry*, New York, Rinehart & Winston, 1968.
48. Karush, F., and Eisen, H.N., *Science* **136**:1032, 1962.
49. Borek, F., Silverstein, A.M., and Gell, P.G.H., *Proc. Soc. Exp. Biol. Med.* **114**:266, 1963.
50. Uhr, J.W., and Pappenheimer, A.M., *J. Exp. Med.* **108**:891, 1958. See also Silverstein, A.M., and Borek, F., *J. Immunol.* **96**:953, 1966.
51. Marrack, P., and Kappler, J., *Adv. Immunol.* **38**:1, 1986.
52. Marchalonis, J.J., Atwell, J.L., and Cone, R.E., *Nature New Biol.* **235**:240, 1972; Cone, R.E., and Marchalonis, J.J., *Biochem. J.* **1430**:345, 1974.
53. See Vitteta, E.S., et al., *J. Exp. Med.* **136**:81, 1972 and *Proc. Natl Acad. Sci. USA* **70**:834, 1973.
54. Feldmann, M., and Nossal, G.J.V., *Transplant. Rev.* **13**:3, 1972, p. 4.
55. I recall having heard this at a meeting in the early 1970s, I think by Felix Haurowitz. I later asked him about it, and while he did not remember it specifically, he admitted that he might well have said it.
56. Mitchison, note 37; see also note 36. A more modern approach to regulatory T cells will be found in Dario, A.A., et al., *Nature Rev. Immunol.* **8**:523, 2008.
57. Zinkernagel, R.M., and Doherty, P.C., *Nature* **248**:701, 1974; *Adv. Immunol.* **27**:51, 1979; Marrack and Kappler, note 51.

16 "Natural" antibodies and "virgin" lymphocytes: the importance of context

...normal serum contains substances which are formed independently of external antigenic stimuli...

Karl Landsteiner[1]

During the early days of any science, before theory can catch up to fact, it is often difficult to understand and explain certain phenomena. Thus, early geologists were at a loss to explain how seashells could be deposited at the top of the Alps in Switzerland; early physicists could not comprehend how X-rays (X = unknown) could penetrate solid matter to expose photographic film; and early physicians could not understand the workings of contagion. In each instance, speculation might run rife until the rational solution was found – sometimes quickly, sometimes long after the organization of the science.

The field of immunology has not been exempt from a similar inability to explain certain findings, and provides two interesting examples. The first – the problem of explaining the presence of certain antibodies in the absence of provable antigenic stimulus – illustrates how the same question can come and go over a long period of time, as the concepts that guide the field change. The second example – whether there exists such an entity as a "virgin" immunocyte – had to wait some eighty years to be posed, until Macfarlane Burnet focused the attention of the field on the cellular dynamics of the immune response. In each case, the final answer would prove more complicated than the original question.

The discussion that follows will attempt to show how the conceptual and technological context of the times may determine how scientists explain certain phenomena, if indeed they permit themselves even to consider them at all.

"Natural" antibodies

The era of Ehrlich's side-chain theory

The first protective antibodies that were identified were those against diphtheria and tetanus toxins; these were shown not only to protect against the two diseases specifically, but even to abort the process once started.[2] During the succeeding decade, a number of other pathogens, and even such benign substances as red blood cells, were found able to elicit the formation of specific antibodies when injected into laboratory animals. But, slowly at first, investigators began finding

A History of Immunology, Second Edition
ISBN: 978-0-12-370586-0

Copyright © 2009, Elsevier Inc.
All rights reserved

specific antibodies in animals not known to have been actively immunized, and antitoxins in humans with no previous history of infection. It was especially telling that many animals possess presumably specific antibodies against the red cells of other species which mediate their hemolysis in conjunction with complement.

In 1900, Karl Landsteiner discovered in humans the presence of anti-erythrocyte antibodies specific for the red cells of other humans;[3] this led to the identification of human blood groups, and such anti-erythrocyte isoantibodies were soon found in other species as well. One might have thought that the "strange" presence of such antibodies would have made people wonder how their production had been stimulated, when it was obvious that no previous specific immunization had taken place. However, this first encounter had been made in the context of an existing theory of antibody formation that made the presence of these "naturally-occurring" antibodies seem fully explicable.

This theoretical setting was Paul Ehrlich's "side-chain theory of antibody formation," discussed in detail in Chapters 3 and 6. For our present purposes it will suffice to mention that the theory, widely accepted in the late 1890s, proposed that each specific antibody began life as the spontaneously formed molecule deposited somehow as a receptor on the cell that had produced it. When this receptor thenceforth reacted with its specific antigen (most often during a natural infection or due to the intercession of an immunologist), it would stimulate the cell to produce an excess of the same antibody which would end up in the circulating blood. It was only natural, therefore, to assume that any antibody found in the unimmunized host must necessarily be a "minor spillover" of that initial production of receptors, and thus a fairly normal occurrence unworthy of special attention. Such a position was so implicit in Ehrlich's theory that few thought even to raise the question.[4]

Thus Landsteiner, discoverer of the ubiquitous presence of blood-group antibodies in "normal" individuals, did not question their spontaneous generation. Even though he was a follower of Jules Bordet and questioned Ehrlich's chemical approach to the specificity of these antibodies, he questioned only the "quality" of these antibodies and not their provenance. Landsteiner would follow Bordet[5] in suggesting that these spontaneously occurring antibodies were not the final product, but merely immature antibodies whose specificity would be "perfected" by interaction with antigen.[6] Thus, so long as Ehrlich's theory of antibody formation held sway, no one thought to question the existence of what would later be termed "natural" antibodies.

The immunochemical era

As we noted previously (and will expand upon in Chapter 17), the period around World War I saw the decline of the medically-oriented immunology of Pasteur, Ehrlich, Behring, and Metchnikoff. The easier vaccines (fowl cholera, rabies, anthrax, and plague) and the useful serotherapies (for diphtheria and tetanus) had been applied, and important new discoveries were slow to appear. This lull led to the rise of a new disciplinary direction that would be dominated for the next fifty years or so by chemically-oriented scientists. These were interested less in where antibodies come from than in the structural and thermodynamic aspects of their interaction with antigen.

As we saw in Chapter 3, the discovery that the immune system could respond not only to dangerous pathogens but also to a host of benign agents like ery-throcytes and simple proteins opened Ehrlich's theory to severe challenge. Why would evolution have favored such meaningless responses? With the decline in this Darwinian theory came, *pari passu*, the disappearance of the belief that antibodies might be spontaneously formed normal products of the body. Even Landsteiner, former believer, came finally to the position that the repertoire of possible specificities was far too great to allow for so many different molecules. Thus a new theory took form; the Lamarckian view that antibodies were common proteins that derived their specificities from interaction with or instruction of some sort by antigen. The two principal instructive theories that saw the light during the 1930s were those of Felix Haurowitz[7] and Linus Pauling.[8] Each modification of the main theme thus retained the notion that specific antibody could only be formed through the intercession of antigen; in antigen's absence, spontaneous formation of its corresponding antibody could not be conceived.

But how to explain the presence in "normal" serum of blood group isoanti-gens, of hemolytic antibodies against the cells of other species, and of modest titers of antibodies against a variety of pathogens, absent previous infection? The answer was in fact that these "spontaneous" or "natural" antibodies could not be explained at all. As is often the case, when a phenomenon cannot be explained, the simplest solution is to pretend it does not exist – and this in fact is what most scientists of the era did.

It must not be thought that the phenomenon of "natural" antibodies was the only one neglected by the immunochemists of the day; they tended to neglect the implications of persisting antibody formation long after antigen had disappeared from the immunized host; they neglected also the secondary antibody response that produced so much more antibody than did the primary response. Even more perplexing to the instructionist theory, but no less neglected by its followers, was the question of how repeated antigenic boosts might result in an ever-increasing affinity of the resulting antibody (later to be termed "affinity maturation"), or the question of how protection could be developed against certain viral infec-tions in the apparent absence of any participation of circulating antibodies?

However, this neglect of the inexplicable is not restricted to any particular group. If the immunochemists permitted themselves to disregard these biological problems, then the early immunobiologists would tend, in their turn, to disre-gard the more problematic chemical ones. Nowhere does the immunobiologists' theoretician-in-chief Macfarlane Burnet take on the important question of the nature of the antibody combining site, nor how it manages to bond to its anti-genic partner.

As we have pointed out elsewhere, no matter what research program or theoretical basis might dominate a science at a given time, there are always those few individuals or groups who are willing to address the difficult problems and unfashionable directions, or to think thoughts that might seem heretical in the context of the prevailing paradigm. Thus, there were some who thought about natural antibodies, and even fewer who actually worked with them. As late as 1966, well before the genetic basis of antibody formation would be clarified, Stephen Boyden would indicate in a broad review of the area[9] that some of these natural antibodies might be stimulated by inapparent exposure or even by cross-reacting substances. The remainder, however, that were also referred to as "normal" antibodies, still posed a serious conceptual problem.

Somewhat related to the problem of natural antibodies (and in a way amusing in retrospect) was the repeated observation that the blood contains appreciably more immunoglobulins than can be accounted for in terms of identifiable antibodies. It was easy to assume that these also were produced spontaneously, but they seemed not to be specific for any particular antigen. However, if, as was increasingly believed, all antibodies are immunoglobulins and all immunoglobulins are antibodies, then one might be justified in calling those whose specificities could not be identified "*non-sense*" antibodies; those whose provenance and specificity were known would appear to "make sense" to the investigator.

It was in this context, then, that the somewhat late-starting Niels Jerne[10] would get his first exposure to immunology at the State Serum Institute in Copenhagen – an event that would have interesting future consequences for the field. Two coincidences seem to have furnished the foundation for Jerne's preoccupation with natural antibodies. In the first place, he was assigned to work on diphtheria toxin–antitoxin reactions, and on the avidity of these anti-toxins, on which he wrote his dissertation.[11] He would note that the antitoxin-like activity found in normal serum appeared to be less avid than that obtained by active immunization, and indeed outlined the phenomenon that would later be known as "affinity maturation." The second chance occurrence came not long after, when a visit to Copenhagen by Gunter Stent and James Watson exposed Jerne to the activities and methods of the phage group. In working on improvements to the assay of phage viruses, Jerne chanced upon the presence of an activating factor in normal horse serum that paralleled precisely the activity of the anti-phage antibodies produced following active immunization with the virus. Here was the proximate stimulus, according to Jerne's later recollection,[12] to his musings on the role of natural antibodies, and thus to his natural selection theory of antibody formation.

The immunobiological era

Throughout the previous chemically-oriented era there had always been the odd individual or laboratory that worked on the more medical, and especially pathological, aspects of the immune response. However, starting in the late 1930s with the work on the apparently autoimmune allergic encephalomyelitis[13] and the genetic basis of the immune response,[14] and later with the rediscovery of autoimmune hemolytic anemias,[15] interest in the medical and even biological implications of the immune response reawakened. The pace quickened with the postulate of immunological tolerance, reports on immunodeficiency diseases and the role of the bursa and thymus, and of the probable immunological basis for the rejection of tissue grafts.[16] Each of these new findings proved either to have no obvious relationship to the prevailing instructionist theory of antibody formation, or indeed directly to contradict its requirements. Thus, the chemically-oriented Lamarckian concept, and even the somewhat more biological form that Burnet had given it,[17] no longer satisfied the requirements posed by the data. Moreover, the recent progress in understanding the structure and function of the genetic basis of protein formation, leading to Francis Crick's dictum that information can only flow from DNA to RNA to protein, made an instructionist mechanism for the induction of antibody formation even more improbable. A new replacement was needed.

Into the breach, in 1955, came Niels Jerne with his natural selection theory of antibody formation.[18] Familiar with the notion of natural antibodies, as we have seen, Jerne followed Ehrlich in suggesting that each specificity of antibody is spontaneously formed and delivered to the bloodstream. There it awaits the appearance of its specific antigen, and when the latter appears and reacts with its partner, the complex returns to the mother cell and somehow stimulates it to form more of the same antibody. The theory lacks elegance, even compared to Ehrlich's selective theory of almost sixty years earlier, but it did serve a double heuristic function; it raised anew the question of natural antibodies (and provided them at long last with a function), and it stimulated David Talmage and Macfarlane Burnet to think about selectionist possibilities as an explanation for the formation of so large a repertoire of specific antibodies.

In fairly short order came Talmage's suggestion[19] and Burnet's somewhat more elaborate concept[20] that antigen *selects* for the production of its corresponding specific antibody. But even more critical was the proposal by each of them that cellular dynamics plays an important role in the antibody response. This was most clearly advanced by Burnet, who not only called for the antibody to function as an Ehrlich-type receptor on the cell involved, but also to have that antigen–receptor interaction lead to a clonal expansion of daughter cells with similar function. With the further elaboration of his theory by Burnet[21] and its supplementation by Joshua Lederberg[22] and Talmage,[23] clonal selection rapidly displaced the earlier instructionist theories of antibody formation.

Resolution of the conceptual problems

The clonal selection theory provided plausible answers to many of the biological aspects of the immune response that instructionists had been unable to address. Not only was it consistent with the developing genetic underpinnings of protein formation in general, and of the large antibody repertoire in particular; it also explained well both the enhanced secondary response and affinity maturation. Indeed, the explanation of these phenomena in the context of a clonal expansion gave substance to the notion of "immunological memory," otherwise imponderable in an instructionist context.

What, then, of natural antibodies? Once again, as in Ehrlich's day, these could be merely a spillover of the normal spontaneous production of samples of the entire antibody repertoire. But was even this conclusion necessary? In addition to the growing impression that many of these "natural" antibodies were in fact the products of subliminal stimuli by cross-reacting antigens, there was yet another explanation that might settle the question permanently. Virologists have long known that viral infection may be accompanied by a substantial polyclonal activation of B cells, with the accompanying production of whatever antibody in their respective genetic programs.[24] But it was becoming increasingly apparent that almost any immune response may be accompanied by both activated cells and lymphokines capable of stimulating a polyclonal expansion of B (and probably also T) cells.[25] Thus, the large "excess" of immunoglobulin in the circulation following specific stimulus might represent a polyclonally-activated sample of the entire past immunological history of the host, and perhaps even a sample of its future capabilities (although there is not yet firm evidence for this latter suggestion). The conclusion seems to be forced, especially by our understanding of the genetic mechanisms at work – that there is no such thing as "nonsense" immunoglobulin. We assume that no matter what combination of gene segments and mutations may appear in a B cell, somewhere there is an antigen that will interact more or less well with its antibody product, whether or not that substance presents itself within the lifetime of the host.

"Virgin" lymphocytes

Leaving behind the antibody molecule, we come now to the changing view of the cell that produces it. Here again we shall see how the conceptual and technological context of each era would affect how, and even whether, the antibody-forming cell would be considered.

Ehrlich's view of the cell

It must be recalled at the outset that little was known in the 1890s about the fine structure and metabolic workings of the mammalian cell. At a time when the immunological repertoire appeared to be limited to only a modest number of pathogens, Ehrlich felt free to suggest that cells (the type not specified) spontaneously produce small amounts of each antibody specificity. These are deposited upon the cell surface (as "side-chains") to await interaction with their specific antigens. Such an interaction would then stimulate the cell to produce large quantities of that specificity, to the exclusion of all the other specificities for which the cell possessed a theoretical competence. It was far too early even to hazard a guess as to how this specific activation and the following antibody formation might occur.

Here, in brief, was the core of Ehrlich's side-chain theory of antibody formation. As is apparent, it is not a cell-selection theory, but a molecule-selection theory in precisely the same sense as was Jerne's natural selection concept. At this stage of the science, no one would yet question either the multi-potentiality of the cell, or whether there existed any qualitative difference between the cell awaiting the interaction with antigen and the same cell after it had been stimulated. All that one supposed at the time was that the former produced only very small samples of specific antibody, whereas the latter sent large amounts *of the same substance* into the circulation.

Ehrlich's theory went into decline in the face of an ever-expanding repertoire that included now not only benign biological molecules and cells, but also large numbers of small chemical molecules attached to protein carriers as haptens. As interest in the theory disappeared, so too did interest in the cell as the seat of antibody formation.

The immunochemical era; instruction

We have covered fully, in previous discussions, why the immunochemists took over from the medically-oriented immunologists, and why a Lamarckian theory of antibody formation should have found favor over Ehrlich's more Darwinian concepts. Now the active factor in antibody formation was no longer the cell, nor even the antibody as co-initiator of the process. The antigen now held center stage, and the cell was relegated to a completely passive role. Its only function was the passive, non-immunological one of producing a steady flow of some type of normal protein whose structure would be altered by antigen during its formation.

Haurowitz's instruction theory had the antigen somehow instructing the mechanism to change the amino acid sequence in order to form a specific site on this otherwise standard protein. Pauling's instruction theory had antigen acting at a later stage, during the final coiling of the polypeptide chain, in order to impress a specific site, as though in a mold. Neither of these ideas endowed the cell with any active contribution to specificity formation, so no believer in instruction (substantially everyone during the 1930s to the 1950s) would raise questions about cells in thinking of the immune response.

Only in the late 1940s would cytologic studies identify the plasma cell as the one involved in antibody formation,[26] and only in the early 1950s would Albert Coons' fluorescent antibody techniques[27] focus attention more closely on what appeared to be a family of lymphoid cells involved in the formation of antibody. But it was still too early to wonder about qualitative differences among these cells, with one exception; the process appeared to involve a major morphological change when what looked like a normal lymphocyte transformed into an anti-body-forming plasma cell.

Clonal selection, genetics, and receptor patterns

With the elaboration of Burnet's theory in the late 1950s[28] (it is really the cell – the clonal precursor – and not the clone that is selected), attention was focused once more on the cell as the active factor in antibody formation. The times were right for such a shift in attention, because the field was in rapid transition during this period. For the reasons already cited, new blood was entering the research laboratories of the discipline. These included physicians and pathologists interested in immunological diseases, anatomists and cytologists interested in the cells and tissues of the immune system, physiologists interested in the mechanisms regulating the immune response, and geneticists interested in how the information for these spontaneously-produced antibodies is encoded in and extracted from the genome.

Apart from the numerous questions that might have been expected from these various specialties, other basic questions about cells were raised by some of the implications of the clonal response. What precisely is the status of a lymphocyte whose receptors are restricted to a single antibody specificity, while it awaits (perhaps in vain) stimulation by antigen? What exactly happens to the resting cell when it is activated to divide and differentiate into a producing cell? By the mid-1960s, discussions of the initial steps in the activation of a clonal precursor began to sound almost like "insemination" by antigen had taken place. One began to hear such terms as "naive" or even "virgin" lymphocyte used to describe the B cell prior to encounter with its specific antigen. And finally, now that it was certain that some of the clonal daughters would become memory cells, was there any difference between the mother cell and her memory-cell daughters in either quality or quantity? Had the original cell (the clonal precursor) indeed lost its "innocence"?

The answer was soon to come from a variety of different directions, and would prove far more complicated than can have been anticipated at the start.[29] First, it was shown that, contrary to the implied views of Ehrlich and Burnet, the antibody produced by the cell to serve as its specific surface receptor differed from that incited by antigenic stimulus. The receptor is formed with an additional "tail" that enables it to atach firmly to the cell membrane (the membrane domain), and that serves also to transmit to the interior the activation signal (via the cytoplasmic domain).

Next, the geneticists showed that a DNA excision takes place not only in the gene segment assortment that determines the V region specificity of the antibody product, but later also as the various daughter cells further differentiate in the immunoglobulin class (isotype) shift that will eventually lead from production of IgM to the several IgG subclasses and to IgE and IgA. Once an excision has occurred, that cell, and any subsequent daughters that it may give rise to, cannot dedifferentiate to form an earlier 5′ isotype, nor return to the pristine state of its ancestral precursor.

Study of the development of both B and T cell lines has shown that at each step of maturation there exists a unique combination of nuclear transcription factors that mediate the functions of the cell. Thus there is a further difference here between the "naive" cell and that following activation, as well as between the resting clonal precursor and its several daughters. Similarly, the developing lymphocyte lines, and especially T cell subsets) are distinguished by unique combinations of cell surface receptors, including the large group of "CD" (clusters of differentiation) molecules. Differences among the T cell subsets may also be characterized by the different spectra of lymphokines that they elaborate to assist in their functions.

Finally, there is the interesting question of what determines the fate of the many daughter cells that arise following the stimulation of a clonal expansion of either B or T cells. Some B cells become relatively short-lived antibody producers, and some T cells become effectors or regulators, while others become long-term memory cells. It is clear from the above discussion that there is a marked difference between the resting clonal precursor B or T cells and any of the daughters that result from antigen activation. To apply the term "virgin" to the precursor is perhaps too anthropomorphic. The term "naive" seems to have become the term of choice, although this is hardly less anthropomorphic than the other.

Notes and references

1. Landsteiner, K., *The Specificity of Serological Reactions*, New York, Dover, 1962, p. 132.
2. Behring, E., and Kitasato, S., *Deutsch. Med. Wochenschr.* **16**:1113, 1890; Behring, E., and Wernicke, E., *Z. Hyg.* **12**:10, 45, 1892.
3. Landsteiner, K., *Zentralbl. Bakteriol.* **27**:357, 1900; *Wien. Klin. Wochenschr.* **14**:1132, 1901.
4. The conceptual problem that natural antibodies posed to those who opposed Ehrlich's theory is well detailed in Keating, P., and Ousman, A., "The problem of natural antibodies 1894–1905," *J. Hist. Biol.* **24**:245, 1991.
5. Bordet, J., *Ann. Inst. Pasteur* **13**:273, 288, 1899.
6. Landsteiner, K., and Reich, M., *Wien klin. Rundschau* **19**:568, 1905. See also Landsteiner, K., in Oppenheimer, C., ed., *Handbuch der Biochemie*, Vol. I, Jena, Fischer, 1910.
7. Breinl, F., and Haurowitz, F., *Z. Physiol. Chemie* **192**:45, 1930.
8. Pauling, L., *J. Am. Chem. Soc.* **62**:2643, 1940; *Science* **92**:77, 1940.
9. Boyden, S.V., "Natural antibodies and the immune response," *Adv. Immunol.* **5**:1, 1966.
10. Having first studied chemistry and then medicine, Jerne would be in his mid-thirties before he began experimental work in immunology.
11. Jerne, N.K., *A Study of Avidity Based on Rabbit Skin Responses to Diphtheria Toxin–Antitoxin Mixtures*, Copenhagen, Munksgaard, 1951.
12. Jerne, N.K., "The natural selection theory of antibody formation: ten years later," in Cairns, J., Stent, G., and Watson, J.D., eds, *Phage and the Origins of Molecular Biology*, Cold Spring Harbor, 1966. But see Thomas Soderqvist's challenge to Jerne's "eureka discovery story," *J. Hist. Biol.* **27**:481, 1994.
13. See Byron Waksman's comprehensive review of the early studies in *Int. Arch. Allergy Appl. Immunol.* **14**: Suppl., 1959.
14. The work was initiated by Gorer, P.A., *J. Pathol. Bacteriol.* **44**:691, 1937; **47**:231, 1938. See also the review on histocompatibility genes, *J. Genet.* **49**:87, 1948.
15. Coombs, R.R.A., Mourant, A.E., and Race, R.R., *Br. J. Exp. Pathol.* **26**:255, 1946.

16. The details on these phenomena and of their collective effect on the field are discussed in Chapter 17.

17. Burnet, F.M., *The Production of Antibodies*. New York, Macmillan, 1941; Burnet, F.M., and Fenner, F., *The Production of Antibodies*, 2nd edn, New York, Macmillan, 1949.

18. Jerne, N.K., *Proc. Natl Acad. Sci. USA* **41**:849, 1955.

19. Talmage, D.W., *Annu. Rev. Med.* **8**, 239, 1957.

20. Burnet, F.M., *Austral. J. Science* **20**:67, 1957.

21. Burnet, F.M., *The Clonal Selection Theory of Acquired Immunity*, Nashville, Vanderbilt University Press, 1959.

22. Lederberg, J., *Science* **129**:1649, 1959.

23. Talmage, D., *Science* **129**:1543, 1959.

24. See, for example, Achmed, R., and Oldstone, M.B.A., in Notkins, A.L., and Oldstone, M.B.A., eds, *Concepts in Viral Pathogenesis*, New York, Springer, 1984, pp. 231–238, and more recently Hunziger, L., et al., *Nature Immunol.* **4**:343, 2003.

25. Silverstein, A.M., and Rose, N.R., *Nature Immunol.* **4**:931, 2003.

26. Fagraeus, A., *Acta Med. Scand.* **Suppl. 204**, 1948.

27. Coons, A.H., Leduc, E.H., and Connolly, J.M., *J. Exp. Med.* **102**:49, 1955.

28. See especially Burnet's *The Clonal Selection Theory*, note 21.

29. One of the best summaries of the development of B and T cell lineages, and of the many different molecules that characterize each stage, can be found in Mak, T., and Saunders, M.E., *The Immune Response: Basic and Clinical Principles*, New York, Elsevier, 2005.

17 The dynamics of conceptual change in immunology

There was no such thing as The Scientific Revolution, and this is a book about it.
                                                                Steven Shapin[1]

Shapin's book was about that major sixteenth- and seventeenth-century upheaval in the way that mankind regarded Nature – the rise of modern science. Historians and sociologists of science debate about whether this should properly be called a "revolution," but no one doubts its significance. In the present chapter, we shall examine two major shifts in the way that immunologists regarded their discipline. However they may be named, each represented an important turning point for the field.

The classical view of scientific progress that was advanced by such analysts as philosopher Karl Popper[2] and sociologist Robert Merton[3] was one of a smooth and progressive evolution toward the ultimate goal – a complete understanding of the physical world. This view was shared by scientists of all types, and the ideas represented by "smooth" and "progressive" were considered to be implicit in the very notion of an evolutionary sequence. But the term "smooth" as an adequate description of the workings of science was soon brought into question from several different directions.[4] Arguing that disagreement among scientists may have even more heuristic value than agreement, they suggest that the resolution of such disputes may cause abrupt shifts rather than smooth transitions in the cognitive content of the scientific enterprise. Indeed, Thomas Kuhn has proposed that radical shifts from former theory and practice may constitute true revolutions, and represent major discontinuities in the development of science.[5] Even in Darwinian evolution the classically ascribed smoothness has been brought into question, as Steven Jay Gould points out in his discussion of the significance of the Burgess shale deposits.[6] As Gould shows, rather than a slow and almost majestic expansion of different life forms there have been violent expansions and contractions in the diversity of species, probably not just once but many times.

We now return to that other adjective usually applied to the historical course of science, "progressive." The term usually implies the more-or-less inexorable forward development of a science along a straight-line track. But as we shall see, twentieth-century immunology appears rather to have involved an ebb and flow, in which some aspects of the science might advance rapidly, others more slowly, and yet others might even retrogress. If "evolution" and "revolution" describe the forward progress of a science, then a third term – "devolution" – may be appropriate to characterize sideways or even reverse changes of direction. It is a reasonable term to describe the atrophy of some aspect of a scientific discipline,

A History of Immunology, Second Edition
ISBN: 978-0-12-370586-0
Copyright © 2009, Elsevier Inc.
All rights reserved

and even the transfer of that aspect out of the mainstream interest and into other hands.

The discussion to follow will thus be concerned in great measure with the rise and fall of research areas and of subdisciplines within the broad field of immunology.[7] We will examine the reasons for the decline of the initially productive research program of immunology early in the twentieth century, and the devolution of several of its components into the hands of "outsiders." This was followed by the appearance of a completely different theoretical framework and an entirely new set of research questions and technical approaches. Whereas the former program had been extrovert, with broad application to and exchange with many fields of biology and medicine, the new program was introvert, asking questions whose answers were of little interest to those outside the field. This new program, involving predominantly chemical approaches to the study of immunity, held sway for some fifty years. Then, a true conceptual and technical revolution altered the course of immunology once again. Perhaps Ludwik Fleck pointed the way, more than seventy years ago, in his book *Genesis and Development of a Scientific Fact*.[8] Implicit in Fleck's description of those leaders who govern a scientific field and determine its values and priorities (what he termed the *Denkkollektiv*) was the possibility that replacement of these arbiters by others with different backgrounds and interests might change the character of the discipline itself. This discussion will illustrate two such *Denkkollektiv* replacements in immunology which parallel the two major transitions that took place in twentieth-century immunology.

The research program of early immunology

During its early years, the research program of immunology was divided among six principal areas, each of which had arisen logically from the germ theory of disease and from developments in public health or, as often happens in science, from chance laboratory observations. While closely interrelated, each component had its own questions and technical approaches.

Preventive immunization

The science of immunology was born in the laboratory of Louis Pasteur, in the context of Pasteur's dedicated commitment to the germ theory of disease. Pasteur's earlier work on the agents responsible for certain diseases in the French silkworm and wine industries had convinced him that each disease is the reproducible result of an infection by a specific microorganism. Moreover, he held not only that spontaneous generation did not exist, but also that these pathogenic agents are constant and specific in their ability to cause a given disease, and cannot undergo transformation to yield some other disease picture. By one of those happy instances of serendipity in science, it was discovered that chickens that had recovered from a mild attack of chicken cholera induced by an attenuated strain were thenceforth protected from challenge with more lethal strains.[9]

This report in 1880 was the first generalization on Edward Jenner's use of cowpox vaccine to protect against smallpox, and opened up an entirely new research program of prophylactic immunization. Pasteur was quick to seize upon these possibilities, as his subsequent work on anthrax, rabies, and other diseases amply testifies. Over the next quarter-century, as the specific pathogens of different diseases were reported with increasing frequency in the journals, scientists throughout the world endeavored to develop their corresponding preventive vaccines, using Pasteurian approaches.

Cellular immunity

The second significant step in the expansion of the immunological research program of the nineteenth century came in 1884, with Ilya Metchnikoff's cellular theory of immunity.[10] Based upon purely Darwinian evolutionary principles, Metchnikoff suggested that the primitive intracellular digestive functions of lower animal forms had persisted in the capacity of the mobile phagocytes of metazoa and higher forms to ingest and digest foreign substances. Metchnikoff proposed that the phagocytic cell is the primary element in *natural* immunity (the first line of defense against infection), and critical also for *acquired* immunity (the heightened protection conferred by preventive immunization or prior infection). Metchnikoff's theory had several far-reaching consequences for biology and medicine. First, it introduced the notion that *inter*specific conflict might contribute as importantly to evolution as the classical Darwinian notion of *intra*specific competition.[11] Here, the struggle for survival was between the infected host and the offending pathogen, with the phagocyte entering the lists as champion of the former.

Another notable contribution of the phagocytic theory was to the field of general pathology. Most believed at the time that inflammation was a damaging component of the disease process itself. Metchnikoff, on the other hand, suggested that the inflammatory response was in fact an evolutionary mechanism designed to protect the organism. Whereas Metchnikoff's idea of the protective role of inflammation eventually triumphed, his cellular theory of immunity stimulated much opposition from those who claimed that humoral (blood-borne) factors were by far the more important. The debate between these two camps over the next two to three decades was fierce,[12] with each side designing experiments to reinforce its own thesis and to show the error of the ways of the opposition. Eventually, Metchnikoff's cellular theory of immunity fell into disfavor early in the last century, not to be revived (in a somewhat different form) for another fifty years or so – but not before its heuristic value had inspired many ingenious experiments and a wealth of important data, and not before Metchnikoff was awarded, with Paul Ehrlich, the Nobel Prize in 1908.

Serotherapy

The third important step in the expansion of the early immunological research program came in 1890 with the demonstration by Behring and Kitasato that preventive immunization with the exotoxins of diphtheria and tetanus organisms resulted in the appearance in the blood of immunized animals of a soluble substance capable of neutralizing these toxins and rendering them innocuous.[13] Moreover, these antitoxins (later generalized with the name antibodies) could be transferred from the blood of an immunized animal to protect a naive recipient from disease. Indeed, it was shown in the case of diphtheria in humans that passive transfer of diphtheria antitoxin might even protect human infants during the early stages of the disease itself. Here was a remarkable new addition to the medical armamentarium, which offered great therapeutic promise in combating a variety of infectious diseases. In the 1890s, the new so-called serotherapy stimulated an explosion of laboratory and clinical experimentation, in recognition of which Behring received the first Nobel Prize in 1901.

It was in connection with his studies of antibodies and his demonstration of how diphtheria toxin and antitoxin preparations might be standardized that Paul Ehrlich devised his side-chain theory of antibody formation.[14] Like Metchnikoff, Ehrlich adopted a Darwinian approach[15] by suggesting that antibodies had evolved as cell receptors, functioning like those receptors necessary to fix nutrients and drugs for their assimilation by the cell. Ehrlich proposed that when these receptors (side-chains) are bound by injected antigen an over-proliferation is stimulated within the cell, resulting in their being cast off into the blood to appear as circulating antibody. This side-chain theory, with its broader implications for how receptors govern *all* types of cellular function, had great influence in pharmacology and in many branches of clinical medicine. Numerous books and reviews appeared during the next decade describing the implications of Ehrlich's side-chain theory for many different branches of clinical medicine. Perhaps most important for the future, Ehrlich attributed the specificity of antibodies to their stereochemical structure, and their interactions with antigen to strictly chemical bonding.

Cytotoxic antibodies

The fourth significant area that occupied early immunologists stemmed from the demonstration by Jules Bordet in 1899 that antibodies specific for erythrocytes could cause their destruction (hemolysis) in conjunction with the nonspecifically-acting serum factor complement.[16] Here was a clear explanation of one of the important mechanisms of protective immunity – the direct destruction of bacterial pathogens through the cooperation of these two immunologic factors.[17] But other far-reaching implications were seen in Bordet's observation. For the first time, the cells and tissues of the immunized host itself were seen possibly to be at risk by an "aberrant" immune response against self components. With little delay, scientists in almost every active laboratory began to immunize experimental animals with suspensions or extracts of almost every tissue or organ in the body, in an attempt to find cytotoxic antibodies that might be responsible for one or another local disease. Soon the journals were filled with reports of such experiments, and indeed much of the 1900 issue of *Annales de l'Institut Pasteur* was devoted to this question.[18] While it was quickly discovered that *xeno*antibodies (those derived by immunizing an animal of another species) and *iso*antibodies (those derived from other members of the same species) were often formed, and might be cytotoxic against the target tissue or organ,[19] *auto*antibodies (obtained by immunizing an animal with its own tissues) were, with few exceptions,[20] rarely produced. Nevertheless, for years thereafter, the possibility was seriously entertained that such cytotoxic antibodies might play an important role in the pathogenesis of a number of diseases, both as pure autoimmune phenomena and as secondary contributors to the lesions seen in such diseases as syphilis and ophthalmitis.[21]

Serodiagnosis

Another consequence of Bordet's observation on the mechanism of immune hemolysis came with the finding that *all* antigen–antibody interactions would result in the nonspecific fixation of complement, and its disappearance from the test mixture.[22] With the rapid development of techniques to measure complement, it was apparent that if a bacterial antigen were available, then the presence or absence of its specific antibody in a patient's serum could be assayed by measuring the effect of such a mixture on a standard amount of complement added to the system. With this, a powerful new tool was added to the arsenal of the student of infectious diseases who could now, in the case of certain diseases, tell whether the patient had previously experienced the disease, and in others determine whether the patient currently had active disease. Occasionally, by studying variations in antibody titer, he might even follow the actual course of a disease process. The first disease to which this new approach was applied was syphilis, in the hands of August von Wassermann and his colleagues in 1906.[23] These serodiagnostic approaches were quickly applied to many other diseases, and the technique and its improvement provided a fertile field of activity for decades to come.

Anaphylaxis and related diseases

A seminal discovery in the history of immunology was made in 1902 by physiologists Paul Portier and Charles Richet.[24] Up until that time, the immune response had been viewed as a purely benign set of mechanisms whose only function was to protect the organism against exogenous pathogens; the work of those searching for cytotoxic antibodies had done little to alter this view. Indeed, it had been found only a few years earlier that an immune response could be stimulated by other than bacterial antigens and toxins. Now came Portier and Richet to demonstrate that even bland substances could, when injected into presensitized individuals, cause severe systemic shock-like symptoms, and even death. They termed this phenomenon *anaphylaxis*, in an attempt to distinguish it from the usual prophylactic results expected of the immune system. Shortly thereafter, Maurice Arthus demonstrated that bland antigens could cause local necrotizing lesions when they react with specific antibody in the skin of test animals – the so-called Arthus phenomenon.[25] Then, in 1906, Clemens von Pirquet and Bela Schick demonstrated that the pathogenesis of so-called serum sickness depends upon an antibody response by the host to the injection of large quantities of foreign protein antigens, such as accompanied the administration of horse antidiphtheria toxin according to Behring's serotherapeutic doctrine.[26]

Here was a group of observations that threatened the very conceptual foundation of immunology, which had held the system to be completely benign and protective. Moreover, it could not be argued that these were only artificial laboratory phenomena; soon thereafter it was demonstrated that two of the significant curses of mankind, hayfever[27] and asthma,[28] also belong to this same group of specific antibody-mediated diseases. It is little wonder, then, that much work was stimulated to clarify the phenomenology of these diseases and the role of the immune response in their pathogenesis, to establish the nature of the antibodies responsible for them, and especially to explain the paradox of a system presumably evolved to protect, somehow giving rise to the very opposite.[29]

The fate of the early research programs

We have seen that during the period 1880 to about 1910, the young and highly productive field of immunology[30] had organized itself predominantly in terms of six major areas of interest: preventive immunization, cellular immunity, serotherapy, antibody-mediated cytotoxicity, serodiagnosis, and anaphylaxis. By the beginning of World War I, while most of its practitioners might not yet have called themselves "immunologists," institutionalization of the discipline had begun in earnest. An institute devoted to its aims had been established for Paul Ehrlich in Frankfurt, and departments and services dedicated to the discipline had been formed within many of the leading research institutions around the world. Sections devoted to one or another component of the immunologic program were to be found at International Congresses of Medicine or Hygiene, and an "invisible college"[31] existed, involving informal exchange among its practitioners. While the pages of the *Annales de l'Institut Pasteur* had long been devoted to immunological reports, the discipline was more formally recognized by the founding in 1908 of the *Zeitschrift für Immunitätsforschung*, and of the American *Journal of Immunology* in 1916. The commonality of interest of this subgroup of scientists and practitioners was recognized, at least in America, by the founding of the American Association of Immunologists in 1913.[32]

Let us now look at developments within each of the components that comprised the early immunological research program. *Preventive immunization* had seen its great victories in the case of chicken cholera, anthrax, rabies, plague, and several other important diseases. But increasingly, pathogenic organisms were being described for which it was proving impossible to prepare efficacious vaccines. These included not only such important agents as the tubercle and leprous bacilli, the cholera vibrio, and the spirochete of syphilis, but also the important group of disease-producing gram-positive organisms, to say nothing of a number of newly described diseases due to viruses and parasites that so ravaged man and animals. Thus, by 1910, the great early promise of Pasteurian immunization was no longer being fulfilled; new successes would thenceforth be few and far between, and achieved only with great difficulty. Work in this area very rapidly left the "classical" immunology laboratory, and was taken over by bacteriologists, virologists, and parasitologists interested more in organisms than in immunologic mechanisms.

The study of *cellular immunology* and of Metchnikoff's phagocytic theory, as we have seen, went into decline early in the century at the hands of proponents of humoralist theories. Cells were much more difficult to work with than humoral antibodies, and no such antibody techniques as agglutination, the antigen–antibody precipitin reaction, immune hemolysis, and the ability to transfer antibody passively from one animal to another existed in the field of cell studies. Indeed, the cell was still considered something of a mystery, whereas Ehrlich's pictures of antibodies and their specific combining sites could almost convince one that the antibody was a "real" entity whose structure and properties were readily understood. For a while it looked as though Almroth Wright might save the day with his theory of opsonins (the collaboration of antibodies to enhance phagocytosis) and program of carefully timed specific immunization to enhance opsonic activity,[33] as so delightfully described by Bernard Shaw in his play *The Doctor's Dilemma*. But the techniques proved difficult and the results variable in practice, so that Wright's approach was rapidly given up as not very useful.

The techniques of *serotherapy* for the prevention or cure of disease suffered a fate similar to that of preventive immunization. After the remarkable demonstration of the efficacy of horse antidiphtheria and antitetanus sera in the treatment of these diseases, no significant further victories were recorded in this area. While laboratories throughout the world continued to produce these two antisera (the Pasteur Institute helped support itself with its stable of immunized horses), interest in this approach waned, since there were so few other significant diseases that were caused by exotoxins and thus amenable to this approach. When, much later, passive transfer of antibody would be employed, it would be by hematologists using human gamma globulin to prevent erythroblastosis fetalis, or by pediatricians employing convalescent sera to deal with poliomyelitis.

As for the interest in *cytotoxic antibodies*, this proved to be ephemeral. Despite all attempts to implicate antitissue and anti-organ antibodies in the pathogenesis of disease, with the exception of anti-erythrocyte antibodies responsible for hemolytic anemias no convincing demonstrations were forthcoming, and immunologists even forgot about Donath and Landsteiner's demonstration of the pathogenesis of paroxysmal cold hemoglobinuria as the possible tip of an autoimmune disease iceberg. By 1912, the study of immune cytotoxic phenomena had left the immunology laboratory, to be pursued only within essentially unrelated clinical specialty areas such as ophthalmology, with its interest in sympathetic ophthalmia and autoimmune disease of the lens. True, the occasional experimental pathologist such as Arnold Rich might study immunocytotoxic events in the pathogenesis of tuberculosis,[34] or the occasional virologist such as Thomas Rivers might demonstrate experimental allergic encephalomyelitis,[35] but these were far out of the current mainstream of immunology, and the results were generally published in other than specifically immunological journals.

Developments within the area of *serodiagnosis* represent a more typical example of disciplinary differentiation for the sociologist of science. These techniques had developed within the very heart of an immunologic enterprise interested in immunity in the infectious diseases, which therefore not only demanded an understanding of disease pathogenesis but also required the ability to diagnose these diseases. Syphilis remained the mainstay of serodiagnostic laboratories, and work to perfect the technique and extend it to other diseases continued throughout the period under discussion. Very quickly, though, the technique became quite routine and *applied*, and immunologists interested in basic mechanisms soon lost interest in the area. Work in this field was taken over by classical bacteriologists, and in fact those who devoted themselves to this and other aspects of serodiagnosis soon began to call themselves "serologists" and worked principally in hospital diagnostic laboratories rather than in those devoted to basic immunologic research.

Soon after their discovery, *anaphylaxis* and its related diseases had also become an intimate concern of immunologic experimentalists. They were interested in the nature of the antibodies responsible for these phenomena, and in the basic mechanisms involved in the diseases which resulted from their action. However, after a short and essentially unsuccessful struggle with the paradox of a system presumably evolved to protect now being demonstrated to *cause* disease, the immunologists soon deserted the field, leaving it to others. In the main, those upon whom these interests devolved were clinicians interested in hayfever and asthma, which had just been identified as "anaphylactic" diseases. In fact, it was this identification that was primarily responsible for the establishment of clinical allergy as a medical subspecialty,[36] and it was primarily in the laboratories of allergists that further progress was realized in sorting out the mechanisms involved and in developing skin tests and therapeutic approaches to the treatment of human allergies. In addition to these, however, the study of anaphylactic and related phenomena was of great interest to physiologists such as Sir Henry Dale,[37] interested in the physiologic mechanisms involved in such diseases, and also to a large group of experimental pathologists interested in the comparative study of the lesions that accompanied these diseases.

The rise of immunochemistry

Thus, the immunologic research program waned in all of its interest areas, so far as the basic scientist was concerned, and several of these areas were taken over by others. Now we shall see how the general field itself experienced a devolution into the hands of a new *Denkkollektiv*. The seeds of the future interest in the chemistry of antigens and antibodies can be traced back to the fertile imagination of Paul Ehrlich. For the first time, Ehrlich's side-chain theory of antibody formation pictured antigen, antibody, and complement as molecules, and their combining sites as stereochemically complementary structures that would account for the specificity of their interactions. At the time, however, little was known of the structure and precise composition of protein molecules, and appropriate techniques were unavailable to translate Ehrlich's theory into laboratory experiments.

It is common in most textbooks to ascribe the paternity of the field of immunochemistry to the famous physical chemist Svante Arrhenius, since he coined the term "immunochemistry" in a famous series of lectures in 1904.[38] Like many another physical scientist, Arrhenius was attracted by the mysteries and by the confusion that existed in biology, and felt that he could bring some order to the chaos by the introduction of the rigorous laws of chemistry and physics. Through his Danish colleague Thorvald Madsen, Arrhenius became interested in the problem of diphtheria toxin–antitoxin titration, and proposed that these interactions are reversible, like the interactions that he had described for weak acids and weak bases that had contributed so much to his earlier fame. But it would probably be erroneous to attribute the fatherhood of the field to Arrhenius, since his contributions were purely theoretical, could not be adequately tested at the time, and had little immediate influence on subsequent events.

Perhaps the true turning point came in 1906, with the demonstration by Obermeyer and Pick that protein antigens could be modified chemically to alter their immunological specificity.[39] For example, when nitrated proteins were employed to immunize animals, the specificity of the resulting antibodies appeared to be directed no longer at the original protein, but rather at the added nitro groups. In an encyclopedic review of this area in 1912,[40] Pick showed that a number of different *synthetic* groupings (called haptens) might be joined to a carrier protein to serve as antigenic determinants. Here was a powerful new tool, with which the small molecules produced in the organic chemistry laboratory could be used to dissect intimately the nature of immunologic specificity and the character of the combining site on antibody. No one exploited this approach more assiduously or to better effect than polymath Karl Landsteiner, who in 1917 published two papers[41] that illustrated the power of this approach, and that helped to define both his own work during the next twenty-seven years and much of the domain of immunochemistry as well. Now the medical significance of antibodies and the biological significance of their formation took a back seat to interest in the chemical nature of antigens and antibodies and the basis of their specificity.

Another approach to the chemistry of antigens and antibodies was opened up in the 1920s by organic chemist Michael Heidelberger. Working initially in the context of a bacteriological laboratory, Heidelberger was able to show that, contrary to the classical view that antibodies could only be formed against protein antigens, the capsular polysaccharides of the pneumococcus could also stimulate a specific antibody response.[42] This led Heidelberger to study the chemical

differences among the polysaccharide antigens of different strains of pneumococcus, in pursuit of which he developed over many years an impressive set of
quantitative techniques that helped establish immunology as a more exact science.

Theory follows mindset

We saw above that during the early biomedical era of immunology, the first
theory of immunity advanced by Ilya Metchnikoff, trained in zoology, followed
strict Darwinian evolutionary principles, and the first theory of antibody
formation proposed by Paul Ehrlich, trained in medicine, was similarly based.
But the chemically-oriented investigators who dominated immunology after
World War I had little interest in the biological basis of immunity. They were,
however, interested in antibodies and their formation, and new theories of
antibody formation were not slow to appear. These new theories no longer
focused on the *function* of antibodies, but rather on their chemical *structure*, and
more specifically on the question of how such a large group of specific molecules
able to interact with an ever-growing universe of potential antigens could
possibly be produced within the vertebrate host. For this was the rock upon
which Ehrlich's side-chain theory had foundered: the improbability that evolution could have accounted for the spontaneous production of so many different
antibodies, the greater portion of which were directed against bland and even
artificial antigens of no obvious evolutionary selective force.

It is not surprising, therefore, that the new chemical theories of antibody
formation were quite Lamarckian in nature; in contrast to the molecules of the
biologist, those of the chemist generally have no evolutionary history behind
them. The most compelling of the new theories to be advanced was that of
biochemist Felix Haurowitz, in 1930.[43] In this, it was proposed that only the
antigen itself might contain all of the information necessary for antibody
formation, and that antigen imposes a complementary structure on a nascent
protein by acting as a template for the synthesis of a unique sequence of amino
acids. This was the first so-called instruction theory of antibody formation. Here
was a ready explanation not only for the tremendous diversity of different
antibodies, but also for how so fine a specificity could be imparted to the antibody molecule. This instructive theory of antibody formation was further refined
in 1940 by chemical physicist Linus Pauling,[44] who proposed that the antigen
serves as a template upon which the nascent amino acid chain coils to form
a protein molecule. It is interesting that, so ingrained in the collective immunological psyche of the times were these chemical ideas, even biologist
Macfarlane Burnet, in his first two theories of antibody formation,[45] felt obliged
to employ Lamarckian instructive approaches.

The scope of the immunochemical research program

The application of synthetic haptens to the study of antibody specificity led to
progress in clarifying the structure of antigen and antibody combining sites, and in
defining the thermodynamic parameters of their interaction.[46] These studies were

facilitated by the development of quantitative techniques for the measurement of these reactions,[47] and by the identification of antibody as a gamma globulin protein,[48] paving the way for the development of chemical purification methods.

The scope of the field of immunology from the 1920s to the early 1960s is perhaps best epitomized by five of the leading books of the period: Well's *The Chemical Aspects of Immunity* in 1924; Marrack's *The Chemistry of Antigens and Antibodies* in 1934; Landsteiner's *The Specificity of Serological Reactions* in 1937; Boyd's *Fundamentals of Immunology* in 1943; and Kabat and Mayer's *Quantitative Immunochemistry* in 1949.[49] These were the reference books from which a generation of young immunologists learned their trade, and little attention was paid in any of them to the biological or medical aspects of the field. If a Max Theiler developed a new vaccine in the mid-1930s against yellow fever,[50] this was of interest only to virologists and students of infectious diseases. If a Hans Zinsser[51] or an Arnold Rich[52] studied allergic reactions to bacteria, or if a Louis Dienes[53] or Simon and Rackemann[54] developed models of delayed hypersensitivity lesions to simple proteins in the 1920s and 1930s, this was only of interest to bacteriologists and experimental pathologists. Finally, if a Thomas Rivers developed an experimental model of allergic encephalomyelitis as early as 1933,[55] this seemed to excite no one at the time. These and other similar excursions into areas of biomedical interest lay out of the mainstream of contemporary immunology, were usually published in "outside" journals, and made little impression upon the governing *Denkkollektiv*. Only a later generation of immunologists more attuned to biological questions would go back to identify these contributions as landmarks in immunological progress.

This is not to suggest that *all* work along the six classical lines described above ceased during the immunochemical era. It has been pointed out that "research areas which have become well established take a long time to die out altogether. There is always *some* work that can be done."[56] Thus, as described above, the clinical allergists gave new life to the study of anaphylactic phenomena by redefining the field along new lines; continued progress was made in the preparation of better toxoids and better modes of immunization; serologists continued to improve and expand the application of serodiagnostic procedures; and, from time to time, an effective vaccine would be developed against one or another disease of man or animals.

The immunobiological revolution

The research program that governed the normative science of the immunochemical era between the 1920s and 1950s produced interesting results. It had gone far to define the chemical nature of both antigens and antibodies, and the precision of their specific interactions. Increasingly, however, biologists working on the fringes of immunology made observations whose explanation was not to be found in the received wisdom of instructionist theories of antibody formation. How, they asked, could antibody formation persist in the apparent absence of antigen? Why should a second exposure to antigen result in an enhanced booster response that is much more productive than is the primary response to antigenic stimulus? How can repeated exposure to antigen change the very *quality* of the antibody, in many instances sharpening its specificity by increasing its affinity for the antigenic determinant employed? Finally, how is it possible that immunity to some viral diseases appears to be unrelated to the presence of circulating antiviral antibodies? These and other biologically-based questions began seriously to challenge the immunochemical paradigm, most notably through the pen of Macfarlane Burnet in his two books on *The Production of Antibodies*, in 1941 and 1949.[57] Burnet complained repeatedly that the chemical theories, while quite elegant, failed to explain the more functional biological aspects of the immune response.

By the 1950s, the stage seemed to be set for a large-scale confrontation such as described by Thomas Kuhn in his book *The Structure of Scientific Revolutions*.[58] On the one hand was the immunochemical tradition, guided by theories that could no longer satisfactorily explain all of the phenomena of the field, and employing approaches that yielded results of increasingly parochial interest and of decreasing marginal value. Challenging this classical tradition was a growing group of biomedical scientists seeking answers to a set of new and important questions that traditional immunochemical theory and practice were ill-prepared to answer.

In the 1940s, Peter Medawar demonstrated that the rejection of tissue transplants was a purely immunologic phenomenon, but one unrelated to humoral antibody.[59] In 1945, Ray Owen described the paradoxical situation of dizygotic twin calves that were incapable of responding to one another's antigens.[60] The explanation of this phenomenon lay in the ontogeny of the immune response in the vertebrate fetus, leading Burnet and Fenner to postulate the existence of a cell-based immunological tolerance[61] – a hypothesis that Peter Medawar (still at the time a zoologist) and colleagues confirmed experimentally,[62] and for which Burnet and Medawar shared the Nobel Prize in 1960. Yet another observation for which no ready explanation was available in classical theory involved the description in the early 1950s of a group of immunological deficiency diseases in man,[63] the explanation of which would go to the very heart of the biological basis of the immune response. Finally, after a hiatus of some forty years or more, interest in autoimmune diseases was re-awakened by new demonstrations of autoimmune hemolytic anemias, experimental and human autoimmune thyroiditis, and allergic encephalomyelitis.[64]

While these new phenomena provided a sufficient basis to question the old values, such questions could only be answered by the development of new methods, and these were rapidly forthcoming. The techniques of immunofluorescence staining[65] and of hemolytic plaque assay[66] permitted the tissue localization and quantitative enumeration of antibody-forming cells. The technique of passive cell transfer,[67] and especially that of cell culture techniques,[68] permitted for the first time the analysis of cell–cell interactions and immunocyte dynamics.

Here was a true revolution in the offing, awaiting only the appearance of a theoretical leader to lead the charge against the old regime and its outmoded paradigm. That theoretician was Burnet. In 1955, Niels Jerne had revived the old

Ehrlich theory of a Darwinian evolution of immunologic capabilities with his natural selection theory of antibody formation.[69] Jerne proposed that the information required for the production of specific antibodies pre-exists within the vertebrate genome, and that antigen does not *instruct* for antibody formation, but rather *selects* for the production of a few among all the possible specificities *already present*. Whereas Jerne's theory was imprecise in many respects, its central feature, that of "natural selection," was adapted by Burnet into his clonal selection theory of antibody formation.[70] This theory defined the immune response in terms of cell receptors and the dynamics of cellular proliferation and differentiation. As further developed by Burnet[71] and refined by David Talmage and Joshua Lederberg,[72] the clonal selection theory not only provided reasonable explanations for all of the hitherto inexplicable biological phenomena, but also served to stimulate a remarkable explosion of experimentation along lines that touched a broad spectrum of biological and medical fields.

Within five years of its introduction, the clonal selection theory had carried the day; the old immunochemical paradigm had been thoroughly overthrown, except in the minds of a few diehard adherents.[73] Most of the young investigators who flocked into this burgeoning field came with prior training in genetics, physiology, experimental pathology, and a variety of clinical disciplines of medicine. Immunology became once again what it had been sixty years earlier: an outward looking discipline with much to offer to and much to gain from a wide variety of interdisciplinary ventures.

In recent years, the new immunology has offered classical evolutionary theory the elegant model of an extremely complicated mechanism that is even able to *anticipate*[74] the appearance of new pathogens, rather than merely slowly adapting its response to their presence.[75] Indeed, this peculiarity of evolution occurred not just once but twice – once for the immunoglobulin B cell receptor[76] and again, somewhat differently, for the T cell receptor![77] It has offered to geneticists the unique example of an immunoglobulin gene superfamily whose components exercise a broad range of interrelated activities extending even beyond the immune response,[78] and whose mechanism for the generation of immunologic diversity has shown how a gene product can be assembled by the variable splicing of many different DNA segments.[79] In its study of lymphokines and cytokines (the hormones of the immune system), modern immunology has offered to the physiologist a variety of examples of how cells may communicate with and influence one another.[80] Finally, the new immunology has assisted many medical subspecialties in defining the pathogenesis of some of their most important diseases, and pointed the way as well to the development of preventive measures or therapeutic modalities to combat these diseases.[81]

Comment

I have attempted here to define three distinct eras in the history of the discipline of immunology. The first, extending from 1880 to about World War I, centered on the new bacteriology and infectious diseases, and had a distinctly medical orientation. Several of the components of the original research program in immunology failed to maintain their original momentum or to fulfill their initial high promise, and went into decline. These include the development of new vaccines, serotherapeutic approaches, the study of cellular immunity, and the study of diseases that might be mediated by cytotoxic antibodies. Two other sub-programs followed a somewhat different course; the study of anaphylaxis and related diseases passed primarily into the hands of clinical allergists, while the development and adaptation of serodiagnostic techniques passed into the hands of the new discipline of serology, both fields moving out of the mainstream of post-World War I immunology.

As interest in the components of the old program was falling away, there developed a new area of interest in immunology. Leadership in the field devolved upon a new group of individuals with a predominantly chemical orientation to the study of antigens and antibodies, who pursued a research program and developed a theoretical base that reflected this orientation well. It may be interesting to examine more closely the forces responsible for this shift in emphasis. When interest in the old areas waned, the medically-oriented practitioners did not switch to more immunochemical lines, but went in other directions. Karl Landsteiner was the only prominent "old-timer" who contributed significantly to the newer immunology, and it was his work that set the tone and attracted the new generation of immunochemists who became the reigning *Denkkollektiv*. A science does not change its precepts and approaches spontaneously; it is moved to the new position by those who explore fertile new areas. This is not to say, however, that there was no longer interesting and important work to be done along the old lines – it was just that such work was no longer "fashionable," as the reception of the work of Dienes, of Rich, of Rivers, and of the early Medawar illustrates.

Whereas the earlier immunological program had interacted extensively with many different fields of biology and medicine, the immunochemical era was characterized by a relative introversion, as compared with the broad influence exerted by the earlier immunological program.[82] We can date this second era from about World War I until the late 1950s and early 60s.

There then occurred an abrupt transition in the field of immunology, which may well be called a scientific revolution in the Kuhnian sense. Since the old theories and old techniques could not satisfactorily explain newer observations, the biologists took over command of the discipline from the chemists. Chemical approaches and chemically-oriented theories were rapidly overthrown by a new *biomedical* paradigm which, guided by the clonal selection theory, now asked a markedly different set of questions, involving the biological basis and biomedical implications of the immune response. Eventually, there occurred a synthesis of the two positions. The chemists (who approached the system by studying the final molecular product) and the biologists (who approached it from the initial cellular interactions) found a common ground in the molecular biology of T and B cell receptors and of lymphokines, and together they have clarified the major questions about antibody formation and cell–cell interactions.[83] The broad implications of this unification into what has been termed the *immune system* are discussed in detail by A.-M. Moulin.[84] The entire process represents an interesting dialectic, involving as it does the early predominantly medical/bacteriological thesis, a chemically-based antithesis, and eventually the modern biomedically-oriented synthesis.[85]

We have concentrated here on devolution and revolution in science. Nevertheless, it will be appreciated that *within* each of the three eras through which immunology passed, the normative science of the period saw the usual evolutionary progression of new techniques and the accretion of new information, all within the context of the then-dominant paradigm.

Notes and references

1. Shapin, S., *The Scientific Revolution*, Chicago, University of Chicago Press, 1996, p. 1.
2. Popper, K., *The Logic of Scientific Discovery*, New York, Basic Books, 1959.
3. Merton, R. K., "The normative structure of science," in *The Sociology of Science*, Chicago, University of Chicago Press, 1973, p. 267 ff.
4. See, for example, historian Thomas Kuhn (*The Structure of Scientific Revolutions*, 2nd edn, Chicago, University of Chicago Press, 1970); philosophers Imre Lakatos ("Falsification and the methodology of scientific research programs," in Lakatos, I., and Musgrave, A., eds, *Criticism and the Growth of Knowledge*, London, Cambridge University Press, 1970) and Paul Feyerabend (*Against Method*, London, Verso, 1978); and sociologist Michael Mulkay ("Sociology of the scientific research community," in Spiegel-Rosing, H., and Price, D.J., eds, *Science, Technology, and Society*, Beverly Hills, Sage, 1977).
5. Kuhn, note 4. But there are many who oppose Kuhn's views, well summarized by L. Laudan in "Two puzzles about science: reflections about some crises in the philosophy and sociology of science," *Minerva* 20:253, 1984.
6. Gould, S.J., *Wonderful Life: The Burgess Shale and the Nature of History*, New York, W.W. Norton, 1989.
7. The rise and fall of disciplines has received much attention in other areas of science. See, for example, *Perspectives on the Emergence of Scientific Disciplines*, Lemaine, G., Macleod, R., Mulkay, M., and Weingard, P., eds, Chicago, Aldine, 1976. Others draw a parallel between scientific disciplines and the evolution of genera and species: Hull, D., "A matter of individuality," *Phil. Sci.*, 45:335, 1978; de Mey, M., *The Cognitive Paradigm: Cognitive Science, a Newly Explored Approach*, Boston, Riedel, 1982. The manner of subdiscipline formation in immunology is considered in Chapter 19.
8. Fleck, L., *Genesis and Development of a Scientific Fact*, Chicago, University of Chicago Press, 1979. The original German version was *Entstehung und Entwicklung einer wissenschaftlichen Tatsache. Einführung in die Lehre vom Denkstil und Denkkollektiv*, Basel, Benno Schwabe, 1935.
9. Pasteur, L., *C. R. Acad. Sci.* 90:239, 952, 1880.
10. Metchnikoff, I.I., *Virchows Arch.* 96:177, 1884. Metchnikoff's theory is most completely expounded in his *Lectures on the Comparative Pathology of*

Inflammation, London, Keegan, Paul, Trench, Trübner, 1893, and in *Immunity in the Infectious Diseases*, New York, Macmillan, 1905.

11. Daniel P. Todes presents a full picture of Metchnikoff's Darwinian thesis in *Darwin without Malthus: The Struggle for Existence in Russian Evolutionary Thought*, New York, Oxford University Press, 1989, pp. 82–103. See also A.I. Tauber, and L. Chernyak, *Metchnikoff and the Origins of Immunology: from Metaphor to Theory*, New York, Oxford University Press, 1991.

12. The dispute between the cellularists and humoralists is described in detail in Chapter 2.

13. Behring, E., and Kitasato, S., *Deutsch. med. Wochenschr.* **16**:1113, 1890; Behring, E., and Wernicke, E., *Z. Hyg.* **12**:10, 45, 1892.

14. Ehrlich, P., *Klin. Jahrb.* **6**:299, 1897.

15. Nowhere in his discussions of the theory does Ehrlich actually acknowledge a Darwinian influence, although such an influence is unmistakable; see Chapter 21.

16. Bordet, J., *Ann. Inst. Pasteur* **12**:688, 1899.

17. Richard Pfeiffer had shown some years earlier that cholera vibrios were destroyed specifically within the immunized host (*Z. Hyg.* **18**:1, 1895).

18. *Ann. Inst. Pasteur* **14**, 1900.

19. Of particular interest in this connection was Karl Landsteiner's observation of isoantibodies in the blood of most humans specific for the erythrocytes of other individuals (Landsteiner, K., *Centrtalbl. Bakteriol. Orig.* **27**:357, 1900), which opened up the field of blood groups, of such great future importance in blood transfusion, immunogenetics, and anthropology. Landsteiner would receive the Nobel Prize in 1930 for this discovery.

20. These exceptions were the finding of experimental spermicidal autoantibodies by S. Metalnikoff (*Ann. Inst. Pasteur* **14**:577, 1900) and the human disease paroxysmal cold hemoglobinuria by J. Donath and K. Landsteiner (*Münch. med. Wochenschr.* **51**:1590, 1904).

21. E. Weil and H. Braun (*Wien. Klin. Wochenschr.* **22**:372, 1909) suggested that an autoimmune response to the tissue *breakdown* products in the syphilitic lesion exacerbates the disease, while the ophthalmologists S. Santucci (*Riv. Ital. Ottal. Roma* **2**:213, 1906) and S. Golowin (*Klin. Monatsbl. Augenheilk.* **47**:150, 1909) suggested that autoantibodies to damaged ocular tissues might account for the pathogenesis of sympathetic ophthalmia.

22. Bordet, J., and Gengou, O., *Ann. Inst. Pasteur* **15**:289, 1901.

23. von Wassermann, A., Neisser, A., Bruck, C., and Schucht, A., *Z. Hyg.* **55**:451, 1906.

24. Portier, P., and Richet, C., *C. R. Soc. Biol.* **54**:170, 1902.

25. Arthus, M., *C. R. Soc. Biol.* **55**:817, 1903.

26. von Pirquet, C., and Schick, B., *Die Serumkrankheit*, Vienna, Deuticke, 1906.

27. Wolff-Eisner, A., *Das Heufieber*, Munich, 1906.

28. Meltzer, S.J., *Trans. Assoc. Am. Physicians* **25**:66, 1910; *J. Am. Med. Assoc.* **55**:1021, 1910.

29. The efforts to come to grips with this paradox are reviewed in Chapter 8.

30. It is always difficult to assign an exact date to the formal admission of a new research area into the ranks of "scientific discipline." If we accept as adequate criteria the existence of a group of individuals who: (1) share the same research interests and methodologies; (2) share a special language peculiar to that area; and (3) utilize and debate one another's results to advance their research program, then clearly immunology had become a discipline during the first decade of the twentieth century.

31. We owe this term to Robert Boyle in the 1640s. It describes informal groups of scientists established for the exchange of information and the advancement of learning.

32. See Silverstein, A.M., "The development of immunology in America," *Fed. Proc.* **46**:240, 1987.

33. Wright, A.E., and Douglas, S.R., *Proc. R. Soc. London Ser. B.* **72**:364, 1903.

34. Rich, A.R., *The Pathogenesis of Tuberculosis*, 2nd edn, Springfield, Charles Thomas, 1951.

35. Rivers, T.M., Schwentker, F.F., and Berry, G.P., *J. Exp. Med.* **58**:39, 1933; Rivers, T.M., and Schwentker, F.F., *J. Exp. Med.* **61**:689, 1935.

36. See, for example, Sheldon Cohen's and other papers on the history of clinical allergy in the fiftieth anniversary issue of *J. Allergy Appl. Immunol.*, **64**:306–474, 1979.

37. Dale, H.H., *J. Pharmacol. Exp. Ther.* **4**:167, 1913; Dale, H.H., and Laidlaw, P.P., *J. Physiol. (Lond.)* **52**:355, 1919.

38. Arrhenius, S., *Immunochemistry*, New York, Macmillan, 1907.

39. Obermeyer, F., and Pick, E.P., *Wien. Klin. Wochenschr.* **19**:327, 1906.

40. Pick, E.P., in Kolle, W., and von Wassermann, A., eds, *Handbuch der Pathogenen Mikroorganismen*, 2nd edn, Jena, Fischer, 1912, Vol. 1, pp. 685–868.

41. Landsteiner, K., and Lampl, H., *Z. Immunitätsforsch.* **26**:258, 293, 1917.

42. Heidelberger, M., *Physiol. Rev.* **7**:107, 1927. This work owes much to the support of Oswald T. Avery, with whom Heidelberger published the initial papers (*J. Exp. Med.* **38**:81, 1923; **42**:367, 709, 1925.

43. Breinl, F., and Haurowitz, F., *Z. Physiol. Chem.* **192**:45, 1930. Other Lamarckian theories had preceeded this one, but they did not enjoy as great an influence. See Chapter 3.

44. Pauling, L., *J. Am. Chem. Soc.* **62**:2643, 1940; *Science* **92**:77, 1940.

45. Burnet, F.M., *The Production of Antibodies*, New York, Macmillan, 1941; Burnet, F.M., and Fenner, F., *The Production of Antibodies*, 2nd edn, New York, Macmillan, 1949.

46. See, for example, Kabat, E.A., *Structural Concepts in Immunology and Immuno-chemistry*, New York, Holt, Rinehart & Winston, 1968, and Pressman, D., and Grossberg, A., *The Structural Basis of Antibody Specificity*, New York, Benjamin, 1968.

47. Kabat, E.A., and Mayer, M.M., *Quantitative Immunochemistry*, Springfield, Charles Thomas, 1949.

48. Tiselius, A., and Kabat, E.A., *J. Exp. Med.*, **69**:119, 1939.

49. Wells, H.G., *The Chemical Aspects of Immunity*, New York, Chemical Catalog Co., 1924; Marrack, J.R., *The Chemistry of Antigens and Antibodies*, London, HMSO, 1934; Landsteiner, K., *The Specificity of Serological Reactions*, Springfield, Charles Thomas, 1936; Kabat and Mayer's *Quantitative Immunochemistry*, note 47; Boyd, W.C., *Fundamentals of Immunology*, New York, Wiley Interscience, 1943. This chemically-oriented text was replaced only in 1963 by one aimed at biologists, J.H. Humphrey and R.G. White's *Immunology for Students of Medicine*, Philadelphia, Davis, 1963.

50. Theiler, M., and Whitman, L., *Am. J. Trop. Dis.* **15**:347, 1935.

51. Zinsser, H., *Bull. NY Acad. Med.* **4**:351, 1928.

52. Rich, note 34.

53. Dienes, L., and Schoenheit, E.W., *Am. Rev. Tuberc.* **20**:92, 1929.

54. Simon, F.A., and Rackemann, F.M., *J. Allergy* **5**:439, 1934.

55. Rivers et al., note 35.
56. From the editors' introduction to *Perspectives on the Emergence of Scientific Disciplines*, note 7, p. 7.
57. Burnet, *The Production of Antibodies*, note 45.
58. Kuhn, note 3.
59. Medawar, P.B., *J. Anat.* **78**:176, 1944; **79**:157, 1945. See also Medawar's *Harvey Lect.*, **52**:144, 1956–57.
60. Owen, R.D., *Science* **102**:400, 1945.
61. Burnet, *The Production of Antibodies*, note 45.
62. Billingham, P.B., Brent, L., and Medawar, P.B., *Nature Lond.* **172**:603, 1953.
63. The first case of agammaglobulinemia was described by Bruton, O.C., *Pediatrics*, **9**:722, 1952; of a defect in T cells by DiGeorge, A.M., *J. Pediatr. (St. Louis)* **67**:907, 1965; and of severe combined immunodeficiency (so-called "Swiss type") by Glanzmann, E., and Riniker, P., *Ann. Pediatr. (Basel)* **175**:1, 1950.
64. Rose, N.R., and Mackay, I.R., eds, *The Autoimmune Diseases*, New York, Academic Press, 1985. The most recent update, the 4th edition of 2006, testifies to the great progress seen in this field.
65. Coons, A.S., Leduc, E.H., and Connolly, J.M., *J. Exp. Med.* **102**:49, 1955.
66. Jerne, N.K., and Nordin, A.A., *Science* **140**:405, 1963.
67. Landsteiner, K., and Chase, M.W., *Proc. Soc. Exp. Biol. Med.* **49**:688, 1942; Mitchison, N.A., *Proc. R. Soc. Lond. B.* **142**:72, 1954; Billingham, R.E., Brent, L., and Medawar, P.B., *ibid.* **143**:43, 1954. See also Claman, H.N., Chaperon, E.A., and Triplett, R.F., *Proc. Soc. Exp. Biol. Med.* **122**:1167, 1966.
68. Mishell, R.I., and Dutton, R.W., *J. Exp. Med.* **126**:423, 1967; Dutton, R.W., *Adv. Immunol.* **6**:253, 1967.
69. Jerne, N.K., *Proc. Natl Acad. Sci. USA* **41**:849, 1955.
70. Burnet, F.M., *Austral. J. Sci.* **20**:67, 1957.
71. Burnet, F.M., *The Clonal Selection Theory of Acquired Immunity*, London, Cambridge University Press, 1959.
72. Talmage, D.W., *Science* **129**:1643, 1959; Lederberg, J., *Science* **129**:1649, 1959.
73. At the Prague conference on *Molecular and Cellular Basis of Antibody Formation* (J. Šterzl, ed., Prague, Czech Academy of Science, 1964) Burnet rose to declare the victory of his clonal selection theory, and few biologists in the audience disagreed. The more chemically oriented were much slower in coming round, and a few (e.g., Felix Haurowitz and Alain Bussard) never conceded. Some identify the 1967 Cold Spring Harbor meeting (*Cold Spring Harbor Symp. Quant. Biol.* **32**, 1967) as the turning point.
74. The term "anticipate" is perhaps too anthropocentric, but a better term does not come to mind. Niels Jerne (personal communication, 1990) points out that one can hardly say that the presence of mutants in a large population of bacteria *anticipates* the arrival of penicillin in explaining the ability to develop a strain with penicillin resistance!
75. Ohno, S., *Perspect. Biol. Med.* **19**:527, 1976; *Progr. Immunol.* **4**:577, 1980.
76. Tonegawa, S., *Nature* **302**:575, 1983.
77. Marrack, P., and Kappler, J., *Science* **238**:1073, 1987.
78. Hood, L., Kronenberg, M., and Hunkapiller, T., *Cell* **40**:225, 1985.
79. Tonegawa, note 76.
80. Oppenheim, J.J., and Shevach, E.M., *Immunophysiology: The Role of Cells and Cytokines in Immunity and Inflammation*, New York, Oxford University Press, 1990.

81. Samter, M., ed., *Immunologic Diseases*, 4th edn, Boston, Little Brown, 1988. See also Chapter 12.
82. It should be noted, however, that the study of antibodies and their specificity throughout this period was intimately connected with physical chemistry, and with developments in the understanding of protein and especially enzyme structure and function. Indeed, L.E. Kay (*Hist. Life Sci.* **11**:211, 1989) suggests that immunology contributed much to the early development of molecular biology.
83. This coming together of chemically- and biologically-oriented immunologists is reminiscent of the so-called evolutionary sythesis (see Mayr, E., and Provine, W.B., *The Evolutionary Synthesis: Perspectives on the Unification of Biology*, Cambridge, Harvard University Press, 1980). In that case, it was the Mendelian geneticists working forward from the genome and the paleontologists working backward from the whole organism and populations who melded their disparate languages and approaches to attack the problems of evolution.
84. Moulin, A.-M., *Rev. Hist. Sci.* **36**:49, 1983; *Hist. Phil. Life Sci.* **11**:221, 1989. See also Moulin's *Le Dernier Langage de la Médicine: Histoire de l'Immunologie de Pasteur au SIDA*, Paris, Presses Universitaire, 1991.
85. I am indebted to Fred Karush for pointing out to me the dialectical nature of this process.

18 Immunology in transition 1951–1972: the role of international meetings and discipline leaders[1]

...these conferences...constitute an experiment in communication.

Maurice Landy[2]

We saw in earlier chapters that the decades that followed World War II witnessed a radical change in the field of immunology. Accompanying a major transition from immunochemical to immunobiological concerns, the discipline became one of the central players in modern biomedical research.[3] The history of that period in immunology has been described in Part I primarily in terms of the conceptual changes that occurred, but the full history of a scientific discipline involves more than an account of conceptual transitions; it has many other dimensions. These include, among others, analyses of the philosophical influences on the scientists,[4] the semantics of the language that they employ,[5] the epistemological basis of their facts,[6] the social influences on their work,[7] the anthropological study of their tribal habits and mores,[8] and the rhetorical strategies that they employ.[9] It also demands a study of the changing structure and of the inner dynamics of the discipline itself – a subject that has increasingly attracted the attention of historians and sociologists of science.[10]

This chapter will extend these more sociological contributions to the development of immunology to include an analysis of disciplinary leadership. It will also cast light on some of the more modern trends in subdiscipline formation. This aspect of the subject will be approached by analyzing the many international meetings that took place during a critical transition period in the development of modern immunology. Scientific meetings are an all-pervasive part of contemporary science[11] and became an integral component at its very beginnings in the seventeenth century, when members of the first learned societies[12] met to spread knowledge among their participants and to further disseminate the new knowledge with the publication of Transactions and Proceedings. Scientific meetings (congresses, symposia, workshops, etc.) have been identified as serving an alternative and less formal channel of communication as compared with professional journals.[13] They have also occasionally been employed as "mirrors" of disciplinary development, or as source material for institutional and disciplinary histories.[14]

The period chosen for this study is the interval from 1951 to 1972, because this was the era of most rapid change in immunology. Changes resulted from,

A History of Immunology, Second Edition
ISBN: 978-0-12-370586-0

Copyright © 2009, Elsevier Inc.
All rights reserved

and also encouraged, the entry into the field of a new generation of investigators. There were other manifestations of this burst of activity in immunology. It was accompanied by a new phase of organization of the scientific discipline, and many new national immunological societies were founded (there are now fifty-four members in the International Union of Immunological Societies).[15] The creation of these societies and the organization of national and international meetings with an explicit immunological agenda is itself a fundamental aspect of the process of discipline formation which will not be explored further here. Many new journals appeared (now numbering well over 300!), reflecting the development of many new immunological subdisciplines. Chairs and departments of immunology were established, and finally, in 1971, the triennial International Congresses of Immunology were initiated.

This chapter will employ a novel approach; the prosopographical analysis (identification by name) of individual members of the community of immunologists who attended these meetings during the period in question. This approach offers the possibility of identifying the leaders of the field and of its subdisciplines (Ludwik Fleck's *Denkkollektiv* – the arbiters of theory and governors of the direction of movement of the field). It also offers clues to when a peripheral field with its own research program and leadership joined the "mainstream" of the parent discipline. In addition, the different types of meetings (from small "by invitation only" to grand international congresses) represent a fascinating spectrum, and some of them have themselves surely influenced the direction and rate of development of the field.

Immunological meetings, 1951–1972

Selection criteria

The broad boundaries of a discipline like immunology make the identification of meetings appropriate for inclusion in this study somewhat ambiguous. A start was made by inventorying the published volumes in the collections of three major libraries – the Welch Library of the Johns Hopkins Medical School in Baltimore, the Danish Science and Medical Library in Copenhagen, and the Library of the Basel Institute of Immunology. As a first approximation, all meeting titles that contain the keyword "immunology" (or more generally "immuno-") will be included, as well as those containing such other keywords as "antigen," "antibody," "allergy," "complement," "transplantation," "histocompatibility," etc. To these are added titles resulting from responses to a questionnaire sent to leading immunologists throughout the world.[16] Our initial survey identified, within the chosen timeframe, over 150 meetings as immunological in the broadest sense; these included basic and clinical immunology, and the border areas between immunology and other disciplines.

Since the purpose of this study is to analyze the disciplinary structure and dynamics of immunology in terms of immunological meetings and their participants (and eventually in terms of their content), many were immediately excluded for lack of published proceedings. Among these are a number of significant series of informal "workshops" which played an important part in defining the several subdisciplines of immunology.[17] Also excluded were all meetings having a primary educational function, such as Summer Schools[18] (although it is sometimes difficult to draw a distinction between such meetings and research conferences). Further, the annual/semiannual meetings of national societies of immunology were excluded, in spite of the fact that these were attended by the vast majority of immunologists in the period under investigation, since lists of participants were not always available. Also excluded were meetings devoted primarily to standardization (for example, of vaccines); to applications of immunological techniques in other fields;[19] or where immunological research problems played only a minor role in the proceedings.[20]

With these exclusions, eighty-eight meetings are included, as listed in Appendix 18.1 to this chapter. It is likely that some meetings applicable to a study of this type have been missed, but the current sample of eighty-eight meetings should represent a good first approximation. As a consequence of an intrinsic international bias in the choice of meetings, leaders of the field are probably overrepresented, since they are more likely than followers to attend (and be invited to) international meetings. However, since much of the analysis is intended precisely to identify these disciplinary leaders, the exclusion of meetings of national immunological societies will likely have little effect on the results.

The taxonomy of meetings

Examination of the meetings chosen for this study shows a great variation in organizational structure and purpose. These variations may depend on size, on exclusivity of participation, and on the avowed intent of the organizers. The taxonomy of scientific meetings appears to represent an important if thus far relatively unexplored area of interest to science studies, but I shall only comment briefly here on some of the more obvious factors involved.

One of the major features of scientific meetings is whether they stand alone or are members of a series. About one-quarter of the meetings included in this study are singular. They most frequently result from the identification of an important or emerging area of the science by a governmental or private institution, a scientific society, or entreprenurial scientists themselves. The meetings are then organized, usually with outside financial support, to record and publicize progress in the field, and occasionally even to provide publicity for the organizer or for his institution.

Among the meeting series included in this study are the transplantation meetings of the New York Academy of Sciences and its organizational successor, the Transplantation Society (series "Transplant"); the immunopathology symposia ("Immunopath"), the Germinal Center series ("Germinal"); the leukocyte culture conferences ("Leukocyte"); the allergology congresses ("Allergy"), and the histocompatibility workshops ("Histocomp"),[21] all devoted to describing and recording progress within a specific subdiscipline of the science. Differently conceived, the Ciba Foundation ("Ciba Found"),[22] the Sanibel Island ("Sanibel"), or the Brook Lodge series ("Brook Lodge") were organized so that each separate meeting was designed to explore in depth a different important evolving area of the discipline. On yet another level, the annual meetings of National Societies, the triennial International Congresses of Immunology ("Intl Congr"), and to an extent the (roughly) decennial Cold Spring Harbor series ("Cold Spring") are meant to summarize the status and current activities of the entire discipline.

In considering meeting size, the Ciba, Sanibel Island, and Brook Lodge meetings often involved only fifteen to thirty participants from among those scientists internationally noted in the specific field of interest, occasionally with invited experts from outside the field for interdisciplinary stimulus. At the next higher level are such meetings as the Prague series ("Prague"), the Germinal Center series, and the later Histocompatibility Testing series, with audiences usually of the order of 100. Above these are such meetings as the Buffalo Convocations ("Buffalo"), the New York Academy of Science transplantation meetings, the Cold Spring Harbor meetings, and the Collegium Internationale Allergolicum series, whose audiences might number several hundred. Finally, there are the National Society meetings (especially in the United States) and the International Congresses, open to the entire world of immunology, with audiences in the thousands.

Another of the structural characteristics of the larger modern meetings involves internal competition among subspecialty areas. While most of the meetings on our list involve single sessions or series of sessions without competition, recent national society meetings and the International Congresses have reflected the great subdivisions within the larger field by holding multiple simultaneous sessions of mini-symposia, workshops, and poster displays.[23]

Yet another meeting variable concerns the intent and aims of the organizers. Thus, the Ciba, Sanibel Island, and Brook Lodge meetings were meant to gather together the world leaders for informal and wide-ranging discussion of a given area, to explore in depth its recent developments and especially its theoretical and practical implications.[24] The histocompatibility workshops were designed at the outset to promote the standardization of reagents and techniques in this technically complicated young field, but soon broadened their scope. The Germinal Center, leukocyte culture, and transplantation series of meetings were intended both to record progress in a subdiscipline as well as to advertise the area broadly. The Buffalo Convocations and Collegium Internationale Allergolicum series were meant not only to record progress in the field, but also to educate their respective memberships on that progress. Finally, The Cold Spring Harbor and International Congress meetings were intended to provide the milestone markers of progress in the entire discipline for all to see.

Exclusivity is yet another variable in the evaluation of meeting structure and function. Most of the smaller meetings (for example, Ciba, Brook Lodge, and many of the early workshops) were by invitation only, the list chosen by the

organizers from among the world leaders in the field of interest. At an intermediate level are the Prague series, the Germinal Center series, the Buffalo
Convocations, and many of the individual symposia, to which many speakers
were invited from among the world leaders, but where the audience might be
open to all comers. Finally, the National Society meetings and International
Congresses represent a mixture, in which the symposia and grand lectures are
presented by invited speakers from among the leaders in the field, but where
poster- and workshop-session presenters are self-selected (although their
abstracts may be screened by a program committee).

The disciplinary leadership of immunology, 1951–1972

Only a few studies have treated scientific meetings quantitatively.[25] More
specifically, no studies so far have applied prosopographical and scientometric
methods to the phenomenon of meetings. To pursue this aim, all individual
participants are analyzed from the list of the eighty-eight immunological meetings described above.

First, does the population of researchers who frequently attend meetings in the
field of immunology, particularly international meetings, constitute the leading
elite of the discipline? If this is the case, the identification of frequent meeting-
goers might be used as a method for mapping the disciplinary elite. For each
selected meeting, the names of all the participants were listed and pooled to
generate a master file of all participants.[26] The pooling procedure is not without
complications, since different individuals may appear under the same name, or
the same individual under different names (for example, are M. Cooper, M.D.
Cooper, and Max Cooper the same person? Is L. Herzenberg the husband or the
wife?), but with very few exceptions one nevertheless may unambiguously
identify a total of 4,806 individuals who have participated in at least one of the
eighty-eight immunological meetings in the period 1951–1972. The records are
assembled in a master text-file in the form of a {4,806 participants; eighty-eight
meetings} matrix.

As expected, the participation in scientific meetings is by no means evenly
distributed. The frequency of attendance at meetings is shown in Table 18.1. The
large majority of participants (3,480 individuals or 72 percent) attended only
one of the eighty-eight meetings in the twenty-two-year period. Only 311
attended at least five meetings (6.5 percent of participants), seventy-nine (1.6
percent) attended at least ten meetings, and only twenty-seven individuals (about
0.5 percent) attended fifteen meetings or more. One single investigator (Robert
A. Good) attended thirty-nine meetings!

One might expect that the higher the frequency of participation, the more
renowned is the researcher. In fact, almost every one of the seventy-nine
researchers that have gone to at least ten meetings are known to historians of
contemporary immunology to be leaders in the field, as having either made
important   discoveries   or   instigated   influential   immunological   research

Table 18.1 Frequency of participation in immunological meetings, 1951–1972

No. of meetings attended (at least)	No. of individuals participating (N)
1	4,806
2	1,326
5	311
10	79
15	27
20	5

programs, or in their capacity as entrepreneurs or scientific gatekeepers.[27] The list of these seventy-nine individuals is provided in Table 18.2.[28]

It may be appropriate to point out here that several of those who figure prominently on the frequency list do so not as leading scientists, but rather as leading meeting organizers. In addition, each meeting organizer might frequently invite particular "favorite scientists" to participate, apart from objective scientific considerations. Again, a leading scientist invited to meetings will often take along a favored colleague or student who might not otherwise have been invited.

Subjective evaluation of the relation between high meeting-frequency and high reputation may be independently assessed by matching the ranking on the meeting list with another accepted indicator of scientific reputation – the number of citations of scientific papers.[29] In general, the frequency with which a researcher went to immunological meetings is clearly correlated with the number of citations of that individual's scientific papers. The citations of the ten most frequent meeting-goers were compared with ten randomly sampled participants at five meetings, and ten others randomly selected from those participants that attended one meeting only. The result strongly indicates that the more frequently a researcher attends immunological meetings the higher is that individual's scientific reputation in the field. The ten most frequent meeting participants (attending eighteen meetings or more) each show some 4,000–5,000 citation equivalents over a twenty-five-year period, whereas researchers who participated in one meeting only rarely have more than a few hundred citation equivalents. Researchers who participated in five meetings display a somewhat more varied pattern.

Hence, we conclude that there is a strong correlation between frequency of participation in immunological meetings and scientific reputation in the field of immunology, for the extreme ends of the meeting frequency spectrum. There are a few significant individual exceptions to this pattern, however. On the one hand there are a few researchers who rank low on the meeting frequency scale, but are generally known as major players in the field of immunology. These may be individuals who just dislike meetings or travel, or those who entered the field late or left it early during the period under study (see note 28). Conversely, several of the more frequent meeting participants are less well known (and less-often cited)

Table 18.2 Most frequent participants, attending ten or more meetings

39	Robert Good	12	James Gowans
21	H. Sherwood Lawrence	12	T. N. Harris
21	Morten Simonsen	12	Maurice Landy
20	Baruj Benacerraf	12	J. J. van Rood (T)
20	Byron Waksman	12	Roy Walford (T)
19	Bernard Amos	11	John Converse (T)
19	Frank Dixon	11	Gustav Dammin (T)
19	Richard Smith	11	Leonard Herzenberg
19	Jonathan Uhr	11	Peter Medawar
18	Rupert Billingham	11	Peter Miescher (I)
18	John Humphrey	11	H. Müller-Eberhard (I)
18	N. Avrion Mitchison	11	Joseph Murray (T)
17	Fritz Bach	11	David Pressman
17	Ruggero Ceppellini	11	Ivan Roitt
17	Felix Milgrom	11	Paul Russell (T)
17	Gustav Nossal	11	Ernst Sorkin
16	Jacques Miller	11	Lewis Thomas
16	Göran Möller	11	John Trentin (T)
16	Jaroslav Šterzl	11	John Turk
15	Michael Feldman	11	D. W. van Bekkum (T)
15	Herman Friedman	11	Robert White
15	Pierre Grabar	10	Max Cooper
15	Milan Hašek	10	Alain de Weck
15	William Hildemann	10	Frank Fitch
15	Peter Perlmann	10	Susanna Harris
15	Jeanette Thorbecke	10	Elvin Kabat
15	William Weigle	10	Henry Kunkel
14	Werner Braun	10	Paul Maurer
14	Paul Teresaki (T)	10	John Merrill (T)
14	Guy Voisin	10	Erna Möller
13	Frank Austen	10	John Najarian (T)
13	Leslie Brent	10	Joost Oppenheim
13	Jean Dausset (T)	10	Zoltan Ovary
13	Felix Rapaport (T)	10	Noel Rose
13	Robert Schwartz	10	Alec Sehon
13	Arthur Silverstein	10	Chandler Stetson
12	Melvin Cohn	10	Zdenek Trnka
12	Joanne Finstad	10	Darcy Wilson
12	Hugh Fudenberg	10	Michael Woodruff
12	Philip Gell		

(T) indicates that most of the meetings attended by the individual were restricted to transplantation; (I) indicates that most of the meetings attended by the individual were restricted to immunopathology.

for their science than for their important roles in disciplinary development: governmental biomedical functionaries (at the National Institutes of Health, etc.), entrepreneurial meeting organizers, "scientific statesmen," "gatekeepers," etc. (for example, Maurice Landy, Lewis Thomas, Howard Goodman, Zdenek Trnka). Finally, a small group of individuals can be identified who rank very highly in citation frequency, but who attended only a single meeting or two; these are scientists famous in fields other than immunology who, for whatever reason, chose to attend an immunologically oriented meeting (for example, Jacques Monod).

Obviously, there is a continuum stretching from researchers with high reputations attending many meetings to more marginal researchers with low reputation in the field – hence, the borderline between a disciplinary elite and a non-elite of immunologists cannot be drawn sharply. For practical purposes, however, we have established limits at the five-, ten-, and fifteen-meeting levels, respectively. We tentatively distinguish four populations of immunologists: a large group of followers that consists of those attending fewer than five meetings; the major professionals in the field consisting of those attending five or more meetings; a disciplinary elite consisting of those attending ten or more meetings; and a core elite consisting of those attending fifteen or more meetings. These definitions will be used in the following section.

The inner dynamics of immunology, 1951–1972

Meeting clusters

We have seen above how data regarding the participation of individuals in immunological meetings can be used to identify leaders in the discipline. It will be obvious also that the timing of initiation of certain meeting series, or of meetings on a given subject matter, tells much about when these subjects (or subdisciplines) attained popularity among immunological researchers. This same material can be used to identify subdisciplinary units by comparing meetings in terms of the overlapping attendance of these same individuals. The technique employed is *cluster analysis*, a widely used method for taxonomic purposes in biological classification, linguistics, sociology, and psychometrics.[30] This statistical approach has been utilized as a routine method for scientiometric purposes, particularly in co-citation analysis,[31] in which articles are classified in pairs on the basis of the similarity between their reference lists.

In this study, meetings are compared with respect to the overlapping participation of individual scientists. Two meetings are said to be more similar than two other meetings if they share a greater overlap of participants. A standard Jaccard similarity measure was chosen together with a standard computer program package for cluster analysis.[32]

Analysis was made with a {meeting:participant} matrix reduced to the 311 researchers who participated in at least five meetings. Each meeting from the selected list of eighty-seven meetings was compared with every other meeting on the list. (Of the original eighty-eight meetings collected for this study, one – the

Second Collegium Allergolicum [Allergy 2-55] – was lost to this analysis because none the 311 individuals had taken part in it). The program starts by grouping together the two meetings with the highest similarity measure (i.e., the greatest overlap of participants), and this meeting-pair is assigned a value. The program continues to group together meeting with meeting, or meeting with meeting-sets of increasing complexity until all meetings have been grouped together in clusters.[33]

The result for meetings attended by the 311 individuals who participated in five or more meetings is shown diagrammatically in Figure 18.1. Those meetings joined by the longest black bars are most similar to one another as measured by overlapping participants.

As may be seen from Figure 18.1, the meetings attended by the 311 "professionals" fall into discrete groups or clusters. Thus, reading from top to bottom, a separate grouping is formed by the leukocyte culture conferences, and another includes the immunopathology symposia. A somewhat larger cluster joins the Prague and Brook Lodge meetings with the 1967 Cold Spring Harbor symposium and the 1971 International Congress. A separate grouping includes the transplantation conferences and histocompatibility workshops, and at the bottom, the allergy congresses form their own cluster.

How can the clustering of meetings be explained? It is surely not geographically based; the Prague meetings organized by Jaroslav Šterzl and Milan Hašek included large numbers of Czechoslovak and Eastern European participants, and yet they cluster with many non-Czechoslovak meetings well beyond the $N = 311$ level. Similarly, the three meetings in Scandinavia (Singular 13-67, Stockholm; Singular 19-69, Lund; and Singular 20-70, Helsinki) had many local participants, but also relate well to other meetings elsewhere, particularly at the higher levels of analysis. Thus, it may be reasonably concluded that most "local" participants would increasingly fail to qualify for analysis as the required number of meetings attended increases.

One might also suspect a generational influence, wherein contemporary meetings might have greater overlap of participants than meetings separated by a larger time-span. Such a generational change might be responsible for the bimodal structure of the Sanibel Island meetings, which at all levels of analysis fall into two subclusters, for the years 1965–1967 and the years 1969–1972. However, the Sanibel Island groupings are perhaps better explained by the decisive change in cognitive content, from developmental biology problems in the three early meetings to the quite different subject of immunoglobulin classes in the later meetings. The relatively minor importance of the generational factor is illustrated by the transplantation meetings held over the eighteen-year period 1954–1972; they cluster together almost independently of chronological position. While the generational factor seems to be of rather minor significance for the major cluster patterns, it may be of some significance within individual subdisciplines where the turnover rate may be high.

If regional bias and contemporaneity may be ruled out as major causes of clustering, then we may assume that the main reason why meetings exhibit

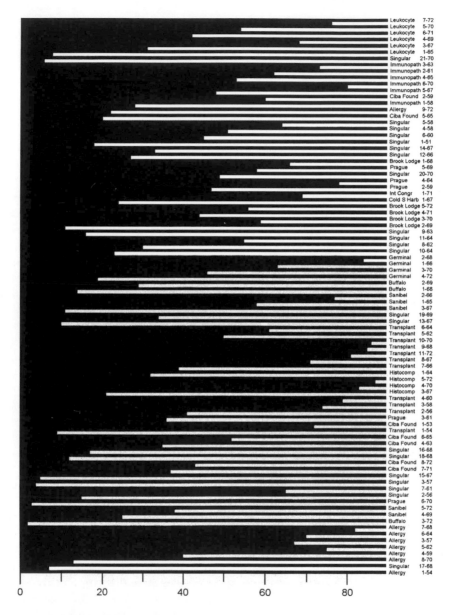

Figure 18.1 Cluster diagram of immunological meetings attended by the 311 individuals who participated in five or more meetings.
From Söderqvist, T., and Silverstein, A.M., *Scientometrics* **30**:243, 1994; Söderqvist, T., and Silverstein, A.M., *Social Studies in Science*, **24**:513, 1994.

similarity in the cluster analysis is that they attract participants with similar scientific interests in response to the aims and program of the meeting. In this sense, scientific subdisciplines can be viewed as analogous to the different traditions in music. One would expect concerts featuring Telemann to cluster more closely to those playing Bach than to those with Wagner programs; all of these would fall into a supercluster of classical music at some distance from the clusters of rock and country music concerts. But there is another factor at work; most concert-goers are self-nominated, whereas most major speakers at scientific meetings are invited by the organizers. This may be either because they are the recognized leaders in the program area, or perhaps because they belong to a small circle of individuals favored by the organizers for other reasons.

The general features of this clustering hold true, with few exceptions, when the eighty-seven meetings are analyzed by zeroing in on the seventy-nine individuals who participated in ten or more meetings (termed the "disciplinary elite"). However, three of these meetings drop out (Allergy 2-55, Allergy 3-57, and Singular 21-70 on leukocytes) because each failed to have a single one of the seventy-nine individuals in attendance. Similarly, reducing the participants to the twenty-seven "core elite" (who attended at least fifteen or more meetings) saw a further six meetings drop out of the list (two further allergy meetings, another small singular meeting (2-56 on development), the sixth leukocyte culture conference, and the two Sanibel Island meetings on immunoglobulins (4-69 and 5-72). It is interesting that as the field narrows down to the most active immunological travelers, it is the early allergy meetings, the two Ciba meetings devoted to immunoglobulins, a leukocyte culture conference, and two small and perhaps more parochial singular meetings that drop out of consideration. This may hint at the relationship of certain subdisciplines to the mainstream interests of the field – a topic that will be explored more fully in the next section.

Subdisciplinary relationships

With the main assumption of scientific relatedness as our point of departure, we will now examine the disciplination of immunology in the period 1951–1972 as reflected by the main international meetings of the period. The First International Congress of Immunology held in Washington DC in 1971 was chosen as the reference point for further discussion. The possibility of holding a congress was suggested in the mid-1960s, formally decided upon by the newly formed International Union of Immunological Societies, and organized by Maurice Landy for the American Association of Immunologists. It was the first international manifestation of the institutionalization of immunology as a scientific discipline. It was a comprehensive meeting; its fifteen sessions and eighty-four workshops were devoted to all of the then-active aspects of immunology.

The 1971 International Congress overlaps considerably (at all levels of analysis) with another meeting that has risen to legendary status in the historical consciousness of immunologists – the Cold Spring Harbor meeting on Antibodies in 1967, supporting the view that the origin, production, structure, and function of antibodies was still the central issue in immunology in the mid-1960s. Centered on these two meetings, a core of closely related immunological meetings may be identified: the Prague meetings; the Brook Lodge series; the Sanibel Island developmental immunology workshops; a small number of individual meetings; and, at a somewhat greater distance, the immunopathology symposia and the germinal center meetings. The overlap of attendees at the three Prague meetings[34] organized by Šterzl (Prague 2-59, 4-64, and 5-69) and the two reference-point meetings further supports the view that the mechanism of antibody formation was at the center of scientific interest among immunologists in the late 1950s and throughout the 1960s.

The homogenous cluster of the germinal center meetings points up an interesting feature of disciplinary relationships. As might be expected, a large number of participants at these meetings came from pathology departments. Curiously, however, cluster analysis shows no significant overlap between these meetings and the immunopathology meetings, many of whose participants also came from pathology departments. This schism is most probably due to the preoccupation of immunopathologists with diseases and their mechanisms, whereas the germinal center meetings dealt more with the structure and function of antibody-producing lymphoid tissues and cells (i.e., anatomy and cytology).

Also belonging to the mainstream of immunological interest is the series of Brook Lodge meetings organized by Maurice Landy from 1968 to 1972. In spite of the variety of issues treated by the five meetings (cellular immunity, surveillance, intervention, and genetic control), the Brook Lodge series nevertheless stands out as a fairly homogenous cluster at all levels of analysis (with the exception of the first meeting on Immunological Tolerance, which clusters more closely with the Prague meetings at the $N = 311$ through $N = 27$ levels of analysis). This confirms our feeling that the organizers had a fairly well-defined purpose for the meetings, and drew the small number of participants from among the most frequent meeting-goers (the "elite").

Somewhat more distantly related is the very homogenous cluster formed by the series of immunopathology symposia, initially organized in 1958 by Peter Miescher and Pierre Grabar. This cluster is quite homogenous at all levels of analysis, reflecting a rather closed community of scientists. It also includes a meeting outside the series, the Ciba meeting (Ciba 2-58) on Cellular Aspects of Immunity, where both Burnet's clonal selection theory and Simonsen's graft-versus-host experiments were broadly discussed for the first time. Closely related to the immunopathology series are two individual meetings on hypersensitivity held in 1958 (Singular 4-58 and 5-58). The increasing number of meetings on such biomedical topics as hypersensitivity, immunopathology, autoimmunity, etc., is perhaps one of the best indicators of the major shift from immunochemistry to immunobiology during this period.

The series of workshops held on Sanibel Island was initiated in 1965 at the instigation of Richard Smith, Robert Good, and Peter Miescher, and was supported by the National Institute of Child Health and Human Development. They gathered a small number of invited participants, initially to discuss and popularize the recent work on developmental immunology. The first three meetings, in 1965 (Phylogeny), 1966 (Ontogeny), and 1967 (Immunodeficiency Diseases), cluster together fairly well at all levels of analysis, and cluster also with the germinal center meetings. The meetings in 1969 (IgA) and 1972 (IgE) left the field of developmental biology and dealt with immunoglobulin isotypes; hence these meetings cluster separately, because the participants were selected from a different subset of the disciplinary leaders.

In addition to the several series of meetings, a number of small but important singular meetings stand out. A small (twenty-eight selected participants) meeting on Regulation of the Antibody Response, organized by Bernhard Cinader in Toronto (Singular 12-68), falls close to the Prague and Brook Lodge meetings, probably because it dealt with problems concerning the regulation of antibody formation. A somewhat larger (eighty-nine participants) meeting on Cell Interactions and Receptor Antibodies in Immune Responses, organized by Olli Mäkelä in Helsinki (Singular 20-70), dealt with the recently discovered functional distinction between B and T lymphocytes. This meeting shows an even greater affinity with other mainline meetings, particularly at the $N = 27$ level, suggesting that this was considered a very hot topic by the core disciplinary elite at the time. Two symposia at Rutgers University, Immunochemical Approaches to Problems in Microbiology, organized by Michael Heidelberger and Otto Plescia (Singular 6-60), and Nucleic Acids in Immunology, organized by Plescia and Werner Braun (Singular 14-67), apparently drew on a similar constituency of immunologists (although, being somewhat more immunochemically oriented, both fall outside the main cluster at the $N = 27$ level). The two meetings on gamma globulins, organized in Sweden (Stockholm, Singular 13-67; and Lund, Singular 19-69) that cluster together up to the $N = 79$ level nevertheless show different affinities to the other meetings of the core group of clusters, presumably reflecting the narrower specialization of their topics.

The Buffalo Convocations (series "Buffalo") were initiated in 1968 by Noel Rose and Felix Milgrom, but cluster analysis shows a striking dissimilarity among the individual meetings, presumably because a different topic was chosen for each meeting. They were attended by a large contingent of local and regional participants, and by a small number of invited international leaders (different for each meeting). This leaves the impression that, in addition to local educational functions, one of the aims of the meetings was to put Buffalo immunology on the world map. The meeting in 1968 clusters weakly with the immunopathology meetings at the $N = 79$ and $N = 27$ levels, whereas the meetings in 1969 and 1972 show much more varied relationships to other meetings.

Most of the meetings discussed thus far reflect a fairly homogenous group of research programs centered on problems of antibody formation, mechanisms of hypersensitivity, and immunopathology. The strong overlap between meetings oriented to problems concerning basic research questions, such as the Prague and Brook Lodge meetings, and meetings seemingly oriented towards more clinical issues, such as the immunopathology symposia, suggests that it is difficult to make a clear distinction between basic science and certain areas of clinical immunology during the period under investigation. But the view of the immunopathology meetings as primarily clinical is somewhat erroneous. Rather, these meetings were devoted principally to basic research on clinically relevant problems and to the establishment of research animal models for these human disease problems, such as immune complex disease, autoimmunity, and the basic mechanisms of allergic diseases. Thus, clinically oriented researchers and those interested in the theoretical problems of antibody formation, T/B cell interaction, immunopathogenetic mechanisms, etc., would likely go to the same meetings.

When subdisciplines join the mainstream

If we are correct in assuming that the interests of those individuals who attended the greatest number of meetings reflect the changing trends of mainstream immunology, then the degree to which they attend certain subdisciplinary meetings should be a measure of the "distance" of that area from the mainstream. In addition, increasing attendance by these disciplinary leaders over time should provide a measure of when the subdiscipline began to join the main current of the field. This conjecture appears to gain support from a consideration of two initially peripheral fields: allergology and transplantation.

The conspicuous cluster near the bottom of the diagram in Figure 18.1 consists primarily of the nine symposia of the Collegium Internationale Allergologicum (series "Allergy") held between 1954 and 1972. Historically, allergy emerged as a distinct clinical discipline after World War I, when anaphylaxis, serum sickness, hayfever, and asthma were recognized as important medical problems unrelated to the growing chemical interests of immunology. This separation of allergy from the core cluster of meetings reflects the fact that most clinical allergologists rarely attended other immunological meetings, and most immunologists did not attend allergology meetings. "Elite" scientists, who are prone to be more specialized, would not usually be invited to or voluntarily attend the allergology meetings.[35] According to Sheldon Cohen, allergology did not contribute substantially, if at all, to the integration of clinical and basic theoretical issues in immunology prior to the 1970s.[36] It is interesting to note that the allergology meeting held in 1972 begins to cluster with the well-defined group of immunopathology meetings – a tendency that increases further with further reductions of the matrix. This appears to reflect the discovery in the late 1960s of the IgE antibody responsible for allergic diseases, and of the beginning elucidation of the immunophysiological mechanisms of allergic reactions. Thus, we see the beginnings of the integration of allergy research into mainstream immunology – a process reflected by the increasing participation of the mainstream leaders at these later meetings.

Yet another fairly well-defined cluster falls outside the core group. This is comprised of the fifteen or so transplantation meetings: the series of seven meetings arranged by the New York Academy of Science between 1954 and 1966; the international Congresses of the Transplant Society held from 1967 on; and the series of histocompatibility workshops, first organized by Bernard Amos in 1964. The transplantation meetings were initiated in the main by plastic surgeons with an interest in transplantation, but basic science reports soon became an integral part of the proceedings. These two series of meetings form a natural cluster through all levels of analysis. Both transplantation series overlap considerably with the Histocompatibility Testing workshops. These small workshop-like meetings were specifically technical in orientation. Their close overlap with the transplantation meetings is explained by the fact that graft rejection, the leading topic of interest to transplant scientists, was early shown to be due to the antigens dealt with in the histocompatibility workshops, and tissue typing of these antigens became increasingly important in organ transplant programs. The proximity to these of the meeting on Mechanisms of Immuno-logical Tolerance, organized by Milan Hašek in Czechoslovakia (Prague 3-61), reflects the growing relationship between research on immunological tolerance and the emerging immunobiological basis for graft rejection. Again, as basic research in transplantation showed graft rejection to be related to cellular immunity, so did this new subdiscipline slowly join the mainstream.[37]

Another well-defined cluster is constituted by the series of annual leucocyte culture conferences, originally concerned primarily with leukocyte (lymphocyte) structure and physiology. The series shows very little overlap with other meet-ings at all levels of analysis, indicating that researchers specializing in these studies did not mix with other immunologists. Thus, this area had not yet, by 1972, joined the immunological mainstream, despite the fact that it eventually became a subject of immense importance for immunological research.

The impact of meetings on disciplination

We have seen above how developments in immunology influenced the nature of the meetings held during the period under study, but it is also reasonable to inquire whether in return some of these meetings themselves might have influ-enced developments in the science. This takes us, of course, away from quan-titative measurements and into the more questionable realm of subjective evaluation, but it may serve a useful purpose to indicate at least a few of the more obvious possibilities.

Three singular meetings during the 1950s helped to focus attention on the growing interest in immunobiology. The first (Singular 1-51) was organized by A. M. Pappenheimer Jr, Merrill Chase, and René Dubos. While still much concerned with classical, chemically-oriented topics, it did introduce several more biological subjects: the passive transfer of delayed hypersensitivity by cells (Chase); the pathogenesis of serum sickness (Janeway); allergic mechanisms (Sherman, Lowell); and applications of labeled antibodies (Coons, Pressman). But perhaps the two most seminal meetings took place in 1958: Lawrence's Cellular and Humoral Aspects of Hypersensitivity States (Singular 4-58) and Schaffer, LoGrippo, and Chase's Mechanisms of Hypersensitivity (Singular 5-58). Here for the first time were assembled all of those disparate investigators who had been working on the biology and pathology of the immune response at what previously had been the fringes of the discipline. The broad overlap of biomedical interests between basic science and clinical researchers became apparent at these exciting meetings, and this sense of a new community of interests was communicated more widely by the many attendees and by the wide circulation of the proceedings.

In a somewhat similar manner, the series of transplantation meetings held by the New York Academy of Sciences and then by the Transplantation Society helped to institutionalize transplantation as an interesting and viable field of study by bringing together clinicians and researchers. More than this, however, these meetings and their proceedings served as the vehicle to attract young students and outside immunologists into the area, thus helping to integrate immunology into transplantation research, and transplantation into the developing currents of immunology.

The Prague meetings organized by Jaroslav Šterzl and Milan Hašek from 1954[38] to 1969 represent an interesting study. Not only did they contribute significantly in focusing early attention on the more biological aspects of immunology; they also played another important role in the institutionalization of immunology in Czechoslovakia. Immunology in Prague has had a long history. Early in the twentieth century, E. Weil and Oskar Bail (at the German University of Prague) contributed significantly to studies of autoimmunity and agressin theory, respectively, and during the 1930s Felix Haurowitz advanced his instructive theory of antibody formation there. However, with the war and subsequent political upheaval, immunology substantially disappeared from the local scene, and Charles University became exclusively Czech. During the 1950s, two young immunologists, Šterzl and Hašek, sought to re-establish immunology in Prague.[39] It is clear that the meetings which they held, attended by many of the world's leading immunologists, not only brought well-justified international recognition to the work of these two laboratories, but also helped to advance the careers of the principals and to garner strong governmental support for their work. However, the political situation that followed the Soviet invasion of Czechoslovakia in 1968 led to a fall from favor of these two investigators, and a decline in the productivity of their laboratories.

It is clear that many of the meetings considered in this study contributed significantly to the internal disciplinization of their respective research areas, and helped to advertise these areas to a wider public. In addition, however, many of them (and especially the smaller ones) saw the presentation of new data and new ideas at the cutting edge of the science, rather than merely serving to summarize past history. Thus, they frequently led the field and stimulated work and thought in new directions. Further, they often provided the setting for the important informal discourse in hallways or over drinks that suggests new experiments, novel collaborations, and the exchange of critical reagents and techniques.

Finally, the Cold Spring Harbor meeting of 1967 and the First International Congress of Immunology stand out as markers of the institutional success of the field, each summarizing an exciting past and pointing to a more exciting future. But the Cold Spring Harbor meeting had an even greater significance; it was the first time since the turn of the century that immunology had been recognized by an "outside" organization as worthy of inclusion in its prestigious meeting and publication series.

Comment

The period 1951–1972 was a time during which a significant cognitive shift occurred in immunology, from chemical to biomedical concerns, and with parallel changes in the research programs involved. The material for the analysis is represented by the subject matter and lists of participants collected from eighty-eight published international meetings on various topics in immunology held during those critical times. The major meeting-goers in the field were identified, and the interrelationships among the various meetings and topics were established by cluster analysis of overlapping attendance.

Like citation analysis,[40] this new method of disciplinary analysis is useful in identifying the scientific leaders in a discipline, and even in the several subdisciplines attached to the mainstream core. However, the approach possesses several additional strengths. It also identifies a significant number of discipline-builders and institutionalizers – those administrators and meeting organizers whose scientific contributions may have been less important to the field than their organizational efforts; further, it adjusts for those famous scientists from other disciplines who may have touched immunology briefly, but whose citations from work in the other discipline might have made them appear more important in this field than is justified.

This new analytical approach appears to be especially useful in the case of emerging new disciplines or those undergoing rapid conceptual change, such as immunology in the 1950s and 1960s. At that time there were very few departments of immunology, and few clearly defined immunological journals. Much of the newer work emerged from departments of microbiology, pathology, pediatrics, medicine, and surgery, and was published in a host of journals not immediately identifiable as immunological. In such instances, literature citation analysis, co-citation analysis, etc., may not provide particularly useful data for the study of discipline formation, but meetings do – particularly international meetings, such as many of those analyzed here. They were often undertaken as conscious attempts to institutionalize disciplinary transitions and new areas of immunology, and more clearly represent the conceptual movement of the field and its changing research programs and changing leadership. Thus the cognitive movement of the field may be defined in terms of the attendance "votes" of the disciplinary (and even subdisciplinary) elites.

One of the most useful results to emerge from this approach has been to provide quantitative data on the *inter*-relationships among the several subdisciplines within the larger science. Once a mainstream core of meetings (topics) has been identified, the statistical program can tell how far from the center a given research area is, and even when in time it joins the mainstream.

Yet another outcome of the analysis of meetings may be to show that they influence the course of the science, as well as merely mirroring it.

A few caveats should be understood in assessing the value of this approach. A number of meetings important to the period under study have been missed, most notably the unpublished series of antibody, complement, and delayed hypersensitivity workshops, as well as the several Gordon Conferences on immunology. Further, the *quality* of participation at a meeting by individuals has not been defined. This may tell much about their standing in the community; obviously, there are recognizable differences between invited keynote and minisymposium speakers, workshop and session chairpersons, and self-volunteered presenters or passive attendees. Similarly, there are hierarchical differences between internationally recognized invitees to small elite meetings and the "junior" participants at larger ones. It is probable, however, that the "disciplinary elite" and "core elite" that are identified in Table 18.2 fairly represent that type of high quality participation.

Notes and references

1. This chapter is the distillation of two earlier studies outlining the usefulness of a prosopographical analysis of the participants at international meetings within the context of the discipline of immunology: Söderqvist, T., and Silverstein, A.M., *Scientometrics* 30:243, 1994; and Söderqvist, T., and Silverstein, A.M., *Social Studies in Science*, 24:513, 1994.
2. Landy, M., in Lawrence, H.S., and Landy, M., eds, *Mediators of Cellular Immunity*, New York, Academic Press, 1969, p. xiii.
3. Note how many clinical and basic science disciplines have spawned subgroups employing the prefix "immuno-" in their names. See also Corbellini, G., *L'evoluzione del Pensiero Immunologico*, Torino, Bollati Boringhieri, 1990, and Moulin, A.-M., *Le Dernier Langage de la Médicine: Histoire de l'Immunologie de Pasteur au Sida*, Paris, Presses Universitaire, 1991.
4. Chernyak, L., and Tauber, A.I., "The idea of immunity: Metchnikoff's metaphysics and science," *J. Hist. Biol.* 23:187, 1990. See also Tauber, A.I., and Chernyak, L., *Metchnikoff and the Origins of Immunology*, New York, Oxford University Press, 1991.
5. See Chapter 14; also Cambrosio, A., Jacobi, D., and Keating, P., *Isis* 84:662, 1983.
6. Fleck, L., *Genesis and Development of a Scientific Fact*, Chicago, University of Chicago Press, 1979 (originally published in German in 1935); Latour, B., and Woolgar, S., *Laboratory Life: The Construction of Scientific Facts*, Princeton, Princeton University Press, 1986.
7. See, for example, the influence of Franco-German enmity on the debate between cellular and humoral immunologists in Chapter 2. Again, the happy offer of a position at the Rockefeller Institute took Karl Landsteiner from a fruitless post-World War I job as a hospital autopsy pathologist in Holland to "write the book" on serological specificity. The preoccupation with health and with "wars" against various diseases after World War II led to a massive infusion of monies for research into various National Institutions and an explosion of progress in many biomedical disciplines.

8. The immunologists at the Walter and Eliza Hall Institute in Melbourne have been studied by Charlesworth, M.J., *Life among the Scientists: An Anthropological Study*, New York, Oxford University Press, 1989. See also Latour and Woolgar, note 6.

9. Söderqvist, T., *The Ecologists: From Merry Naturalists to Saviours of the Nation*, Stockholm, Almqvist and Wicksell, 1986.

10. See, for example, Lemaine, G., Macleod, R., Mulkay, M., and Weingard, P., eds, *Perspectives on the Emergence of Scientific Disciplines*, Chicago, Aldine, 1976; Darden, L., and Maull, N., *Phil. Sci.* 44:43, 1979; Hull, D., *Phil. Sci.* 45:335, 1978; Timothy Lenoir, "The discipline of nature and the nature of discipline," in Messer-Davidow, E., and Shumway, D., eds, *Knowledges: Historical and Critical Studies in Disciplinarity*, Charlottesville, University of Virginia Press, 1993.

11. For example, the number of international scientific congresses increased from one or two per year in the 1850s to some thirty per year by the end of the century (*Les Congrès Internationaux de 1681 á 1899: Liste Complète*, Paris, Union des Associations Internationales, 1960) quoted from Schroeder-Gudehus, B., in Spiegel-Rösing, I., and de Solla Price, D., eds, *Science, Technology and Society*, Sage, 1977, Ch. 13.

12. For example, the Lincei (1603) and Cimento (1657) in Italy; the Royal Society (1662) in England; the Academia Naturae Curiosi (1662) in Germany; and the Académie des Sciences (1666) in France.

13. Orr, R.H., Coyl, E.B., and Leeds, A.A., "Trends in oral communication among biomedical scientists: meetings and travel", *Federation Proc.* 23:1146–1154, 1964; Garvey, W.D., Lin, N., Nelson, C.E., and Tomita, K., "Research studies in patterns of scientific communication: II, The role of the national meeting in scientific and technical communication," in Garvey, W.D., ed., *Communication: The Essence of Science*, Oxford, Pergamon, 1979, pp. 184–201. Warren Hagstrom discusses the function of meetings for informal scientific recognition, and supports the view, also frequently made by scientists, that the major reason scientists go to meetings is to meet colleagues face-to-face for informal recognition (Hagstrom, W.O., *The Scientific Community*, New York: Basic Books, 1965, pp. 29–33).

14. Howard-Jones, N., *The Scientific Background of the International Sanitary Conferences 1851–1938*, Geneva, World Health Organization, 1975; Brobeck, J., Reynolds, O., and Appel, T., *History of the American Physiological Society: The First Century*, Bethesda, American Physiological Society, 1987, pp. 315–332; Woodford, F.P., *The Ciba Foundation: An Analytical History 1949–1974*, Amsterdam, Elsevier, 1974; Paterson, E.M., *History of the National Conference of Tuberculosis Workers 1909–1955*, New York, National Tuberculosis Association, 1956.

15. As mentioned above, the American society was founded in 1913. The British society was founded in 1959. In the late 1960s and 1970s a flood of new national societies were formed. The International Union of Immunological Societies was formed at a meeting in 1969 of representatives from eleven national immunological societies, to coordinate all international activities and to sponsor the International Congresses of Immunology (minutes from meeting of the IUIS, 5 May 1969); for a historical sketch of the IUIS, see Cinader, B., *Immunol. Today* 13, 323, 1992.

16. From fifty-eight questionnaires sent out, forty responses were received!

17. For example, the Antibody Workshops were initiated in the late 1950s (for a historical sketch, see Porter, R.R., "Antibody structure and the antibody workshops 1958–1965," *Perspect. Biol. Med.* 29:161, 1986); the Delayed Hypersensitivity Workshops were organized in the early 1960s by Merrill Chase; the

Complement Workshops met first in the late 1960s; and the Allergy Round Tables had an even longer history (recounted by Tuft, L., *NER Allergy Proc.* 6:279, 1985). Those Gordon Conferences devoted to immunology fulfilled both research and educational functions, but were excluded on the same grounds.

18. For example, several Summer Schools were organized by Alec Sehon in Canada starting in 1966; The American Association of Immunologists has run a series of Summer Schools since 1966; other National Societies have organized similar programs; and the Federation of American Societies for Experimental Biology (FASEB) has devoted a number of its summer courses to immunological topics.

19. Thus, such meetings as "Forensic Immunology, Medicine, Pathology, and Toxicology" (London, 1963) and "Immunity in Viral and Rickettsial Diseases" (Israel, 1972) were excluded.

20. Examples of such meetings excluded from the study include: US Public Health Service Immunization Conferences devoted to governmental immunization programs; "Forensic Immunology," "Immunology of the Liver," and "Immunoassay of Hormones," etc., in which immunological problems did not figure prominently; and "Immunity in Viral and Rickettsial Diseases," devoted primarily to the pathogens themselves. In addition, we have restricted our analysis of allergology meetings to the series of Collegium Internationale Allergologicum; exclusion of the seven congresses of the International Association of Allergology 1951–1970 should not change the result of the analysis significantly.

21. Much of the early history of histocompatibility antigens and of the workshop series may be found in Teresaki, P.I., ed., *History of HLA: 10 Recollections*, Los Angeles, UCLA Press, 1990, and Amos, D.B., "Fundamental antigens of HLA," *Human Immunol.* 30:236, 1991.

22. For a history of the Ciba Foundation meetings and their stated purpose, see Woodford, note 14.

23. In the late 1950s and early 1960s, the Annual Meeting of the American Association of Immunologists sat in single session over four to five days; one heard everything going on in the field and had the impression that one could comprehend the entire scope of the science. This is now virtually impossible; the programs of the Annual Meetings of National Societies and of the International Congresses of Immunology may list multiple grand- and mini-symposia, numerous special lectures, and often well over 100 workshop/poster sessions, all over a period of perhaps one week. In addition, numerous satellite symposia on specific subdisciplinary topics are held apart from, but in conjunction with, the Congresses.

24. Maurice Landy (personal communication, 1992), organizer of the Brook Lodge meetings, has emphasized the importance of the small informal discussion meeting of international leaders in helping to clarify the issues and indicate the new research directions of emerging areas of interest.

25. Fye lists attendance at the regular meetings of the American Physiological Society, 1887–1899, in absolute numbers and percentage of members only (Fye, W.B., *The Development of American Physiology: Scientific Medicine in the Nineteenth Century*, Baltimore, Johns Hopkins University Press, 1987, Appendix 3). Jackson and Saunders, in their chapter on meetings in the history of the American Physiological Society, list attendance by numbers and as the percentage of total membership numbers; and distribution by specialty of volunteered papers, and by mode of delivery (Jackson J.M., and Saunders, J.F., "Spring and Fall Scientific Meetings," in Brobeck, Reynolds, and Appel, note 14, pp. 315–332.

26. Usually proceedings list "contributors" – i.e., persons who both attended and gave a paper at the meeting. Sometimes the proceedings list all persons attending the meeting. In a few cases, however, the proceedings list all authors of papers, without giving any information on whether only one or all of the authors actually participated in the meeting. In these cases, all authors are included into the master matrix as "participants." The result is a slight over-representation of individuals who may not have physically attended the meeting.

27. Lenoir's distinction (note 10) between founders of cognitive research programs vs founders of disciplines (discipline builders, entrepreneurs, and scientific gatekeepers) has proved useful for the present purpose.

28. The names of a few prominent immunologists are missing from this list. Some (such as Nobelists Niels K. Jerne and Macfarlane Burnet) were not avid meeting-goers; some (such as Frank Putnam, Michael Sela, and Nobelists George Snell, Rodney Porter, and Gerald Edelman) may have worked in areas somewhat apart from the developing biomedical mainstream; others may have attained prominence in the field only after the period under consideration. Other noted immunologists (for example, David Talmage, Albert Coons, Olli Mäkelä, Herman Eisen, Hugh McDevitt, Manfred Mayer, and Alfred Nisonoff, among others) are missing from the list for reasons unknown to us.

29. Garfield, E., "Citation indexes for science," *Science* **122**:108, 1955; Garfield, E., Sher, L.H., and Torpie, R.J., *The Use of Citation Data in Writing the History of Science*, Philadelphia, Institute of Scientific Information, 1964. See also Price, D.J. de S., *Science Since Babylon*, New Haven, Yale University Press, 1961; Price, D.J. de S., *Little Science, Big Science*, New York, Columbia University Press, 1963.

30. For a general introduction, see, Gordon, A.D., *Classification*, London, Chapman and Hall, 1981, particularly Chapters 2 and 3. A more advanced introduction can be found in Kaufman L., and Rousseeuw, P.J., *Finding Groups in Data; An Introduction to Cluster Analysis*, New York, John Wiley & Sons, 1990.

31. Small, H.G., "A co-citation model of a scientific specialty; a longitudinal study of collagen research," *Social Studies Sci.* **7**:139, 1977.

32. Cf. Gordon, note 30, Chapter 2. Lynn Gale, of the Center for Advanced Study in the Behavioral Sciences, Stanford University, generously performed the cluster analysis for this study.

33. Different cluster procedures can be used; only the Average Linkage Cluster procedure gave both good resolution and non-arbitrary clustering.

34. The Prague meetings, like some others, were celebrated as social as well as scientific events; one was disinclined to reject an invitation, and might forego some other meeting in favor of these, if financially limited.

35. In fact, many of these allergology meetings did invite one or two "mainstream scientists," presumably for the education of their membership.

36. Sheldon Cohen, personal communication, November 1991.

37. Indeed, Peter Medawar and his colleagues, working in a zoology department, did not think of themselves as immunologists until the late 1950s or early 1960s (personal communications from Rupert Billingham, 1991, and Leslie Brent, 1992).

38. The data for the 1954 meeting have not been available.

39. Silverstein, A.M., "Immunology in Prague: critical contributions to a biological revolution," *Folia Microbiol.* **30**:179, 1985.

40. See, for example, Garfield and Garfield, Sher, and Torpie, note 29.

Appendix 18.1
Immunological meetings, 1951–1972

Meeting notations are provided for ease of identification in Figure 18.1. The name refers to the meeting series; the first number gives the position in the series, and the suffix number the meeting year.

Allergy: Collegium Internationale Allergolicum series
Brook Lodge: Brook Lodge series
Buffalo: Buffalo Convocations
Ciba Found: Ciba Foundation series
Cold Spring: Cold Spring Harbor series
Germinal: Germinal Center series
Histocomp: Histocompatibility Workshops
Immunopath: Immunopathology Symposia
Int Congr: International Congress of Immunology
Leukocyte: Leukocyte Culture series
Prague: Prague series
Sanibel: Sanibel Island series
Singular: Singular meetings outside series
Transplant: New York Academy of Science transplantation meetings and its successor, the
    International Congresses of the Transplantation Society

1951

Singular 1-51: *The Nature and Significance of the Antibody Response* (New York),
    A. M. Pappenheimer, ed., New York, Columbia Univ. Press, 1953.

1953

Ciba Found 1-53: *Preservation and Transplantation of Normal Tissues* (London),
    G.E.W. Wolstenholme and M. P. Cameron, eds., Boston, Little Brown, 1954.

1954

Transplant 1-54: 'The relation of immunology to tissue homotransplantation' (NY Acad.
    Sci.), *Ann. NY Acad. Sci.* 59:277–466, 1955.
Allergy 1-54: 'Aspects of allergy research' (London), *Int. Arch. Allergy Appl. Immunol.*
    6, 193, 1955.

1955

Allergy 2-55: 'Migraine and vascular allergy' (Basel), *Int. Arch. Allergy Appl. Immunol.*
    7:193, 1955.

1956

Transplant 2-56: 'Second tissue homotransplantation conference' (NY Acad. Science), *Ann. NY Acad. Sci.* **64**:735–1073, 1957.

Singular 2-56: *Immunology and Development* (Bar Harbor), M.V. Edds, ed., Chicago, Univ. Chicago Press, 1956.

1957

Allergy 3-57: 'Third Symposium, Collegium Internationale Allergolicum' (London), *Int. Arch. Allergy Appl. Immunol.* **11**:1, 1957.

Singular 3-57: *Allergic Encephalomyelitis* (Bethesda), M. W. Kies and E. C. Alvord, eds., Springfield, Charles C. Thomas, 1959.

1958

Singular 4-58: *Cellular and Humoral Aspects of Hypersensitivity States* (New York), H. S. Lawrence, ed., New York, Hoeber-Harper, 1959.

Singular 5-58: *Mechanisms of Hypersensitivity* (Detroit, Henry Ford Hosp. Int. Symp.), J. Shaffer, G. A. LoGrippo, and M. W. Chase, eds., Boston, Little Brown, 1959.

Transplant 3-58: 'Third tissue homotransplantation conference' (NY Acad. Sci.), *Ann. NY Acad. Sci.* **73**:1, 1958.

Immunopath 1-58: *Immunopathology-Immunopathologie* (Seelisberg), P. Grabar and P. Miescher, eds., Basel. Benno Schwabe, 1959.

1959

Allergy 4-59: 'Fourth Symposium, Collegium Internationale Allergolicum' (Rome), *Int. Arch. Allergy Appl. Immunol.* **18**:1–236, 1961.

Prague 2-59: *Mechanisms of Antibody Formation* (Prague), M. Holub and L. Jarošková, eds., Prague, Czechoslovak Academy of Sciences, 1960.

Ciba Found 2-59: *Cellular Aspects of Immunity* (Royaumont), G.E.W. Wolstenholme and M. O'Connor, eds., Boston, Little Brown, 1960.

1960

Transplant 4-60: 'Fourth tissue homotransplantation conference' (NY Acad. Sci.), *Ann. NY Acad. Sci.* **87**:1, 1960.

Singular 6-60: *Immunochemical Approaches to Problems in Microbiology* (Rutgers, NJ), M. Heidelberger and O. Plescia, eds., New Brunswick, Institute of Microbiology, 1961.

1961

Singular 7-61: *International Symposium on Tissue Transplantation* (Santiago, Chile), A.P. Christoffanini and G. Hoecker, eds., Santiago, Universidad de Chile, 1962.

Prague 3-61: *Mechanisms of Immunological Tolerance* (Liblice), M. Hašek, A. Lengerová, and M. Vojtísková, eds., Prague, Czech Academy of Sciences, 1962.

Immunopath 2-61: *Mechanism of Cell and Tissue Damage Produced by Immune Reactions*, (Brook Lodge), P. Grabar and P. Miescher, eds., New York, Grune & Stratton, 1962.

1962

Allergy 5-62: 'Fifth Symposium, Collegium Internationale Allergolicum' (Freiburg), *Int. Arch. Allergy Appl. Immunol.*, **22**:69, 1963.

Transplant 5-62: 'Fifth tissue homotransplantation conference' (NY Acad. Sci.), *Ann. NY Acad. Sci.* **99**:335, 1962.

Singular 8-62: *The Thymus in Immunobiology* (Minneapolis), R. A. Good and A. E. Gabrielson, eds., New York, Hoeber-Harper, 1964.

1963

Singular 9-63: *Cell Bound Antibodies* (Washington, DC), B. Amos and H. Koprowski, eds, Philadelphia, Wistar Institute Press, 1963.

Ciba Found 4-63: *The Immunologically Competent Cell* (London), G.E.W. Wolstenholme and J. Knight, eds., Little Brown, Boston, 1963.

Immunopath 3-63: *Immunopathology* (La Jolla), P. Grabar and P. A. Miescher, eds., Basel, Benno Schwabe, 1963.

1964

Prague 4-64: *Molecular and Cellular Basis of Antibody Formation* (Prague), J. Šterzl, ed., Prague, Czechoslovak Academy of Sciences, 1965.

Singular 10-64: *The Thymus* (Philadelphia), V. Defendi and D. Metcalf, eds., Philadelphia, Wistar Inst. Press, 1964.

Singular 11-64: 'Autoimmunity: experimental and clinical aspects' (NY Acad. Sci.), *Ann. NY Acad. Sci.* **124**:1, 1965.

Transplant 6-64: 'Sixth International Transplantation Conference' (NY Acad. Sci.), *Ann. NY Acad. Sci.* **120**:1, 1964.

Allergy 6-64: 'Sixth Symposium, Collegium Internationale Allergolicum' (London), *Int. Arch. Allergy Appl. Immunol.* **28**:1, 1964.

Histocomp 1-64: *Histocompatibility Testing, 1964* (Durham, NC), D. B. Amos, ed., Publication No. 1229, National Academy of Sciences, NRC, Washington, DC.

1965

Sanibel 1-65: *Phylogeny of Immunity* (Sanibel Island, Florida), R. T. Smith, P. A. Miescher, and R. A. Good, eds., Gainesville, Univ. Florida Press, 1966.

Histocomp 2-65: *Histocompatibility Testing, 1965* (Leiden), J.J. van Rood, ed., Copenhagen, Munksgaard, 1966.

Ciba Found 5-65: *Complement* (London), G.E.W. Wolstenholme and J. Knight, eds., Boston, Little Brown, 1965.

Ciba Found 6-65: *The Thymus; Experimental and Clinical Studies* (London), G.E.W. Wolstenholme and R. Porter, eds., Boston, Little Brown, 1966.

Immunopath 4-65: *Immunopathology* (Monte Carlo), P. Grabar, ed., New York, Grune & Stratton, 1965.

Leukocyte 1-65: *Leukocyte Culture Workshop* (Washington, DC), (Abstracts unpublished: bound vol. at National Library of Medicine, Bethesda, Maryland).

1966

Transplant 7-66: 'Seventh International Transplantation Conference' (NY Acad. Sci.), *Ann. NY Acad. Sci.* **129**:884, 1966.

Germinal 1-66: *Germinal Centers in Immune Responses* (Bern), H. Cottier, N. Odartchenko, R. Schindler, and C. C. Congden, eds., New York, Springer, 1967.

Sanibel 2-66: *Ontogeny of Immunity* (Sanibel Island, Florida), R. T. Smith, R. A. Good, and P. A. Miescher, eds., Jacksonville, Univ. of Florida Press, 1967.

Singular 12-66: *Regulation of the Antibody Response* (Toronto), B. Cinader, ed., 2nd edn, Springfield, Thomas, 1971.

1967

Singular 13-67: *Gamma Globulins – Structure and Control of Biosynthesis* (Stockholm), J. Killander, ed., New York, Interscience, 1967.

Singular 14-67: *Nucleic Acids in Immunology* (New Brunswick), O.J. Plescia and W. Braun, eds., New York, Springer, 1968.

Singular 15-67: *Differentiation and Immunology*, K.B. Warren, ed., New York, Academic Press, 1968.

Leukocyte 3-67: *Proceedings of the Third Leukocyte Culture Conference* (Iowa City), W. O. Rieke, ed., New York, Appleton-Century Crofts, 1969.

Sanibel 3-67: *Immunologic Deficiency Diseases of Man* (Sanibel Island, Florida), D. Bergsma, ed., New York, The National Foundation, 1967.

Histocomp 3-67: *Histocompatibility Testing, 1967* (Turin), R. Ceppellini, ed., Baltimore, Williams & Wilkins, and Copenhagen, Munksgaard, 1968.

Immunopath 5-67: *Immunopathology* (Punta Ala), P. A. Miescher and P. Grabar, eds., New York, Grune & Stratton, 1967.

Transplant 8-67: *Advance in Transplantation* (First International Congress of the Transplantation Society, Paris), J. Dausset, J. Hamburger, and G. Mathé, eds., Baltimore, Williams & Wilkins. 1968.

Cold Spring 1-67: *Antibodies* (New York), Cold Spring Harbor Laboratories, 1967.

1968

Buffalo 1-68: *1st International Convocation on Immunology* (Buffalo), N. R. Rose and
F. Milgrom, eds., Basel, Karger, 1969.
Germinal 2-68: *Lymphatic Tissue and Germinal Centers in Immune Responses* (Padua),
L. Fiore-Donati and M. G. Hanna, eds., New York, Plenum, 1969.
Transplant 9-68: 'Proceedings of the Second International Congress of the Trans-
plantation Society' (New York), *Transplantation Proc.* 1:1, 1969.
Allergy 7-68: 'Seventh Symposium, Collegium Internationale Allergolicum' *Int. Arch.
Allergy Appl. Immunol.* 36:1, 1969.
Brook Lodge 1-68: *Immunological Tolerance* (Brook Lodge), M. Landy and W. Braun,
eds. New York, Academic Press, 1969.
Singular 16-68: *The Immune Response and its Suppression* (Davos), E. Sorkin, ed.,
New York, Karger, 1969.
Singular 17-68: *Current Problems in Immunology* (Grosse Ledder), O. Westphal,
H.E. Bock, and E. Grundmann, eds., New York, Springer, 1969.
Singular 18-68: *Organ Transplantation Today* (Amsterdam), N.A. Mitchison, J.M.
Greep, and J.C.M. Hattinga-Verschure, eds., Baltimore, Williams & Wilkins, 1969.

1969

Singular 19-69: *Human Anti-human Gammaglobulins: Their Specificity and Function*
(Lund), E. Grubb and G. Samuelsson, eds., New York, Pergamon, 1971.
Buffalo 2-69: *Cellular Interactions in the Immune Response* (Buffalo), S. Cohen,
G. Cudkowicz, and R. T. McCluskey, eds., Basel, Karger, 1971.
Brook Lodge 2-69: *Mediators of Cellular Immunity* (Brook Lodge) H. S. Lawrence and
M. Landy, eds., New York, Academic Press, 1969.
Sanibel 4-69: *The Secretory Immunologic System* (Vero Beach), D. H. Dayton, P. A.
Small, R. M. Chanock, H. E. Kaufman, and T. B. Tomasi, eds., Washington, DC,
US Government Printing Office, 1970.
Prague 5-69: *Developmental Aspects of Antibody Formation and Structure* (Prague),
1969, J. Šterzl and I. Říha, eds., Prague, Academia, 1970.
Leukocyte 4-69: *Proceedings of Fourth Annual Leukocyte Culture Conference* (Hanover,
N.H.), O. R. McIntyre, ed., New York, Appleton Century Crofts, 1971.

1970

Leukocyte 5-70: *Proceedings of the Fifth Annual Leukocyte Culture Conference*
(Ottowa), J. E. Harris, ed., New York, Academic Press, 1970.
Immunopath 6-70: *Immunopathology* (Grindelwald), P. A. Miescher, ed., New York,
Grune & Stratton, 1971.
Prague 6-70: *Immunogenetics of the H-2 System* (Liblice), A. Lengerová and
M. Vojtísková, eds., Basel, Karger, 1971.
Histocomp 4-70: *Histocompatibility Testing, 1970* (Los Angeles), P. Teresaki, ed.,
Baltimore, Williams & Wilkins and Copenhagen, Munksgaard, 1971.

Transplant 10-70: 'Proceedings of the Third International Congress of the Transplantation Society' (The Hague), *Transplantation Proc.*, 3:1, 1971.
Allergy 8-70: 'Eighth Symposium, Collegium Internationale Allergolicum' (Montreux), *Int. Arch. Allergy Appl. Immunol.* 41:1–236, 1970.
Singular 20-70: *Cell Interactions and Receptor Antibodies in Immune Responses* (Helsinki), O. Mäkelä, A. Cross, and T. U. Kosunen, eds., New York, Academic Press, 1971.
Singular 21-70: *Role of Lymphocytes and Macrophages in the Immune Response* (Munich), D.C. Dumonde, ed., Berlin, Springer, 1971.
Germinal 3-70: *Morphologic and Functional Aspects of Immunity* (Uppsala), K. Lindahl-Kiessling, G. Alm, and M. G. Hanna, eds., New York, Plenum, 1971.
Brook Lodge 3-70: *Immune Surveillance* (Brook Lodge), R. T. Smith and M. Landy, eds., New York, Academic Press, 1970.

1971

Int Congr 1-71: *Progress in Immunology* (Washington, DC), D. B. Amos, ed., New York, Academic Press, 1971.
Brook Lodge 4-71: *Immunologic Intervention* (Brook Lodge), J. W. Uhr and M. Landy, eds., New York, Academic Press, 1971.
Ciba Found 7-71: *Ontogeny of Acquired Immunity* (London), R. Porter and J. Knight, eds., New York, Elsevier, 1972.
Leukocyte 6-71: *Proceedings of the Sixth Leukocyte Culture Conference* (San Juan Islands, Washington), M. R. Schwartz, ed., New York, Academic Press, 1972.

1972

Leukocyte 7-72: *Proceedings of the Seventh Leukocyte Culture Conference* (Quebec), F. Daguillard, ed., New York, Academic Press, 1973.
Ciba Found 8-72: *Corneal Graft Failure* (London), R. Porter and J. Knight, eds., New York, Elsevier, 1973.
Brook Lodge 5-72: *Genetic Control of Immune Responsiveness* (Brook Lodge), M. Landy and H. McDevitt, eds., New York, Academic Press, 1972.
Germinal 4-72: *Microenvironmental Aspects of Immunity* (Dubrovnik), B. D. Janković and K. Janković, eds., New York, Plenum, 1973.
Buffalo 3-72: *Specific Receptors of Antibodies, Antigens, and Cells* (Buffalo), D. Pressman *et al.*, eds., Basel, Karger, 1973.
Histocomp 5-72: *Histocompatibility Testing, 1972* (Evian), J. Dausset, ed., Baltimore, Williams & Wilkins and Copenhagen, Munksgaard, 1973.
Sanibel 5-72: *The Biological Role of the Immunoglobulin E System* (Vero Beach), K. Ishizaka and D. H. Dayton, eds., Washington, DC, US Govt Printing Office 1973.
Allergy 9-72: 'Ninth Symposium, Collegium Internationale Allergolicum' (London), *Int. Arch. Allergy Appl. Immunol.* 45:1–329, 1973.
Transplant 11-72: 'Proceedings of the Fourth International Congress of the Transplantation Society' (San Francisco), *Transplantation Proc.*, 5:1, 1973.

19 The emergence of subdisciplines

Disciplines are the infrastructure of science...

Timothy Lenoir[1]

Students of the sociology of science have long been interested in the nature of scientific disciplines, in how new disciplines are formed, and in how established disciplines differentiate to form further subspecialty areas.[2] In some instances, the intellectual and even technical content of such subspecialty areas derives from developments within the parent discipline itself – a new discovery, a new technology, or the extension of a research program across the boundary to another discipline (thus, for example, immunophysiology). In other instances, the program may be borrowed almost fully formed from the cognitive content of some other rapidly moving and highly popular discipline (for example, ocular immunology). Sometimes these subdisciplines remain within the mainstream of the parent science (molecular immunology); in others, they may assume a more-or-less independent existence (clinical allergy). Occasionally, a technical advance may seem to constitute a subdiscipline, but finds a wide diffusion among many fields. Immunohistochemistry is an excellent example of this; it would be difficult to find the field of biology or medicine in which the use of labeled antibodies as tracers has not found important application.

In this respect, the field of immunology should be of special interest, since during its 130 years it has interacted with and affected the course of many biological and medical disciplines, and spawned many new specialties and subspecialties. The manifestations of such new areas usually take the form of their own conferences, specialty textbooks, and journals, and eventually the formation of their own professional societies. One need only list the many disciplines whose names modify the word *immunology* or are prefixed by *immuno-*. Among the former we have *Cellular, Molecular, Veterinary, Pediatric, Cancer, Neuro-, Ocular* and many more. Among the latter we find *-pathology, -biology, -physiology, -genetics, -pharmacology, -histochemistry* and again many more.

We saw in Chapter 9 how the study of allergy and hypersensitivity arose within the context of immunology and then established itself as an independent clinical field. Again, Chapter 11 touched upon the coming together of immunologists with tissue transplant surgeons.[3] Similarly, Chapter 17 reviewed briefly how serological assays for antigen and antibody, developed initially within the confines of the immunological research laboratory, moved to the hospital diagnostic laboratory under the rubric "serology." And in Chapter 18 we saw how the very titles of international meetings and symposia might confirm the emergence of new subdisciplines. In this chapter we will examine several examples of subdiscipline formation in order to illustrate further the various

A History of Immunology, Second Edition
ISBN: 978-0-12-370586-0

Copyright © 2009, Elsevier Inc.
All rights reserved

ways that they develop, how these specialties may manifest themselves, and the variety of their relationships with the parent discipline of immunology.

Ocular immunology

Ophthalmology, one of the earliest of the medical specialties, was influenced directly by two of the six components of the developing research program of the young field of immunology.[4] These were applied in turn to the explanation of their own clinical disease problems. The first of these was the "doctrine" of cytotoxic antibodies. Richard Pfeiffer showed in 1894 that specific antibodies could mediate the destruction of bacteria,[5] and Jules Bordet showed in 1899[6] that antibodies could destroy (hemolyse) erythrocytes with the help of the nonspecifically-acting serum factor complement. Pandora's Box had been opened. Investigators everywhere tested extracts of other tissues and organs in an attempt to account for the tissue destruction seen in so many diseases of unknown etiology and pathogenesis.[7]

The second important component of the early immunological research program that influenced ophthalmology was that of anaphylaxis and other related disease mechanisms. In 1902, Paul Portier and Charles Richet reported that animals could be so sensitized by an antigen that a second challenge with that antigen would lead to shock-like symptoms and even death.[8] Shortly thereafter, Maurice Arthus demonstrated that bland antigens injected repeatedly into the skin could cause local necrotizing lesions.[9] Finally, in 1906, Clemens von Pirquet and Bela Schick showed that the pathogenesis of human serum sickness involves an antibody response in the host to the injection of large quantities of foreign protein antigens.[10] Anaphylaxis the word and anaphylaxis the concept became so popular that they penetrated all branches of medicine. This spread of interest in "anaphylaxis" was accelerated when it became apparent that those scourges of mankind, hayfever and asthma, were also related to similar mechanisms.[11]

Ophthalmologists very quickly seized upon the concepts and techniques of the early immunologists, and slowly ocular immunology developed as a subdiscipline within the larger field of ophthalmology. Of equal interest is the fact that while many other clinical fields of medicine attempted to integrate the turn-of-century immunological excitement into their clinical and research programs, ophthalmology shares only with hematology a continuation of that interest even during the half-century or so when mainstream immunology abandoned its biomedical concerns in favor of more parochial immunochemical approaches. The interaction of these two disciplines accelerated when the immunobiological revolution got fully under way in the early 1960s.

Autoimmune diseases

Paul Ehrlich's inability to detect damaging autoantibodies led to his famous dictum of *horror autotoxicus*[12] that seemed to rule out even the theoretical possibility of such an event. Ophthalmologists, however, appeared instead to believe the contradictory demonstration by Donath and Landsteiner in 1904 of an autoimmune hemolytic anemia.[13] They took seriously the possibility that cytotoxic antibodies might be responsible for some of the ocular diseases whose origin they could not explain.

Sympathetic ophthalmia

Ophthalmologists were not long in responding to the doctrine of cytotoxic antibodies. Several suggestions were made[14] that sympathetic ophthalmia might be caused by the formation of such antitissue antibodies. They postulated that the immune system might respond to damaged ocular tissue in the traumatized eye, and the antibodies thus produced would then attack the hitherto normal contralateral eye. It was then that the famous ophthalmologist Anton Elschnig of Prague entered the fray. In the first such collaboration, Elschnig and the prominent Prague immunologist Edmund Weil became the leading advocates of an autoimmune pathogenesis of sympathetic ophthalmia.[15] They spoke of a systemic "hypersensitivity" induced by antigens released from the wounded eye. Thus, the mildest disturbance in the second eye might lead to inflammation and blindness.

The prevailing thought for several decades was that the pathogenesis of sympathetic ophthalmia was based upon an autoimmune response to uveal pigment.[16] One of the chief obstacles to progress in understanding the etiology and pathogenesis of sympathetic ophthalmia was the absence of a satisfactory animal model of the disease. Then, Freund's adjuvant was described in 1942[17] – a technique that not only permitted the development of a promising animal model in the guinea pig,[18] but also advanced the cause of so many other autoimmune disease models.

It was shown some years later that retinal extracts are much more efficient in producing autoimmune disease than uveal extracts,[19] and this initiated the search for the organ-specific antigens involved. Three such antigens were found initially, localized by immunofluorescent analysis to the outer segments of the retina.[20] Then, simultaneously, two different laboratories isolated and identified a soluble retinal antigen (S-antigen) able to induce autoimmune disease in animals.[21] Since then, a number of other organ-specific antigens have been implicated, including an interphotoreceptor retinoid-binding protein (IRBP)[22] and even the visual pigment itself, rhodopsin.[23] By varying the dosage and sensitizing regimen, it has been possible to alter the previously-observed chronic inflammatory picture to that of a granulomatous form more typical of human sympathetic ophthalmia.[24] All of the most recent data on mechanisms of host response in this experimental model attest to its autoimmune basis and to its close relationship with human sympathetic ophthalmia.

The lens

A new chapter in the history of immunology was opened by Paul Uhlenhuth when he reported in 1903 that the antigens of the lens of the eye are organ-specific.[25] This was the first intimation that unique antigens might exist within a single organ and, further, that they might be shared among widely divergent species.[26] It was then shown that these lens antigens could mediate both active and passive anaphylaxis in test animals.[27] It was only when Krusius demonstrated in 1910 that rupture of the lens capsule in a *normal* guinea pig would both sensitize the animal and also serve as the disease-producing challenge[28] that the true implications of this system for ocular disease became apparent. In a perceptive review of the field,[29] Römer and Gebb concluded that the lens is indeed self and that it can stimulate an immune response. However, being firm adherents to the theories of Paul Ehrlich, they had perforce to conclude that the lens must somehow elude the normal workings of the host's immune response. Here was an early foretaste of the later idea of the *sequestered antigen*.

Interest in lens-induced immunogenic disease continued. The clinical condition was named *endophthalmitis phacoanaphylactica* in 1922,[30] and these investigators also reported positive skin test responses in patients with this disease, using proteins extracted from the lens.[31] Perhaps the most significant contribution to our understanding of this autoimmune process was the production of an experimental disease in rats that is histopathologically similar to that seen in the human.[32] It was also possible to transfer the disease adoptively, using immune serum. There seems to be little evidence that T cell mechanisms contribute significantly to the pathogenesis of this disease process.

Sjögren's syndrome

Sjögren's syndrome is a later addition to the list of autoimmune diseases of the eye. It was first described clinically in 1933,[33] and autoantibodies were only reported in 1958.[34] It is a disease characterized by an autoimmune attack on the lacrimal and salivary glands, involving a chronic inflammatory destruction of acinar cells and ductular epithelium.[35] This results in the vexing problem of dry eyes and a dry mouth. Recent experiments appear to confirm the existence of lacrimal gland-specific antigens, able to induce autoimmune dacryoadenitis in the rat without cross-reacting involvement of the salivary gland.[36] In these animal models of Sjögren's syndrome, T cell-mediated immunity appears to be the dominant mechanism involved.[37]

Allergic diseases

Uveitis

From the very outset, it had been demonstrated that the eye shares in the general hypersensitivity of the immunized or infected host,[38] and that such sensitization can be induced also by intraocular administration of antigen. However, it remained for Sattler[39] to show in 1909 that bland antigen introduced into the vitreous of the rabbit eye would cause, in addition, a *local ocular hypersensitivity* – an observation extended by the elegant studies of the Seegals.[40] Such a sensitized eye will respond for many months thereafter with an acute anterior uveitis when specific antigen is introduced intravenously, or even by feeding. Here was a possible animal model of human recurrent nongranulomatous anterior uveitis that stimulated later workers to investigate its pathogenesis.

Among the results of these studies was the finding that the tissues of the eye can support the local formation of antibody, much like a regional lymph node. This understanding emerged initially from studies of equine periodic ophthalmia, an ocular infection due to leptospira. It was shown by Goldmann and Witmer that so efficient is the eye in producing specific antibody that the high serum titers of antileptospiral antibodies found in these horses could have originated *solely* from local formation in the infected eyes.[41] It was suggested later that the inflammatory response accompanying intraocular antibody formation might be the primary pathogenetic contributor to recurrent anterior uveitis, due to the persistence of specific memory cells within the uveal tract of the sensitized eye.[42] In line with developments in the general immunopathology of inflammation, it has been shown that lymphokines play an active role in the mediation of uveal inflammatory responses.[43]

The cornea

It was Karl Wessely[44] who first noticed that the injection of antigen intra-stromally in the central cornea of a sensitized rabbit would produce an opaque ring of interstitial keratitis. This observation attracted much attention, especially on the part of Aurel von Szily, who devoted an entire book[45] to the study of this phenomenon. In a model collaboration between immunopathologists and ophthalmologists, Germuth, Maumenee, and coworkers,[46] employing modern techniques such as immunofluorescent histochemical staining, proved that the "Wessely ring" was in fact a true intracorneal Arthus reaction – i.e., an immune complex deposit with local fixation of complement and the release of pharma-cologic agents that attract polymorphonuclear leukocytes to the site. In a similar interdisciplinary collaboration, Waksman and Bullington had earlier demon-strated an equivalent Arthus reaction within the uveal tract.[47]

It is interesting that during those early decades when investigators puzzled over the difference between antibody-mediated "immediate" hypersensitivities and tuberculin-type "delayed" reactions, the cornea was employed to distinguish between them. Antigen injected into the center of the avascular cornea would stimulate keratitis where antibody is involved, but fail to do so in tuberculin-type hypersensitivities.

The conjunctiva

We saw earlier (Chapter 9) that the conjunctiva may become inflamed by instillation of tuberculin in the sensitized individual. With the finding that hayfever and asthma are also immunologic ("anaphylactic") reactions, it became apparent that immunogenic conjunctivitis was a more general phenomenon, and ophthalmic clinicians became more interested in its pathogenesis and treatment.[48] Perhaps the most interesting development along these lines was the suggestion that a chronic immune response to the antigens of *Chlamydia trachomatis* might account for the primary lesions seen in trachoma.[49] Trachoma, one of the major causes of blindness throughout the world, is characterized by the exuberant development of germinal centers beneath the infected conjunctival epithelium; the conjunctival sac has been compared to a lymph node cut open, where antigenic stimuli enter through an overlying epithelium.[50]

Corneal transplantation

The history of attempts to transplant cornea to restore sight is a long one,[51] but it was only around the beginning of the twentieth century that technical improvements in trephines and sutures permitted the ophthalmic surgeon to begin to realize consistent success in this venture, employing what we now call allogeneic tissue. When, as frequently happened, a graft failed, it was usually attributed to some unknown physiological factor. Despite the fact that the immunologic "laws of transplantation" were well worked out by tumor transplanters early in the twentieth century,[52] the information appears to have been unknown outside that field.

It remained for Peter Medawar to clarify the immunologic basis of allograft rejection in his elegant series of studies in the 1940s,[53] and these immediately caught the attention of transplant surgeons everywhere. It was A. E. Maumenee who was chiefly responsible for bringing these immunological findings to the attention of ophthalmologists, and for bringing to the attention of the transplant immunologists the special characteristics of the cornea that allowed keratoplasty to succeed, while most other tissue and organ grafts failed.[54] Again, the important mechanism responsible for the occasional graft rejection was shown to be the cellular immune response mediated by effector lymphocytes, and the high success rate of allokeratoplasty was shown to reside in an immunological privilege of the cornea, involving both the afferent and efferent limbs of the immune response, due to the avascularity and lack of lymphatic drainage in the cornea.[55] This view is supported by the observation that increased vascularization of the graft predisposes to rejection; the privilege residing mainly in the absence of visiting lymphocytes.

Privileged sites

The extensive studies of H. S. N. Greene on the transplantation of tumors and endocrine tissues into the anterior chamber of the eye[56] had implied that this site might enjoy a degree of immunological privilege, although this privilege appears not to be absolute.[57] We saw above that the privilege of the cornea was attributed to its avascularity and to its lack of lymphatic drainage, and this was elegantly reinforced by the demonstration by Barker and Billingham that transplants survive when grafted onto a pedicle of skin deprived of lymphatics.[58]

Perhaps related to this phenomenon are the studies of Streilein and his collaborators, who suggest that when antigens are presented to the immune system through the anterior chamber of the eye, an "anterior chamber-associated immune deviation" (ACAID) results from the induction of suppressor T cells.[59] It is possible that the slow exit of a well-filtered antigen from the eye into the circulation may mimic the ability to induce tolerance by intravascular administration of ultracentrifuged (aggregate-free) soluble proteins.

Institutionalization of the discipline

It is always difficult to assign an exact date to the formal establishment of a scientific discipline. In the case of ocular immunology/immunopathology, the components of a specific research program (involving sympathetic ophthalmia, lens-induced disease, "anaphylactic" keratitis and conjunctivitis, etc.) had already been identified by 1912. A small interacting community of clinical and laboratory researchers had thus been formed. The first special monograph devoted to ocular immunopathology appeared in 1914,[60] and other texts and monographs followed, especially after the 1960s.[61] Departments of ophthalmology throughout the world added immunological research laboratories to their facilities.[62] These ocular immunology units soon began to hire basic science faculty members – a trend that accelerated after World War II, with the expansion of interest in all biomedical specialties.[63] The establishment of the National Eye Institute at the US National Institutes of Health in Bethesda, Maryland, in 1968 testifies to the growing scientific and political strength of ophthalmic and vision research, from which research in ocular immunology also benefited immeasurably.

One of the hallmarks of disciplinary institutionalization is the development of informal and then formal networks of individuals with common scientific interests. In 1966 an Ocular Immunology and Microbiology Discussion Group was organized in San Francisco, which soon served as the model for the Section on Ocular Immunology and Microbiology when the Association for Research in Vision and Ophthalmology (ARVO) was reorganized into disciplinary sections in 1968.[64] The ARVO Section meets annually, and has grown from a few dozen persons at the start to a current membership of almost 600, more than 50 percent of whom are basic scientists.

One of the more significant organizational developments in this field was the founding in 1974 of a series of quadrennial International Symposia on the Immunology and Immunopathology of the Eye. Six such symposia were held, of which the sessions and resulting publications[65] helped to consolidate this specialized community of scientists.

Also functioning to define the scope of the field was a set of workshops sponsored by the National Eye Institute.[66] Immunology had been identified by the Eye Institute as one of the principal research areas for future exploitation, and these meetings were designed to define the current status of the field and to identify new and fruitful approaches. Finally, as the ultimate mark of rise of a scientific discipline, two journals have been established to record progress in the field.[67]

Pediatric immunology

We noted above that ocular immunology had its start in the hands of young ophthalmologists who saw in the exciting new findings of immunology the opportunity to stake out for themselves interesting and rewarding careers. In contrast to this, the first stirrings of interest in what would later become pediatric immunology took place within the very bosom of the expanding young field of immunology. It was Paul Ehrlich, working in Robert Koch's Institute in Berlin, who would undertake the first landmark experiments in this area. Curiously, the initial stimulus for these imaginative experiments[68] came less from an interest in fetal and neonatal immunology than from the desire to settle one of the many nature/nurture disputes of the day.

Paul Ehrlich's studies

It was a widespread belief, in the latter part of the nineteenth century, that disease and even acquired immunity could be transferred genetically from father to offspring via an "altered zygote." This mechanism was especially implicated in discussions of cases of congenital syphilis.[69] Indeed, Dr Arthur Conan Doyle wrote a short story in 1894 involving the transmission of this disease over three generations, from grandfather to father to son.[70] Then, following the demonstration of the efficacy of passively administered antitoxin in curing diphtheria in children, it was widely noted that the newborn was resistant to disease and might even have antitoxin in its blood. Some immediately suggested that here was further proof of the genetic transmission of immunity from parent to offspring. Paul Ehrlich, a closet Darwinist, refused to believe this Lamarckian explanation, and set out to disprove it.

Having previously demonstrated his experimental expertise in several quite elegant studies on immunity to the plant toxins ricin and abrin,[71] Ehrlich applied the same system to a convincing explanation of the phenomenon of neonatal immunity. After reviewing earlier discussions of the question, he introduced his report thus:[72]

> three different possibilities present themselves which, differing in principle, must also be separately treated. The immunity of the offspring can be effected by: 1) inheritance in the ontogenetic sense; 2) the transfer of maternal antibody; and 3) the direct intrauterine influence on the fetal tissues by the immunizing agent. ...I have been able to succeed in finding a simple research plan which made it possible to establish in each instance the mechanism of inherited immunity.

Passive transfer of maternal antibody

Ehrlich published the first results of these studies in mice in 1892 in a paper entitled "On immunity by inheritance and suckling." The elegance of this "simple research plan" will quickly become apparent. Ehrlich first tested the offspring of immune fathers and normal mothers, and found that they showed no protection. By contrast, the offspring of normal fathers and immune mothers were almost routinely protected. In addition, he determined the duration of this immunity; the newborn mice enjoyed almost complete protection from these toxins for about the first six weeks postpartum. The immunity then waned over the succeeding few weeks, and had completely disappeared by the third month of life. He could conclude, therefore, that:

> We can presently distinguish two types of immunity, the first of which may be termed active and the second passive. ...It is not to be doubted that the immunity that we have observed in the offspring of immune mothers...depends on the transfer of maternal antibody.

Using now the offspring of non-immune parents derived from ricin- or abrin-immune grandparents, Ehrlich showed that no immunity to these toxins was transmitted to the second generation. But Ehrlich did not yet know how and when antibody is transferred from mother to offspring. After determining that the newborn of an immune mother has protection at birth, he could conclude that there must be passive, transplacental transfer of maternal antibody *in utero*. But he went further, and showed that the neonate of a normal mother given to suckle to an immune foster mother would receive protection, whereas the neonate of an immune mother given to suckle to a normal foster mother would lose its protection rapidly. Ehrlich could now conclude "that the young come into the world endowed with maternal antibody. ...On the other hand...my experiments show with every certainty that milk...supplies antibody to the suckling young."

Ehrlich then raised the question of the origin of milk antibodies. The solution to this was extraordinarily simple. He transferred horse anti-tetanus serum passively to a nonimmune nursing mother, and demonstrated the appearance of complete immunity in the suckling young within the next twenty-four hours.

Ehrlich next discussed the curious and hitherto unknown fact that the antibodies in question appear to pass unchanged through the intestinal wall of the newborn into its circulation. As he put it, "More wonderful, however, is the fact...." Then, always interested in the practical, Ehrlich began a long discussion of the clinical implications of his findings, and concluded, "Thus, mothers milk is the most ideal food for the newborn." In a later paper with Hübener, Ehrlich would extend these same findings to the tetanus system, using suckling guinea pigs as well as mice.[73]

Curiously, these studies of the transmission of immunity to the young constitute almost the only instance in which Ehrlich's scientific activities failed to exert a lasting effect on a biomedical science. The work appears to have been substantially forgotten,[74] and the burst of research activity on maternal-fetal/newborn immunological relationships in the period after World War II received its impetus from other sources. Only then did the pediatric community begin to take a serious interest in the expanding implications of the new immunobiology for their discipline.

Pediatricians take notice

During the first half of the twentieth century, as the practice of preventive vaccinations spread – especially against the dreaded childhood disease diphtheria – it was reported that the immune response of neonates and young children is deficient; they demonstrate little or no "natural" antibody, and respond poorly if at all to diphtheria toxoid injections.[75] Given the demonstrable immaturity of the mammalian fetus and neonate in so many other areas of physiology and behavior, the finding of an immaturity in the immune response was not unexpected.

This latter view was given a theoretical underpinning by Macfarlane Burnet in the second edition of his *The Production of Antibodies*.[76] In explanation of the absence of rampant autoimmune disease in the vertebrate host, a mechanism seemed to be required to permit "self–nonself discrimination" to take place during the maturation of the immune response. This seemed to require a lengthy period of nonreactivity during which the native antigens of the fetus would acquire "self-markers" to exempt them from attack, resulting in what would later be called "immunological tolerance." Burnet's later clonal selection theory,[77] and its concept of "clonal deletion" during a putative null period during intrauterine life, reinforced the view that the mammalian fetus and neonate are unable to respond to any antigenic stimulus.

In the years following World War II, information on the timing of maturation of the immune response saw conflicting data appear. Those interested in pediatric vaccines continued to report poor (if any) response to the standard immunizations.[78] This appeared to be supported by observations suggesting that human fetuses have immature lymphoid development and do not form plasma cells.[79] But these were apparently normal and unstimulated fetuses. When, however, the placental protection is compromised by passage of such infectious agents as *Treponema pallidum* or *Toxoplasma gondii*, then the lesions produced by these congenital infections may be characterized by abundant plasma cell development and antibody formation as early as mid-gestation.[80]

The principal observations that attracted the interest of the pediatric profession to things immunological, though, were reports of the existence in young children of immunological deficiency diseases. The first such report to attract attention was that of military pediatrician Colonel Ogden Bruton, who reported a case of agammaglobulinemia in 1952, involving an inability to form antibodies.[81] This called the attention of the field to an earlier report by Glanzmann and Riniker[82] of an apparent immunodeficiency disease characterized by lymphopenia and susceptibility to fatal infectious processes. Then came the description of a primary immunodeficiency disease that would prove to involve both the humoral and the cellular arms of the immune response.[83] This was initially called "Swiss-type agammaglobulinemia," but when its full nature became apparent it was renamed "severe combined immunodeficiency (SCID)" by a World Health Organization committee. In time SCID proved to be a heterogeneous group of genetic diseases, including an X-linked form,[84] an autosomal recessive form,[85] and the adenosine deaminase-deficient form.[86] The early history of these deficiency diseases is well recounted by Rosen.[87] If the Bruton case involved only the humoral immune system, and the Swiss type (SCID) involved both the humoral and cellular components, then it remained for DiGeorge to complete the picture with a report of an immunodeficiency that involved only a deficit in the T cell function.[88]

A second, related discovery during this same period proved to be equally significant to the field of pediatrics. This was the discovery by Miller and others of the immunological significance of the thymus.[89] This organ had always been a mystery to pediatricians; it is very prominent in early life, and then involutes after subserving its heretofore mysterious function. Now that function would be made clear – to mediate the maturation of the T cell components of the immune response.

It was all of these reports that stimulated pediatrician Robert A. Good to engage himself in both the clinical and the basic science aspects of pediatric immunology. He was soon to become one of the foremost members of this expanding group from his position at the University of Minnesota. There, with his students – most notably Max Cooper – Good contributed importantly to defining the research program of this expanding subdiscipline. (Good's students would eventually number some 300!) The early interests of the group included studies on the function of the thymus, the division of labor between thymus and avian bursa of Fabricius/mammalian bone marrow in forming what would later be called the T cell and B cell systems, and in demonstrating the immunological implications of several additional genetic syndromes.[90] Equally important was the group in Boston, led by Charles Janeway and his students David Gitlin, Fred Rosen, Ralph Wedgwood, and Walter Hitzig. These investigators made many significant contributions to our understanding of immunologic deficiency diseases. It is safe to say that the field of pediatric immunology would scarcely justify the name, were it not for the scientific "children" and "grandchildren" of Good and Janeway.

Finally, there is another congenital disease of pediatric interest; its description and the clarification of its etiology and pathogenesis actually preceded those of the other syndromes described above. It is erythroblastosis fetalis, an intra-uterine hemolytic disease characterized at birth by the appearance of anemia with many immature red blood cells. It was only with the discovery of the Rh factor and its specific hemagglutinin by Landsteiner and Wiener[91] that the process could be explained.[92] An Rh− mother carrying a Rh+ fetus may respond, especially after multiple pregnancies, to the transplacental passage of erythrocytes with the production of specific antibodies. These antibodies, crossing the placenta back to the fetal circulation, will destroy the fetal red cells, thus causing an immunogenic hemolytic anemia.

Immunologists take notice

The immunological community was as quick to respond to the exciting findings of the 1940s and 1950s as were the pediatricians. Indeed, it was often difficult to determine the primary disciplinary allegiance of some of the players in the field. Burnet's concept of immunologic tolerance would soon be confirmed experimentally by the newborn-mouse experiments of Medawar's group,[93] who still called themselves zoologists in 1953! Expanding work on the implications of neonatal thymectomy and bursectomy contributed importantly to the elaboration of the dual workings of the immune system – the T and B cell responses. They would also shed light on the mechanisms by which tolerance is attained, including the many steps in the developmental pathways of the several lymphocyte subsets.[94]

With the growing appreciation of the fact that clonal selection is based upon the DNA encoding for the large repertoire of specific responses, many studies of fetal and neonatal immune responses were undertaken to establish the nature of the early response.[95] It was found that the young of many species, and even fetuses *in utero*, are able to mount an effective immune response; that there is a seemingly programmed stepwise maturation of these responses; and that the early responses are polyclonal, consisting of cross-reacting specificities worthy of the adult. Each of these results would have an important bearing on the development of the clinical subdiscipline. The excitement that these observations generated helped to attract young scientists into the field, while the results profoundly affected the approaches to diagnosis and therapy in the pediatric clinic.

Institutionalization of the discipline

We have noted that little attention was paid by pediatricians to Paul Ehrlich's studies in this area, and interest in preventive immunizations against childhood diseases rested primarily with the public health rather than with the pediatric community. Thus, prior to World War II it would not be possible to identify any significant manifestations of what would later be called "pediatric immunology."

However, with the expansion of interest in this field during the 1950s and 1960s, and especially with the seeding of medical schools and children's hospitals (primarily by those who had trained with Good and Janeway), the discipline began to organize itself. The increasing clinical importance of immunology led to the establishment of immunology/allergy divisions within departments of pediatrics, first in Minneapolis under Good, in Boston under Rosen, in Seattle under Wedgwood, and at Duke in Durham under Rebecca Buckley. Soon the trend spread to other major schools and hospitals in the United States and in Europe. With the increase in the funding of the National Institutes of Health, training programs in pediatric immunology or allergy/immunology expanded to several dozens.

Meanwhile, such organizations as the American Association of Immunologists, the American Academy of Allergy, Asthma, and Immunology, the Society for Pediatric Research, and the Clinical Research Society began to devote special sessions to topics of intimate interest to pediatric immunology. International symposia devoted exclusively to the ontogeny of the immune response were organized. Some were devoted predominantly to basic immunological studies,[96] while others were more clinically oriented.[97] Soon monographs and texts devoted to the subject[98] would furnish the *vade mecums* and reference guides for the next generation.

Finally there came those two most prominent features that attest to the existence of a valid scientific discipline: its own scientific societies and its own journals. Among the former we may note The Academy of Pediatrics Section on Allergy & Immunology, The International Society of Developmental and Comparative Immunology, The European Society of Pediatric Allergy and Clinical Immunology, The European Society for Immunodeficiencies, and The European Society for Paediatric Haematology and Immunology. Among the latter we find: *Pediatric Allergy & Immunology, Developmental Immunology*; *Clinical and Developmental Immunology*; *Pediatric Asthma, Allergy, and Immunology*; *Developmental & Comparative Immunology*; *Japanese Journal of Pediatric Allergy and Clinical Immunology*; and the *Egyptian Journal of Pediatric Allergy and Immunology*.

Immunophysiology

In our discussion of the development of ocular immunology, we saw that much of the progress in the field was made by ophthalmologists applying immuno-logical concepts and techniques to problems of clinical interest. In considering pediatric immunology, we found that the honors were shared almost equally between the two parent disciplines. Immunologists were interested in what the ontogeny of the response could tell them about its inner workings, while pedi-atricians sought information that would help them to protect the newborn from infections and to devise therapies for congenital deficiency diseases. Now, at the other end of this spectrum, we will see that it was primarily immunologists seeking to dissect the immune response and its controls who led the way in what we now call "immunophysiology."

The term *immunophysiology* is of recent vintage; the early days of immu-nology saw concepts that only implied a relationship between the immune response and general physiological mechanisms. The term suggests a recognition of several intimate relationships: (1) that the immune response obeys the grand laws of general physiology; (2) that specificity, the hallmark of every immuno-logic event, can be identified somewhere in all immunophysiologic processes; and (3) that inflammation (with its implications for allergic and other immu-nopathologic processes) is an integral component of immunologic reactions. These relationships have not always been acknowledged, and indeed were sometimes hotly contested.

Several early conceptual problems inhibited the association of immunologic reactions with the workings of general physiology. One was the notion that immunity (a term with benign and protective connotations) might itself give rise to harmful inflammatory disease;[99] this was strongly contested for many decades. Another was the attempt to restrict immunologic reactions to those involving specificity-controlled agents only. Eventually, however, immunologists came to recognize and accept the Janus-like aspect of the response, and that the collaboration of specific and nonspecific components is essential to almost all immunologic reactions.[100]

Immunity as a physiological process: early stirrings

We saw in Chapter 1 that the earliest medical views on acquired immunity were physiological in nature. Rhazes, in the tenth century, thought that smallpox (then accepted as a benign childhood disease) actually assists in the normal process of ridding the blood of children of its "excess moisture," thus changing it to the "drier" adult state that is immune to subsequent infection.[101] Similarly, the sixteenth-century Italian physician Fracastoro saw the symptoms of smallpox as the salutary means by which a "menstrual contaminant" (with which all are tainted at birth) is discharged from the body. Once gone, the disease could not recur because the menstrual contaminant was now missing; thus, the individual would enjoy lifelong immunity.[102]

Each of these early theories of immunity is couched in terms of normal physiologic mechanisms, as that physiology was understood at the time. These and explanations of acquired immunity as a physiological depletion remind us that Louis Pasteur himself initially sought to explain immunity in terms of the depletion (during the response to an initial infection or immunization) of some trace nutrients critical to the growth of the pathogen.[103]

In his proposal that phagocytic cells play the central role in immunity, Ilya Metchnikoff also saw normal physiology at work.[104] He viewed the function of these cells as a Darwinian extension of their earlier role in mediating the nutrition of the organism. Now they would ingest and thus neutralize pathogenic organisms and toxins. All immunologists know the fairy-tale version that appeared in the biography written by Metchnikoff's wife Olga,[105] describing his "Eureka moment" when he saw the phagocytes surround the rose thorn inserted into the starfish larva. This version stands as the founding myth of immunophysiology.

Paul Ehrlich's side-chain theory also sought to explain antibody formation as a normal bodily function. He proposed that antibodies are nothing more than cell receptors for bacterial toxins, analogous to those required to transact such other cell functions such as nutrition and drug interactions.[106] Antibodies are thus normally present, and appear in excess in the blood only when antigen stimulates their overproduction by the appropriate cells.

Both Metchnikoff's phagocytic theory and Ehrlich's side-chain theory were attempts to integrate immunity into the general biology of the times. Despite their opposing positions (see Chapter 2), both theories saw the immune response as merely one component of the evolution of a larger physiologic system that serves the needs of the body's economy. This view, however, would not survive for long. Both theories would fall out of favor with the scientific community, contributing to an almost half-century separation of immunology from the mainstream of physiology and medicine. Phagocytosis as an explanation for acquired immunity gave way to the growing popularity of humoral antibodies, and suffered also for an apparent lack of the specificity that seemed to define immunity reactions. Ehrlich's theory of antibody formation fell out of favor for other reasons. With the demonstrations that antibodies can be produced against an increasingly large universe of antigens and even synthetic haptens, it soon appeared inconceivable that such diversity could pre-exist in the normal physiologic functions of cells. Immunology would soon lose its close connection to physiology and medicine, to become a more chemically-oriented science with a Lamarckian rather than a Darwinian theoretical base.

Allergy and its mediators

The discoveries of anaphylaxis, the Arthus reaction, and serum sickness during the early years of the twentieth century were described above. Here were harmful disease states that seemed to involve those same factors that provide for protective immunity. Was it possible that these hitherto benign mechanisms might also be harmful? All of these conditions would eventually be subsumed under the general category of "allergic reactions."

Henry Dale's extensive work on the *in vitro* system that compared the contraction of uterine strips from sensitized guinea pigs either by histamine or by specific antigen (called "the Schultz–Dale phenomenon") contributed much to the understanding of the mechanism of anaphylactic shock.[107] This relationship was strengthened by the demonstration by Dale and Laidlaw that a syndrome analogous to anaphylactic shock could be induced by the intravenous injection of histamine, a normal constituent of the body.[108] With the eventual demonstration that a specific IgE antibody–allergen interaction can mediate the degranulation of mast cells and basophils with the release of many different active molecules, including histamine,[109] Dale's contribution to the pharmacology of immediate hypersensitivity reactions can be appreciated. Here was a mechanism whereby a modest specific stimulus might be amplified greatly to produce a profound reaction, through the intercession of nonspecific factors.

Anaphylatoxins

The discovery of anaphylaxis (and related pathologic conditions caused by specific antibodies) stimulated an intense search for the mechanisms involved. One of the key observations in this field, made as early as 1909 by Friedemann,[110] was that the characteristic symptoms of acute anaphylactic shock could be elicited in guinea pigs by injection of antigen–antibody mixtures preincubated with fresh serum. The interpretation of these results was that complement is fixed to the immune complex, activated, and exercises its putative enzymatic power to engage in proteolysis, the breakdown products of which would constitute toxic substances or anaphylatoxins. Indeed, even substances that could fix complement nonspecifically, such as kaolin, starch, and agar, were able to stimulate anaphylatoxin formation. Here was a specific immunologic trigger that would initiate a nonspecific train of pharmacologic events, in this case contributing not to protection from disease, but to its development.

Complement and its products

In 1888, George Nuttall described a naturally occurring bactericidal property present in normal serum.[111] Hans Buchner observed that this property disappears whenever the serum is heated, and he hypothesized the existence of a thermolabile substance (which he named "alexin," later to be renamed "complement" by Ehrlich) that acts in concert with specific antibacterial antibodies.[112] The cooperation between these two serum components was still more impressively demonstrated by Richard Pfeiffer, with the antibody-mediated lysis of cholera vibrios.[113] Jules Bordet concluded that "the intense vibriocidal power, as it presents in the serum of immunized individuals, is due to the action of two substances, the first... endowed with the characteristic of specificity,...the second nonspecific."[114] It was based upon his studies of immune hemolysis that Bordet could later set out the major premise of immunophysiology, almost echoing his mentor Metchnikoff. He characterized immunity as "a primordial function that would exist no less if there were no pathogenic germs on the surface of the earth, but which is admirably adapted...to the protective role that it was in a position to fulfill."[115]

Four components of complement were described between 1907 and 1938, based upon rather crude fractionation and treatment procedures, but further analysis was hindered by the inadequacy of contemporary biochemical tools and biological assays. In the 1950s, Louis Pillemer showed that the third component of complement (C3) was not absorbed but rather inhibited by zymosan, a substance heretofore used for C3 inactivation. From these experiments, Pillemer inferred the existence of a new pathway for the activation of complement that bypassed the initial stages (C1, 2, 4) classically triggered by immune complexes.[116] Pillemer linked this new mechanism, which he termed the "properdin pathway," to natural immunity. He suggested that it was the absence of properdin that was responsible for the poor bactericidal properties of some sera. This concept was not accepted by many workers, who were troubled by the apparently nonspecific mode of action of properdin and by the exaggerated claims for its importance.[117] When cellular immunologists made the acceptance of nonantibody (nonspecific) factors respectable, Pillemer's work on properdin was revisited.[118]

With the demonstration of the many components that participate in the complement cascade,[119] and of the pharmacologic activities of the byproducts of C3 and C5 activation, the old concept of anaphylatoxins found verification and even new significance. The importance of complement was further enhanced by the demonstration of its important role in the pathogenesis of many types of immunogenic inflammation.[120]

Immunity without antibodies

Cellular immunity: the return of the receptor

With the growing understanding that so many "allergic" reactions depend upon identifiable circulating antibodies, the differences between these and the tuberculin and related reactions challenged the canonical wisdom of the field. As described in Chapter 2, not only were there differences in the timing of development of these responses (thus the term "delayed hypersensitivity"), many investigators also sought in vain for the antibodies presumed to be responsible.[121] Unlike the "immediate" hypersensitivities, these could not be transferred with serum.

It was only after 1933 that Karl Landsteiner studied cutaneous sensitization and contact dermatitis reactions in the context of his classical studies on hapten specificity. Here, too, some of these appeared to develop without the participation of antibodies. Only when Landsteiner's collaborator, Merrill Chase, worked out a suitable technical approach was an answer to this conundrum found. Contact dermatitis was shown to be transferable to naive recipients using peritoneal cells from sensitized donors.[122] (It was really Chase's obstinacy that had been determinant; perhaps it is giving too much credit to Landsteiner, who scarcely needs it, to assign to him "a great role in the renaissance of cellular studies."[123])

The establishment of the ability to transfer specific cutaneous sensitivity with cells had enormous theoretical consequences. This new category, at first relegated to the bottom of the list of "antigen–antibody reactions" in general pictures of immunity,[124] emerged as an attractive model for various intriguing phenomena. It was soon applied to tuberculin hypersensitivity, and finally to allograft rejection, to resistance to certain viral infections, and to a variety of autoimmune processes. In short, it provided the cornerstone for the new subdiscipline of cellular immunology.

Immunocyte receptors

In his side-chain theory, Ehrlich had endowed all cells with a universal competence for the production of receptors that would mediate antibody formation. As he had predicted, it would eventually be demonstrated that the receptor on an antibody-forming cell – B cell – is a (slightly modified) immunoglobulin of the selected specificity.[125] With the discovery of the T cell system responsible for "delayed-type" hypersensitivities, these cells were also shown to bear surface membrane receptors (TCR) that mediate their specific functions.[126]

Thus were the several aspects of the immune response restored to membership in the family of receptor-mediated physiological functions.

From MIF to lymphokines

Rich and Lewis showed in 1932 that the normal emigration of leukocytes from a spleen explant taken from a tuberculous animal could be inhibited by the addition of tuberculin.[127] They thought that the tuberculin killed the migrating cells specifically. However, thirty years later George and Vaughan showed that peritoneal macrophages were inhibited from migrating out of capillary tubes in culture, rather than being killed.[128] It remained for David[129] and, independently, Bloom and Bennett[130] to explain the phenomenon. They demonstrated that the specific interaction of antigen with sensitized lymphocytes results in the formation of an active substance that functions to inhibit the movement of macrophages. It was shown further that only a very small proportion of sensitized lymphocytes is required, and that a cell-free supernatant from sensitized lymph node cells incubated with antigen will serve to inhibit normal peritoneal cell migration. Bloom named this substance *migration-inhibition factor* (MIF). These experiments reinforced the conviction that only a few sensitized cells may act as a trigger to provoke an entire cascade of reactions, including those classically described as inflammation. These reactions were mediated by molecules that acted nonspecifically; they were clearly not specific antibodies. In 1969, Dudley Dumonde suggested the term "lymphokines" for these substances.[131]

Pandora's Box had now been opened: the quest for factors was under way. Ultimately, the idea emerged that these factors, secreted by lymphocytes as effectors, might also provide the key for the ignition of inflammatory reactions. Further, they might play a major role in regulating the immune response as well. The interplay between antigen-specific cells and potentiating factors was unmistakable. The intensified search for new factors, and for the cell receptors that trigger their release, confirmed that a large number of such molecules have evolved to serve a variety of physiological functions within the context of the immune system. Indeed, lymphocyte subsets might even be characterized by the lymphokine spectrum that they release, or to which they respond. It was soon recognized that lymphokines are but a subset of a larger functional group – factors that any cell might release (cytokines) to mediate the intercellular communications necessary for its function.

A new generation of immunologists now recognized that, within the general context of specificity, nonspecific interactions might contribute importantly to immunologic phenomena. This cooperation between different molecules (and even different cell lines) is now one of the principal features of the full version of the immune system.

Institutionalization

It is not surprising that the movement toward the institutionalization of what we have termed "immunophysiology" should have lagged behind the two other subdisciplines described above. Whereas these two straddled the boundary between a basic and a clinical science, immunophysiology arose and continued to function well within the expanding research program of immunological laboratories. However, the chief reason is that the main content and even the name of the discipline had to await the discovery of lymphokines and the appreciation of their broad significance. Thereafter, the pace quickened significantly.

Soon symposium proceedings and monographs appeared to summarize advances in the field; the catalog of the National Library of Medicine lists well over 350 with "lymphokine," "cytokine," or "immunophysiology" in the title.[132] Laboratories devoted to immunophysiology followed soon after; among the first were those at the Dana Farber Cancer Center in Boston, at the Institute of Physiology and Pharmacology in Marburg, Germany, and at the National Institutes of Health in Bethesda, Maryland. There was even a "Laboratory of Integrative Immunophysiology" in the Department of Agriculture at the University of Illinois at Urbana. The Pasteur Institute in Paris has organized a formal course in immunophysiology.

The discipline saw its ultimate validation in the formation of two formal organizations: The International Cytokine Society and The International Society for Interferon & Cytokine Research, the latter with its own organ, the *Journal of Interferon and Cytokine Research*. Indeed, the catalog of the National Library of Medicine lists eight journals with the keyword "Cytokine" in the title, and four with the keyword "lymphokine."

Comment

Three examples of subdiscipline formation within the broader field of immunology have been chosen to illustrate some of the disparate factors that might contribute to the definition of a research/clinical specialty. On the one hand there is the interplay between the clinic and the basic science laboratory. Ocular immunology was initiated by ophthalmic clinicians who saw immediately that developments in the young field of immunology (anaphylaxis, immune complex disease, autoimmunity) might provide explanations for some of their troubling clinical problems.

With pediatric immunology, the credits were fairly equally divided; clinicians worried about the ability of neonates to respond to pediatric immunizations, and sought explanations for a variety of immunological deficiency diseases, mostly affecting young children. Basic immunologists, for their part, saw in Burnet's clonal selection theory a call to study the ontogeny of the immune response and, later, the maturational steps of B and T lymphocyte subsets. As for immunophysiology, it must be admitted that the hint of physiological thinking underlay the work of Metchnikoff, Ehrlich, and Bordet. But there was little by way of formal development along those lines, despite the early work on histamine and on complement-based anaphylatoxins. Then a single discovery – that of MIF – started a chain reaction that was almost entirely restricted to basic researchers for a long time. Only later would clinicians borrow some of these findings to develop therapeutic modalities, primarily in the area of immunoregulation.

A second sociological factor, which somewhat parallels the first, involves the institutional affiliations of the workers in each of these fields. Those medically trained individuals who work in ocular immunology will invariably work in eye-associated institutions, and will identify with ophthalmology. And, while basic scientists in ocular immunology may maintain one foot in each field, they also will almost invariably be found in ophthalmic institutions. In pediatric immunology, again the field is split; those who study developmental immunology will generally emphasize clearly either the "pediatric" or the "immunology" in identifying themselves. This emphasis will usually determine whether they work in a department of pediatrics or of immunology. The situation regarding immunophysiology is much clearer; these workers will almost invariably call themselves "immunologists" rather than "physiologists," and pursue their interests almost entirely in immunological institutions.

What is so interesting sociologically about a field like immunology, which touches so many other disciplines, is that it is often difficult to determine where the primary allegiance of an investigator lies. Some of the most important advances in immunology are currently being reported by those who may not even think of themselves as immunologists, and who choose to publish in other than "immunological" journals!

Notes and references

1. Lenoir, T., *Instituting Science: The Cultural Production of Scientific Disciplines*, Stanford, Stanford University Press, 1997, p. 46.
2. See, for example, Shapere, D., in Shapere, D., ed., *Reason and the Search for Knowledge*, Dordrecht, Riedel, 1984; Lemaine, G., Macleod, R., Mulkay, M., and Weingard, P., eds, *Perspectives on the Emergence of Scientific Disciplines*, Chicago, Aldine, 1976; Darden L., and Maull, N., *Phil. Sci.* **45**:43, 1979; Hull, D., *Phil. Sci.* **45**:335, 1978; and Lenoir, note 1.
3. See also Brent, L., *A History of Transplantation Immunology*, San Diego, Academic Press, 1997.
4. The six principal components of the early immunological research program are discussed in Chapter 17. These were: preventive immunization, cellular (phagocytic) immunity, serotherapy, cytotoxic antibodies, serodiagnosis, and anaphylaxis and related phenomena.
5. Pfeiffer, R., *Z. Hyg.* **18**:1, 1894.
6. Bordet, J., *Ann. Inst. Pasteur* **12**:688, 1899.
7. See, for example, *Ann. Inst. Pasteur* **14**, 1900.
8. Portier, P., and Richet, C., *C. R. Soc. Biol.* **54**:170, 1902.
9. Arthus, M., *C. R. Soc. Biol.* **55**:817, 1903.
10. von Pirquet, C., and Schick, B., *Die Serumkrankheit*, Vienna, Deuticke, 1906.
11. Wolff-Eisner, A., *Das Heufieber*, Munich, 1906; Meltzer, S., *J. Am. Med. Assoc.* **55**:1021, 1910.
12. Ehrlich, P., Verh. 73, *Ges. Deutsch. Naturforsch. Aerzte*, 1901. Reprinted in *The Collected Papers of Paul Ehrlich*, Vol. 2, Pergamon, New York, 1957, p. 298. See also Chapter 8.
13. Donath J., and Landsteiner, K., *Münch. med. Wochenschr.* **51**:1590, 1904.

14. Santucci, S., *Riv. Ital. Ottal. Roma* **2**:213, 1906; Golowin, S., *Russky Vratch*, No. 22, May 29, 1904, and *Klin. Monatsbl. Augenheilk.* **47**, 150, 1909. A similar suggestion was made for the pathogenesis of syphilis, involving the participation of "autoantibodies" directed against the tissue breakdown products associated with this disease; Weil, E., and Braun, H., *Wien. klin. Wochenschr.* **20**:527, 1907; **22**:372, 1909.

15. Elschnig, A., *von Graefes Arch. Ophthalmol.* **75**:459, 1910; **76**:509, 1910; **78**:549, 1911; and **79**:428, 1911.

16. Woods, A.C., *Arch. Ophthalmol.* **45**:557, 1916; **46**:8, 503, 1917; **47**:161, 1918; Friedenwald, J.S., *Am. J. Ophthalmol.* **17**:1008, 1934.

17. Freund, J., and McDermott, K., *Proc. Soc. Exp. Biol. Med.* **49**:548, 1942.

18. Collins, R.C., *Am. J. Ophthalmol.* **32**:1687, 1949; **36**:150, 1953.

19. Wacker, W.B., and Lipton, M.M., *Nature* **206**:253, 1965; *J. Immunol.* **101**:151, 1968.

20. Kalsow, C.M., and Wacker, W.B., *Intl Arch. Allergy Appl. Immunol.* **44**:11, 1973; **48**:287, 1975.

21. Wacker, W.B., et al., *J. Immunol.* **119**:1949, 1977; Dorey, C., and Faure, J.-P., *Ann. Immunol. Inst. Pasteur* **128**:229, 1977; de Kozak, Y., et al., *Curr. Eye Res.* **1**:327, 1981.

22. Gery, I., Mochizuki, M., and Nussenblatt, R.B., *Progr. Retinal Res.* **5**:75, 1986.

23. Marak, G.E., et al., *Ophthalmic Res.* **12**:165, 1980; Meyers-Elliot, R.H., et al., *Clin. Immunol. Immunopathol.* **27**:81, 1983.

24. Rao, N.A., Wacker, W.B., and Marak, G.E., *Arch. Opthalmol.* **97**:1954, 1979.

25. Uhlenhuth, P., in *Festschrift zum 60 Geburtstag von Robert Koch*, Jena, Fischer, 1903, pp. 49–74.

26. Following the path laid out by Nuttall, G. (*Blood Immunity and Blood Relation-ships*, Cambridge, Cambridge University Press, 1904), who used cross-reactions among the protein antigens of different species to measure interspecies relationships, Halbert and Manski (*Progr. Allergy* **7**:107, 1963) used lens proteins to show species relationships across the entire vertebrate spectrum.

27. Kraus, R., Doerr, R., and Sohma, M., *Wien. klin. Wochenschr.* **21**:1084, 1908; Andrejew, P., and Uhlenhuth, P., *Arb. Kaiserl. Gesundheitsamte* **30**:450, 1908.

28. Krusius, F.F., *Arch. Augenheilk.* **67**:6, 1910.

29. Römer, P., and Gebb, H., *von Graefes Arch. Ophthalmol.* **81**:367, 387, 1912.

30. Verhoeff, F.H., and Lemoine, A.N., *Acta Intl Congr. Ophthalmol. (Washington)*, **1**:234, 1922.

31. Verhoeff, F., and Lemoine, A.N., *Am. J. Ophthalmol.* **5**:737, 1922. (It was not then considered dangerous or unethical to inject a "self-antigen" into a patient!)

32. Marak, G.E., et al., *Exp. Eye Res.* **19**:311, 1974; Marak, G.E, Font, R.L., and Alepa, F.P., *Mod. Probl. Ophthalmol.* **16**:75, 1976; *idem*, *Ophthalmic Res.* **9**:162, 1977.

33. Sjögren, H., *Acta Ophthalmol.* **Suppl. 2**:1, 1933.

34. Jones, B.R., *Lancet* **ii**:773, 1958.

35. Manoussakis, M.N., and Moutsopoulos, H.M., in Rose, N.R., and Mackay, I.R., eds, *The Autoimmune Diseases*, 4th edn, New York, Academic Press, 1985, pp. 401–416.

36. Mizejewski, G.J., *Experientia* **34**:1093, 1978; Liu, S.H., Prendergast, R.A., and Silverstein, A.M., *Invest. Ophthalmol. Vis. Sci.* **28**:270, 1987.

37. Jabs D.A., and Prendergast, R.A., *Invest. Ophthalmol. Vis. Sci.* **29**:1437, 1988.

38. Nicolle, M., and Abt, G., *Ann. Inst. Pasteur* **22**:132, 1908.
39. Sattler, C.H., *Arch. Augenheilk.* **64**:390, 1909.
40. Seegal, D., and Seegal, B.H., *Proc. Soc. Exp. Biol. Med.* **27**:390, 1930; *J. Exp. Med.* **54**:265, 1931.
41. Goldmann, H., and Witmer, R.H., *Ophthalmologica*, Basel **127**:323, 1954; Witmer, R.H., *Am. J. Ophthalmol.* **37**:243, 1953; *Arch. Ophthalmol.* **53**:811, 1955.
42. Silverstein, A.M., in Maumenee, A.E., and Silverstein, A.M., eds, *Immunopathology of Uveitis*, Baltimore, Williams & Wilkens, 1964, pp. 83–110.
43. Chandler, J.W., Heise, E.R., and Weiser, R.S., *Invest. Ophthalmol.* **12**:400, 1973; Liu, S.H., Prendergast, R.A., and Silverstein, A.M., *Invest. Ophthalmol. Vis. Sci.* **24**:361, 1983.
44. Wessely, K., *Münch. med. Wochenschr.* **58**:1713, 1911.
45. von Szily, A., *Die Anaphylaxie in der Augenheilkunde*, Stuttgart, Ferdinand Enke, 1914.
46. Germuth, F.G., et al., *Am. J. Ophthalmol.* **46**:282, 1959; *J. Exp. Med.* **115**:919, 1962.
47. Waksman, B.H., and Bullington, S.J., J. *Immunol.* **76**:441, 1956.
48. Lagrange, H., and Delthil, S., *Les Conjonctivites de Nature Anaphylactique*, Paris, Doin, 1932. Advances in the field up to 1958 are summarized in Theodore, F.H., and Schlossman, A., *Ocular Allergy*, Baltimore, Williams & Wilkins, 1958. The work relating allergic conjunctivitis to IgE-mediated mechanisms is reviewed by Allansmith, M.R., *The Eye and Immunology*, St Louis, Mosby, 1982. This stimulated research on the possible immunopathogenesis of such diseases as vernal catarrh, a seasonal conjunctivitis most often associated with pollen allergy (Allansmith, M.R., and Frick, O.L., *J. Allergy* **34**:535, 1963).
49. Dhermy, F., Coscas, G., Nataf, R., and Levaditi, J.C., *Rev. Intl Trachome* **44**:295, 1968.
50. Jones, B.R., cited in Duke-Elder, S., *System of Ophthalmology*, Vol. 8, London, Kimpton, 1965, pp. 4–5; see also Silverstein, A.M., and Prendergast, R.A., in Lindahl-Kiessling, et al., eds, *Morphological and Functional Aspects of Immunity*, New York, Plenum, 1971, p. 583.
51. See, for example, Leigh, A.G., *Corneal Transplantation*, Oxford, Blackwell, 1966, pp. 1–5.
52. See Chapter 11.
53. Medawar, P.B., *J. Anat.* **78**:176, 1944; **79**:157, 1945; see also Medawar's *Harvey Lect.* **52**:144, 1956–57.
54. Maumenee, A.E., *Ann. NY Acad. Sci.* **59**:453, 1955; see also Paufique, L., Sourdille, G.F., and Offret, G., *Les Greffes de la Cornée (Kératoplasties)*, Paris, Masson, 1948.
55. Billingham, R.E., and Boswell, T., *Proc. R. Soc. London (Biol.)* **141**:392, 1953; Khodadoust, A.A., and Silverstein, A.M., *Invest. Ophthalmol.* **11**:137, 1972.
56. Greene, H.S.N., Cancer Res. **2**:669, 1942; **7**:491, 1947. A more extensive discussion of this phenomenon will be found in Woodruff, M.F.A., *The Transplantation of Tissues and Organs*, Springfield, Charles C. Thomas, 1960. Medawar had observed earlier (*Br. J. Exp. Pathol.* **29**:58, 1948) that allogeneic skin grafts would survive in the anterior chamber only so long as they remained unvascularized. See also Brent, note 3.
57. Raju, S., and Grogan, J.B., *Transplantation* **7**:475, 1969; Franklin, R.M., and Prendergast, R.A., *J. Immunol.* **104**:463, 1970. See also Niederkorn, J.Y., *Adv. Immunol.* **48**:191, 1990.

58. Barker, C.F., and Billingham, R.E., in Dausset, J., et al., ed., *Advances in Transplantation*, Baltimore, Williams & Wilkins, 1968, p. 25.
59. Kaplan H.S., and Streilein, J.W., *J. Immunol.* **118**:809, 1977; **120**:689, 1978; Niederkorn, J.Y., Streilein, J.W., and Shadduck, J.A., *Invest. Ophthalmol. Vis. Sci.* **20**:355, 1980; and Streilein-Stein, J., and Streilein, J.W., *Intl Rev. Immunol.* **21**:123, 2002.
60. von Szily, note 45. Much later came Böke, W., *Immunpathologie des Auges*, Basel, Karger, 1968; Rahi, A.H.S., and Garner, A., *Immunopathology of the Eye*, Oxford, Blackwell, 1976; Faure, J.-P., Bloch-Michel, E., Le Hoang, P., and Vadot, E., *Immunopathologie de l'Oeil*, Paris, Masson, 1988.
61. See, for example, Woods, A.C., *Allergy and Immunity in Ophthalmology*, Baltimore, Johns Hopkins Press, 1933; *Ocular Allergy*, note 48; Campinchi, R., et al., *L'Uvéite, Phénomènes Immunologiques et Allergiques*, Paris, Masson, 1970; Friedlaender, M.H., *Allergy and Immunology of the Eye*, New York, Harper & Row, 1979; Smolin, G., and O'Connor, G.R., *Ocular Immunology*, 2nd edn, Boston, Little Brown, 1986.
62. I count at least five such laboratories in Germany and one in Prague prior to World War I. Between the Wars, two were added in the United States – one at Johns Hopkins University and one at Columbia University. After World War II, an additional six ocular immunology laboratories were opened in the United States, and others in at least twelve other countries throughout the world.
63. An endowed Research Chair in Ophthalmic Immunology was created in 1964 at the Wilmer Ophthalmological Institute, The Johns Hopkins University School of Medicine, Baltimore.
64. Henkind, P., "History of the Association for Research in Vision and Ophthalmology," *Invest. Ophthalmol. Vis. Sci.* **17**(Suppl.):2, 1978.
65. Strasbourg, 1974: Böke, W., and Luntz, M.H., eds, *Ocular Immune Responses*, Basel, Karger, 1976; San Francisco, 1978: Silverstein, A.M., and O'Connor, G.R., eds, *Immunology and Immunopathology of the Eye*, New York, Masson, 1979; Seattle, 1982: O'Connor, G.R., and Chandler, J.W., eds, *Advances in Immunology and Immunopathology of the Eye*, New York, Masson, 1985; Padova, 1986: Secchi, A.G., and Fregona, I.A., eds, *Modern Trends in Immunology and Immunopathology of the Eye*, Milan, Masson, 1989; Tokyo, 1990: Usui, M., Ohno, S., and Aoki, K., eds, *Ocular Immunology Today*, Amsterdam, Excerpta Medica, 1990; Washington, DC., 1994: Nussenblatt, R.B., et al., eds, *Advances in Ocular Immunology*, New York, Elsevier, 1994.
66. *Immunology of the Eye, I. Immunogenetics and Transplantation Immunity*, Steinberg, G.M., Gery, I., and Nussenblatt, R.B., eds, 1980; *II. Autoimmune Phenomena and Ocular Disorders*, Helmsen, R.J., Suran, A., Gery, I., and Nussenblatt, R.B., eds, 1981; *III. Immunological Aspects of Ocular Diseases*, Suran, A., Gery, I., and Nussenblatt, R.B., eds, 1981, Washington DC, Information Retrieval.
67. *Ocular Immunology and Inflammation* and *Ophthalmic Immunology and Inflammation*.
68. For a full description of the experimental details of Ehrlich's work in this area, see Silverstein, A.M., *Nature Immunol.* **1**:93, 2000, and *Paul Ehrlich's Receptor Immunology*, San Diego, Academic Press, 2002, pp. 27–40.
69. See Fournier, A., *L'hérédité Syphilitique*, Paris, Masson, 1891; Hutchinson, J., *Syphilis*, 3rd edn, London, Cassell, 1893.

70. Doyle, A.C., "The Third Generation," reprinted in Rodin, A.E., and Key, J.D., *Conan Doyle's Tales of Medical Humanism and Values: Round the Red Lamp*, Malabar, The Spermaceti Press, 1992. The implications of these concepts are discussed in detail in Silverstein, A.M., and Ruggere, C., "Dr Arthur Conan Doyle and the case of congenital syphilis," *Perspect. Biol. Med.* **49**:209, 2006.

71. Ehrlich, P., *Deutsch. med. Wochenschr.* **17**, 976, 1218, 1891.

72. Ehrlich, P., *Z. Hygiene* **12**:183, 1892.

73. Ehrlich, P., and Hübener, W., *Deutsch. med. Wochenschr.* **18**:511, 1892.

74. Ehrlich's contributions to this field are scarcely mentioned in definitive works on the subject (such as Brambell, F.W.R., *The Transmission of Passive Immunity from Mother to Young*, Amsterdam, North Holland, 1970; Hemmings, W.A., ed., *Maternofoetal Transmission of Immunoglobulins*, Cambridge, Cambridge University Press, 1976), and not at all in more general summaries (for example, Solomon, J.B., *Foetal and Neonatal Immunology*, Amsterdam, North Holland, 1971; Loke, Y.W., *Immunology and Immunopathology of the Human Foetal–Maternal Interaction*, Amsterdam, North Holland, 1978).

75. Topley, W.W.G., and Wilson, G.S., *Principles of Bacteriology and Immunity*, 2nd edn, Baltimore, William Wood, 1936, p. 855.

76. Burnet, F.M., and Fenner, F., *The Production of Antibodies*, 2nd edn, New York, Macmillan, 1949.

77. Burnet, F.M., *Austral. J. Sci.* **20**:67. 1957; *The Clonal Selection Theory of Acquired Immunity*, London, Cambridge University Press, 1959.

78. Osborn, J.J., Dancis, J., and Julia, J.F., *Pediatrics* **9**:736, 1952; Smith, R.T., and Bridges, R.A., *J. Exp. Med.* **108**:227, 1958.

79. Bridges, R.A., Condie, R.M., Zak, S.J., and Good, R.A., *J. Lab. Clin. Med.* **53**:331, 1959; Black, M.M., and Speer, F.D., *Blood* **14**:848, 1959.

80. Silverstein, A.M., and Lukes, R.J., *Lab. Invest.* **11**:918, 1962; Silverstein, A.M., *Nature* **194**:196, 1962.

81. Bruton, O.C., *Pediatrics* **9**:722, 1952.

82. Glanzmann, E., and Riniker, P., *Ann. Pediat.* **175**:1, 1950.

83. Tobler, R., and Cottier, H., *Helv. Pediat. Acta* **13**:313, 1958; Hitzig, W.H., et al., *Helv. Pediat. Acta* **13**:551, 1958.

84. Gitlin, D., and Craig, G.M., *Pediatrics* **32**:517, 1963.

85. Hitzig, W.H., et al., *J. Pediat.* **78**:968, 1971.

86. Giblett, E.R., et al., *Lancet* **2**:1067, 1972.

87. Rosen, F.S., *Immunol. Rev.* **178**:8, 2000. A more recent history of the entire field will be found in Stiehm, E.R., and Johnston, R.B., *Pediat. Res.* **57**:458, 2005.

88. DiGeorge, A.M., *J. Pediat.* **67**:907, 1965.

89. Miller, J.F.A.P., *Lancet* **2**:748, 1961. See also Janković, B.D., Waksman, B.H., and Arnason, B.G., *J. Exp. Med.* **116**:159, 1962, and Good, R.A., et al., *J. Exp. Med.* **116**:773, 1962.

90. A review of much of the early work from Good's laboratory will be found in Bergsma, D., et al., eds, *Immunological Deficiency Diseases in Man*, New York, The National Foundation, 1967.

91. Landsteiner, K., and Wiener, A.S., *Proc. Soc. Exp. Biol. Med.* **43**:223, 1940; *J. Exp. Med.* **74**:309, 1941.

92. Levine, P., et al., *J. Am. Med. Assoc.* **116**:825, 1941; *Am. J. Obstet. Gynecol.* **42**:925, 1941. See also Wiener, A.S., and Sonn, E.B., *Am. J. Dis. Child.* **68**:317, 1944.

93. Billingham, R.E., Brent, L., and Medawar, P.B., *Nature* **172**:603, 1953.

94. See, for example, the chapters on B cell and T cell development in Mak, T.W., and Saunders, M.E., *Immune Response: Basic and Clinical Principles*, Burlington, Elsevier, 2006.

95. See Chapter 4. For earlier summaries of the ontogenetic studies, see Silverstein, A.M., *Science* **144**:1423, 1964; Sterzl, J., and Silverstein, A.M., *Advances Immunol.* **6**:337, 1967.

96. Thus, Smith, R.T., Good, R.A., and Miescher, P.A., eds, *Ontogeny of Immunity*, Gainesville, University of Florida Press, 1967; Warren, K.B., ed., *Differentiation and Immunology*, New York, Academic Press, 1968; *Ontogeny of Acquired Immunity*, Ciba Foundation Symposium, Amsterdam, Elsevier, 1972; and Cooper, M.D., and Dayton, D.H., eds, *Development of Host Defenses*, New York, Raven Press, 1977.

97. For example, *Immunological Deficiency Diseases of Man*, note 90; Hodes, H., and Kagan, B.M., *Symposium on Pediatric Immunology*, New York, Science and Medicine Publishing, 1979.

98. Stiehm, E.R., Ochs, H.D., and Winkelstein, J.A., *Immunological Disorders in Infants and Children*, Philadelphia, W.B. Saunders, 1973 (now in its fifth edition, 2004); Spirer, Z., Roifman, C.M., and Bransky, D., *Pediatric Immunology*, Basel, Karger, 1993; Ochs, H.D., Smith, C.I.E., and Puck, J.M., eds, *Primary Immuno-deficiency Diseases: A Molecular and Genetic Approach*, New York, Oxford University Press, 1999; Wolf, R.L., *Essential Pediatric Allergy, Asthma, and Immunology*, New York, McGraw-Hill, 2004.

99. Peter Medawar has commented on "the philological chamber of horrors..." (Medawar P.B., *Aristotle to Zoos*, Cambridge, Harvard University Press, 1983, p. 257). Even the very name of our discipline has been taxed by Gell and Coombs in "Immunology, too late to change this word; we shall have to think straight in spite of it" (Gell, P.G.H., and Coombs, R.R.A., *Clinical Aspects of Immunology*, Oxford, Blackwell, 1967, p. xxii).

100. Cunningham A.J., *Curr. Top. Microbiol. Immunol.* **67**:97, 1974.

101. Rhazes, *A Treatise on the Smallpox and Measles* (translated by W.A. Greenhill), London, Syndenham Society, 1848.

102. Fracastoro, G., *De Contagiosis Morbis et Eorum Curatione* (translated by W.C. Wright), New York, Putnam, 1930.

103. Pasteur, L., Chamberland, C., and Roux, E., *C. R. Acad. Sci.* **90**:239, 952, 1880.

104. Metchnikoff, I., *Virchows Arch.* **96**:177, 1884. Metchnikoff's clearest exposition of the Darwinian source of his theory can be found in his 1892 book *Leçons sur la Pathologie Comparée de l'Inflammation*, Paris, Masson.

105. Metchnikoff, O., *Life of Elie Metchnikoff*, Boston, Houghton-Mifflin, 1921.

106. Enrlich, P., *Klin. Jahrb.* **60**:299, 1997 (English translation in *Collected Papers*, Vol. 1, p. 107, note 12).

107. Dale, H.H., *J. Pharmacol. Exp. Ther.* **4**:167, 1913.

108. Dale, H.H., and Laidlaw, P.P., *J. Physiol.* **52**:355, 1919.

109. See, for example, Kay, A.B., Austen, K.F., and Lichtenstein, L.M., *Asthma: Physiology, Immunopharmacology, and Treatment*, Orlando, Academic Press, 1984.

110. Friedemann, U., *Z. Immunitätsf.* **2**:591, 1909. See also Friedberger, E., *Berl. Klin. Wochenschr.* **47**:1490, 1910,

111. Nuttall, G., *Z. Hyg.* **4**:353, 1888.

112. Buchner, H., *Zentralbl. Bakt.* **6**:561, 1889.

113. Pfeiffer, note 5.

114. Bordet, note 6.
115. Bordet, J., *Traité de l'Immunité dans les Maladies Infectieuses*, Paris, Masson, 1920, p. 695.
116. Pillemer, L., et al., *Science* **120**:279, 1954; Pillemer, L., *Trans. NY Acad. Sci.* **17**:526, 1955.
117. For a history of this dispute, see Ratnoff, W.D., *Perspect. Biol. Med.* **23**:638, 1979/1980.
118. For example, Götze, G., and Müller-Eberhard, H.J., *Advances Immunol.* **34**:1, 1976; Fearon, D.T., *Crit. Rev. Immunol.* **1**:1, 1979. Fearon, D.T., and Austin, K.F., *New Engl. J. Med.* **303**:259, 1980.
119. Mayer, M.M., *Proc. Natl Acad. Sci. USA*, **69**:2954, 1972; Mayer, M.M., *Complement* **1**:2, 1984.
120. Shin, H.S., et al., *Science* **162**:361, 1968; Lichtenstein, L.M., et al., *Immunology* **16**:327, 1969; Hugli, T.E., and Müller-Eberhard, H.J., *Adv. Immunol.* **26**:1, 1978.
121. Most notable among these was Arnold Rich; see his *The Pathogenesis of Tuberculosis*, Springfield, Charles C. Thomas, 1942. See also Chapter 15.
122. Landsteiner, K., and Chase, M.W., *Proc. Soc. Exp. Biol. Med.* **49**:688, 1942. See also Chase, M.W., *Harvey Lect.* **1**:169, 1967.
123. Talmage, D.W., *Annu. Rev. Immunol.* **4**:1, 1986, p. 4.
124. See, for example, Miescher, P., and Vorlaender, K., *Immunopathologie Clinique et Expérimentale*, Paris, Flammarion, 1959, p. 60.
125. Reth, M., *Annu. Rev. Immunol.* **10**:97, 1992.
126. Marrack, P., and Kappler, J., *Adv. Immunol.* **38**:1, 1986.
127. Rich, A.R., and Lewis, M.R., *Bull. Johns Hopkins Hosp.* **50**:115, 1932.
128. George, M., and Vaughan, J.H., *Proc. Soc. Exp. Biol. Med.* **111**:514, 1962. Waksman, B.H. and Matoltsy, M. had shown earlier (*J. Immunol.* **81**:220, 1958) that in fact the macrophages were not killed by antigen.
129. David, J.R., *Proc. Natl Acad. Sci. USA* **56**:72, 1966.
130. Bloom, B.R., and Bennett, B., *Science* **153**:80, 1962.
131. Dumonde, D.C., et al., *Nature* **224**:38, 1969.
132. Typical of the early offerings are: *Biology of the Lymphokines* (Cohen, S., et al., eds, New York, Academic Press, 1979); *Human Lymphokines* (Khan, A., and Hill, N.O., eds, New York, Academic Press, 1982); *Immunophysiology: the Role of Cells and Cytokines in Immunity and Inflammation* (Oppenheim, J.J., and Shevach, E.M., eds, New York, Oxford, 1990); *Immunophysiology of the Gut* (Walker, W.A., Harmatz, P.R., and Wershil, B.K., eds, New York, Academic Press, 1993).

20 Immune hemolysis: on the value of experimental systems

Philosophers of the biomedical sciences have long been interested in how scientific knowledge is constructed, how it relates to theory, and how it radiates within the sciences and may even find practical application outside these bounds. There are many dimensions to this question. On the one hand it is proposed that knowledge and theory are purely social constructs, only approachable by the study of scientists, their ideas, and their institutional settings. At the other end of the spectrum are those who maintain that only an understanding of the technological core of a science (i.e., the techniques used and the data that result) is necessary for a reconstruction of its knowledge base. In the middle is the suggestion that both theory and practice contribute, and that there are boundary negotiations between the two which are important for an understanding of how science progresses.[1]

Because of the central role that immunology has assumed among modern biomedical sciences,[2] it offers several good examples of this interaction that combine both historical and ethnographic approaches. Thus, Cambrosio and Keating have studied hybridoma technology[3] and the bases for the identification of lymphocyte subsets[4] from this point of view, and show clearly how data and perceptions interact in the development of research programs. In this chapter we examine the interaction of technique and fact with theory, in terms of the consequences of the discovery of the phenomenon of immune hemolysis. This study highlights again the many important threads that may move back and forth between theory and practice, in a complicated but productive and ever-broadening dialectic which serves to expand the horizons of the science.

Background to the discovery

Earlier chapters have shown how immunology was born in the context of the new bacteriology of the late nineteenth century; from the outset, it was devoted almost exclusively to problems of infectious diseases. Its initial research programs were concerned with both practical and theoretical aspects of this new approach to disease prevention. On the practical side were various extensions of Louis Pasteur's demonstration that acquired immunity might be induced by vaccination with attenuated pathogens. Important also were studies to improve upon Behring and Kitasato's demonstration of the serotherapeutic effect of the passive transfer of antitoxins for the treatment of tetanus and diphtheria.

On the conceptual side was the question of the mechanism whereby the body defends itself against infection. In a wide-ranging debate, Elie Metchnikoff in

A History of Immunology, Second Edition
ISBN: 978-0-12-370586-0

Copyright © 2009, Elsevier Inc.
All rights reserved

France claimed that cellular (phagocytic) mechanisms are all-important, whereas Paul Ehrlich and Richard Pfeiffer in Germany held that the active factor is humoral antibody. The debate stimulated much important work on both sides,[5] but always with the implicit assumption that the immune response was devoted entirely to bacterial pathogens and their noxious products.

In 1894, Richard Pfeiffer showed that cholera vibrios can be lysed *in vivo* by passively administered anticholera antibodies.[6] The following year Jules Bordet, in Metchnikoff's laboratory, demonstrated the same effect *in vitro*, with the additional finding that the lytic effect depends upon the action of two components: thermostable antibody, and thermolabile alexin (complement).[7] Then, in 1898, Belfanti and Carbone found that animals were able to form antibodies which would agglutinate and mediate hemolysis of foreign erythrocytes.[8] The importance of this observation was recognized immediately by Bordet, who quickly showed that the mechanism responsible for immune hemolysis is precisely the same as that for immune bacteriolysis – i.e., the combined action of specific antibody and complement.[9]

The discovery of immune hemolysis had profound implications for the further development of immunology. First and foremost, it showed conclusively that the immune response is not limited to bacterial pathogens and their toxic products, thus requiring a major readjustment of most current ideas about the evolution of this "protective" response of the body.[10] Secondly, the fact that a destructive immune response might be mounted against an important *cellular* component of the body implied that other tissues and organs might be similarly attacked by their respective antibodies, with interesting implications for the pathogenesis of a number of *non*infectious diseases. Thirdly, *the experimental system itself* served as the vehicle for a lively debate about the mechanism of action of antibody and complement, from which emerged a number of significant facts and expanded concepts that would add to the perceived importance of immunology in turn-of-the-century medicine. Finally, the technique would find important practical application in the diagnosis of disease, and soon would move from the research laboratory to a more clinical setting.

In the discussion that follows, we shall explore the several consequences of the discovery of immune hemolysis in greater detail. Not only did this new experimental system lead to an expansion of the early research program of immunology; its pursuit also resulted in a number of important conceptual and practical advances.

Conceptual consequences

The Ehrlich–Bordet debate on mechanisms

We saw, in Chapter 6, the first fruits of Bordet's discovery of immune hemolysis – Paul Ehrlich's immediate recognition that this new experimental system would provide the perfect vehicle for experiments to help to validate his theories on the nature and mode of action of antibodies. His 1897 theory on how antibodies are formed, the famous "side-chain" theory,[11] held that antibodies are naturally occurring cell receptors whose stereochemical structures match exactly the complementary structures on their respective antigens. The theory proposed that this "fit" would permit a tight chemical bond to form between the two members.

Since his theory was then under attack, Ehrlich used the hemolytic system not only to substantiate his earlier claims, but to expand the concept as well. Between 1899 and 1901 he undertook a remarkable series of experiments on immune hemolysis, with his associate Julius Morgenroth.[12] Since complement was also involved in the process of hemolysis (in addition to antibody and the antigenic erythrocyte), Ehrlich also postulated a specific stereochemical interaction involving the active site on the complement "molecule" and its "receptor." Thus, the work on immune hemolysis permitted Ehrlich to elaborate further his side-chain theory with the addition of new components and mechanisms – additions which then suggested further experiments and yet further elaborations in an ever-expanding interplay between data and theory. The discovery of immune hemolysis had demonstrated its first heuristic value.

Ehrlich's ideas did not sit well with Bordet, who favored a less elaborate explanation of the mode of action of antibody and complement based upon colloidal adsorption – i.e., a physical rather than chemical interaction. Ehrlich and Bordet thereupon commenced a lengthy dispute that went on for many years, each devising experiments based upon immune hemolysis to support his own view and to cast doubt on that of the opposition. Even the languages employed in the laboratories and publications of the protagonists semantically reflected their differing views on mechanism; Ehrlich employed the terms *Komplement*, and *Ambozeptor* or *Zwischenkörper* for antibody, while Bordet preferred to speak of *alexine* and *substance sensibilisatrice*.[13] Regardless of the outcome of this great dispute (and both Ehrlich and Bordet were ultimately proved correct in part), it stimulated a wealth of new and important data while it lasted.

Cytotoxic antibodies

Bordet's elaborate experiments on the phenomenon of immune hemolysis excited a wave of interest across Europe. If cytotoxic antibodies could be formed against *one* of the body's cell types, why not against others? Herein, perhaps, lay the explanation for the pathogenesis of many of the diseases for which no reasonable cause was then known. In numerous laboratories (and especially at the Pasteur Institute), basic researchers and clinicians ground and extracted almost every tissue and organ in the body to determine whether they would stimulate the formation of specific cytotoxic antibodies when injected into laboratory animals. In the main, these efforts were successful! In addition to destructive antibodies against erythrocytes, spermotoxins, neurotoxins, leukotoxins, hepatotoxins, and many others were described.[14]

It was a rare medical subspecialty that did not see in this new approach the solution to some of its most difficult disease problems. Perhaps the best example of this involves the ophthalmologists' attempt to explain the pathogenesis of sympathetic ophthalmia, an inflammatory disease that might affect a normal eye long after a penetrating injury to the patient's other eye. As early as 1904, the Russian Golowin had suggested that sympathetic ophthalmia might be caused by the formation of cytotoxic antibodies against ocular tissues, citing Bordet's studies to justify the thesis.[15] This idea was developed independently by Santucci in 1907,[16] who based it upon experiments showing that rabbits injected with emulsified ocular tissues would develop endophthalmitis. He suggested that the resorption of damaged tissues in the first, wounded eye would initiate the formation of "cyclotoxins" that might later attack the ciliary body of the hitherto normal fellow eye. The approach was taken up by numerous ophthalmic investigators over the next eighty years – a persistent attack that ultimately led to validation of the hypothesis and the identification of retinal proteins as the inciting antigens.[17] It must be admitted, however, that ophthalmology was the only clinical discipline to benefit from this approach during the early years; others made the effort, but the various cytotoxins could not be identified with disease, and they quickly lost interest. They would not realize the importance of cytotoxic antibodies (and cells) until the renaissance of interest in autoimmunity following World War II.

Horror autotoxicus *and the question of autoimmunity*

During the course of his investigations on the mechanism of immune hemolysis, Ehrlich asked whether goats, which so easily produce hemolysins against the red cells of other species, might also be able to form such products against the cells of other goats and even against their own erythrocytes. Many tests showed that while animals could form *iso*antibodies against the red cells of other members of their species, they were never observed to form *auto*antibodies against their own cells. This led him to postulate the concept of *horror autotoxicus*, stating that "it would be dysteleologic in the extreme" were the organism able to mount an immune and damaging response to its own substance."[18] Ehrlich would repeat this dictum frequently over subsequent years, with the authority of a world-famous leader in the field, and no one who subscribed to his teachings would dare to contradict it. But Karl Landsteiner worked in Vienna, far from Ehrlich's base at Frankfurt am Main, and subscribed to the ideas of Bordet. In this, he had been influenced by the vehement anti-Ehrlich ideas of his mentor Max Gruber.[19]

Bordet's original work on hemolysis had interested Landsteiner in anti-erythrocyte antibodies, and he quickly published several studies on hemagglutination and immune hemolysis,[20] including his landmark discovery of human blood groups (see below). Here was an unencumbered mind prepared for radical speculation in this area. When clinician Julius Donath came to the more senior Landsteiner for help in sorting out the pathogenesis of paroxysmal cold hemoglobinuria (PKH), Landsteiner was both familiar with the experimental system and uninhibited by Ehrlich's "law." He designed experiments using the serum and cells from PKH patients and normal controls, and proved beyond question that PKH is an autoimmune disease.[21] Unfortunately, most of the rest of the world believed in Ehrlich, and the broad significance of the PKH finding was substantially lost for half a century.[22]

Effects on Metchnikoff's phagocytic theory

It is ironic that one of the more significant nails in the coffin of Metchnikoff's phagocytic theory of immunity should have been driven by his own most famous student, Jules Bordet. Even before the discovery of immune hemolysis in 1899, the cellular theory had suffered severe attack by the proponents of a humoral explanation of immunity. Behring and Kitasato's discovery of antitoxic antibodies which protect against diphtheria and tetanus, and Pfeiffer's demonstration that antibody mediates the lysis of cholera organisms (to which Bordet contributed in no small measure, as indicated above), could not be gainsaid. When Bordet's finding (that antibody and complement mediate hemolysis) started the wide search for other cytotoxic antibodies, attention was further diverted from cellular mechanisms.

This diversion was magnified when Ehrlich seized upon the Bordet finding and used it to extend and perfect his humoralist side-chain theory; shortly thereafter, Ehrlich published pictorial representations of these humoral entities that seemed to lend credibility to their importance.[23] Now, in addition to the ease of working with antibodies, employing such techniques as hemolysis, bacterial agglutination, and the precipitin reaction (in contrast to the difficulties involved in working with cells), one actually felt that one could "see" the antibody itself in terms of Ehrlich's cartoons. As a result, the cellularist sun went into decline while that of the humoralists rose brightly, as is testified to by the vote of scientists around the world in their choice of research problems. By the time that the Swedish Academy voted to award the 1908 Nobel Prize in Medicine jointly to Ehrlich and Metchnikoff, the latter's theory had substantially become history rather than current belief.

Practical consequences of immune hemolysis

The Wassermann reaction and serodiagnosis

During the course of his extensive attempts to show that Ehrlich was wrong in thinking that there were multiple complements of differing specificities, Bordet was able to show that even the antibody-sensitized stromata of hemolyzed erythrocytes would deplete a serum of *all* of its complement.[24] It was then but a short step for Bordet and Octave Gengou to demonstrate that *any* antigen–antibody interaction would fix complement, thus depriving it of its ability further to mediate the immune hemolysis of a standard dose of indicator-sensitized erythrocytes.[25] Here, then, was a system that would permit the measurement of any mixture involving an antigen and its specific antibody.

These results suggested to many investigators that if one had a known antiserum, one could use it and the hemolytic indicator system to test for antigen. Alternatively, if a known antigen were available, one could similarly test for specific antibody. For the first time, a sensitive and specific serodiagnostic test for infectious disease seemed possible. The first applications of this approach involved reports of the serodiagnosis of tuberculosis, independently by Wassermann and Bruck[26] and by Citron.[27] Then Wassermann and coworkers[28] (and independently, Detré[29]) reported application of this technique to the diagnosis of syphilis, and a new industry was born. This diagnostic technique was considered so important that it soon moved out of the research laboratory and into clinical hospital practice, where it was for a long time one of the mainstays of the clinical laboratory. The complement fixation test, along with several other immunodiagnostic procedures, became the domain of a new specialty that called itself "serology," and over the next half-century the journals were filled with articles describing modifications of the complement fixation test and its application to other diseases.

The discovery of blood groups

During the late 1890s, the discovery of the phenomena of hemagglutination and of immune hemolysis (and especially the work of Bordet, and of Ehrlich and Morgenroth) focused the attention of many investigators on the erythrocyte as a convenient tool for productive study. Many laboratories took up the new subject, including that of Karl Landsteiner. However, Landsteiner was less interested in the antibodies found in the serum of immunized subjects than in those present in normal individuals. He felt, with Bordet, that they were not the same, and that these so-called "natural" antibodies were at best the precursors of the immune bodies.[30] There is the feeling, in reviewing Landsteiner's studies on hemagglutination and hemolysis over the ensuing decade, that most of them were aimed at testing and verifying this hypothesis, and at challenging Ehrlich's theories at the same time.

Whatever the case, this great admirer of Bordet wasted no time in devoting himself to the study of anti-erythrocyte antibodies, and within a year of Bordet's publication had sent in several reports of his work.[31] These included studies not only of natural antibodies to xenogeneic erythrocytes in the serum of several species, but also of natural isoantibodies in the sera of animals and, finally, of man.[32] These Landsteiner found in abundance and, with the experimental care typical of the man, he tested the sera of a number of his laboratory colleagues against a library of red cells from the same individuals. He was able, from these results, to describe three different erythrocyte groups,[33] the first two of which contained their own distinctive surface antigens while the third possessed neither antigen. These later became types A, B, and O; the fourth type, later known as AB, was reported by Landsteiner's assistants the following year.[34] The subsequent story of Landsteiner's discovery of other blood groups (M, N, and P), and later of the Rhesus system and the implications of this for the pathogenesis of erythroblastosis fetalis, are too well known to justify further elaboration here.

If Bordet's immune hemolysis stimulated the discovery of blood groups, then surely that discovery led to the rapid application of blood typing for trans-fusions, anthropological studies,[35] and forensic purposes.[36] The practical value of these approaches are still being realized now, 100 years later!

Comments

We have seen how the chance discovery of a phenomenon, the immune hemo-lysis of erythrocytes by specific antibody and nonspecific complement, struck a responsive chord among researchers in the nascent field of immunology at the end of the nineteenth century. The observation was rapidly converted to a full experimental system whose implications broadly expanded the range of contemporary theory and research programs. Not only did this new experi-mental approach lead directly to new avenues of research and practical appli-cation (such as the search for cytotoxic antibodies as the cause of disease, and the development of serodiagnostic procedures to detect infectious disease), but it had interesting indirect effects as well. By highlighting still further the ease of working with antibodies, it provided significant support to the humoralist overthrow of Metchnikoff's cellular theory of immunity. Again, simply by calling attention to the existence of anti-erythrocyte antibodies, it stimulated many different investigations and helped to prepare the mind of investigators for particular interpretations of their results. It is likely that without the stimulus of the work of Bordet and of Ehrlich and Morgenroth on immune hemolysis, Landsteiner's discovery of the ABO blood groups might have been much delayed. Again, without this earlier work, it is unlikely that the experiments of Donath and Landsteiner, leading to an autoimmune concept of the pathogenesis of paroxysmal cold hemoglobinuria, would have been designed at the time and carried out so convincingly.

Most interesting is the demonstration of the continuing border negotiations between theory and practice revealed by an analysis of the uses to which Bordet and Ehrlich put this new experimental system. Time after time an experimental result would suggest a modification of theory, which in turn would suggest a new hypothesis and a new experiment, in an ever-expanding spiral which contributed much to the growing body of immunological knowledge as well as to the growing acceptance by the medical community of immunology as an important new discipline. It was not by chance that three of the first eight Nobel Prizes in Physiology or Medicine were awarded to those associated with the field: von Behring, 1901; Koch, 1905; and Ehrlich and Metchnikoff, 1908. Charles Richet would receive the Prize in 1913 for the discovery of anaphylaxis, Bordet (belatedly) in 1919, and Landsteiner (for blood groups) in 1930. Nor was it surprising that the burst of activity, stimulated in no small part by the discovery of immune hemolysis, would lead to the founding of the first journals in the field, the *Zeitschrift für Immunitätsforschung* in 1908 and the *Journal of Immunology* in 1916, and of the first professional society in the field, the American Association of Immunologists, in 1913.

Notes and references

1. Galison, P., *Image and Logic: The Material Culture of Microphysics*, Chicago, University of Chicago Press, 1997, Chapter 9.
2. See, for example, Moulin, A.-M., *Le Dernier Langage de la Médicine: Histoire de l'Immunologie de Pasteur au Sida*, Paris, Presses Universitaires, 1991; Tauber, A.I., and Chernyak, L., *Metchnikoff and the Origins of Immunology*, New York, Oxford University Press, 1991; Corbellini, G., *L'Evoluzione del Pensiero Immunologico*, Turin, Bollati Boringhieri, 1990; Mazumdar, P.M.H., *Species and Specificity: An Interpretation of the History of Immunology*, Cambridge, Cambridge University Press, 1995; and Brent, L., *A History of Transplantation Immunology*, London, Academic Press, 1997.
3. Cambrosio, A., and Keating, P., "Between fact and technique: the beginnings of hybridoma technology," *J. Hist. Biol.* **25**:175, 1992.
4. Cambrosio, A., and Keating, P., "A matter of FACS: constituting novel entities in immunology," *Med. Anthropol. Q.* **6**:362, 1992.
5. The nature and heuristic value of the debate between cellularists and humoralists is discussed in Chapter 2; see also Silverstein, A.M., "The Pasteur Institute and the Birth of Immunology: The Great Immunological Debates," in Cazenave, P.-A., and Talwar, G.P., eds, *Immunology: Pasteur's Heritage*, New Delhi, Wiley Eastern, 1991, pp. 11–20.
6. Pfeiffer, R., and Issaeff, V., *Z. Hygiene* **17**:355, 1894; Pfeiffer, R., *Z. Hygiene* **18**:1, 1894.
7. Bordet, J., *Ann. Soc. R. Sci. Med.* **4**:455, 1895.
8. Belfanti, S., and Carbone, T., *Giorn. Accad. Med. Torino* **46**:321, 1898.
9. Bordet, J., *Ann. Inst. Pasteur* **12**:688, 1899.
10. An implicit teleology had hitherto suggested that the immune response had evolved to protect the body, and that its functions were directed solely against pathogenic organisms and their associated toxins.
11. Ehrlich, P., "Die Wertbemessung des Diphtherieheilserums und deren Theoretische Grundlagen," *Klin. Jahrb.* **60**:299, 1897 (English translation in Himmelweit, F., ed., *The Collected Papers of Paul Ehrlich*, Vol. II, London, Pergamon, 1957, pp. 86–106).
12. Ehrlich, P., and Morgenroth, J., "Zur Theorie der Lysinwirkung," in *Collected Papers*, pp. 143–149; idem, "Ueber Hämolysine," Part II, pp. 156–164; Part III, pp. 196–204; Part IV, pp. 213–223; Part V, pp. 234–245; Part VI, pp. 256–277.
13. For a detailed account of these mutually incommensurable languages and their consequences, see Chapter 14.
14. A number of the early findings are discussed and/or reviewed in Vol. 14 of *Ann. Inst. Pasteur*, 1900. See also the review by Hans Sachs, in *Handbuch der Technik und Methodik der Immunitätsforschung*, Vol. 2, Jena, Fischer, 1909, p. 186 ff.
15. Golowin, S., "Importance of cellular poisons in the pathology of the eye and especially in the pathogenesis of sympathetic inflammation", *Russky Vratch* **3**:802, 1904; idem, *Klin. Monatsbl. Augenheilk.* **47**:150, 1909.

16. Santucci, S., "L'Oftalmia Simpatica in Relazione alla Teoria delle Citotossine," *Ann. Ottalmol. Pavia* **36**:244, 1907.

17. For a more complete account of this long series of investigations, see Silverstein, A.M., "Ocular immunology: on the birth of a new discipline," *Cell. Immunol.* **136**:504, 1991.

18. Ehrlich, P., *Deutsch. med. Wochenschr.* **27**:865, 888, 913, 1901.

19. For a discussion of the Ehrlich–Gruber debate, see Chapter 6.

20. Landsteiner, K., *Zentralbl. Bakt. Orig.* **25**:546, 1899; Landsteiner, K., *Zentralbl. Bakt. Orig.* **27**:357, 1900; Donath, J., and Landsteiner, K., *Wien klin. Wochenschr.* **14**:713, 1901.

21. Donath, J., and Landsteiner, K., "Ueber Paroxysmale Hämoglobinurie," *Münch. med. Wochenschr.* **51**:1590, 1904.

22. Indeed, it can reasonably be suspected that this concept, originating from so distinguished a source, was responsible in good part for the almost sixty-year delay in the general appreciation that autoimmune diseases were possible. I note elsewhere (Chapter 8) that Ernest Witebsky, a scientific grandson of Ehrlich and firm adherent of the master's teachings, long refused to believe even his own results on the discovery of experimental autoimmune thyroiditis.

23. Ehrlich, P., "Croonian Lecture: On Immunity with Special Reference to Cell Life," *Proc. R. Soc. London*, **66**:424, 1900 (reprinted in *Collected Papers*, Vol. II, pp. 178–195). See also Alberto Cambrosio, Daniel Jacobi, and Peter Keating's discussion of the influence of these pictorializations, in *Usages de l'Image au XIX Siècle*, Paris, Creaphis, 1991, and Cambrosio, A., Jacobi, D., and Keating, P., "Ehrlich's 'beautiful pictures' and the controversial beginnings of immunological imagery," *Isis* **84**:662, 1993.

24. Bordet, J., *Ann. Inst. Pasteur* **14**:257, 1900.

25. Bordet, J., and Gengou, O., *Ann. Inst. Pasteur* **15**:289, 1901.

26. von Wassermann, A, and Bruck, C., *Deutsch. med. Wochenschr.* **32**:449, 1906.

27. Citron, J., *Zentralbl. Bakt.* **41**:230, 1906.

28. von Wassermann, A., Neisser, A., and Bruck, C., *Deutsch. med. Wochenschr.* **32**:745, 1906.

29. Detré, L, *Wien. klin. Wochenschr.* **19**:619, 1906.

30. A more detailed discussion of Landsteiner's views and of his opposition to Ehrlich's theories may be found in Chapter 14.

31. Karl Landsteiner's complete bibliography may be found in the appendix to his *The Specificity of Serological Reactions*, New York, Dover, 1962, prepared by Landsteiner's long-time collaborator Merrill Chase as a posthumous tribute.

32. The first brief report, apparently the result of a chance observation, was that of Landsteiner, note 20 (1900).

33. Landsteiner, K., *Wien. klin. Wochenschr.* **14**:1132, 1901.

34. von Decastello, A., and Stürli, A., *Münch. med. Wochenschr.* **49**:1090, 1902.

35. von Dungern, F., and Hirszfeld, L., *Z. Immunitätsforsch.* **4**:531, **6**:284, 1910. See also Hirszfeld, L., *Ergeb. Hyg. Bakteriol.* **8**:367, 1926.

36. Lattes, L., *Arch. Antropol. Crim. Psych. Med. Leg.* **36**:4, 1915; *Arch. Ital. Biol.* **64**:3, 1915. See also Lattes, L., *Individuality of the Blood* (English translation), London, Oxford University Press, 1932.

21 Darwinism and immunology: from Metchnikoff to Burnet

I have indeed dared to put forward a new theory of inflammation, only because I felt that I had Darwin's great conception as a solid foundation to build upon...
Elie Metchnikoff, 1892[1]

It is the rare branch of the biological or medical sciences that has not been affected by Charles Darwin's theory of evolution. However, the theory is not monolithic; it is really a set of five interlaced subtheories, as Ernst Mayr has pointed out:[2] evolution *per se* (i.e., change); common descent; gradualism; speciation; and natural selection. These are not of equal importance when we consider the influence of Darwinism on any particular scientific specialty. In the late nineteenth century, those scientists most interested in Darwinian thought were:[3]

- geologists, interested in the time-span of change, and the gradualist contradiction of saltationist and cataclysmic theories
- paleontologists, interested in the contradictions that common descent and speciation posed for essentialist views on the fixity of species
- zoologists and botanists, interested in the morphological relations among taxa, usually on the higher levels of whole organisms or major systems and functions.

In general, physiologists interested in the cellular and molecular mechanisms of function paid little overt attention to evolutionary theory.[4] Their references to evolution were in the main implicit, and made in the context of organs and systems rather than of mechanisms.[5] Other biomedical sciences paid little attention to Darwinian precepts. Only in the field of pathology was the inheritance of such acquired characteristics as malformations viewed as a significant component of the evolutionary process, most notably by Rudolph Virchow, the father of cellular pathology.[6] The inheritance of acquired characteristics was called "soft" inheritance, and was associated with the name of Lamarck; it would eventually be viewed as a heresy in evolutionary thought. In the nineteenth century, though, this was not viewed as anti-Darwinian, since Darwin himself had believed initially in soft inheritance. Only in the next century did the movement called "Neodarwinism" purify the theory by ruling out this type of inheritance.[7]

It is the purpose of this chapter to examine the history of Darwinian influences on immunological thought during the formative years of the discipline. It may be noteworthy that the first discussion devoted broadly to this question was that of Alain Bussard's "Darwinisme et Immunologie," presented by this multi-talented individual to a meeting of philosophers in 1982.[8] But those who knew him will

A History of Immunology, Second Edition
ISBN: 978-0-12-370586-0

Copyright © 2009, Elsevier Inc.
All rights reserved

recall that Bussard remained throughout his career an unreconstructed immunochemist, always searching for an instructionist alternative to Burnet's Darwinian clonal selection theory. He seems, in this discussion, to perpetuate the traditional nineteenth- and early twentieth-century French distaste for Darwinism,[9] and indeed repeats somewhat ruefully that "one of my Anglosaxon colleagues [noted that] there is always present in the heart of a Frenchman a slumbering Lamarckian."

The struggle for existence

Darwin's notion of "the survival of the fittest" was expressed in terms of a Malthusian contest[10] among individuals *within* a species, selection favoring those best able to compete. In 1882, Virchow's student Paul Grawitz advanced a theory of acquired immunity based upon the idea that disease represents an *inter*-species struggle between the cells of the host and the parasite.[11] He supposed that infection or active immunization would specifically "energize" host cells to battle more efficiently – a vital quality that he thought would be inherited by later generations of cells. It was Ilya Metchnikoff, with his phagocytic theory of immunity, who gave full voice to the suggestion that the critical struggle in disease is *between* different species; the immune response represents the principal weapon used by the host to combat the pathogenic organism actively.[12]

Metchnikoff was initially critical of the Malthusian basis of Darwin's theory,[13] but his study of embryology and the evolution of the process of digestion convinced him of the importance of Darwinian concepts, so that when he first observed phagocytosis[14] he was quick to give it a Darwinian interpretation.[15] He suggested that vertebrate phagocytic cells are the remnants in higher animals of the original primitive intracellular digestive process of lower organisms. These cells had then assumed the added function in higher organisms of digesting effete elements and noxious foreign invaders. Thus, the host is now an active participant in the outcome of an *inter*-specific struggle for survival between pathogen and host, and the phagocytes have evolved to become the principal factors in immunity to infection. (The earlier size disparity that pictured the conflict as between microbe and man was now recast more appropriately as the more balanced struggle between microbe and phagocytic cell.)

It is interesting that Louis Pasteur, very much the religious conservative and at least a passive anti-Darwinist,[16] would invite Metchnikoff to become a Chief of Service at his new research Institute in Paris.[17] It was, curiously, the expatriate Metchnikoff who would represent France at the Fiftieth Anniversary Jubilee celebration of Darwin's *Origin* in 1909 in Cambridge, England.

Paul Ehrlich's only mention of Darwinism came in his 1892 study of the transmission of immunity from mother to offspring.[18] Here, he mentions that

"It is generally accepted that acquired characteristics are not inherited, *in contrast to the original Darwinian theory*" (my italics), and he will demonstrate this with an *experimentum crucis*[19] that shows that all neonatal acquired immunity is passively transferred, and nothing is inherited as an intrinsic quality by the newborn from an immune mother or father.[20]

But Ehrlich's 1897 side-chain theory of antibody formation[21] was implicitly Darwinian. He implied that specific receptors for nutrients, toxins, etc. had evolved to permit the cell to engage in its normal physiological processes. These naturally occurring receptors are normally present as "side-chains" on the cell surface; when antigen interacts with its specific receptor on a cell, it stimulates that cell to overproduce these receptors and to release them into the blood as circulating antibodies. Interestingly, when the antibody repertoire size was seen to be very large, it was a Darwinian argument from the pen of Ehrlich's opponent Max Gruber that helped to sink the side-chain theory. Gruber asked how in evolution there could have developed so many side-chains (antibody receptors) specific for substances that might never be seen by the body. What is the selective force that acts to preserve these unlikely specificities?[22]

William Welch, of Johns Hopkins, was, at the turn of the century, America's leading pathologist.[23] He had studied in Europe, and was interested in the new bacteriology and immunology as well as in the mechanisms of tissue damage that accompany infectious diseases. Indeed, from time to time he took it upon himself to review for his contemporaries the recent advances in bacteriology and immunity research.[24] In the context of one such review, presented as the Huxley Lecture for 1902 at the Charing Cross Hospital Medical School in London,[25] Welch advanced a most curious theory of immunological function and of immunopathology. He suggested that bacteria might have evolved an immune defense mechanism for their protection analogous to that of their vertebrate opponents!

Welch first reviewed Metchnikoff's phagocytic theory (which did not receive his enthusiastic support) and then Ehrlich's side-chain theory of receptors (which did). He pointed out that while certain disease agents inflict their damage by means of toxins, others do not. Thus, the host employs antitoxins to protect itself against the exotoxins of diphtheria and tetanus. However, against such organisms as the cholera bacillus it responds with the production of bacteriolytic antibodies directed against the pathogen itself rather than against its products, as Pfeiffer had shown.[26] Then the anti-cell phenomenon was generalized; Bordet showed that antibodies could be raised that are able to destroy erythrocytes,[27] Metalnikoff demonstrated antibodies against spermatozoa,[28] and soon investigators were experimenting with the production of antibodies that might destroy the cells of any and all tissues and organs;[29] these they called *cytotoxins*.

Welch seized upon this phenomenon of cytotoxin formation, and applied it to the question of the struggle between host and parasite. He suggested that if the host had evolved receptors (cytolytic antibodies) specific for invading pathogens to help in its defense, might not the pathogen also have evolved antibody receptors to attack the cells of the infected host, thus assisting it in the struggle? As Welch summarized this theory of infectious disease pathogenesis:[30]

> *the struggle between the bacteria and the body cells in infections may be conceived as an immunizing contest in which each participant is stimulated by its opponent to the production of cytotoxins hostile to the other, and thereby endeavors to make itself immune against its antagonists.*

Not only was this an extension of immunological theory, but it served also as a convenient way for pathologist Welch to explain many of the damaging features of certain infectious processes that hitherto could not be ascribed to the action of the usual toxin molecules. Although Welch suggests in this talk that some preliminary data from his laboratory appear to support the existence of what he termed "bacteriogenic cytotoxins," he seems thereafter to have reconsidered the theory, since nowhere later does he revisit it.

If indeed the pathogen may not develop its own defensive immune system, it is still not without certain other modes of self-protection. Thus, in the contest between species, both sides may evolve a variety of protective mechanisms. Rolf Zinkernagel has pointed out that,[31] in the co-evolution between parasite and vertebrate host, a balance of forces is generally arrived at to assure the survival of both parties.

Immunochemistry: immunology without Darwin

Just as Metchnikoff's cellular theory of immunity gave way in the 1890s to repeated demonstrations of the importance of humoral antibodies, so too did Ehrlich's side-chain theory lose adherents a decade later when challenged by the implications of an expanding repertoire of antibody specificities. With the demise of these biological theories, Darwinian influences on immunologic thought also disappeared, not to be revived for half a century or more. Immunology seemed to lose its medical orientation.[32] Antitoxic serotherapy could not be extended beyond diphtheria and tetanus; there were few new vaccine successes against the remaining important infectious diseases; and work on anaphylaxis, asthma, and hayfever was pretty much taken over by the new field of clinical allergy. It was, insofar as theory and practice in biology and medicine were concerned, a time of relative stagnation.

Into the breach stepped more chemically-oriented investigators, impressed by the demonstration that antigens could be chemically manipulated. Here was a way to study chemical structure and function, and the stoichiometry and thermodynamics of the antigen–antibody interaction. The new emphasis was on *immunochemistry*, a term coined by Svante Arrhenius,[33] while studies on immunopathology, autoimmunity, etc. now fell out of the mainstream. This new approach was fortified by Landsteiner's impressive studies on serological specificity,[34] and by Heidelberger's introduction of quantitative methods.[35]

The change in orientation of the field was accompanied by an analogous shift in theory. If there seemed to be no way that Darwinian concepts could explain the generation of such a large diversity of immunological specificities, then obviously the information for their formation must arise from outside the host, and antigen appeared to be the only candidate to mediate this acquired characteristic. It is not that chemically-oriented immunologists of the period consciously adopted a Lamarckian point of view, but rather that chemists are not accustomed to thinking in evolutionary terms. Their molecules have no built-in history, beyond the simple fact of their formation, and no internal program built up over time. Thus, a chemical approach to a theory of antibody formation was perforce *non*-Darwinian, and might even appear to be Lamarckian.

Now the host was once again pictured to be passive, and the antigen served as a template to impress upon a nascent antibody molecule the structure determining specificity. In Felix Haurowitz's scheme[36] the antigen was supposed somehow to control the order of addition of amino acids to the growing peptide chain, thus determining specificity. In Linus Pauling's view,[37] the nascent globulin would fold itself around the antigen template, in that way assuming a configuration that would assure a tight fit and thus specificity. Only with Macfarlane Burnet's 1941 adaptive enzyme theory of antibody formation[38] and his 1949 indirect template theory[39] was a Lamarckian inheritance of these acquired characteristics postulated explicitly, where acquired information might descend from mother somatic cells to their daughters.

Immunobiology: Darwin returns to immunology

Chemical approaches dominated immunology for almost half a century, but during the late 1940s and 1950s a group of phenomena began to challenge the field. These included tissue transplantation, immunological deficiency diseases, autoimmune diseases, and immunological tolerance, each of which posed questions that the current immunochemical paradigm could not answer. In addition, there was now a new generation of investigators, who approached immunology differently and were not wedded to its former preconceptions; they came from various clinical disciplines, from experimental pathology, and from such basic sciences as genetics and physiology, etc.

Macfarlane Burnet, the archtypical biologist, was a virologist who had studied the evolution of influenza serotypes, the competitive struggles for survival between host and parasite, and clonal phenomena in tissue culture and in the reproduction of bacteria. He had early on developed an interest in the immune response, and speculated freely on possible mechanisms for the formation of antibodies, as we have seen. It will be noted that even in his early instructionist theories, Burnet placed emphasis on the importance of cell population dynamics for any theory of antibody formation.

Sparked by Niels Jerne's 1955 publication of a natural selection theory of antibody formation,[40] both David Talmage[41] and Burnet[42] postulated that the most important role in antibody formation was played by cells actively selected by antigen to proliferate and differentiate. The theory of clonal selection was most fully elaborated by Burnet in 1959,[43] and was given further genetic support by Joshua Lederberg[44] and statistical support by Talmage.[45] Burnet would repeatedly refer to his theory as "Darwinian" and "evolutionary," and indeed it was.

The clonal selection theory advanced three major propositions:[46]

1. The capacity to produce antibody is encoded in the genome, with the repertoire expanded by somatic mutational events
2. An antibody-forming cell is restricted to one (or very few) specificities, and puts specific antigen-reactive receptors on its cell membrane;
3. The cells selected by antigen are stimulated, inducing antibody formation and clonal proliferation – daughter cells inherit the properties of their progenitors.

Clonal selection rapidly found favor among all save a few diehard immuno-chemists. It appeared to explain well such *biological* phenomena as the long-term continuation of antibody formation; no longer was persisting antigen necessary. It furnished an explanation for the more rapid and heightened secondary booster response by the presence of primed memory cells. It explained affinity maturation as the favored selection of those clonal precursors with a better fit for antigen. Finally, it sought to explain immunological tolerance in terms of fetal clonal deletion, and autoimmunity in terms of "forbidden clones."

Darwinism triumphant

We noted in the discussion above how the ebb and flow of Darwinian thought in immunology followed so precisely the alternating dominance of biologists and chemists in the field. Now, however, modern immunogenetics has demonstrated the evolution of the immunoglobulin superfamily of genes,[47] presumably derived originally from primitive cell adhesion molecules.[48] We understand whence (if not precisely how) the minigenes arose that combine to form immunoglobulin and T cell receptors, and that these complicated mechanisms were a unique type of evolution, designed to prepare the host to deal with unanticipated future challenges (rather than having to adapt slowly to each new environmental change).[49]

This does not mean, of course, that all questions relating to the evolution of acquired immune responses have been answered. We do not know yet what specificities are encoded in the germline, nor by what selective forces they are conserved.[50] There is still debate about whether evolution has provided for the regulation of immune responses in terms of idiotypic networks or by means of complicated sets of regulatory cellular or molecular on/off signals. Among the few who have attempted to formulate a comprehensive view of the immune system in evolutionary terms are Rodney Langman and Melvin Cohn[51] from one direction, and Irun Cohen from another.[52] Even here, however, their suggestion that the underlying evolutionary driving force has been the discrimination of self from non-self has been roundly questioned.[53]

Immunology is not unique in facing further gaps in our understanding of the evolution of its mechanisms; every area of Darwinian evolution has sparked continuing controversy. Thus, there is still dispute about mechanisms of speciation, about whether some evolution is punctuated rather than smoothly gradual, and especially about how natural selection operates on phenotypes (such as antibody formation) that are governed by multigenic processes.

Notes and references

1. Metchnikoff, E., *Lectures on the Comparative Pathology of Inflammation*, New York, Dover, 1968, p. xvii (reprint of the 1893 English translation).
2. Mayr, E., in Kohn, D., ed., *The Darwinian Heritage*, Princeton, Princeton University Press, 1985, pp. 755–772.
3. See, for example, Hull, D.L., in *The Darwinian Heritage*, note 2, pp. 773–812.
4. For a discussion of Darwin and the physiologists, see French, R.D., *J. Hist. Biol.* 3:253, 1970.
5. Geison, G., *Michael Foster and the Cambridge School of Physiologists*, Princeton, Princeton University Press, 1978, pp. 340–344.
6. Churchill, F.B., *J. Hist. Med.* 31:117, 1976.
7. See Ernst Mayr's prologue to Mayr, E., and Provine, W.B., eds, *The Evolutionary Synthesis*, Cambridge, Harvard University Press, 1980, pp. 1–48.
8. Bussard, A., *Bull. Soc. Franç. Philosophie* 77:1, 1982.
9. Conry, Y., *L'Introduction du Darwinisme en France au XIX Siècle*, Paris, Vrin, 1974, p. 231; Corsi, P., in *The Darwinian Heritage*, pp. 698–711, note 3.
10. Malthus had made the dire prediction that population would increase faster than the food supply, leading to fierce competition among the members of a species.
11. Grawitz, P., "Die Theorie der Schutzimpfung," *Arch. Pathol. Anat. Physiol.* 84:87, 1881, p. 106 ff.
12. See Tauber, A.I., and Chernyak, L., *Metchnikoff and the Origins of Immunology: from Metaphor to Theory*, New York, Oxford University Press, 1991, pp. 101–153; see also Tauber, A.I., *The Immune Self: Theory or Metaphor?*, Cambridge, Cambridge University Press, 1994.
13. Todes, D.P., "Metchnikoff, Darwinism, and the phagocytic theory," in *Darwin without Malthus: The Struggle for Existence in Russian Evolutionary Thought*, Oxford, Oxford University Press, 1989, pp. 82–103; Tauber, A.I., and Chernyak, L., "The problem of evolution in Metchnikoff's works," in *Metchnikoff and the Origins of Immunology*, note 12, pp. 68–100.
14. Metchnikoff, I., *Archiv. Pathol. Anat.* 96:177, 1884. The full range of Metchnikoff's Darwinian contributions is covered in Gourko, H., Williamson, D.T., and Tauber, A.I., eds, *The Evolutionary Biology Papers of Elie Metchnikoff*, Boston, Dordrecht, 2000.
15. Metchnikoff's early elaboration was in the context of the phagocyte's contribution to a protective inflammation, in *Lectures on the Comparative Pathology of Inflammation*, New York, Dover reprint, 1968 (first published in French in 1892). Metchnikoff later extended it more broadly as the leading player in defense against

all infectious diseases in *Immunity in the Infectious Diseases*, New York, Johnson reprint (first published in French in 1901).

16. In 1877, Pasteur wrote a letter to C. James, author of a scurrilous attack on Darwin's evolution entitled *Du Darwinisme ou l'Homme Singe*. "I read enough to be charmed by your very elegant and clear style, and by the high purpose and healthy doctrines that have inspired this excellent book," cited in Conry, Y., *L'Introduction du Darwinisme*, note 9, p. 231.

17. Moulin, A.-M., *Le Dernier Langage de la Médicine: Histoire de l'immunologie de Pasteur au Sida*, Paris, Presses Universitaires, 1991, p. 49 ff.

18. Ehrlich, P., *Z. Hyg. Infektkr.* **12**:183, 1892; English translation in *Collected Papers of Paul Ehrlich*, Vol. II, London, Pergamon, 1957, pp. 31–44.

19. These experiments are discussed in Silverstein, A.M., *Nature Immunol.* **1**:93, 2000, and in Silverstein, A.M., *Paul Ehrlich's Receptor Immunology: The Magnificent Obsession*, New York, Academic Press, 2002, pp. 27–39.

20. Some still believed in the inheritance of acquired characteristics, including S. Arloing, who said in his book *Les Virus* (Paris, Alcan, 1891, p. 285) "when [the father] participates in the act of insemination, he may transmit a part of the immunity which he enjoys. The male ovule...has been dynamically modified...it carries a vaccinating substance which spreads in all of the cells of the embryo and fetus."

21. Ehrlich, P., *Klin. Jahrb.* **60**:299, 1897; English translation in *Collected Papers of Paul Ehrlich*, Vol. 2, pp. 107–125. These concepts were more fully elaborated several years later in Ehrlich's Croonian Lecture before the Royal Society, *Proc. R. Soc. London* **66**:424, 1900, and in *Collected Papers*, Vol. 2, pp. 178–195.

22. von Gruber, M., *Münch. med. Wochenschr.* **48**:1214, 1901; *Wien. Klin. Wochenschr.* **16**:791, 1903.

23. Fleming, D., *William H. Welch and the Rise of Modern Medicine*, Boston, Little Brown, 1954.

24. Welch W.H., *Papers and Addresses by William Henry Welch*, Vol. II, *Bacteriology*, Baltimore, Johns Hopkins Press, 1920.

25. Welch, W.H., *Br. Med. J.* ii:1105, 1902.

26. Pfeiffer, R., *Z. Hyg. Infektkr.* **18**:1, 1894.

27. Bordet, J., *Ann. Inst. Pasteur* **12**:688, 1899.

28. Metalnikoff, S., *Ann. Inst. Pasteur* **14**:577, 1900.

29. See, for example, Metchnikoff's review of the field of cytotoxins in *Ann. Inst. Pasteur* **14**:369, 1900.

30. Welch, W.H., note 25, page 1109.

31. Zinkernagel, R.M., "Immunity taught by viruses," *Science* **271**:173, 1996. See also the criticism of this article by McKean, G.A., Nunney, L., and Zuk, M., "Immunology taught by Darwin," *Science* **272**:634, 1996, and Zinkernagel's response in *Science* **272**:635, 1996.

32. The transition from medical to chemical orientation is discussed in Chapter 17.

33. Arrhenius, S., *Immunochemistry*, New York, Macmillan, 1907.

34. Landsteiner, K., *The Specificity of Serological Reactions*, New York, Dover, 1962, a reprint of the second edition, published by the Harvard University Press in 1945.

35. See Kabat, E.A., and Mayer, M.M., *Quantitative Immunochemistry*, Springfield, Thomas, 1948.

36. Breinl, F., and Haurowitz, F., *Z. Physiol. Chem.* **192**:45, 1930.

37. Pauling, L., *J. Am. Chem. Soc.* **62**:2643, 1940.

38. Burnet, F.M., *The Production of Antibodies*, Melbourne, Macmillan, 1941.
39. Burnet, F.M., and Fenner, F., *The Production of Antibodies*, 2nd edn, Melbourne, Macmillan, 1949.
40. Jerne, N.K., *Proc. Natl Acad. Sci. USA* **41**:849, 1955.
41. Talmage, D.W., *Annu. Rev. Med.* **8**:239, 1957.
42. Burnet, F.M., *Austral. J. Sci.* **20**:67, 1957.
43. Burnet, F.M., *The Clonal Selection Theory of Antibody Formation*, Cambridge, Cambridge University Press, 1959.
44. Lederberg, J., *Science* **129**:1649, 1959.
45. Talmage, D.W., *Science* **129**:1643, 1959.
46. The theory has sometimes been attacked because Burnet was wrong about some of its ancillary propositions, most notably about the mechanism and significance of immunological tolerance. This debate is discussed in Chapter 5.
47. Williams, A.F., *Immunol. Today* **8**:298, 1987; Hunkapiller, T., and Hood, L., *Adv. Immunol.* **44**:1, 1989.
48. Edelman, G.M., "CAMs and Igs: cell adhesion and the evolutionary origins of immunity," *Immunol. Rev.* **100**:9, 1987; Matsunaga, T., and Mori, N., *Scand. J. Immunol.* **25**:485, 1987.
49. See Susumo Ohno's speculation on what he termed "Promethian evolution," *Perspect. Biol. Med.* **19**:527, 1976; Ohno, S., et al., *Progr. Immunol.* **4**:577, 1980.
50. See the stimulating discussion by Cohn, M., Langman, R.E., and Geckeler, W., *Progr. Immunol.* **4**:153, 1980.
51. See Langman, R.E., *The Immune System*, San Diego, Academic Press, 1989, and Melvin Cohn's introduction therein, pp. xii–xlvi.
52. Cohen, I., *Tending Adam's Garden: Evolving the Cognitive Immune Self*, New York, Academic Press, 2000.
53. See Chapter 5, and the discussions entitled "Self–nonself discrimination revisited" in *Seminars Immunol.* **12**:159, 2000.

22 The end of immunology?

As this younger generation of professionals is pressing rapidly toward the definitive solution of the antibody problem, we older amateurs had perhaps better sit back, waiting for the End.

Niels Jerne, 1967[1]

For the discipline of immunology, the scientific and even social event of the year 1967 was the International Symposium on Antibodies,[2] which convened at Cold Spring Harbor, Long Island, during a sunny week in June. Most of the world's leading immunologists came to discuss their most recent findings, and the Proceedings were attended by the electric excitement that characterizes a science in which progress is almost breathtakingly rapid. It seemed indeed as though a watershed had been reached, or that the divide was at least in sight. All the important conceptual questions of the past eighty years appeared for the first time to be answerable.

The meeting was opened by Sir Macfarlane Burnet, who declared that a new paradigm now directed the development of hypothesis and the design of experiments in immunology. In less than a decade, Burnet's clonal selection theory of antibody formation, based upon genetic control of antibody specificity, had deposed all earlier instruction theories. In fact, Burnet had more than hinted two years earlier[3] that his theory had furnished the approaches that would soon solve all outstanding problems in immunology (if indeed that had not already occurred!).

The progress reported during the course of the Symposium seemed to foretell the rapid solution of the most pressing problems in immunology. On the molecular level, immunoglobulin chain sequencing had reached the point where the complete structure of the immunoglobulin molecule was in sight, and the genetic basis for the specificity repertoire and the recognition of antigenicity seemed finally to be within reach. Delineation of the steps in antibody biosynthesis appeared well in hand, as did the structure of the antigen-specific lymphocyte receptor. On the cellular level, much information was available on the differentiation pathways of immunocytes, and on the collaboration of macrophages. Although T and B lymphocytes had not yet been named, studies of the differences between delayed hypersensitivity and antibody formation, and especially the newer information on the role of the thymus and of the avian bursa and the mammalian bone marrow, presaged the identification of interacting lymphocyte subsets. One could be justifiably pleased, in June of 1967, that important answers were arriving with impressive rapidity.

The Symposium closed with a masterful summary by Niels Jerne, entitled "Waiting for the End." He rightly drew attention to the triumph of the clonal selection theory, and implied that *cis*- (cellular) immunologists working forward

A History of Immunology, Second Edition
ISBN: 978-0-12-370586-0

Copyright © 2009, Elsevier Inc.
All rights reserved

from the first interaction of antigen with cell, and *trans-* (molecular) immunologists working backward from the structure of the antibody molecule, were very close to meeting one another in between – at which point all but the minor details would presumably have been settled. As Jerne implied in his final paragraph, cited above, the definitive solution of "the antibody problem" was near at hand. Jerne would later confirm this, at a meeting in Australia organized in Burnet's honor. He declared that immunology had been "solved" in 1957 with the publication of Burnet's clonal selection theory. By this time (1969), the title of Jerne's talk –"The complete solution of immunology"[4] – appeared to many in this rapidly expanding field to be accurate, but sounded like hyperbole to some others.

This declaration – that it was pretty much all over in immunology – calls to mind a similar, earlier pronouncement. In 1930, Dr Gerald Webb, one of the founders and the first President of the American Association of Immunologists (1913–1915), resigned from the group. He had been one of the young American students of Almroth Wright in England, who, back home in America, had banded together in 1913 to institutionalize their exciting young discipline. Now, for Webb, the excitement had subsided, and he could say that he "had lost interest because he could not see that it [the Society] was doing much to advance its science."[5] When a science appears unexciting, so also must its professional society.

Declarations of "the end" in other fields

Immunology is not the only discipline that has seen such pessimistic conclusions. However, history suggests that predictions about "definitive solutions" of scientific questions should be made with extreme caution. For some 2,000 years in the case of Aristotle, and almost 1,500 years in the case of Galen, the Western world appeared to have concluded that little could be added to the natural history writings of the former, nor to the medical writings of the latter, save perhaps for trivial scholastic disputation.[6] In another instance, the chief engineer of the Roman army wrote some 2,000 years ago that he could not envisage any improvement in instruments of warfare over the Roman short sword, the javelin, and the catapulta machine that had conquered the world.[7] One can imagine the same statement about "the ultimate weapon" being made repeatedly in the future, about the longbow, the arquebus, the repeating rifle, poison gas, and ultimately the hydrogen bomb. It was often predicted that these "ultimate weapons" would surely render warfare no longer possible!

While such declarations have been made in many fields,[8] perhaps none has seen quite such explicit ones as the discipline of physics.[9] During the eighteenth century, Newtonian physics was taught as the ultimate attainable truth and, following the advances of the nineteenth century, America's leading physicist Albert Michelson would say in 1894 that "it seems probable that most of the grand underlying principles have been firmly established, and that further advances are to be sought chiefly in the rigorous application of these principles..."[10]

Lord Kelvin, then the world's leading physicist, is reputed also to have said a bit earlier that it was pretty much all over in contemporary physics, and that "the future truths of physical science are to be looked for in the sixth place of decimals."[11] As Derek Price so aptly put it in his *Science Since Babylon*: "by about 1890, all natural phenomena had been divided and ruled, and only unimportant problems remained.... It was obviously reasonable to believe that finality was just around the corner."[12]

Among the many other examples that might be given, we may cite an essay by the great pathologist Rudolf Virchow in 1877, some eighteen years after the publication of his landmark book *Cellular Pathology*.[13] Virchow seemed to imply that pathology had solved its major problems, and there remained only a "mopping up," when he wrote:[14]

> *What efforts had to be made...to assign every phenomenon...to its proper place. And yet we seem to have succeeded in bringing firm order out of seeming chaos; the thousands of individual facts have been comprehended in a few well established laws and made easily accessible to the understanding of the younger generation in the new order.*

This view – that the pace of new discoveries was rapidly slowing and that mankind's knowledge was approaching an upper limit – has recurred often in history. It may remind us of the apocryphal story of the Director of the US Patent Office in the 1850s, who was reputed to have quit his job and suggested to Congress that his office be closed, since there would soon be nothing left to invent![15]

Silent decisions to leave immunology

We have thus far considered only explicit declarations of the end of a discipline. However, not every famous scientist who felt that he had "done it all" would declare to the world that it was all over in his discipline. Many, having achieved their ambitions within a field, would, with minimal fanfare, move on to new disciplines whose challenges seemed only to await their arrival. Some would visit the new field only briefly, soon to return to their original occupations; others would make the shift a permanent one. We will examine briefly several examples of scientists who have moved from one field to another, to point out some of the variations that may occur.

Paul Ehrlich

We come now to the interesting case of Paul Ehrlich, who, when he entered immunology in 1890, had already become world famous for his contributions to histology, cell physiology, and hematology.[16] In 1891 Ehrlich directed his attention to the exciting young field of immunology, and in a series of outstanding contributions to experimental immunology[17] he taught the world how to measure accurately both toxins and antitoxins. He provided a theoretical basis for the understanding of the formation and interactions of antibody – his side-chain theory of 1897.[18] In it, and in his famous experiments on immune hemolysis,[19] he not only explained *why* and *how* antibodies are formed, but even provided pictures to illustrate their chemical specificity and mode of action.[20] Ehrlich's ideas and experimental results would sweep the world and assure him the Nobel Prize in 1908.[21]

By 1901, however, Ehrlich must have decided that there was little about antibodies, complement, and the immune response left to explain – his work seemed to have answered most of the important questions. As he had done previously after contributing so much to histology and hematology, Ehrlich would move to another field – in this instance to experimental tumor research. Thenceforth, he would return to immunology only to defend his theory and data from attack, and that with reluctance. Eventually, he would move on to the development of therapeutic agents against various diseases, and would give the world the first scientific chemotherapy and Salvarsan treatment of syphilis. For this, it is almost generally agreed that he might well have received a second Nobel Prize, had he lived long enough.[22]

Gerald Edelman

Edelman trained in immunology, devoting himself to the study of the structure of the antibody molecule. In a series of landmark studies, he showed that myeloma proteins could be reductively cleaved into heavy and light chains, and that Bence Jones proteins were in fact immunoglobulin light chains. He and his colleagues further defined the different physiologic roles played by the several domains of immunoglobulins. This approach culminated in his establishment of the first complete amino acid sequence of a large protein molecule, an immunoglobulin.[23] For this, Edelman shared the 1972 Nobel Prize with Rodney Porter.

With what appeared to be the "solution to the antibody problem," Edelman then branched out broadly,[24] taking the Darwinian concepts of immunoglobulin evolution to point out the importance of cell adhesion molecules in morphogenesis[25] and to discuss what he suggested was the similar Darwinian evolution of neural function.[26] It is noteworthy that, like Francis Crick (see below), Edelman moved to neurobiology as the most challenging area, and to La Jolla, California, as the best place to pursue these interests.

Silent decisions to join immunology

Linus Pauling

Pauling was the *Wunderkind* of chemical physics during the 1920s and 1930s, bringing quantum mechanics to the solution of the nature of the chemical bonds that determine inorganic crystal structure.[27] After receiving the first Langmuir Prize of the American Chemical Society and becoming the youngest individual ever elected to the National Academy of Sciences, he seemed to tire of crystals, and sought to apply his ideas to large, biologically significant molecules. He started with hemoglobin, and made notable contributions; foremost among these was, with Harvey Itano and others, the definition of the molecular-genetic defect in sickle cell hemoglobin.[28] In 1940, Pauling advanced a theory of antibody formation based upon the instruction by antigen of how the nascent globulin peptide chain was to be coiled to confer specificity.[29] Applying his knowledge of bond angles and sizes to a variety of proteins, in 1950, he came up with the solution to the structure of the polypeptide chain, involving an alpha-helix held together by hydrogen bonds.[30] It was this application of physical principles to the protein molecule that won him the Nobel Prize in 1954. He then entered the race to solve the structure of DNA, but lost out to Watson and Crick – apparently because the X-ray data available to him were inferior to theirs.[31] Thenceforth, Pauling seemed to lose interest in this area too, and became an activist in the peace movement (for which he won a second Nobel Prize in 1962), and a promoter of what he termed "orthomolecular medicine" and "orthomolecular psychiatry."[32]

Pascual Jordan

Jordan was a German nuclear physicist who had contributed importantly to the development of quantum mechanics and to Heisenberg's uncertainty principle. He also speculated on the philosophical meaning of recent developments in physics.[33] In the late 1930s he turned to biology, and attempted to apply quantum-mechanical reasoning to the problem of specificity, and formulated thereon a theory of antibody formation and the mechanism of allergic reactions.[34] The theory attracted little attention among immunologists already strongly committed to the earlier instruction theories of Haurowitz and Pauling; Jordan presumably returned to physics, since his voice was heard no further in biology. Curiously, Jordan's theory is known primarily because Linus Pauling and Max Delbrück subjected it to scathing criticism. They pointed out that it was the physics that was wrong, and thus the biology could not be correct.[35]

Leo Szilard

Szilard was a well-trained and highly productive nuclear physicist who, in part due to the post-war status of theoretical physics and also to his distaste for the atom and hydrogen bomb programs, turned his attention to biology in the late 1940s. In the late 1950s Szilard could be found on the boardwalk in Atlantic City on a sunny afternoon, outside the headquarters of the Immunology Society. There he would interview targeted immunologists. If not fully satisfied, he would invite one or another of them to dinner and a further cross-examination, at his apartment in Washington, DC.[36] He became interested in the genetic control of protein formation in general, and of antibody formation in particular.[37] In the exciting Lwoff–Monod–Jacob days of enzyme induction and enzyme repression,

Szilard put together a theory of antibody formation in these terms,[38] better known for the elegance of its argument than for its heuristic power.

Other examples of discipline shift

It will be obvious to the reader that immunology is not the only field of science that has witnessed either an egress from or an immigration into the field by those seeking either stimulation or laurels elsewhere. Examination of a few of the more prominent examples will shed additional light on the why and wherefore of this phenomenon.

Max Delbrück

Delbrück came up in physics, in the exciting era of relativity and quantum mechanics. He had studied with Max Born and then with Niels Bohr, and, according to Lily Kay, "found the prospect of conventional research in physics [in the early 1930s] dull and uninspiring."[39] Attracted by Bohr's interest in biology, and with the support of the Rockefeller Foundation,[40] Delbrück attempted to apply the complementarity principle of physics to biology – in fact, he tried throughout his career to demonstrate that biology is a branch of theoretical physics.[41] After studying radiation-induced mutations in drosophila, he assisted at the birth of molecular biology with his studies of molecular genetics employing bacteriophages, which won him the Nobel Prize in 1969. However, as early as 1953, following the Watson–Crick double helix, which appeared to many to be the climax achievement of molecular biology, "Delbrück began to lose interest in phage and in genetics, just as he had lost interest in the anti-climactic physics research of the 1930s."[42] Here is the case of a very bright young man who had done little in the field in which he had initially been trained, but excelled after moving to another. In fact, Delbrück never did consider that he had moved from physics to biology; rather, he felt that he was bringing biology into physics, where it belonged.[43]

Francis Crick

Crick had been trained in physics, but, like Delbrück fifteen years earlier, chose to devote himself to biology. Max Perutz introduced him to Lawrence Bragg, and he was given a place in the Cavendish Laboratory at Cambridge – a hotbed of activity in the X-ray analysis of protein structure. Accidentally falling in with the visiting James Watson, they chose to concentrate on the structure of DNA – a happy choice, as the success of the double helix and the 1962 Nobel Prize testify. Deciding that the 1966 Cold Spring Harbor meeting on the genetic code marked the end of classical molecular biology, to which Crick himself had contributed handsomely, he (along with Sydney Brenner) decided "that it was time to move on to new fields. We selected embryology..."[44] Unable to gain a foothold in the biochemistry of his chosen problem, he eventually moved (in 1976) to the Salk Institute in Southern California, where he devoted himself to problems in neurobiology until his death, at the age of eighty-eight, in 2004.

Cyril Hinshelwood

Hinshelwood was a physical chemist who had contributed importantly to the field of chemical kinetics during the 1920s and 1930s.[45] He made significant contributions to the understanding of the function of catalysts, and shared the Nobel Prize for Chemistry in 1956. During the latter part of the 1930s, he moved from the study of purely chemical reactions to the field of bacteriology. Here, he applied his former approaches to the study of the kinetics of physiologic processes in bacterial cells. He proposed that the regulation of bacterial growth and function was due to the balanced action of a network of interdependent enzymes.[46] It is unclear to what extent Hinshelwood's change of research substrate was due to a push away from the old field or, alternatively, to a pull towards the new one. One must, however, suspect that there came a time when he began to lose interest in his former occupation.

Comment

Why would anyone declare that "it is all over" in a scientific discipline, and that "the end is near"? One reason, as we have seen, is that progress may have slowed temporarily, as seems to have been the case in 1890s physics for Albert Michelson, in 1920s immunology for Gerald Webb, and in 1930s physics for Max Delbrück. A second reason is that a leader in the field, after making significant contributions to its progress, may decide (somewhat egotistically) that he has pretty much solved all the important problems in the field, and only a "mopping up" of details remains. This may have been the case with Lord Kelvin, Rudolf Virchow, and Linus Pauling, and was surely the case with Paul Ehrlich, Francis Crick, Gerald Edelman, Niels Jerne and Macfarlane Burnet.

An analysis of the many instances of predictions of "the end" discloses a very significant error common to almost all predictors; they assume that the discipline (or field, or science in general) is bounded by the facts and theories *as they are known at the time*. But Frontinus could not have imagined that gunpowder would enter into warfare; Lord Kelvin could not envisage the development of quantum mechanics, the discovery of X-rays and radioactivity, or Einstein's relativity. Neither Burnet nor Jerne could know yet that there was a second complete system of immune responses, the T cell system, in addition to the more familiar antibody system, or that immunoregulation would soon assume such great importance.

There is a further subset of this latter phenomenon. A prominent scientist may decide that he has written *finis* on that portion of the discipline that interests him, and decide that there is nothing else interesting to do – this with complete disregard for any other exciting activity in the larger field. This seems to have been the case with Crick, who was interested first in the structure of DNA and then in defining the genetic code; when the latter was solved, he left the field despite the growing ferment in the applications of molecular biology to an ever-growing number of important biological problems. The same estimate can be made about Edelman, whose primary interest was in molecules; with the solution of the nature and origins of immunoglobulins, he left the field and its exciting ferment about the role of cells in the process.

Among those scientists who decided that the time had come to change disciplines, it is worth noting which of them achieved significant success in their new fields, and why. Although the returns were not positive in every case, we may safely conclude that this further success can be claimed for Paul Ehrlich, for Max Delbrück, for Linus Pauling, for Cyril Hinshelwood, and for Gerald Edelman. I believe that the reason for success in each instance lies in the fact that each of them brought from the former discipline a generally useful approach with which to confront the new one. Ehrlich carried over his concept of specific receptors from immunology to the design of chemotherapeutic agents. Delbrück looked for the biological equivalents of the elementary particles and complementarity of theoretical physics, and developed the bacteriophage approach to molecular genetics. Pauling applied his knowledge of the chemical bond to biological macromolecules. Hinshelwood, in his turn, applied his knowledge of chemical kinetics to bacterial physiology. Finally, Edelman applied the Darwinism current in immunological thought to the broader question of the evolution of molecular controls of histogenesis. Where the scientist moves to another field, and attempts to utilize *its* concepts and approaches rather than those of his former discipline, the success rate seems somewhat more equivocal. (Indeed, the story here is undoubtedly weighted in favor of those who succeeded, for the failures often escape notice.)

We may note, finally, that the phenomenon of the declaration of "the end of…" is not limited to science itself. There seemed to be a recent epidemic of such predictions in a broad range of fields at the end of the twentieth century, most seemingly unrelated to any expectation of a millennial tragedy. It may be significant that a printout of book titles beginning with *The end of…* from a prominent Internet source[47] listed many thousands of items. While there are many duplicate entries in the list (for example, multiple editions of the same volume), many fictional works with that initial title, and of course many historical works describing the end of something that in fact ended in the past, not a few describe the end of something that would surely occur sometime in the near future. One notes such recent titles as *The End of History*, *The End of Ideology*, *The End of Nature*, *The End of Affluence*, *The End of Architecture*, and *The End of the Nation State*.[48]

An epistemological postscript

Two views of the future of the scientific enterprise were advanced in America after World War II. The optimistic one was expressed by Vannevar Bush in his manifesto *Science, the Endless Frontier*.[49] Bush claimed that each new discovery would raise more questions than it answered, and that the scientific horizon would recede endlessly. The pessimistic viewpoint was advanced by AAAS President Bentley Glass, who suggested that the rate of new discoveries was slowing, and that we had attained a "Golden Age" in biology; the science was approaching asymptotically the upper limit of its possible accomplishments.[50] More recently, it has been suggested that we are approaching the "final solution" of physics, when the *The Theory of Everything* would be contained in a single equation linking together gravity, the weak and strong nuclear forces, and electromagnetism.[51] This type of closure of science in general has most recently been advanced by John Horgan in *The End of Science*.[52]

An interesting epistemological variant of this somewhat pessimistic view is embodied in the suggestion that there is in fact something beyond the restricting envelope, *but it is unknowable;*[53] thus the field is equally restricted and subject to an imminent closure. This is the position taken by David Lindley, among others, in *The End of Physics*.[54] He argues that a "Grand Unifying Theory" in physics is unattainable,[55] and that we already know most of the knowable! As Lindley concludes:[56]

> *This theory of everything, this myth, will indeed spell the end of physics. It will be the end not because physics has at last been able to explain everything in the universe, but because physics has reached the end of all the things it has the power to explain.*

In a sense, this is also the implicit position taken by Peter Medawar in his book *The Limits of Science*.[57] Medawar proposes that "there is no limit upon the power of science to answer questions of the kind science *can* answer." (To this writer, this smacks of tautology!) It will be apparent that this view holds that "unknowability" is inherent in and limited by the very nature of the facts/phenomena themselves. But a different view is that the position of the restricting envelope of knowability is imposed not by the inherent quality of the facts themselves, but only by the limitations of the mind of man to comprehend them.[58]

Finally, a somewhat more pragmatic view of the situation is advanced by Nicholas Rescher in his *Scientific Progress*.[59] Rescher suggests that in the end, it may only be budgetary considerations that limit scientific progress. Lindley has also suggested that we may be approaching the limits in cost and size of the instruments required to plumb further the depths of particle physics – society cannot afford accelerators the size of Switzerland![60]

Notes and references

1. Jerne, N.K., "Summary: waiting for the end," in *Antibodies: Cold Spring Harbor Symp. Quant. Biol.* 32:591–603, 1967.
2. *Antibodies*, note 1.

3. This suggestion was made at an international symposium held in Prague in 1964, on "Molecular and Cellular Aspects of Antibody Formation."

4. Jerne, N.K., "The complete solution of immunology," *Austral. Ann. Med.* **18**: 345–348, 1969.

5. Clapesattle, H., *Dr Webb of Colorado Springs*, Boulder, Colorado Association University Press, 1984, p. 401.

6. Castiglioni, A., *A History of Medicine*, 2nd edn, New York, Knopf, 1947, p. 221; Leicester, H.M., *Development of Biochemical Concepts from Ancient to Modern Times*, Cambridge, Harvard University Press, 1974, Chapters 2 and 7.

7. Sextus Julius Frontinus, *Strategematicon [The Strategems]*, quoted in James, P., and Thorpe, N., *Ancient Inventions*, New York, Ballantine Books, 1994, pp. 207–208.

8. For a more extended discussion, see Silverstein, A.M., *Hist. Sci.* **37**:407, 1999.

9. See, for example, Lindley, D., *The End of Physics*, New York, Basic Books, 1993, and Horgan, J., *The End of Science*, New York, Broadway Books, 1997.

10. Michelson, A.A., in a speech quoted in part in *Physics Today*, April 1968, p. 9.

11. The quotation appears in Michelson's speech, note 10. He attributed it to "an eminent physicist," and Robert Milliken suggested that Michelson meant Kelvin, but the attribution has never been substantiated.

12. Price, J.D. de S., *Science Since Babylon*, New Haven, Yale University Press, 1961, p. 37.

13. Virchow, R., *Die Cellularpathologie in ihrer Begründung auf physiologische und pathologische Gewebelehre*, Hirschwald, Berlin, 1858; English edition, *Cellular Pathology*, Dover, New York, 1971.

14. Virchow, R., "Standpoints in scientific medicine," in Rather, L.J., ed., *Disease, Life, and Man: Selected Essays by Rudolf Virchow*, Stanford, Stanford University Press, 1958, p. 143. Virchow had written an essay with the same title thirty years earlier (in Rather, p. 26), in which he had outlined many of the outstanding problems in pathology, whose solution had presumably now been attained by the cellular approach.

15. See Koshland, D.E., *Science* **267**:1575, 1995. The true story of the affair is told by Jeffrey, E., *J. Patent Off. Soc.* July, 1940, 479–481.

16. Ehrlich's activities in these different fields are variously described in the numerous biographies – see, for example, those by his secretary Martha Marquardt, *Paul Ehrlich* (New York, Schuman, 1951); by his colleague A. Lazarus, *Paul Ehrlich* (Vienna, 1922); by Heinrich Satter to accompany the 1962 Paul Ehrlich Prize, *Paul Ehrlich: Begründer der modernen Chemotherapie* (München, 1962); and by Ernst Bäumler, *Paul Ehrlich, Scientist for Life* (New York, Holmes & Neier, 1984). The best review of the full scope of Ehrlich's science will be found in the Festschrift prepared in honor of his sixtieth birthday, *Paul Ehrlich: Eine Darstellung seines wissenschaftlichen Wirkens* (Jena, Fischer, 1914).

17. Silverstein, A.M., *Cell. Immunol.* **194**:213, 1999; *Nature Immunol.* **1**:93, 2000; and *Paul Ehrlich's Receptor Immunology: The Magnificent Obsession*, New York, Academic Press, 2002.

18. Ehrlich, P., *Klin. Jahrbuch* **6**:299, 1897; English translation in *The Collected Papers of Paul Ehrlich*, Vol. II, London, Pergamon, 1957, pp. 107–125.

19. The English translations of the six hemolysis papers by Ehrlich, P., and Morgenroth, J. are in Ehrlich's *Collected Papers* Vol. II, pp. 150–155; 165–172; 205–212; 224–233; 246–255; and 278–297. See also Ehrlich's Croonian Lecture, *Proc. R. Soc. London* **66**:424, 1900.

20. See also Chapter 6.

21. Ehrlich shared the 1908 Nobel Prize with Elie Metchnikoff, both for their impressive contributions to immunology.

22. Ehrlich died in 1915, after having been nominated for a second Nobel Prize. It has been suggested by G. Liljestrand (in *Nobel. The Man and his Prizes*, Stockholm, The Nobel Foundation, 1950, pp. 135–316) that, had Ehrlich survived, he might well have been honored again.

23. Edelman, G.M., *Proc. Natl Acad. Sci. USA* **63**:78, 1969.

24. Edelman may have been one whose shift to a new field was not made quietly. He is quoted in a *NY Times* magazine cover story in 1988 (cited by Horgan, J., *The End of Science*, New York, Broadway Books, 1996, p. 165) as suggesting that his Nobel Prize-winning work on the structure of the antibody molecule had "solved" immunology. "Before I came to it, there was darkness – afterwards there was light."

25. See Edelman, G.M., "CAMS and Igs: cell adhesion and the evolutionary origins of immunity," *Immunol. Rev.*, **100**:11, 1987; Edelman, G.M., and Thiery, J.-P., eds, *The Cell in Contact: Adhesions and Junctions as Morphogenetic Determinants*, New York, Wiley, 1985.

26. Edelman, G.M., *Neural Darwinism: The Theory of Neuronal Group Selection* (New York, 1987). Among his many other books on neurobiology are *Topobiology*, New York, Basic Books, 1988; *Bright Air, Brilliant Fire: On the Matter of the Mind*, New York, Basic Books, 1992; *Wider than the Sky: The Phenomenal Gift of Consciousness*, New Haven, Yale University Press, 2004; and *Second Nature: Brain Science and Human Knowledge*, New Haven, Yale University Press, 2007.

27. Pauling published his trail-blazing ideas first in 1931 (*J. Am. Chem. Soc.* **53**:1367, 1931). He extended and formalized the approach in *The Nature of the Chemical Bond*, Ithaca, Cornell, 1939.

28. Pauling L., et al., "Sickle cell anemia, a molecular disease," *Science* **110**:64, 1949.

29. Pauling, L., "A theory of the structure and process of formation of antibodies," *J. Am. Chem. Soc.* **62**:2643, 1940.

30. Pauling, L., Corey, R., and Bransom, H.R., *Proc. Natl Acad. Sci. USA* **37**:205, 1951.

31. Judson, H., *The Eighth Day of Creation*, New York, Simon & Schuster, 1979, p. 152 ff.

32. See Serafini, A., *Linus Pauling: A Man and his Science*, New York, Simon & Schuster, 1989.

33. See, for example, Bayler, R.H., *From Positivism to Organism: Pascual Jordan's Interpretation of Modern Physics in Cultural Context*, Thesis, Harvard University, 1994.

34. Jordan, P., "Heuristische Theorie der Immunisierungs- und Anaphylaxie-Erscheinungen," *Z. Immunitätsf.*, **97**:330, 1940.

35. Pauling L., and Delbrück, M., "The nature of the intermolecular forces operative in biological processes," *Science* **92**:77, 1940.

36. The author, together with Jonathan Uhr, was subjected to this (actually quite fascinating) set of interviews.

37. See Lanouette, W., *Genius in the Shadows: A Biography of Leo Szilard*, New York, Scribner, 1992.

38. Leo Szilard, "The molecular basis of antibody formation," *Proc. Natl Acad. Sci. USA* **46**:293, 1960.

39. Kay, L., "Conceptual models and analytical tools: the biology of physicist Max Delbrück," *J. Hist. Biol.* **18**:207, 1985.

40. The important role that the Rockefeller Foundation played in the movement of physical scientists into biology is discussed by Kohler, R.E., in Reingold, N., ed., *The Sciences in the American Context*, Washington, DC, Smithsonian Institute Press, 1979, and by Abir-Am, P., *Soc. Stud. Sci.* **12**:341, 1982.

41. Kay, L.E., "The secret of life: Niels Bohr's influence on the biology program of Max Delbrück," *Riv. Storia Scienza* **2**:487, 1985. See also Olby, R., *The Path to the Double Helix*, Seattle, University of Washington Press, 1974; Judson, *The Eighth Day*, note 31; and Kay, L.E., *The Molecular Vision of Life*, Oxford, Oxford University Press, 1996.

42. Kay, "Conceptual models," note 39, p. 245.

43. Delbrück never gave up considering himself a physicist. Thus, in 1949 he made a speech entitled "A Physicist looks at Biology," reprinted in Cairns, J., Stent, G., and Watson, J., eds, *Phage and the Origins of Molecular Biology*, Cold Spring Harbor, Cold Spring Harbor Laboratory, 1966, p. 22. He was still a physicist when later he wrote "A physicist's renewed look at biology: twenty years later," *Science* **168**:1312, 1970.

44. Crick, F., *What Mad Pursuits: A Personal View of Scientific Discovery*, New York, Basic Books, 1988, p. 143 ff.

45. Hinshelwood, C., *Kinetics of Chemical Change in Gaseous Systems*, Oxford, Oxford University Press, 1926; Hinshelwood, C., and Williamson, A.T., *The Reaction between Hydrogen and Oxygen*, Oxford, Oxford University Press, 1934.

46. Hinshelwood, C., *The Chemical Kinetics of the Bacterial Cell*, Oxford, Oxford University Press, 1946; Hinshelwood, C., and Dean, A.C.R., *Growth, Function, and Regulation in Bacterial Cells*, Oxford, Oxford University Press, 1966.

47. The Internet source consulted during February, 2008 was Amazon.com. There are quite a few pertinent titles starting with *The Death of…*, but this list is more devoted to detective novels than to serious predictions of the imminent demise of a scientific discipline.

48. Fukuyama, F., *The End of History and the Last Man*, New York, Free Press, 1992; Bell, D., *The End of Ideology*, Glencoe, Free Press, 1962; McKibben, B., *The End of Nature*, New York, Random House, 1990; Jeffrey Madrick, *The End of Affluence*, New York, Random House, 1995; Noever, P., ed., *The End of Architecture*, Munich, Prestel, 1993; Ohmae, K., *The End of the Nation State*, New York, Simon & Schuster, 1995.

49. Bush, V., *Science: The Endless Frontier*, Washington DC, US Government Printing Office, 1945. A similar view has been advanced by John Maddox, former editor of *Nature*, in *What Remains to be Discovered*, New York, Free Press, 1998.

50. Glass, B., "Science: endless horizon or golden age?" *Science* **171**:23, 1971; also Glass, *Q. Rev. Biol.* **54**:31, 1979. Gunther Stent had earlier declared the end of all progress in *The Coming of the Golden Age: A View of the End of Progress*, New York, Natural History Press, 1969. In his *Science Since Babylon* (note 12), Derek Price, while not calling for an end of science, had earlier predicted an approaching asymptote in the scientific enterprise based upon purely demographic and financial considerations. The logarithmic expansion experienced since the founding of the Royal Society in 1660 could not continue unabated, otherwise the entire population (and the entire national budget) would soon be devoted to scientific pursuits.

51. Stephen Hawking would entitle his 1980 inaugural lecture as the Lucasian Professor of Mathematics at Cambridge "Is the end of theoretical physics in sight?" *Phys. Bull.* **January**, pp. 15–17, 1981.

52. Horgan, J., note 24. Horgan and John Maddox (note 49) have debated the issue hotly in several venues; see, for example, *The New York Times*, November 10, 1998, p. D5. See also Harwitt, M., *Cosmic Discovery*, Basic Books, 1981.

53. Emil du Bois-Reymond speculated as early as the 1870s on the limits of science, "Über die Grenzen des Naturerkennens," in *Reden von Emil Du Bois-Reymond*, Vol. I, Leipzig, 1912, pp. 441–473. He later suggested that some of Nature's most fundamental problems were in fact insoluble, in "Die Sieben Welträtsel," *Reden*, Vol. II, pp. 65–98.

54. Lindley, D., *The End of Physics: The Myth of a Unified Theory*, New York, Basic Books, 1993.

55. In this connection, see Nobel Prize winner Steven Weinberg's *Dreams of a Final Theory*, New York, Pantheon, 1994.

56. Lindley, *End of Physics*, note 54, p. 255. Entering variously into this position are Werner Heisenberg's Uncertainty Principle of 1927, and Kurt Gödel's Incompleteness Theorem of 1931.

57. Medawar, P.B., *The Limits of Science*, New York, Harper & Row, 1984.

58. Just as Noam Chomsky suggests that our brain hardware imposes restrictions on the form of our language, so historian J.B. Bury, in *The Idea of Progress: An Inquiry into its Origins and Growth*, New York, Dover, 1932, p. 4, raised the possibility that we shall "soon reach a point in our knowledge of nature beyond which the human intellect is unqualified to pass."

59. Nicholas Rescher, *Scientific Progress*, Pittsburgh, University of Pittsburgh Press, 1977, pp. 7–15.

60. Lindley, *End of Physics*, note 54, p. 113. Lindley was here extending the point made earlier by Price (see note 50). Many in the physics community felt betrayed by the American Congress when it cancelled the expensive superconducting supercollider.

Appendix A
The calendar of immunologic progress

A1. Epochs in immunology

Most fields of science, even those born in the present century, go through periods of high excitement and productivity interspersed with periods of relative inactivity. The active periods, and the renaissance that may follow a period of doldrums, are often stimulated by a single individual – an Aristotle, a Newton, or an Einstein – with a startling new observation or with a theoretical construct that opens up new vistas. Similarly, the quiet periods in the life of a science (its "Dark Ages") usually occur when the old theories and the old observations have been substantially milked dry of their useful consequences; much of the activity in the field is devoted to repeating and refining the already well-worked phenomenologies, while awaiting the spark that will ignite a renaissance. As the foregoing pages have made abundantly clear, the field of immunology has witnessed three such epochs in the first hundred years of its existence.

The "Golden Age" of bacteriology

Immunology was born of the dramatic success of Louis Pasteur's germ theory of disease, with the significant assistance of Robert Koch. With the realization that each infectious disease has a specific and identifiable etiology, a happy laboratory accident permitted Pasteur (he of the prepared mind) to discover that specific acquired immunity can be induced by the use of bacterial cultures whose virulence has been attenuated. This was accomplished in 1880, to protect chickens against the *choléra des poules*, and was quickly followed by similar observations on anthrax, rabies, and numerous other diseases for which the agents responsible were being reported with ever-increasing frequency. Here were the methodology and the rationale that made a science of immunology, and that were able at last to explain the eighty-year-old observation of Edward Jenner.

With the discovery by von Behring and Kitasato that active *and even passive* immunity could be induced to protect against diphtheria and tetanus toxins, it appeared for a time that no infectious disease could resist the armamentarium of the new immunology – prophylactic vaccination and therapeutic serum treatment. All that would be necessary would be to identify the pathogen using Koch's postulates, attenuate it, and vaccinate – or, alternatively, identify its toxin and manufacture (in horses) its antidote. The euphoria was great, as was the activity in every laboratory, each anxious to be the first to claim a cure for one or another of the many diseases that had plagued mankind up to that point, but whose evils would soon be erased with these new approaches. And indeed, many were the accomplishments of this Golden Age. The plague bacillus was identified, and a reasonable vaccine produced. The cholera organism was identified, and a useful vaccine appeared imminent. It seemed only a matter of time before those great scourges of mankind – tuberculosis, leprosy, typhus, typhoid fever, syphilis, etc. – would be brought under control.

This hubris lasted less than a quarter-century. Despite the ready availability of well-characterized etiologic agents, prophylactic immunization just did not seem to function as expected in such diseases as tuberculosis. Again, it became apparent that while one could kill cholera vibrios *in vitro* and *in vivo* with appropriate antisera, these did not seem to protect against the gastrointestinal ravages of the clinical disease. In other instances, such as syphilis, the pathogen was identified but could not be grown in culture, and efforts at preventive vaccination came to naught. In numerous other diseases (later shown to be caused by various viruses and rickettsiae), the nature of the agents themselves remained a mystery. Rabies proved to be the exception, and not the rule. Thus it was that by the first decade of the last century the easy victories over disease belonged mostly to the past, and the more difficult problems of tuberculosis, syphilis, yellow fever, trypanosomiasis, and a host of others began to appear insurmountable. The once-promising young field of immunology now appeared unlikely to satisfy all that had been expected of it.

The "Dark Ages" of immunochemistry

It is perhaps too harsh a verdict to call the fifty-odd years between about 1910 and 1960 the Dark Ages. Work continued in immunology, and some of it proved to be extremely useful. But it was activity of a different sort than that witnessed earlier. As pointed out in various chapters of this book, the immunologic concerns with disease that typified the bacteriologic era slowed, and became more sporadic. Improvements were made in the use of diphtheria toxoid, and in the quantitative measurement of toxins and antitoxins. It took some twenty-five years to develop a vaccine against yellow fever, and some thirty years to work out the problems of preventive immunization against pneumococcal pneumonia – an advance perhaps unfortunately aborted by the advent of antibiotics.

It was during this period that immunology passed from the hands of the bacteriologist into those of the chemist. The humoral theory of immunity, involving molecules and stereochemical structures, had replaced Metchnikoff's essentially biologic theory of cellular (phagocytic) immunity. The emphasis was now on the study of antibodies and antigens, their specificities, and the modes of their interactions. While impressive progress was made in this area by Karl Landsteiner and others, it appeared slow and incremental, and brought with it few important generalizations. Others devoted themselves to the problem of quantitation of the precipitin and the agglutinin reactions, to complement fixation and immune hemolysis, and to improvements in the various serodiagnostic tests then available. Since the mainstream of immunology was now dominated by chemical approaches, it is no wonder that the theories of antibody formation then extant had the antigen chemically instruct the formation of the antibody combining site. In addition, the immunologic leaders of the day appeared content to leave the more biologically-oriented aspects of immunology to others; allergy was left to the clinicians, and any excursions into questions of immunopathology were left to experimental pathologists, and little remarked in the mainstream of immunology.

Perhaps the best way to characterize this era in immunology, and contemporary perceptions of it, is by citing the actions and the words of Dr Gerald Webb, one of the founders and the first President of the American Association of Immunologists. Shortly after the start of the great depression in the early 1930s, Dr Webb resigned his membership of the Immunology Society. This was admittedly due in part to financial problems, but also "he had lost interest because he could not see that it [the Society] was doing much to advance its science." By implication, it was the science itself that was not moving appreciably, and therefore was no longer exciting.

The "Renaissance" of immunobiology

The ten years following World War II saw a number of reports, mostly by outsiders, that would soon unsettle the reigning paradigm of immunology. These were predominantly biological observations that were difficult to integrate into the old way of thinking. Peter Medawar showed that the rejection of skin grafts is due to a genetically controlled immunologic mechanism unlike the standard antigen–antibody interaction then in vogue; Ray Owen reported on the existence of cattle chimeras, which led to Burnet's prediction and Billingham, Brent, and Medawar's confirmation of the existence of immunologic tolerance – a notion incompatible with previous theory; Ogden Bruton reported on agammaglobulinemia in children; and Jacques Miller, Byron Waksman, and Robert Good demonstrated the importance of the thymus in immunological functions – all of which opened up new avenues for immunologic research.

None of these phenomena could be easily integrated into previous notions of the structure and functions of the immune apparatus, and they cried out for a new conceptual synthesis. This was quickly forthcoming, in Niels Jerne's natural selection theory of antibody formation and in Macfarlane Burnet's clonal selection theory of antibody formation. Taken together, these new phenomena and new concepts led to a radical shift in the direction of immunology, away from the old chemical pursuits and toward more biological directions. As a result, cellular immunology, immunopathology, autoimmunity, immunogenetics, and many other biological avenues began to flourish, after fifty years of relative neglect. This is not to say that immunochemistry dropped out of the picture; on the contrary, work on the structure of the immunoglobulin molecule and the molecular basis of immunologic specificity attest to the great progress in this area. However, these advances depended less on following the old pathways than through participation in the modern revolution in molecular biology.

A2. Seminal discoveries

1714   Report to Royal Society on oriental practice of variolation for smallpox – Emanuele Timoni (*Phil. Trans. R. Soc.* **29**:72)

1721   First immunologic clinical trial, for smallpox variolation – Sir Hans Sloane (*Phil. Trans. R. Soc.* **49**:516, 1756)

1798   Cowpox vaccination – Edward Jenner (*An Inquiry...*, London, Sampson Low)

1880   First modern controlled experiment in immunology – Louis Pasteur (*C. R. Acad. Sci.* **90**:239, 952)

1884   Phagocytic theory of immunity – Ilya Metchnikoff (*Virchows Arch.* **96**:177)

1888   Isolation of diphtheria toxin – Emile Roux and Alexandre Yersin (*Ann. Inst. Pasteur* **2**:629)

1888   "Natural" immunity – George Nuttall (*Z. Hyg.* **4**:353)

1889   Discovery of complement (alexin) – Hans Buchner(*Zentralbl. Bakteriol.* **6**:561)

1890   Antitoxic antibodies and passive transfer – Emil von Behring and Shibasaburo Kitasato (*Deutsch. med. Wochenschr.* **16**:1113)

1891   Koch phenomen and tuberculin skin test – Robert Koch (*Deutsch. med. Wochenschr.* **17**:101)

1891   Immunity to plant toxins ricin and abrin – Paul Ehrlich (*Deutsch. med. Wochenschr.* **17**:976, 1218)

1894   Immune bacteriolysis, "Pfeiffer phenomenon" – Richard Pfeiffer (*Z. Hyg.* **18**:1)

1896   Bacterial agglutination – Max von Gruber and Herbert E. Durham (*Münch. med. Wochenschr.* **43**:285)

1897   Quantitative titration of diphtheria toxin and antitoxin – Paul Ehrlich (*Klin. Jahrb.* **6**:299)

1897   Side-chain theory of antibody formation – Paul Ehrlich (*Klin. Jahrb.* **6**:299)

1897   Precipitin reaction – Rudolph Kraus (*Wien. klin. Wochenschr.* **10**:736)

1898   Immune hemolysis – S. Belfanti and T. Carbone (*Giorn. R. Acad. Torino* **46**:321); Jules Bordet (*Ann. Inst. Pasteur* **12**:688)

1899   Antitissue (antisperm) antibodies – Karl Landsteiner (*Zentralbl. Bakt.* **16**:13)

1900   Blood groups (ABO) – Karl Landsteiner (*Centralbl. Bakt. Orig.* **27**:357)

1901   Complement fixation – Jules Bordet and Octave Gengou (*Ann. Inst. Pasteur* **15**:289)

1902   "Danysz phenomenon" – Jean Danysz (*Ann. Inst. Pasteur* **16**:331)

1902   Anaphylaxis – Paul Portier and Charles Richet (*C. R. Soc. Biol.* **54**:170)

1903   The Arthus reaction – Maurice Arthus (*C. R. Soc. Biol.* **55**:817)

1903   Opsonins – Almroth Wright and S.R. Douglas (*Proc. R. Soc. Ser. B.* **72**:364).

1903   Organ specificity of antigens – Paul Uhlenhuth (in *Festschrift zum 60 Geburtstag von Robert Koch*, Fischer, Jena)

1904   First autoimmune disease, paroxysmal cold hemoglobinuria – Julius Donath and Karl Landsteiner (*Münch. med. Wochenschr.* **51**:1590)

1906   Specificity of chemically treated antigens – Friedrich Obermayer and Ernst Pick (*Wien. klin.Wochenschr.* **19**:327)

1906   Serodiagnosis of syphilis – A. von Wassermann, A. Neisser, C. Bruck, and A. Schucht (*Z. Hyg.* **55**:451)

1906   Serum sickness – Clemens von Pirquet and Bela Schick (*Die Serumkrankheit*, Deuticke, Vienna)

1907   Defines "immunochemistry" – Svante Arrhenius (*Immunochemistry*, New York, Macmillan)

1910   "Schultz–Dale phenomenon" – W.H. Schultz (*J. Pharmacol. Exp. Ther.* **2**:221); H.H. Dale (*J. Pharmacol. Exp. Ther.* **4**:167, 1913)

1910   Autoantigenicity of lens protein – F.F. Krusius (*Arch. Augenheilk.* **67**:6)

1920   Hapten inhibition – Karl Landsteiner (*Biochem. Z.* **104**:280)

1921   Passive transfer of allergy – Carl Prausnitz and Heinz Küstner (*Zentralbl. Bakt.* **86**:160)

1923   Pneumococcal polysaccharides – Michael Heidelberger and Oswald T. Avery (*J. Exp. Med.* **38**:73)

1926   MNP blood groups – Karl Landsteiner and Philip Levine (*Proc. Soc. Exp. Biol. Med.* **24**:600, 941)

1929   Quantitative precipitin reaction – Michael Heidelberger and Forrest E. Kendall (*J. Exp. Med.* **50**:809)

1929   Delayed hypersensitivity to simple proteins – Louis Dienes and E.W. Schoenheit (*Am. Rev. Tuberc.* **20**:92)

1930   Instruction theory of antibody formation – F. Breinl and Felix Haurowitz (*Z. Physiol. Chem.***192**:45)

1934   "Jones–Mote phenomenon" – T.D. Jones and J.R. Mote (*N. Engl. J. Med.* **210**:120)

1937   Histocompatibility complex – Peter A. Gorer (*J. Pathol. Bacteriol.* **44**: 691); George D. Snell (*J. Genet.* **49**:87, 1948)

1939   Antibodies as globulins – Arne Tiselius and Elvin Kabat (*J. Exp. Med.* **69**:119)

1940   Rh blood group system – Karl Landsteiner and Alexander Wiener (*Proc. Soc. Exp. Biol. Med.* **43**:223)

1942   Passive cell transfer of delayed hypersensitivity – Karl Landsteiner and Merrill Chase (*Proc. Soc. Exp. Biol. Med.* **49**:688)

1942   Freund's adjuvant – Jules Freund and K. McDermott (*Proc. Soc. Exp. Biol. Med.* **49**:548)

1944   Transplantation immunology – Peter Medawar (*J. Anat.* **78**:176)

1945   Chimerism in cattle – Ray D. Owen (*Science* **102**:400)

1948   Agar gel immunodiffusion – J. Oudin (*Ann. Inst. Pasteur* **75**:30); Örjan Ouchterlony (*Acta Pathol. Microbiol. Scand.* **26**:537, 1949)

1948   Role of the plasma cell – Astrid Fagraeus (*Acta Med. Scand.* **Suppl. No. 64**)

1950   Enzymatic cleavage of antibody molecule – Rodney R. Porter (*Biochem. J.* **46**:479)

1952   Agammaglobulinemia – Ogden C. Bruton (*Pediatrics* **9**:722)

1953   Immunoelectrophoresis – Pierre Grabar and Curtis Williams (*Biochem. Biophys. Acta* **10**:193; **17**:65,1955)

1953   Immunologic tolerance – Rupert E. Billingham, Leslie Brent, and Peter B. Medawar (*Nature* **172**:603)

1954   Transfer factor – H. Sherwood Lawrence (*J. Clin. Invest.* **33**:951)

1955   Natural selection theory of antibody formation – Niels K. Jerne (*Proc. Natl Acad. Sci. USA* **41**:849)

1955   Fluorescent antibody immunohistochemistry – Albert Coons, E.H. Leduc, and J.M. Connolly (*J. Exp. Med.* **102**:49)

1956   Bursa of Fabricius – Bruce Glick, T.S. Chang, and R.G. Jaap (*Poultry Sci.* **35**:224)

1956   Allotypes – J. Oudin (*C.R. Acad. Sci.* **242**:2606); R. Grubb (*Acta Pathol. Microbiol. Scand.* **39**:195)

1956   Size of antibody combining site – Elvin Kabat (*J. Immunol.* **77**:377; *ibid.* **97**:1, 1966)
1957   Interferon – Alik Isaacs and Jean Lindemann (*Proc. R. Soc. Lond. (B)* **147**:258)
1957   Clonal selection theory of antibody formation – F. Macfarlane Burnet (*Austral. J. Sci.* **20**:67)
1959   IgA – Joseph Heremans, M.T. Heremans, and H.W. Schultze (*Clin. Chim. Acta* **4**:96)
1960   Radioimmunoassay – Rosalyn S. Yalow and S.A. Berson (*J. Clin. Invest.* **39**:1157)
1961   Role of the thymus – Jacques Miller (*Lancet* ii:748)
1961   Immunoglobulin light and heavy chains – Gerald Edelman and M.D. Poulik (*J. Exp. Med.* **113**:861)
1963   Hemolytic plaque assay – Niels Jerne and Albert Nordin (*Science* **140**:405)
1963   Idiotypes – Henry Kunkel, M. Mannik, and R. Williams (*Science* **140**:1218); Jacques Oudin and M. Michel (*C. R. Acad. Sci.* **257**:805); Philip Gell and Andrew Kelus (*Nature*, London, **201**:687, 1964)
1966   Lymphokines (MIF) – Barry Bloom and B. Bennet (*Science* **153**:80); John David (*Proc. Natl Acad. Sci. USA* **56**:72)
1966   T cell subsets – Henry Claman, E.A. Chaperon, and R.F. Triplett (*Proc. Soc. Exp. Biol. Med.* **122**:1167); G.F. Mitchell and J.F.A.P. Miller (*J. Exp. Med.* **128**:801, 821, 1968)
1966   IgE – Kimishige Ishizaka and Teruko Ishizaka (*J. Immunol.* **97**:75, 840; *J. Allergy* **37**:165, 336)
1969   Helper T cells – N. Avrion Mitchinson (in *Immunological Tolerance*, New York, Academic Press, p. 149)
1970   Hypervariable regions of Ig – T.T. Wu and Elvin Kabat (*J. Exp. Med.* **132**:211)
1970   Immunoglobulin domains – Gerald Edelman (*Biochemistry* **9**:3197)
1970   Suppressor T cells – Richard Gershon and K. Kondo (*Immunology* **18**:723)
1974   Idiotype networks – Niels Jerne (*Ann. Immunol.* **1250**:373)
1975   Hybridomas – Georges Köhler and Cesar Milstein (*Nature* **256**:495)
1975   MHC restriction – Rolf Zinkernagel and Peter Doherty (*J. Exp. Med.* **141**:1427)

A3: Important books in immunology, 1892–1968

(These titles provide a good indication of the changing directions and new developments in the field. After 1968, the volume became overwhelming.)

1892   *Die Blutserumtherapie*, Emil von Behring, Leipzig, Georg Thieme
1892   *Leçons sur la Pathologie Comparée de l'Inflammation*, Elie I. Metchnikoff, Paris, Masson
1895   *Immunity: Protective Inoculations in Infectious Diseases and Serum-Therapy*, George M. Sternberg, New York, William and Wood
1901   *L'immunité dans les Maladies Infectieuses*, Elie Metchnikoff, Paris, Masson
1902   *Immunität und Immunisierung. Eine medicinisch-historische Studie*, Ludwig Hopf, Tübingen, Franz Pietzcker
1902   *Ehrlichs Seitenkettentheorie and Ihre Anwendung auf die künstlichen Immunisierungsprozesse*, Ludwig Aschoff, Jena, Fischer
1903   *Die Antikörper*, Emil von Dungern, Jena, Fischer

1904   *Immune Sera, Hemolysins, Cytotoxins, and Precipitins*, August von Wasserman, New York, John Wiley

1904   *Toxine und Antitoxine*, Karl Oppenheimer, Jena, Fischer

1904   *Gesammelte Arbeiten Zur Immunitätsforschung*, Paul Ehrlich, Berlin, August Hirschwald (English edition, New York, Wiley, 1906)

1904   *Blood Immunity and Blood Relationship*, George H.F. Nuttall, Cambridge, The University Press

1904   *Die Ehrlich'sche Seitenkettentheorie und ihre Bedeutung für die medizinischen Wissenschaften*, Paul Römer, Wien, Alfred Hölder

1906   *Das Heufieber*, Alfred Wolff-Eisner, Munich, 1906

1906   *Die Serumkrankheit*, Clemens von Pirquet and Bela Schick, Vienna, Deuticke

1907   *Immunochemistry*, Svante A. Arrhenius, New York, Macmillan

1909   *Handbuch der Technik und Methodik der Immunitätsforschung*, R. Kraus and C. Levaditi, eds, Jena, Fischer

1909   *Studies in Immunity*, Robert Muir, London, Henry Frowde

1909   *Die Serodiagnose der Syphilis*, Carl Bruck, Berlin, Springer

1909   *Die Krise in der Immunitätsforschung*, E. Sauerbeck, Leipzig, Klinkhardt

1909   *Studies in Immunity*, Jules Bordet (Collected Papers), translated by Frederick P. Gay, New York, John Wiley & Sons

1909   *Studies on Immunization and their Application to the Treatment of Bacterial Infection*, Almroth E. Wright, London, Constable

1910   *Klinische Immunitätslehre und Serodiagnostik*, Alfred Wolff-Eisner, Jena, Fischer

1911   *Allergie*, Clemens von Pirquet, Chicago, American Medical Association, Chicago

1914   *Handbuch der Immunitätsforschung und Experimentellen Therapie*, R. Kraus and C. Levaditi, eds, Jena, Fischer

1914   *Infection and Resistance*, Hans Zinsser, New York, Macmillan

1914   *Die Anaphylaxie in der Augenheilkunde*, Aurel von Szily, Stuttgart, Ferdinand Enke

1917   *Anaphylaxie et Antianaphylaxie: Bases Experimentale*, Alexandre Besredka, Paris, Masson

1920   *Traité de L'immunité dans les Maladies Infectieuses*, Jules Bordet, Paris, Masson

1921   *De L'anaphylaxie à L'immunité*, Maurice Arthus, Paris, Masson

1921   *Studien über die Überempfindlichkeit*, Carl Prausnitz and Heinz Küstner, Jena, Fischer

1924   *Immunity in Natural Infectious Diseases*, Felix D'Hérelle, Baltimore, Williams & Wilkins

1924   *The Chemical Aspects of Immunity*, H. Gideon Wells, New York, Chemical Catalog Co.

1927   *La Vaccination Préventive Contre la Tuberculose par le "B.C.G."*, Albert Calmette, C. Guérin, A. Bouquet, and L. Négre, Paris, Masson

1928   *Etudes Sur L'immunité dans les Maladies Infectieuses*, Alexandre Besredka, Paris, Masson

1928   *Konstitutionsserologie und Blutgruppenforschung*, Ludwig Hirszfeld, Berlin, Springer

1929   *The Immunology of Parasitic Infections*, William H. Taliaferro, New York, Century

1930   *Le Choc Anaphylactique et la Principe de la Désensibilisation*, Alexandre Besredka, Paris, Masson

1931 *Asthma and Hay Fever in Theory and Practice*, A.F. Coca, M. Walzer, and A.A. Thommen, Springfield, Charles C. Thomas

1933 *Die Spezifizität der Serologischen Reaktionen*, Karl Landsteiner, Berlin, Springer

1933 *Allergy and Immunity in Ophthalmology*, Alan C. Woods, Baltimore, Johns Hopkins Press

1934 *The Chemistry of Antigens and Antibodies*, John R. Marrack, London, HMSO

1936 *The Principles of Bacteriology and Immunology*, 2nd edn, W.W.C. Topley and G.S. Wilson, Baltimore, W. Wood

1936 *Tissue Immunity*, Reuben L. Kahn, Springfield, Charles C. Thomas

1936 *The Specificity of Serological Reactions* (English revised edition), K. Landsteiner, Springfield, Charles C. Thomas

1937 *Les Immunités Locales*, Alexandre Besredka, Paris, Masson

1937 *The Phenomenon of Local Tissue Reactivity and its Immunological and Clinical Significance*, Gregory Schwartzman, Oxford, University Press

1938 *The History of Bacteriology*, William Bulloch, Oxford, Oxford University Press

1941 *The Production of Antibodies. A Review and a Theoretical Discussion*, F.M. Burnet, M. Freeman, A.V. Jackson, and D. Lush, Melbourne, Macmillan

1943 *Fundamentals of Immunology*, William C. Boyd, New York, Interscience

1943 *Allergy, Anaphylaxis and Immunotherapy*, Bret Ratner, Baltimore, Williams & Wilkins, Baltimore

1944 *The Pathogenesis of Tuberculosis*, Arnold Rich, Springfield, Charles C. Thomas

1945 *Immunocatalysis*, M.G. Sevag, Springfield, Charles C. Thomas

1948 *Antibody Production in Relation to the Development of Plasma Cells*, Astrid Fagraeus, Stockholm, Esselte Aktiebolag (Suppl. 204 to *Acta Med. Scand.*)

1948 *Experimental Immunochemistry*, Elvin A. Kabat and Manfred M. Mayer, Springfield, Charles C. Thomas

1949 *The Production of Antibodies*, 2nd edn, F.M. Burnet and F. Fenner, Melbourne, Macmillan

1951 *Antibodies and Embryos*, F.W.R. Brambell, W.A. Hemmings, and M. Henderson, London, Constable

1951 *Blood Transfusion in Clinical Medicine*, P.L. Mollison, Oxford, Blackwell

1951 *A Study of Avidity Based on Rabbit Skin Responses to Diphtheria Toxin-Antitoxin Mixtures*, Niels K. Jerne, Copenhagen, Munksgaard

1953 *Immunity, Hypersensitivity, Serology*, Sidney Raffel, New York, Appleton and Century Croft

1954 "The relation of immunology to tissue homotransplantation," *Ann. NY Acad. Sci.* **59**:277-465

1956 *Blood Group Substances: Their Chemistry and Immunochemistry*, Elvin A. Kabat, New York, Academic Press

1956 *Immunohématologie Biologique et Clinique*, Jean Dausset, Paris, Flammarion

1957 *Collected Papers of Paul Ehrlich*, Vol. II, New York, Pergamon Press

1957 *Immunpathologie in Klinik und Forschung, und das Problem der Autoantikörper*, P. Miescher and K.O. Vorlaender, eds, Stuttgart, Georg Thieme

1958 *1st International Symposium on Immunopathology*, Pierre Grabar and Peter Miescher, eds, Basel, Benno Schwabe

1959 *Mechanisms of Hypersensitivity* (Henry Ford Hospital International Symposium), J.H. Shaffer, G.A. LoGrippo, and M.W. Chase, eds, Boston, Little Brown

1959 "Experimental allergic encephalomyelitis," Byron, H. Waksman, Basel, Karger (Suppl. to *Intl Arch. Allergy Appl. Immunol.* **XIV**), Basel

1959   *The Clonal Selection Theory of Acquired Immunity*, F. Macfarlane Burnet,
       Cambridge, The University Press
1959   *Cellular and Humoral Aspects of Hypersensitive States*, H. Sherwood Lawrence,
       ed., New York, Hoeber
1960   *Mechanisms of Antibody Formation*, M. Holub and L. Jarošková, eds, Prague,
       Czechoslovak Academy of Sciences
1960   *Analyse Immuno-électrophorétique*, P. Grabar and P. Burtin, Paris, Masson
1960   *The Transplantation of Tissues and Organs*, Michael F.A. Woodruff, Springfield,
       Charles C. Thomas
1960   *Cellular Aspects of Immunity* (Ciba Symposium), G.E.W. Wolstenholme and
       M. O'Connor, eds, Boston, Little Brown
1962   "Tumor immunity," D.B. Amos, ed., *Ann. NY Acad. Sci.* **101**
1962   *Mechanisms of Immunological Tolerance*, M. Hašek and A. Lengerová, eds,
       Prague, Czechoslovak Academy of Sciences
1962   *Introduction to Immunochemical Specificity*, William C. Boyd, New York,
       Interscience
1963   *Clinical Aspects of Immunology*, P.G.H. Gell and R.R.A. Coombs, eds, Oxford,
       Blackwell Scientific
1963   *Autoimmune Diseases*, F. Macfarlane Burnet and I.R. Mackay, Springfield,
       Charles Thomas
1963   *Immunology for Students of Medicine*, J.H. Humphrey and R.G. White,
       Philadelphia, F.A. Davis
1964   *The Thymus in Immunobiology*, Robert A. Good and Ann B. Gabrielson, eds,
       New York, Hoeber
1965   *Autoimmunity and Disease*, L.E. Glynn and E.L. Holborow, Philadelphia,
       Davis
1965   *Immunologic Diseases*, Max Samter, ed., Boston, Little Brown
1965   *Molecular and Cellular Basis of Antibody Formation*, J. Šterzl, ed., Prague,
       Czechoslovak Academy of Sciences
1965   *Complement* (Ciba Symposium), G.E.W. Wolstenholme and J. Knight, eds,
       Boston, Little Brown
1966   *Immunotolerance to Simple Chemicals*, Alain de Weck and J.R. Frey, New York,
       Elsevier
1966   *Phylogeny of Immunity*, R.T. Smith, P. Miescher, and R.A. Good, eds, Gaines-
       ville, University of Florida Press
1966   *The Thymus* (Ciba Symposium), G.E.W. Wolstenholme and R. Porter, eds,
       Boston, Little Brown
1967   *Delayed Hypersensitivity*, J.L. Turk, Amsterdam, North Holland
1967   *Gamma Globulin Structure and Control of Biosynthesis* (Proceedings of the
       Third Nobel Symposium), J. Killander, ed., Stockholm, Almgvist and Wiksells
1967   *Antibodies*, Cold Spring Harbor Symposium on Quantitative Biology, New York,
       Cold Spring Harbor
1967   *Natural and Acquired Immunological Unresponsiveness*, William O. Weigle,
       Cleveland, World Publishing Co.
1967   *Germinal Centers in Immune Responses*, H. Cottier, N. Odartchenko, R.
       Schindler, and C.C. Congdon, eds, New York, Springer
1967   *The Gamma Globulins*, Charles A. Janeway, ed., Boston, Little Brown
1968   *The Structural Basis of Antibody Specificity*, David Pressman and Alan
       Grossberg, New York, Benjamin

1968 *Structural Concepts in Immunology and Immunochemistry*, Elvin A. Kabat, New York, Rinehart and Winston

1968 *Handbook of Immunodiffusion and Immunoelectrophoresis*, Ö. Ouchterlony, Ann Arbor, Science Publishers

1968 *Monoclonal and Polyclonal Hypergammaglobulinemia*, Jan G. Waldenström, Nashville, Vanderbilt University Press

1968 *Immunologic Deficiency Diseases in Man*, D. Bergsma, ed., New York, The National Foundation

Appendix B
Nobel Prize highlights in immunology

1901. The first Nobel Prize in Medicine was awarded to **EMIL von BEHRING** [1854–1917]. von Behring studied under Robert Koch at Koch's Institute in Berlin. Following Löffler's isolation of the diphtheria bacillus in 1883 and the identification of diphtheria exotoxin by Roux and Yersin in 1888, von Behring, with his colleagues Kitasato and Wernicke, showed in 1890–1892 that diphtheria and tetanus immunity were due to the formation of circulating antitoxins. He showed that passive administration of antitoxin serum to diseased patients might effect a cure, thus opening the way for serum immunotherapy in a number of diseases. His citation read: "For his work on serum therapy, especially its application against diphtheria, by which he has opened a new road in the domain of medical science and thereby placed in the hands of the physician a victorious weapon against illness and death."

1905. To **ROBERT KOCH** [1843–1910], "for his investigations and discoveries in regard to tuberculosis." Koch had been a small-town physician in Germany, when his private investigations on the life cycle of the anthrax bacillus and the etiology of anthrax excited the medical profession in 1876. He was given first a laboratory and then an institute in Berlin, and it was there, with the help of a distinguished series of students, that he made bacteriology a true science by his development of stringent bacterial isolation and culture techniques, and by his emphasis on the famous Koch postulates for proof of etiology. Koch devoted himself to the study of a number of different diseases, but it was his identification of the tubercle bacillus and of tuberculin, and his continuing devotion to the study of tuberculosis, that earned him the Nobel Prize. Both the immunodiagnostic tuberculin reaction and the "Koch phenomenon," involving the excessive dermal reaction to tubercle bacilli in the skin of sensitized animals, played a major role in the later elucidation of the mechanisms of cellular immunity.

1908. The prize this year was shared by **ELIE METCHNIKOFF** [1845–1916] and **PAUL EHRLICH** [1854–1915], "in recognition for their work on immunity." Metchnikoff was born in the Russian Ukraine, and studied zoology with an emphasis on comparative embryology. In 1884, working in a marine biology laboratory in Italy, he made the initial observations on the phagocytic cells of starfish larvae that provided the basis for his cellular (phagocytic) theory of immunity. When Metchnikoff left Russia for political reasons, Pasteur offered him a position at his new institute in Paris, where Metchnikoff devoted the rest of his life to an impressive series of investigations in support of his phagocytic theory, and to its vigorous defense from the many attacks by those who favored the view that immunity was based upon humoral (i.e., antibody–complement) mechanisms.

Paul Ehrlich was born in Strehlen, Germany. He studied medicine, and early became interested in the staining reactions of cells in tissues, devising some of the most useful stains for the tubercle bacillus and for blood leukocytes. In 1890, he became an assistant to Koch at the Institute for Infectious Diseases, where he commenced his immunologic studies. Following early work on the antibody response to the plant toxins abrin and ricin, Ehrlich made his most notable early contribution to immunology in 1897, with publication of his paper describing the first practical method for standardization of diphtheria toxin and antitoxin preparations. This same publication also contained the outline of his famous side-chain theory of antibody formation, which greatly influenced immunologic theories for several decades. With Julius Morgenroth, he published an important series of papers on the mechanism of immune hemolysis. Shortly after the turn of the century, Ehrlich gave up most of his activities in immunology to pursue his interests in the chemical treatment of disease, making important discoveries in the treatment of trypanosomiasis and of syphilis (Salvarsan – the "magic bullet"), and helping to found scientific pharmacology.

1913. To CHARLES RICHET [1850–1935], "for his work on anaphylaxis." Richet was a Parisian who studied medicine and became especially interested in physiology. It was these interests which led him to study the physiologic effects on mammals of marine invertebrate poisons, while cruising on the yacht of the Prince of Monaco. With his colleague Paul Portier he discovered the phenomenon of anaphylaxis, dependent not upon the toxic properties of the substance injected, but only upon its function as an antigen in the previously sensitized animal. In so doing, he opened up a new and at the time surprising vista in medicine, by showing that the "protective" mechanisms of immunity might also function to cause disease. The later demonstration of the relationship between experimental anaphylaxis and other more familiar human allergies made this observation clinically as well as theoretically important to immunology.

1919. To JULES BORDET [1870–1961], "for his studies in regard to immunity." Bordet was a Belgian physician who went to study with Metchnikoff at the Pasteur Institute in Paris at the age of twenty-four. He made important early contributions to an understanding of the mechanism of complement-mediated bacteriolysis, and in 1899 discovered the phenomenon of specific hemolysis. Shortly thereafter, in collaboration with his assistant and brother-in-law Octave Gengou, Bordet described the phenomenon of complement fixation and its diagnostic possibilities. This soon developed into a powerful tool in the diagnosis of infectious diseases, most notably in the hands of August von Wasserman and his colleagues, in their complement-fixation test for syphilis. Bordet made many other important contributions to immunology, and is known also for his famous debates with Ehrlich on the nature of antigen–antibody–complement interactions.

1930. To KARL LANDSTEINER [1868–1943], "for his discovery of the human blood groups." Landsteiner was a Viennese pathologist who developed a keen interest in structural organic chemistry before embarking on a career in immunology. From the very outset, Landsteiner seemed always to choose important areas in which to work, or to make important those subjects to which he turned his attention. In early studies of anti-erythrocyte antibodies, he described in 1901 the set of human isoagglutinins which now comprise the ABO system of blood groups. In 1926, Landsteiner and Philip Levine discovered the MNP system, and in 1946, with Albert Wiener, the Rhesus system of blood groups. He was the first to demonstrate that poliomyelitis could be produced in non-human primates, and one of the first to make the same observation for syphilis. During World War I he became interested in the antibody response to chemically defined haptens, and over the next quarter-century, primarily at the Rockefeller Institute in New York, he contributed impressively to an understanding of the chemical basis for antigen–antibody interactions, as summarized in his famous book *The Specificity of Serologic Reactions*. While acknowledging the importance of his discovery of blood groups, Landsteiner is said to have felt that his Nobel Prize should rather have been awarded for his work on antibody–hapten interactions.

1951. To **MAX THEILER** [1899–1972], "for his development of vaccines against yellow fever." Theiler was a South African who studied medicine in Britain and then moved to the United States in 1922, first to the School of Tropical Medicine at Harvard and then to the Rockefeller Institute in New York. It was he who showed that yellow fever was caused by a filterable virus, and his description of the mouse protection test (in which serum antibody mixed with the virus protects a mouse from the lethal effects of intracerebral inoculation) provided a very important tool for epidemiologic and other studies of yellow fever. In the late 1930s, he succeeded in developing attenuated strains of yellow fever virus by serial passage *in vitro* in mouse and chick embryo tissue cultures. By these means, strains were developed that retained their immunogenicity, but which were devoid of pathogenicity – the basis of the current yellow fever vaccines.

1957. To **DANIEL BOVET** [1907–1992], Swiss physiologist and pharmacologist, "for his development of antihistamines in the treatment of allergy." The discovery of the Schultz–Dale phenomenon, in which a strip of sensitized uterine tissue could be caused to contract under the action of antigen, provided a useful *in vitro* model for allergic reactions and for the clarification of the physiologic mechanisms involved. This led to the finding that histamine is the most significant agent released in anaphylaxis, along with serotonin and other active substances. Bovet must have been exposed to immunology and allergy while working under Emile Roux at the Pasteur Institute in Paris, and published extensively on the response of the autonomic nervous system to various chemicals. It was this that led him to a study of agents which might counter the effects of histamine, and from this emerged the drugs which were to prove so useful in the treatment of asthma and hayfever. Even had he not become famous for his work on antihistamines, his South-American adventures with curare and its mode of action, and his development of curare-like relaxants, tranquilizing drugs, and anesthetics would have given him a secure place in the annals of medicine.

1960. To F. **MACFARLANE BURNET** [1899–1985] and **PETER B. MEDAWAR** [1915–1987], "for the discovery of acquired immunological tolerance." World War II stimulated basic research in a number of areas of science, among them being the search to improve the survivability of skin and other tissue grafts on burn and wound victims, and to explain their rejection. Medawar, a Briton who had trained in zoology and pathology at Oxford, was interested in tissue repair, and thus in problems of tissue transplantation. His initial work established conclusively that the rejection of foreign skin grafts followed all of the rules of immunologic specificity, and was in fact based upon the same mechanisms responsible for protection against bacterial and viral infections. The follow-up work which he and a series of distinguished students (most notably Rupert Billingham and Leslie Brent) undertook firmly established transplantation immunobiology as an important subdiscipline, yielding many later dividends in the field of clinical organ transplantation. In 1945–1947, Ray Owen reported the curious observation that dizygotic cattle twins which had shared the same circulatory system *in utero* had become blood-cell chimeras, unable to respond immunologically to one another's antigens. This observation was seized upon by the Australian physician-virologist Macfarlane Burnet, not only a productive investigator but also a wide-ranging theoretician. Burnet had published, in 1941, a stimulating book on *The Production of Antibodies*, and was now preparing a revision of this book with his colleague Frank Fenner. The new book (1949) not only proposed a novel indirect template theory of antibody formation, but also provided a theoretical explanation for Owen's findings. Burnet and Fenner suggested that immunologic responses arise fairly late in embryonic life, and involve a cataloguing by a system of "self-markers" of those antigens then present, to which the host thenceforth would be tolerant, and unable to respond immunologically. Any antigens not so catalogued would be "non-self," and could later stimulate an active immune response. The suggestion was made that any antigen introduced during this critical period would be adopted as self, would induce tolerance, and thus would be unable later to activate the immunologic apparatus. These concepts were further developed by Burnet in his clonal selection theory of antibody formation. Burnet and Fenner's suggestion on tolerance was put to the test by Medawar and his colleagues, and in 1953 they provided ample confirmation of the Burnet–Fenner hypothesis, using inbred strains of mice – a phenomenon to which Medawar gave the name *acquired immunological tolerance*.

1972. To **RODNEY R. PORTER** [1917–1985] of Oxford University and **GER-ALD M. EDELMAN** [1929–] of the Rockefeller University, for their work on the chemical structure of antibodies. The demonstration by A. Tiselius and E. A. Kabat that antibodies are high molecular weight gamma globulins made it clear that it would be extremely difficult to define chemically the basis for either their primary immunologic specificity, or their secondary biological functions. Porter undertook to cleave the antibody molecule with enzymes in an attempt to obtain smaller, active fragments, and succeeded in 1958 in accounting for the entire molecule in terms of papain cleavage into two identical Fab fragments and a third Fc fragment; the former contained the antibody-binding sites, while the latter was responsible for the secondary biological activity of antibodies. Edelman then showed that homogeneous myeloma globulin could be reductively cleaved into its component polypeptide chains, comprising both light (L) and heavy (H) chains. He also showed that the L chains of different guinea pig antibodies had different electrophoretic mobility patterns, and further that the Bence Jones protein of multiple myeloma was similar to the L chains of antibody. Porter and his colleagues next demonstrated that the immunoglobulin molecule was composed of two light and two heavy chains, leading to the now-accepted model for IgG. The isolation of immunoglobulin chains and fragments now permitted an approach to their primary amino acid sequencing, and this was hotly pursued in the laboratories of Porter, Edelman, and many other investigators. From this work emerged an understanding of the existence of both variable and constant regions on the L and H chains, and the ability to compare primary sequences among different antibody specificities, different isotypes, and even different species. Finally, in 1969, Edelman and his coworkers succeeded in working out the primary sequence of an entire immunoglobulin molecule, helping to define not only the location of the active site, but also the location of the "domains" responsible for the secondary biologic activities of antibodies.

1977. To **ROSALYN YALOW** [1921–], "for the development of radio-immuno-assays of peptide hormones" (shared with **ROGER GUILLEMIN** and **ANDREW SCHALLY,** "for their discoveries concerning the peptide hormone production of the brain"). Beginning in the early 1950s, Yalow and her long-term collaborator Solomon Berson investigated the causes of insulin resistance in diabetes. They discovered that diabetics treated with insulin formed antibodies specific for this insulin, but their initial attempt to publish this important observation was rejected, in the belief that so small a molecule was incapable of being immunogenic. Berson (who died in 1972) and Yalow then showed that the addition of increasing amounts of unlabeled insulin to an immune complex of anti-insulin and its radiolabeled antigen resulted in a measurable displacement of the labeled insulin. This discovery formed the basis of the first radioimmu-noassay of a hormone, capable of estimating nanogram or even picogram quantities. Since then, this assay system has been applied to other hormones and biologically active substances, and has become a valuable tool for much basic and clinical research. It was this technique that contributed importantly to the isolation and characterization of hypothalamic hormones by Guillemin and Schally.

1980. To **BARUJ BENACERRAF** [1920–], **JEAN DAUSSET** [1916–2006], and **GEORGE SNELL** [1903–1996], "for their work on genetically determined structures of the cell surface that regulate immunologic reactions." The demonstration that the ability of mice to reject tumors was genetically determined stimulated geneticist Snell to search for methods to study the genes responsible for this phenomenon. This led Snell, in the mid-1940s, "to invent the idea of congenic mice" – animals that are bred to be genetically identical except at a single locus or genetic region. In collaboration with Peter Gorer, Snell identified a locus important for allograft rejection, designated H (histocompatibility)-2, and subsequently shown to be a complex of many closely linked genes, with many different alleles occurring at each locus. The work of numerous investigators has since contributed to a better understanding of the composition and many of the functions of this complicated stretch of DNA, now called the *major histocompatibility complex* (MHC). In the 1950s, Jean Dausset of France found iso-antibodies against leukocyte antigens in the blood of transfusion recipients, helping to demonstrate the analogy between the H-2 complex of the mouse and the human leukocyte antigen (HLA) system in man, and providing a powerful tool to define individual HLA antigens. In 1965, Dausset and his coworkers described a system of some ten human antigens encoded for in the histocompatibility complex, containing "sub-loci," each of which specified a limited number of antigenic alleles. It was this approach that finally opened the way for the definition and genetic location of those major and minor antigens responsible for histoincompatibility. However, the importance of the genes in the HLA and H-2 complexes had thus far been restricted to the somewhat unphysiologic practices of tissue and blood transplantation. It remained for Benacerraf and his coworkers to demonstrate that many of the genes located within the MHC may also control active immune responses to various antigenic stimuli. Utilizing simple antigens such as synthetic polypeptides, Benacerraf, McDevitt, and others found that the ability of an animal to respond immunologically to a given antigen was controlled by specific genes – called Ir (for immune response) genes – subsequently shown by others to reside within the I region of the MHC. Since then, work in Benacerraf's and in many other laboratories has shown the importance of I region genes in controlling the intercommunication among immunocytes responsible for the regulation of the immune response, and the importance of some MHC genes in predisposing for certain chronic diseases.

1984. The prize this year was shared between **CESAR MILSTEIN** [1927–2002] and **GEORGES F. KÖHLER** [1946–1995] for development of the technique of monoclonal antibody formation, and **NIELS K. JERNE** [1912–1994] for his theoretical contributions that have shaped our concept of the immune system. Henry Kunkel and coworkers showed in 1955 that myeloma tumors produced monoclonal antibodies, and Michael Potter showed in 1962 that such plasma cell tumors could be induced in mice, while others were able to adapt such tumors to grow indefinitely in culture. In 1974 Köhler started a postdoctoral fellowship in Milstein's laboratory in Cambridge, and the two undertook to immortalize antibody-forming cells by fusing them with myelomas, in order to study the genetic basis of antibody diversity. It was hoped that the tumor cell would endow the otherwise short-lived antibody-forming cell with the capacity for long-term survival in the resulting hybrid (called a hybridoma). The key to success in this venture was the development of a selective technique to recover only fused cells, employing a mutant myeloma cell-line deficient in the enzyme hypoxanthine phosphoribosyl transferase. Without this enzyme, the cells would die in a medium containing hypoxanthine, aminopterine, and thymidine (HAT), but the hybrid cells would survive and could be selected, since the normal antibody-forming cell component of the hybrid would contribute the enzyme required. Isolation of a hybridoma clone would thus yield large quantities of monoclonal antibodies specific for a single antigenic determinant. The availability of such pure reagents has provided one of the most powerful tools of the current revolution in molecular biology, and has opened up new avenues of investigation in many basic and clinical sciences.

Niels Jerne's contributions to immunology are almost too numerous to record, and his influence on the field is impossible to exaggerate. While still a student, Jerne made important observations on the avidity of antibodies, and on the changing quality of antibodies in response to successive booster immunizations. In 1963, Jerne and Albert Nordin described the hemolytic plaque assay method for enumerating antibody-forming cells – a technique that would receive broad application in studies of the cellular events underlying the antibody response. But it was Jerne's theoretical contributions that helped to bring immunology and immunologists to their current important position in the biomedical sciences. In 1955, Jerne was the first modern scientist to challenge the then-current instructive theories of antibody formation, by proposing a selective theory in which antigen functions to *select* specifically from a *pre-existing* repertoire of antibody-forming capabilities. While the particulars of Jerne's hypothesis might require correction, his theory served as the critical stimulus to Macfarlane Burnet's clonal selection theory of antibody formation. In 1971, Jerne made another conceptual leap to explain the development of the repertoire of T cell specificities. He postulated that the principal driving force which stimulates lymphocytes to divide and mutate at a high rate in the thymus is the individual's major histocompatibility complex antigens. Once again, a Jerne formulation served as an important stimulus to experimental and conceptual progress in the field. The third (and perhaps the most profound) of Jerne's theories was his idiotype network theory of 1974. Jacques Oudin, Henry Kunkel, and Philip Gell had previously shown that the antibody-combining site possesses unique antigenic determinants (idiotypes). Jerne's proposal was that the balanced production of a cascading network of idiotypes and anti-idiotypes might constitute one of the principal regulatory mechanisms governing the immune response. This theory has important implications for the physiology of the immune system, and for its regulation of such pathological states as autoimmunity.

1987. To **SUSUMU TONEGAWA** [1939–] for his work on the molecular biology of immunoglobulin genes, demonstrating how antibody diversity is generated. It had always been difficult to believe that all of the genes required to generate the remarkable diversity of antibodies could be present in the germline; thus, most investigators favored a somatic mutation theory acting upon only a very limited number of germline genes. In 1965, Dreyer and Bennett proposed that less DNA might be required if multiple variable region genes could combine with a single constant region gene for a given isotype (the two gene–one polypeptide theory). This speculation was confirmed in 1976 by Tonegawa and Hozumi, by showing that C region and V region genes were separate in embryonic DNA. Tonegawa, Gilbert, and Maxam then showed that combination of these two genes in differentiated cells still involves their separation by a noncoding DNA sequence (an "intron"). It was further found by Tonegawa, and also by Philip Leder and his colleagues, that the variable region polypeptide chain contains more amino acids than is encoded in the V region DNA of the light chain, suggesting that yet another DNA segment might be needed to encode the complete variable region. Tonegawa and his colleagues soon located the missing DNA segment, which was designated J (for joining). Thus, the combination of a single constant region segment with one of several J segments and one of many V region segments would suffice to generate a wide range of different light chains. In studying the assembly of the genes for the heavy chains of antibody, it was found by both Tonegawa and Leroy Hood that now three separate DNA segments must be joined to complete the sequence for the heavy chain variable region. In addition to the V and J segments, a third group of DNA segments, termed D (for diversity), was involved. As well as the ability to choose among multiple V, D, and J elements, additional variability is introduced into the heavy chain by permitting splicing in the middle of a triplet codon, resulting in a translation shift. The demonstration by Hood and others that mutations in these gene segments could also occur appeared finally to complete the picture of how the immense repertoire of antibody specificities is generated. Tonegawa's work, and that of others, has had important implications in other areas, including the structure and formation of T cell receptors and the DNA rearrangements that might be responsible for lymphomas and leukemias.

1996. The prize was awarded to **PETER DOHERTY** (1940–) and **ROLF ZINKERNAGEL** (1944–) for their demonstration of the MHC restriction of cytotoxic T cell recognition of viral antigens on infected cells. The 1950s were years of great ferment in immunology. On the one hand, Burnet's notion of immunological tolerance focused attention on the "immunological self" – a term that would eventually take on more than just metaphorical implications. On the other hand, the differences between the familiar functions of antibody and the role of cells in delayed-type hypersensitivity and in destroying the cells bearing allogeneic transplantation antigens would lead to the separation of B and T cell lineages and functions. In 1954, Mitchison speculated that cellular recognition of skin-sensitizing antigens occurred only when these were present on the surface membranes of autochthonous cells, thus mimicking foreign transplantation antigens. It was the suggestion by Lewis Thomas that these cellular mechanisms had evolved to control tumor formation (immunological surveillance) that led Sherwood Lawrence to extend the idea to include protection from all intracellular parasites. In an imaginative conceptual leap, Lawrence went on to propose that immunological recognition might involve the parasitic (viral) antigen only in association with a self-antigen – the so-called *self plus X* hypothesis. The scene now moves forward to the early 1970s, when prospective surgeon Zinkernagel stopped in Canberra on his *Wanderjahr* and was, for lack of adequate space, assigned to share the same lab with prospective veterinarian Doherty. The project upon which they collaborated was then a highly popular one; it involved the study of the mechanism of damage caused by virus-immune cytotoxic cells in lymphocytic choriomeningitis virus (LCMV) disease of mice. It had recently been reported that many aspects of the immune response to certain antigens was controlled by *immune response genes*, known to be a part of the major histocompatibility complex (MHC) and, more specifically, that susceptibility to LCMV disease might be related to the particular MHC of the infected mouse. Doherty and Zinkernagel chose for their experiments an *in vitro* system that measures the ability of virus-immune effector cells to destroy virus-infected target cells. When the two cell types originated in the same strain of mice, death of the infected cells was efficient. When, however, the virus-specific cytotoxic cells and the virus-infected target cells came from mice of different MHC haplotypes, destruction of the targets generally failed to take place. The investigators were then able to conclude that the effector cell must recognize two signals on a virus-infected cell, one derived from the virus and the other the MHC molecule normally present on the cell. It was subsequently found that the T cell receptor is so constructed that it is able to bind tightly to a polypeptide breakdown product of the antigen, which lies in a special cleft within the MHC molecule. Thus, T cell recognition is truly "in the context of self."

Appendix C
Biographical dictionary

(This list has been restricted to those who contributed significantly to the development of immunology prior to the early 1960s.)

ACKROYD, John Fletcher [1914–2002]. MB, Bristol, 1938; DSc, London, 1956. Worked at St Mary's Hospital, London. Described autoimmune thrombocytopenic (Sedormid) purpura. Book: *Symposium on Immunological Methods*, London, 1964. Obit.: *R. Coll. Physicians Munks Roll* No. 5345.

ARRHENIUS, Svante [1859–1927]. Born in Vik (Uppsala). Studied physics at Uppsala. Director, Nobel Inst., Physical Chemistry, 1905. Theory of electrolyte dissociation; suggested reversibility of antigen–antibody complexes; coined the term "immunochemistry." Nobel Prize for Chemistry, 1903. Book: *Immunochemistry*, New York, Macmillan, 1907. Biogs: E.H. Riesenfeld, *Svante Arrhenius*, Leipzig, Akademische Verlag, 1931; *Dict. Sci. Biog.* 1:296, 1970. Obit.: *J. Chem. Soc.* 1:1380, 1928.

ARTHUS, Nicolas Maurice [1862–1945]. Born in Angers. Taught physiology, Fribourg, 1895; Univ. Lausanne, 1907; Inst. Pasteur, Lille, 1920; Univ. Marseilles. Studied physiological effects of venoms; discovered local anaphylactic reaction (Arthus phenomenon), 1903. Book: *De l'Anaphylaxie à l'Immunité*, Paris, Masson, 1921. Obit.: *Bull. Acad. Méd. Paris* 129:374, 1945.

ASCHOFF, Ludwig [1866–1942]. Born in Berlin. Educated at Bonn, Berlin, and Strasbourg. Professor of Pathology at Freiburg. Developed concept of reticuloendothelial system. Book: *Ehrlichs Seitenkettentheorie...*, Jena, Fischer, 1902. Obit.: *J. Pathol. Bacteriol.* 55:229, 1943.

ASKONAS, Birgitta A. [1923–]. PhD, Cambridge. Natl Inst. Med. Research, 1953; Head, Immunology Division, 1976. Many contributions to mechanism of antibody formation, immunoregulation, viral immunology.

BAIL, Oscar [1869–1927]. Born in Tillisch, Germany. Studied medicine in Vienna. Professor of Hygiene at the German University of Prague. Advanced theory of aggressins, and early instructive theory of antibody formation. Obit.: *Z. Immunitätsforsch.* 55:i–iv, 1928.

BEHRING, Emil Adolph von [1854–1917]. Born in Deutsch-Eylau. Studied medicine in Berlin. Entered Army Medical Corps; assistant in Koch's Institute, 1889; Prof. of Hygiene in Halle, 1894; Prof. in Marburg from 1895. Established Behringwerke, commercial firm for production of antitoxins. Discovered diphtheria and tetanus antitoxins with Kitasato, 1890. Nobel Prize, 1901; ennobled, 1901. Books: *Die Blutserumtherapie*, Leipzig, Thieme, 1902; *Gesammelte Abhandlungen*, Bonn, Marcus and Webers, 1915. Biogs; H. Zeiss and R. Bieling, *Behring, Gestalt und Werk*, Berlin, Schultz, 1940; P. Schaaf, *Emil von Behring* *zum Gedächtnis...*, Marburg, 1942; *Dict. Sci. Biog.* 1:574, 1970. Obit.: *Berl. klin. Wochenschr.* 54:471, 1917.

BENACERRAF, Baruj [1920–]. Born in Caracas. Studied at New York, Virginia, and Paris (with Halpern). Prof. Pathology, NY Univ., 1958; Chief, Immunology Lab., NIH, 1968; Prof. Pathology, Harvard, 1970. Numerous contributions, including carrier effect in delayed hypersensitivity; lymphocyte subsets; Ir genes and immunogenetics of MHC. Nobel Prize, 1980. Book: *Textbook of Immunology* (with Unanue), Baltimore, Williams & Wilkins, 1979.

BESREDKA, Alexandre [1870–1940]. Born in Odessa. MD, Paris, 1897. Worked at Pasteur Institute under Metchnikoff. Studied anaphylaxis and anti-anaphylaxis; concept of local immunity. Books: *Anaphylaxie et Anti-anaphylaxie*, 1918; *Histoire d'une Idée: L'oeuvre de Metchnikoff*, 1921; *Etudes sur l'Immunité dans les Maladies Infectieuses*, 1928. Obit.: *Rev. Path. Comparée* 40:112, 1940.

BILLINGHAM, Rupert Everett [1921–2002]. Born in Warminster, Wiltshire, England. Educated at Oxford. University College, London (with Medawar), 1951; Wistar Inst., Philadelphia, 1957; Chairman, Dept Cell Biol., Univ. Texas, Dallas, 1971. Many contributions to transplantation biology; established (with Medawar and Brent) immunologic tolerance; privileged sites for transplantation; maternal–fetal immunologic relationships. Books: *Immunobiology of Transplantation*, 1971; *Immunobiology of Mammalian Reproduction*, 1976. Obit.: *Biog. Mem. Fell. R. Soc.* 51:33–50.

BORDET, Jules Jean Baptiste Vincent [1870–1961]. Born in Soignies, Belgium. Studied medicine, Univ. of Brussels. Préparateur in Metchnikoff's laboratory, 1894–1901; founded Institut Pasteur of Brussels, 1901; Prof. Bacteriology in Brussels, 1907. Discovered immune hemolysis, 1899; complement fixation with Gengou, 1901; disputed with Ehrlich about nature and mode of action of antibodies and complement. Nobel Prize, 1919. Book: *Traité de l'Immunité dans les Maladies Infectieuses*, 1920. Biogs: De Kruif, P., *Men Against Death*, New York, Harcourt Brace, 1932; *Dict. Sci. Biog.* 2:300, 1970; see *Volume Jubilaire de Jules Bordet, Ann. Inst. Pasteur* 79, 1950. Obit.: *Ann. Inst. Pasteur* 101:1, 1961.

BOVET, Daniel [1907–1992]. Born in Neuchatel. DSc, Univ. Geneva, 1929. Worked at Inst. Pasteur, 1929; organized Lab. of Therapeutic Chemistry, Inst. Superiore de Sanitá, Rome, 1948. Studied pharmacology of nervous system, action of curare, therapy of allergies, etc. Nobel Prize, 1957. Books: *Structure Chimique et Activité Pharmacodynamique des Médicaments du Système Nerveux Végetatif*, 1948; *Curare and Curare-like Agents*, 1959.

BOYD, William Clouser [1903–1982]. Born in Dearborn, Missouri. Educated at Harvard and Boston Univ. Prof. Immunochemistry, Boston Univ. Worked in blood groups; lectins; physical anthropology. Books: *Fundamentals of Immunology*, 1943; *Genetics and the Races of Man*, 1950; *Introduction to Immunochemical Specificity*, 1962. Obit.: *NY Times*, Feb. 23, 1983.

BRAMBELL, Francis William Rogers [1901–1970]. Born in Sandy Cove, Ireland. Educated at Dublin and London. Prof. Zoology, Bangor, Wales. Studied transfer of antibodies from mother to fetus. Books: *Development of Sex in Vertebrates*, 1930; *Antibodies and Embryos*, 1951. Biog.: *Biog. Mem. Fellows R. Soc.* **19**:129, 1973. Obit.: *Nature* **228**:694, 1970.

BRENT, Leslie Baruch [1925–]. Born in Köslin, Germany. PhD, London, 1954. Worked with Medawar and Billingham, Univ. College, London, 1954; Natl Inst. Med. Res., 1962; Prof. Zoology, Univ. Southampton, 1965; Prof. Immunology, St Mary's, 1969. Many contributions to transplantation biology; established immunological tolerance (with Medawar and Billingham); graft-versus-host disease; histocompatibility antigens; immunogenetics. Book: *History of Transplantation Immunology*, San Diego, Academic Press, 1997.

BRUCK, Carl [1879–1944]. Born in Glatz, Germany. Studied at Munich and Berlin. Breslau Univ., 1908; Director, Dept Dermatology, Altona City Hospital. Worked in syphilology; developed complement fixation test for syphilis (with Wassermann and Neisser). Obit.: *Arch. Dermatol. Syphilol.* **73**:426, 1950.

BRUTON, Ogden Carr [1908–2003]. Born in Mt Gilead, N. Carolina. MD, Vanderbilt Univ., 1933. US Army pediatrician. Described first case of human agammaglobulinemia.

BUCHNER, Hans [1850–1902]. Born in Munich. Bacteriologist and immunologist. Educated in Munich and Leipzig. Military surgeon, ultimately Surgeon General. Professor of Hygiene in Munich, 1894. Discovered complement and studied bactericidal action of normal serum; leading proponent of humoral theory of immunity. Obit.: *Münch. med. Wochenschr.* **49**:844, 1902.

BURNET, Frank Macfarlane [1899–1985]. Born in Traralgon, Victoria, Australia. MD, Melbourne, 1923. Director, Walter and Eliza Hall Inst., 1944. Important contributions to virology; theoretical immunology; clonal selection theory of antibody formation; immunological tolerance. Nobel Prize, 1960. Books: *Production of Antibodies* (with Fenner), 1949; *Natural History of Infectious Diseases*, 1953; *Clonal Selection Theory of Antibody Formation*, 1959; *Autoimmune Diseases* (with Mackay), 1963; *Cellular Immunology*, 1969. Autobiog.: *Changing Patterns*, 1969. Biog.: C. Sexton, *The Seeds of Time*, 1991. Obit.: *Nature* **317**:108, 1985.

BUSSARD, Alain [1917–]. Born in Paris. Director, Cellular Immunology Lab., Inst. Pasteur, Paris. Many contributions to immunologic tolerance; molecular immunology; nature of antigenicity. Books: *La Tolérance Acquise et la Tolérance Naturelle à l'Egard de Substances Antigèniques Définies*, Paris, CNRS, 1963; *Antigènicité*, Paris, Flammarion, 1963.

CALMETTE, Albert [1863–1933]. Born in Nice. Educated at Clermont-Ferrand and Paris. Surgeon in naval and colonial services. Founded Institut Pasteur in Saigon. Sub-director, Institut Pasteur, Paris. With Guérin, discovered BCG, and worked on snake venoms and plague serum. Biog.: Noel Bernard, *La Vie et l'Oeuvre d'Albert Calmette*, Paris, 1961. Obit.: *Ann. Inst. Pasteur* **51**:559, 1933.

CAMPBELL, Dan H. [1907–1974]. Born in Fremont, Ohio. PhD, Univ. Chicago, 1935. Worked at Cal. Inst. Technol. with Pauling. Many contributions to immunochemistry and antibody specificity; studied persistence of antigen. Book: *Methods in Immunology*, 1963. Obit.: *Immunochemistry* **12**:439, 1975.

CANTACUZENE, Jean [1863–1934]. Born in Bucharest. MD, Paris, 1894. Student of Metchnikoff at Pasteur Inst.; Prof. Exper. Med., Bucharest, 1902. Worked on role of phagocytes in immunity. Biog.: *Homage à la Mémoire du Prof. Cantacuzène*, Paris, 1934. Obit.: *Bull. Acad. Méd. Paris* **111**:884, 1934.

CHASE, Merrill [1905–2008]. Born in Providence, Rhode Island. PhD, Brown Univ. Assistant to Landsteiner, and then Prof., Rockefeller Inst., 1965. Many contributions to delayed hypersensitivity and contact dermatitis; first passive transfer of tuberculin and contact hypersensitivity; mechanism of action of adjuvants; quantitative methods. Obit.: *NY Times*, April 19, 2008.

COCA, Arthur Fernandez [1875–1959]. Born in Philadelphia. MD, Univ. Penn.; studied at Heidelberg. Cornell, 1910; Prof. Columbia, 1932; Lederle Labs., 1931–1948. Many contributions to allergy with Robert A. Cooke; classification of human allergies; named atopic antibodies; isolation and use of allergens. Books: *Essentials of Immunology for Medical Students*, 1925; *Asthma and Hay Fever in Theory and Practice* (with Walzer and Thommen), 1931. Founding editor, *J. Immunol.*, 1916–1947. Obit.: *J. Am. Med. Soc.* **172**:835, 1959.

COHN, Ferdinand Julius [1828–1898]. Born in Breslau. Educated at Breslau and Berlin. Prof. Botany, Breslau. Leading botanist; argued for constancy of bacterial species; supported Pasteur against spontaneous generation; "discovered" Koch. Biogs: Pauline Cohn, *Blätter der Erinnerung*, Breslau, 1901; *Dict. Sci. Biog.* **3**:336, 1971. Obit.: *Munch. med. Wochenschr.* **45**:1005, 1898.

COHNHEIM, Julius [1839–1884]. Born in Demmin, Pomerania. Prof. in Kiel, 1868; Breslau, 1872; Leipzig, 1878. Eminent pathologist; proposed vascular theory of inflammation. Book: *Lectures on General Pathology*, London, New Sydenham Soc., 1889. Obit.: *Ges. Abhandl., Berlin* 1885, pp. vii–li.

COOKE, Robert Anderson [1880–1960]. Born in Holmdel, New Jersey. MD, Columbia, 1904. Prof. Allergy and Immunology, Cornell, 1920. Classification of human allergies; developed skin test and desensitization methods; helped found allergy societies. Obit.: *Ann. Allergy* **21**:107, 1963.

COOMBS, Robert Royston Amos (Robin) [1921–2006]. PhD, Cambridge, 1947. Prof. Pathology, Cambridge Univ. Developed Coombs test for autoimmune hemolytic anemia; many contributions to immunohematology, immunopathology, and serology. Books: *The Serology of Conglutination and its Relation to Disease*, 1961; *Clinical Aspects of Immunology* (with Gell), 1963. Obit.: *NY Times*, March 27, 2006.

COONS, Albert Hewett [1912–1978]. Born in Gloversville, New York. MD, Harvard, 1937. Professor, Harvard Univ. Developed fluorescent antibody immunohistochemistry. Lasker Medal, 1959; Ehrlich Prize, 1961; Behring Prize, 1966. Obit.: *J. Histochem. Cytochem.* **27**:1117, 1979; see *Ann. NY Acad. Sci.* **420**:6, 1983.

DALE, Henry Hallett [1875–1968]. Born in London. MD, 1909; studied with Starling and Ehrlich. Nat. Inst. Med. Res., 1914–1942; Prof. Chemistry, Davy-Faraday Lab., 1942. Discovered histamine; Schultz–Dale test for anaphylaxis; important contributions to chemical transmission of nerve impulses. Nobel Prize, 1935. Biogs: See *Ciba Foundation Symposium on Histamine: Festschrift in Honour of Sir Henry Dale*, Boston, 1956; *Dict. Sci. Biog.* **15**:104, 1978. Obit.: *Br. Med. J.* **3**:318, 1968.

DAMESHEK, William [1900–1969]. Born in Voronezh, Russia. MD, Harvard, 1923. Prof. Hematology, Tufts Univ., Boston. Leading hematologist; described autoimmune hemolytic anemias; agranulocytosis. Editor-in-Chief of *Blood*. Biog.: See issue of *Blood* in honor of Dameshek's sixtieth birthday, **15**: May, 1960. Obit.: *Blood* **35**:1, 1970.

DANYSZ, Jan (Jean) [1860–1928]. Born in Poland. Worked at Institut Pasteur, Paris. Discovered Danysz phenomenon of neutralization of diphtheria toxin; discovered virus for destroying rodents; later worked on chemotherapy. Obit.: *Bull. Inst. Pasteur* **26**:97, 1928.

DAUSSET, Jean Baptiste Gabriel [1916–2006]. Born in Toulouse. MD, Paris. Director, Natl Blood Transfusion Center, 1950; Director, Immunogenetics and Transplantation, Inst. Nat. Santé, 1968. Many contributions to immunogenetics and transplantation; major histocompatibility complex. Nobel Prize, 1980. Books: *Immunohématologie, Biologique et Clinique*, 1956; *HLA and Disease* (with Sveljgaard), 1977.

DIENES, Louis Ladislaus [1885–1974]. Born in Tokay, Hungary. MD, Budapest, 1908; studied in France and Germany. Director, von Ruck Research Lab. Tuberculosis, 1921; bacteriologist, Mass. General Hospital, 1930. Studied bacterial allergy; pleuropneumonia-like organisms; demonstrated production of tuberculin-type hypersensitivity to bland proteins. See Festschrift: *Symposium on Mycoplasma and L Forms of Bacteria, in honor of Louis Dienes*, New York, Gordon Breach, 1971. Obit.: *J. Infect. Dis.* **130**:89, 1974.

DIXON, Frank James [1920–2008]. Born in St Paul, Minnesota. MD, Univ. Minn., 1944. Washington Univ., 1948; Univ. Pittsburgh, 1951; Director, Scripps Inst. Res. Foundation, La Jolla, 1974. Many contributions to antibody formation; immunopathology; pathogenesis of immune complex diseases. Obit.: *Los Angeles Times*, Feb. 10, 2008.

DOERR, Robert [1871–1950]. Born in Tesco, Hungary. MD, Vienna, 1897. Worked with Paltauf at Vienna Univ.; Director, Hygienic Inst. Basel, 1919. Worked on viral immunology, experimental anaphylaxis, and human allergies.

Biog.: See Festschrift on seventieth birthday, *Schweiz. Z. Pathol. Bakteriol.* **4**: 1941. Obit.: *Wien. klin. Wochenschr.* **64**:129, 1952.

DONATH, Julius [1870–1950]. MD, Vienna, 1895; assistant to Nothnagel, 1898. Dept Head, Vienna Merchants' Hospital, 1910–38; Israelite Comm. Hospital, Vienna, 1938. Described (with Landsteiner) the first autoimmune disease, paroxysmal cold hemoglobinuria, 1904. Biog.: Speiser and Smekal's *Karl Landsteiner,* Bruder Hollinek, Vienna, 1975, p. 126.

DONIACH, Deborah [1912–2004]. MB, London, 1945. Prof. Clinical Immunology, Middlesex Hospital. First demonstrated (with Roitt and others) autoantibodies in Hashimoto's disease; many contributions to study of autoimmune thyroid and liver diseases. Obit.: *Br. Med. J.* **328**:351, 2004.

DOUGLAS, Stewart Ranken [1871–1936]. Born in Caterham, Surrey, England. MB, St Barts, London; Indian Medical Service (Army); worked under Wright at St Mary's, London; Director, Bacteriology Dept, Natl Inst. Med. Res., London. Discovered and developed (with Wright) opsonins and related serotherapy. Obit.: *Lancet* **1**:229, 1936.

DUCLAUX, Emile [1840–1904]. Born in Aurillac, France. Ecole Normale, Paris; préparateur under Pasteur. Prof. Biological Chemistry, Sorbonne; succeeded Pasteur as Director, Pasteur Inst. Assisted Pasteur in many immunologic discoveries. Biog.: *Dict. Sci. Biog.* **4**:210, 1971. Obit.: *Ann. Inst. Pasteur* **18**:273, 337, 1904.

DUNGERN, Emil Freiherr von [1867–1961]. MD, Munich, 1892; studied also in Freiburg and Pasteur Inst. Paris. Head, Biochem. Dept, Heidelberg Inst. Cancer Res. 1906; Director, Hamburg/Eppendorff Cancer Inst., 1913. Discovered (with Hirszfeld) heredity of blood groups; defined fourth group as AB and Landsteiner's third group as O.

DURHAM, Herbert Edward [1866–1945]. Born in London. Educated at Cambridge, Guy's Hospital; worked with von Gruber in Vienna. Discovered (with Gruber) bacterial agglutination; later worked on yellow fever and beriberi. Obit.: *Lancet* **2**:654, 1945.

EDELMAN, Gerald Maurice [1929–]. Born in New York. MD, Penn., PhD, Rockefeller Inst. with Kunkel; Prof. Rockefeller Inst., 1966. Immunoglobulin chains; domains; total amino acid sequence of Ig; role of cell adhesion molecules; neurobiology. Nobel Prize, 1972. Books: *Neural Darwinism,* 1987; *Topobiology,* 1988; *The Remembered Present: A Biological Theory of Consciousness,* 1989; *Morphoregulatory Molecules,* 1990; *Bright Air, Brilliant Fire: On the Matter of the Mind,* 1992; *Consciousness,* 2000; *Wider than the Sky: The Phenomenal Gift of Consciousness,* 2004.

EHRICH, William Ernst [1900–1967]. Born in Dahmen, Germany. MD, Rostock, 1924. Studied pathology at Rockefeller Inst, 1926; privatdozent Rostock, 1931; Univ. Penn., 1936. Worked on antibody formation; lymphoid tissue responses; renal diseases. Obit.: *Verh. Deut. Ges. Pathol.* **54**:589, 1970.

EHRLICH, Paul [1854–1915]. Born in Strehlen, Silesia. Educated at Breslau and Strassburg. Assistant in Frerich's Clinic, 1878; assistant in Koch's Inst., 1890; Director, Inst. für Serumprufung, Steglitz, 1896; Director, Inst. für exper. Therapie, Frankfurt am Main, 1899. Developed histologic stains; many contributions to hematology; side-chain theory of antibody formation; developed assays for diphtheria toxin and antitoxin; mechanism of immune hemolysis; developed chemotherapeutic theory and practice; important early contributions to cancer research. Nobel Prize, 1908. Books: *Collected Studies on Immunity*, 1906; *Collected Papers of Paul Ehrlich* (three vols), 1957. Biogs: Martha Marquardt, *Paul Ehrlich als Mensch und Arbeiter*, Berlin, 1924; A.M. Silverstein, *Paul Ehrlich's Receptor Immunology: The Magnificent Obsession*, 2002; *Dict. Sci. Biog.* 4:295, 1971; (and many others). Obit.: See centenary tribute, *Bull. NY Acad. Med.* **30**:968, 1954.

EISEN, Herman Nathaniel [1918–]. Born in New York. MD, NY Univ. Columbia Univ. P & S, 1944; Prof. Washington Univ., 1955; Prof. MIT, 1973. Developed equilibrium dialysis (with Karush); many contributions to mechanism of contact dermatitis; antigen recognition; antibody structure and function; cytotoxic T cells. Books: *Microbiology* (with Davis and Dulbecco), 1967; *Immunology*, 1974; *General Immunology*, 1990.

ELSCHNIG, Anton Phillip [1863–1939]. MD, Graz. Univ. Vienna, 1895; Prof. Ophthalmology, German Univ. Prague, 1907. Advanced theory of autoimmune pathogenesis of sympathetic ophthalmia; one of leading ophthalmologists of his day. Obit.: *Am. J. Ophthalmol.* **23**:214, 1940.

FAGRAEUS-WALLBOM, Astrid Elsa [1913–1997]. PhD, Stockholm, 1948. Chief, Virus Dept, Natl Bacteriol. Lab., 1962; Prof. Immunology, Karolinska Inst., 1965. First clear demonstration (in her doctoral thesis) that antibodies are produced in plasma cells; worked in clinical immunology; cell membrane antigens. Book: *Antibody Production in Relation to the Development of Plasma Cells*, Stockholm, 1948. Obit.: *Scand. J. Immunol.* **47**:91, 1998.

FENNER, Frank John [1914–]. Born in Ballaret, Australia. Studied at Adelaide Univ. and Hall Inst. with Burnet. Director, John Curtin School Med. Res., 1967; Director, Center for Resource and Environmental Systems, 1973. Worked in virology; host–parasite interactions. Book: *The Production of Antibodies*, 2nd edn (with Burnet), Melbourne, 1949.

FORSSMAN, Magnus John Karl August [1868–1947]. Born in Kalmar, Sweden. Educated at Lund and Stockholm. Prof., Lund. Discovered heterophile antigen (named Forssman antigen). Monographs: "Die heterogenetischen Antigene, besonders die sog. Forssman Antigene und ihre Antikörper," in Vol. 3, Kolle and Wassermann's *Handbuch der pathogenen Mikroorganismen*, 1928, pp. 469–526; "Heterogenetic Antigens and Antibodies", *Acta Pathol. Microbiol. Scand.* Suppl **16**. Obit.: *Acta Pathol. Microbiol. Scand.* **25**:513, 1948.

FRACASTORO, Girolamo [1478?–1553]. Born in Verona. Studied at Padova. Physician, astronomer, geographer, poet, and humanist. Advanced theory of contagion; early theory of acquired immunity; gave syphilis its name. Books: *Syphilis sive Morbus Gallicus*, 1530; *De Sympathia et Antipathia Rerum* and *De Contagione*, 1546. Biog.: *Dict. Sci. Biog.* 5:104, 1972; see *Ann. Med. History* 1:1, 1917.

FREUND, Jules [1890–1960]. Born in Budapest. MD, Budapest, 1913. Prof. Preventive Med., Budapest, 1917; Henry Phipps Inst., Penn., 1926; Cornell, 1932; Bureau of Labs., NY, 1938; NIH, 1957. Studied allergic encephalomyelitis; antibody formation; developed Freund's adjuvant. Lasker Award, 1959. Obit.: *Lancet* 1:1031, 1960.

FRIEDBERGER, Ernst [1875–1932]. Born in Giessen. Educated in Berlin, Giessen, Munich, and Würzburg. Taught at Univ. Königsberg, 1903; Berlin Univ., 1913; Director, Prussian Res. Inst. for Hygiene and Immunology, 1926. Many contributions to allergy research. Obit.: *Z. Immunitätsf.* 73:i, 1932.

GAFFKY, Georg Theodore August [1850–1918]. Born in Hanover. Educated in Army Medical Dept, Berlin; assistant to Koch. Prof. Hygiene, Giessen, 1888; Head, German Plague Comm., India, 1897; Director, Inst. für Infektionskrankheiten, Berlin. First to cultivate typhoid bacillus; fought Pasteur about anthrax (with Koch). Biog.: *Dict. Sci. Biog.* 5:219, 1972. Obit.: *Berlin klin. Wochenschr.* 55:1062, 1918.

GAY, Frederick Parker [1874–1939]. Born in Boston. MD, Johns Hopkins; worked under Bordet in Paris. Prof. Bacteriology, Columbia Univ., 1923. Studied immune hemolysis and anaphylaxis; translated and popularized Bordet. Book: *Agents of Disease and Host Resistance*, London, 1935. Obit.: *Science* 90:290, 1939.

GELL, Philip George Houthem [1914–2001]. MB, Cambridge, 1940. Prof. Exper. Pathology, Birmingham. Studied histopathology and specificity of delayed hypersensitivity; described carrier effect (with Benacerraf); codiscoverer of idiotypy; viral immunology. Book: *Clinical Aspects of Immunology* (with Coombs), 1962. Obits: *Biog. Mem. Fellows R. Soc.* 49:163, 2003; *Cell. Immunol.* 213:1, 2001.

GENGOU, Octave [1875–1959]. Worked with Bordet in Paris. Studied immune hemolysis; discovered (with Bordet) complement fixation test. Obit.: *Mem. Acad. R. Belg.* 7:79, 1969.

GLICK, Bruce [1927–]. Born in Pittsburgh. PhD, Ohio State, 1955. Prof., Mississippi State Univ. Discovered role of thymus in antibody formation (while a graduate student).

GOOD, Robert Alan [1922–2003]. Born in Crosby, Minnesota. MD, Univ. Minnesota. Prof. Microbiology and Pediatrics, Minnesota; Director of Research, Sloan-Kettering Inst., NY, 1973; Prof., Univ. of Oklahoma, 1982; Chairman Pediatrics, Univ. S. Florida, 1985. Studied ontogeny and phylogeny of immune response; role of thymus and bursa of Fabricius; clinical and experimental immunodeficiency diseases. Lasker Award, 1970. Robert A. Good Immunology

Society formed in his honor (see *Immunl. Res.* 40:49, 2008). Books: *The Thymus in Immunobiology*, 1964; *Phylogeny of Immunity*, 1966; edited many others. Obits: *NY Times* June 18, 2003; *Devel. Compar. Immunol.* **28**:371, 2004

GORER, Peter Alfred [1907–1961]. Born in London. MD, Guy's Hospital, 1932; studied with J.B.S. Haldane, Univ. College, London. Prof. Pathology, Guy's Hospital. Important contributions to transplantation genetics; discovered antigen II associated with tumor rejection; identified (with Snell) the mouse H2 complex. Biog.: see *The Gorer Symposium*, P.B. Medawar and T. Lehner, eds, Oxford, Blackwell, 1985. Obits: *Biog. Mem. Fell. R. Soc.* 7:95, 1961; *Lancet* 2:1120, 1961.

GOWANS, James [1924–]. Born in Sheffield, England. MB, London, 1947; PhD, Oxford, 1953. Director, MRC Cell. Immunobiology Unit, Oxford, 1963; Secretary, Medical Research Council UK, 1977. Major contributions to knowledge of recirculation and function of lymphocytes.

GRABAR, Pierre [1898–1986]. Born in Kiev. Educated in Strasbourg and Paris. Chef de Service, Pasteur Inst., 1938; Director, Natl Center for Scientific Research, Paris, 1961. Studied antigen–antibody reactions; developed immunoelectrophoresis (with Williams); "carrier" theory of antibody function. Book: *Analyse Immuno-électrophoretique* (with Burtin), 1960. Obit.: *Bull. Acad. Nat. Méd.* 170:635, 1986.

GRUBER, Max von [1853–1927]. Born in Vienna. Educated at Univ. of Vienna. Prof., Graz, 1884; Vienna, 1887; Munich, 1902. Studied antigen–antibody reactions; serology; discovered bacterial agglutination (with Durham); fought with Ehrlich on nature of specificity and of antigen–antibody reaction. Biog.: *Dict. Sci. Biog.* 5:563, 1972. Obit.: *Wien klin. Wochenschr.* **40**:1304, 1927.

HAFFKINE, Waldemar [1860–1930]. Born in Odessa. Educated at Odessa Univ.; assistant in Pasteur Inst., Paris, 1888. Founded Government Research Lab. (now Haffkine Inst.) in Bombay. Developed and introduced prophylactic immunization for cholera and plague. Biog.: *Dict. Sci. Biog.* **6**:11, 1972. Obit.: *Br. Med. J.* 2:801, 1930.

HAŠEK, Milan [1925–1985]. Born in Prague. PhD, Prague Univ., 1955. Chairman, Dept Exper. Biol. Genetics, later an Inst. of Czech Academy of Sciences, 1961. Developed technique of chick embryo parabiosis; demonstrated immunologic tolerance; many contributions to transplantation biology. Obit.: *Immunogenetics* **21**:105, 1985.

HAUROWITZ, Felix [1896–1987]. Born in Prague. Educated at German Univ., Prague; worked on protein chemistry with Breinl at Prague; Univ. Istanbul, 1939; Prof. Biochemistry, Indiana Univ., 1958. Leading protein chemist; worked on hemoglobin chemistry; immunochemistry; proposed (with Breinl) instruction theory of antibody formation, 1930. Books: *Chemistry and Biology of Proteins*, New York, Academic Press, 1950; *Immunochemistry and the Biosynthesis of Antibodies*, New York, Interscience, 1968. Obit.: *NY Times*, Dec. 6, 1967.

HEIDELBERGER, Michael [1888–1991]. Born in New York. PhD, Columbia Univ. and Federal Polytechnic Inst., Zurich. Rockefeller Inst, 1912; Columbia P & S, 1927; Rutgers Univ. Inst. of Microbiology, 1955; NY Univ., 1964. Isolation and immunochemistry of pneumococcal polysaccharides; quantitative immunochemistry. Lasker Award; National Medal for Science; Behring Award; Pasteur Medal (Sweden); Légion d'Honneur. Autobiog.: "Reminiscences: A Pure Organic Chemist's Downward Path," *Ann. Rev. Microbiol.* **31**:1, 1977; *Ann. Rev. Biochem.* **48**:1, 1979; *Immunol. Rev.* **82**:7, 1984; **83**:5, 1985. Biog.: *Hospital Practice*, Oct., 1983, pp. 214–230. Obits: *J. Immunol.* **148**:301, 1991; *NY Times* June 27, 1991.

HEKTOEN, Ludvig [1863–1951]. Born in Westby, Wisconsin. Studied at Chicago, Uppsala, Prague, and Berlin. Prof. Pathology, Univ. Chicago, 1901; Director, McCormick Inst. Infect. Diseases, Chicago. Studied immunochemistry of thyroid and other tissue proteins; mechanisms of immunity. Biog.: *Dict. Sci. Biog.* **6**:232, 1972. Obit.: *Arch. Pathol.* **52**:390, 1951.

HEREMANS, Joseph F. [1927–1975]. MD, Louvain, 1952; PhD, 1960. Worked with Grabar, Kunkel, and Waldenstrom. Prof. Internal Med. and Immunochemistry, Louvain, 1965. Played a major role in isolation and characterization of IgA; studied secretory immune system. Book: *Molecular Biology of Human Proteins* (with Schultze), 1966. Obit.: *Eur. J. Immunol.* **6**:1, 1976.

HERICOURT, Jules [1850–1933]. Worked with Richet on serotherapy of cancer; anaphylaxis. Book: *La Sérotherapie: Historique, Etat Actuelle*, Paris, Rueff, 1899.

HUMPHREY, John Herbert [1916–1987]. Born in West Byfleet, Surrey, England. Studied at Cambridge and Univ. College Medical School, London. Lister Inst. Preventive Med., 1941; Middlesex Hospital, 1942; Med. Res. Council, 1946; Natl Inst. Med. Res., 1949; Prof. Royal Postgrad. School of Medicine, 1976. Numerous contributions to antibody formation, immunochemistry, immunopathology. Book: *Immunology for Students of Medicine* (with White), 1963. Obits: *Biog. Mem. Fell. R. Soc.* **36**:274, 1990; *Scand. J. Immunol.* **27**:617, 1988.

JENNER, Edward [1749–1823]. Born in Berkeley, Gloucestershire, England. Apprenticed in Surgery, 1762; studied under John Hunter, 1770; MD, St. Andrew's, 1792. Practiced medicine privately; many studies in zoology and animal behavior; discovered application of cowpox (vaccinia) virus for preventive vaccination for smallpox. Books: *Inquiry into the Cause and Effects of the Variolae Vaccinae*, 1798; *On the Influence of Artificial Eruptions in Certain Diseases*, 1822. Biogs: John Baron, *The Life of Edward Jenner*, 1827; William LeFanu, *Biobibliography of Edward Jenner*, Harvey & Blythe, London, 1951; *Dict. Sci. Biog.* **7**:95, 1973.

JERNE, Niels Kaj [1911–1994]. Born in London. Studied at Leiden and Copenhagen. Danish State Serum Inst., 1943; WHO, Geneva, 1956; Prof. Pathology, Univ. Pittsburg, 1962; Goethe Univ., Frankfurt, 1966; Director, Basel

Inst. for Immunology, 1969. Studied antibody formation and avidity; natural selection theory of antibody formation, 1955; hemolytic plaque assay for antibody, 1963; idiotype network theory, 1974. Nobel Prize, 1984. Biog.: T. Söderqvist, *Science as Biography: The Troubled Life of Niels Jerne*, Yale Press, 2003. Obits: *Br. Med. J.* 309:1434, 1994; *Biog. Mem. Fell. R. Soc.* 43:236, 1997.

KABAT, Elvin Abraham [1914–2000]. Born in New York. PhD, Columbia P & S (with Heidelberger); studied with Tiselius at Uppsala. Instructor in Pathology, Cornell, 1938; Prof. Microbiology, Columbia, 1941. Separated globulins by electrophoresis (with Tiselius); demonstrated 7S and 19S gamma globulins; studied anticarbohydrate antibodies; size of the antibody combining site; demonstrated (with Wu) hypervariable regions on Ig chains; proposed minigene hypothesis. Books: *Experimental Immunochemistry* (with Mayer), 1948; *Blood Group Substances: Their Chemistry and Immunochemistry*, 1956; *Structural Concepts in Immunology and Immunochemistry*, 1968. Autobiog.: *Ann. Rev. Immunol.* 1:1, 1983. Biog.: see Schlossman and Benacerraf, *Molecular Immunol.* 21:1009, 1984. Obits: *Nature* 407:315, 2000; *NY Times*, June 22, 2000.

KAHN, Reuben Leon [1887–1979] Born in Swir (Kovno), Lithuania. Studied at Valparaiso; DSc, NY Univ., 1916. Michigan Dept Health, 1920; Prof. Bacteriology, Univ. Michigan, 1928; Prof. Serology, 1951; Prof. Microbiology, Howard Univ., 1968. Many contributions to serology of syphilis; Kahn test. Biog.: *J. Natl Med. Assoc.* 63:388, 1971.

KALLOS, Paul [1902–1988]. Born in Budapest. MD, Peso, 1929. Lab. Chief, Tuberculosis San., 1924; Research Assoc., Leipzig, 1929, Nürnburg, 1931, Uppsala, 1934; Director, Immunol. Res., Wenner–Gren Inst., Stockholm, 1937. Worked on allergy and tuberculosis. Edited many journals and publications on allergy. Founded *Progr. Allergy*, 1939.

KARUSH, Fred [1914–1994]. Born in Chicago. PhD, Chicago, 1938. Chemist, Dupont, 1944; Univ. Penn., 1950, Prof. Immunochemistry, 1957. Studied protein interactions; thermodynamics of antigen–antibody interactions; developed (with Eisen) equilibrium dialysis; many contributions to molecular immunology.

KITASATO, Shibasaburo [1852–1931]. Born in Oguni, Kumamoto, Japan. Studied at Imperial Univ., Tokyo; with Koch at Koch's Inst., Berlin, 1885. Founded Inst. for Infectious Diseases (later Kitasato Inst.), Tokyo. Discovered (with Behring) tetanus antitoxin, 1890; discovered plague bacillus (independently of Yersin), 1894. Ennobled to rank of Baron. Biogs: M. Miyajima, *Robert Koch and Shibasaburo Kitasato*, Sonor, Geneva, 1931; *Dict. Sci. Biog.* 7:391, 1973. Obit.: *J. Pathol. Bacteriol.* 34:597, 1931.

KOCH, Heinrich Hermann Robert [1843–1910]. Born in Clausthal, Germany. MD, Göttingen, 1866. Practiced medicine; Kreisphysikus, Wollstein, 1872; founded School of Bacteriology at Gesundheitsamt, Berlin; Director, Inst. für Infektionskrankheiten (Koch's Inst.), Berlin. Working alone as district physician, discovered tubercle bacillus, 1882; developed techniques for pure culture of bacteria; discovered cholera vibrio; fought with Pasteur about anthrax; discovered tuberculin. Nobel Prize, 1905. Biogs: Bruno Heymann, *Robert Koch*, Leipzig, 1932; *Dict. Sci. Biog.* 7:420, 1973. Obit.: *Br. Med. J.* 1:1386, 1910.

KRAUS, Rudolph [1868–1932]. Born in Jungbunzlau, Bohmen. MD, German Univ., Prague, 1893. Assistant to Paltauf in Vienna; Director, Bacteriologic Inst., Buenos Aires; State Serum Inst., Sao Paulo; State Serotherapeutic Inst., Vienna. Discovered precipitin reaction; edited (with Uhlenhuth and then Levaditi) *Handbuch der Immunitätsforschung und experimentelle Therapie*, 1907, 1914. Biog.: *Wien. med. Wochenschr.* 118:869, 1968. Obit.: *Z. Immunitätsf.* 76:i, 1932.

KUNKEL, Henry George [1916–1983]. Born in New York. MD, Johns Hopkins, 1942. Prof. Medicine, Rockefeller Inst. Many contributions to basic and clinical immunology, including: myeloma proteins as immunoglobulins; structure and genetics of Ig; human allotypy; discovered idiotypy (independently of Oudin and Gell); rheumatoid factor as autoantibody; discovered IgA. Lasker Award; Gairdner Award. Obit.: *J. Exp. Med.* **161**:869, 1985.

LANDSTEINER, Karl [1868–1943]. Born in Vienna. MD, Vienna; assistant to Gruber. Worked in pathologic anatomy and experimental pathology, Vienna; Ziekenhuis, The Hague, 1919; Prof., Rockefeller Inst., 1922. Discovered ABO and other blood groups; discovered Rh factor; argued nature and origin of antibodies with Ehrlich; demonstrated poliomyelitis infection in monkeys; perfected the use of artificial haptens for the study of antibody specificity. Nobel Prize, 1930. Book: *Die Spezifizität der serologischen Reaktionen*, 1933 (English editions, 1936, 1945). Biogs: P. Speiser and F.G. Smekal, *Karl Landsteiner*, Vienna, 1975; *Dict. Sci. Biog.* 7:622, 1973. Obit.: *J. Immunol.* 48:1, 1944.

LAWRENCE, Henry Sherwood [1916–2004]. Born in New York. MD, NY Univ., 1943. Head, Infectious Dis. and Immunology Unit, NY Univ., 1959; Prof. Medicine, 1961. Discovered transfer factor; studied contact dermatitis and delayed hypersensitivity. Founded *Cell. Immunol.* Book: *Cellular and Humoral Aspects of Delayed Hypersensitivity States*, 1959. Obit.: *NY Times*, April 8, 2004.

LEDERBERG, Joshua [1925–2008]. Born in Montclair, New Jersey. PhD, Yale (with Tatum), 1947. Univ. Wisconsin, 1947; Stanford Univ., 1954; President, Rockefeller Univ. Showed sexual reproduction in bacteria (with Tatum); demonstrated transduction of genetic material in bacteria; contributed to formulation of clonal selection theory of antibody formation. Nobel Prize, 1958. Obit.: *NY Times*, Feb. 6, 2008.

LEPOW, Irwin Howard [1923–1984]. Born in New York. PhD, Case-Western Reserve, 1951; MD, 1958; studied with Pillemer. Case-Western Reserve, 1951; Chairman Pathology, Univ. Connecticut, 1967; President, Sterling Winthrop Res. Inst., 1978. Studied properdin; complement; mechanisms of immunologic injury. Obit.: *J. Immunol.* 134:1320, 1985.

LEVADITI, Constantin [1874–1953]. Born in Romania. Educated in Paris. Pasteur Inst., Paris, 1901; Prof., 1924. Many contributions to syphilology and immunology; stain for *T. pallidum*; worked with Landsteiner on polio and scarlet fever; edited (with Kraus) *Handbuch der Immunitätsforschung und experimentelle Therapie*, 1914. Biog.: *Dict. Sci. Biog.* 8:273, 1973. Obit.: *Ann. Inst. Pasteur* 85:535, 1953.

LEVINE, Philip [1900–1987]. Born in Kletsk, Russia. MD, Cornell, 1923. Rockefeller Inst., 1925; Univ. Wisconsin, 1932; Beth Israel Hospital, Newark, 1935; Ortho Res. Foundation, 1944. Discovered (with Landsteiner) M, N, and P blood groups; many contributions to blood groups and immunohematology. See tribute to Dr Philip Levine, *Am. J. Clin. Path.* 74:368, 1980.

LINDENMANN, Jean [1924–]. Born in Zagreb. Educated at Zürich. Univ. Florida, 1962; Zurich Univ., 1964. Discovered interferon (with Isaacs); advanced early network theory of idiotype–anti-idiotype; viral immunology; history of immunology.

MACKANESS, George Bellamy [1922–2003]. Born in Sydney. MD, Univ. Sydney, 1945; PhD, Oxford, 1953. Austral. Natl Univ., 1954; Univ. Adelaide, 1962; Director, Trudeau Inst., 1965; President, Squibb Inst. Med. Res., 1976. Studied role of macrophages in cellular immunity; tumor immunology. Obit.: *Tuberculosis* 87:391, 2007.

MACKAY, Ian Reay [1922–]. MD, Univ. Melbourne. Walter and Eliza Hall Inst., 1955; Chairman, Clinical Res. Unit. Work on autoimmune diseases; immuno-pathology; clinical immunology. Books: *Autoimmune Diseases* (with Burnet), 1963; *The Human Thymus*, 1969; *The Autoimmune Diseases* (with Rose), 1985.

MADSEN, Thorvald [1870–1957]. Born in Frederickberg, Denmark. Educated Copenhagen Univ., 1893; worked with Ehrlich in Frankfurt and at Pasteur Inst., Paris. Director, Statens Serum Inst., Copenhagen, 1902; President, Hygienic Commission, League of Nations, 1921–1940. Studied diphtheria toxin–antitoxin interactions and assay; advanced (with Arrhenius) theory of reversibility of Ag–Ab binding. See *Liber Gratulorius in Honorem Thorvald Madsen*, Copenhagen, Munksgaard, 1930. Obit.: *Br. Med. J.* 1:1010, 1957.

MANWARING, Wilfred Hamilton [1871–1960]. Born in Ashland, Virginia. MD, Hopkins, 1904; studied at Berlin, Frankfurt, and London. Assistant in Pathology, Univ. Chicago, 1904; Univ. Indiana, 1905; Rockefeller Inst., 1910; Prof. Bacteriology and Exper. Pathology, Stanford, 1913. Studied antibody formation; allergy; fought against Ehrlich's theories; attempted to revive antigen-incorporation theory of antibody formation.

MARQUARDT, Martha [1875–1956]. Long-time secretary to Ehrlich; helped to preserve Ehrlich papers; published biography *Paul Ehrlich als Mensch und Arbeiter*, 1924 (English editions 1949, 1951). Obit.: *Br. Med. J.* 2:181, 307, 1956.

MARRACK, John Richardson [1889–1976]. MB, Cambridge, 1912; MD, 1923. Prof. Chemical Pathology, Cambridge, London Hospital; Univ. Texas, Houston (after retirement). Proposed bivalency of antibody; lattice theory of antigen–antibody complex formation; used colored dyes to label antibodies, 1934. Book: *The Chemistry of Antigens and Antibodies*, 1934. Obit.: *Lancet* 2:378, 1976.

MASUGI, Matazo [1896–1947]. MD, Univ. Tokyo, 1921; studied with Aschoff and Rössle in Germany. Prof. Pathology, Chiba Med. College, 1927. Described relationship between monocyte and histiocyte; produced experimental immunogenic kidney disease (Masugi nephritis). Obit.: *Acta Pediat. Jap.* 68:587, 1964.

MATHE, Georges [1922–]. Born in Lermages, France. Educated Univ. Paris. Univ. Paris, 1952; Prof., Cancer Research, 1956; Head, Dept Hematology, Gustav-Roussy, 1961; Director, Inst. Cancerologie et Immunogénétique, 1965. Many contributions to cancer immunology and immunotherapy; bone marrow transplantation.

MATHER, Cotton [1663–1728]. Born in Boston. MA, Harvard, 1681. New England divine. Promoted inoculation against smallpox; advanced theory of acquired immunity, 1724. Book: *The Angel of Bethesda*, 1724. FRS. Biog.: I. Bernard Cohen, *Cotton Mather and American Science and Medicine*, two vols, 1980.

MAYER, Manfred Martin [1916–1984]. Born in Frankfurt. PhD, Columbia, 1942. Assistant to Heidelberger, Columbia; Prof. Immunology, Johns Hopkins, 1946. Isolated and characterized complement components; mechanism of complement action; discovered *T. pallidum* immobilization test for syphilis (with Nelson); lymphotoxins. Book: *Experimental Immunochemistry* (with Kabat), 1948. Obit.: *J. Immunol.* 134:655, 1985.

MEDAWAR, Peter Brian [1915–1987]. Born in Rio de Janeiro. PhD, Oxford, 1935. Lecturer in Zoology, Oxford, 1938; Prof. Zoology, Univ. Birmingham, 1947; Prof., Univ. College London, 1951; Director, Med. Res. Council, 1962; Clin. Res. Ctr, Northwick Park, 1971. Founded modern transplantation immunobiology; proved immunological tolerance (with Billingham and Brent); developed antilymphocyte serum. Copley Medal, Royal Soc.; Nobel Prize, 1960. Books: *The Life Science* (with J.S. Medawar), 1977; *Advice to a Young Scientist*, 1979; *Aristotle to Zoos*, 1983 (and others). Autobiog.: *Memoir of a Thinking Radish*, Oxford, University Press, 1986. Festschrift, sixty-fifth birthday, *Cell. Immunol.* 62:231, 1981. Obit.: *Lancet*, Oct. 17, 1987, p. 923.

MELTZER, Samuel James [1851–1920]. Born in Ponovyezh, Russia. MD, Univ. Berlin, 1882; worked with Welch at Bellevue. Head of Physiology, Rockefeller Inst., 1904. Identified asthma as manifestation of anaphylaxis. Biogs: *J. Allergy* 35:215, 1964; *Dict. Sci. Biog.* 9:265, 1974. Obit.: *Proc. Soc. Exp. Biol. Med.* 18(Suppl.), 1921.

METALNIKOFF, Sergei [1870–1946]. Worked with Metchnikoff at Pasteur Inst., Paris; Prof. Zoology, St Petersburg; returned to France, 1919. Studied auto-antisperm antibodies; neurologic aspects of immune response. Book: *Rôle du Système Nerveuse et des Facteurs Biologiques et Psychiques dans l'Immunité*, Paris, Masson, 1934. Obit.: *Ann. Inst. Pasteur* 72:860, 1946.

METCHNIKOFF, Ilya Ilich (Elie) [1845–1916]. Born in Ivanovska, Ukraine. PhD, Univ. Odessa; studied with Siebold, Leukhart, and Kovalevsky. Prof. Zoology, Univ. Odessa, 1867; Chef de Service, Pasteur Inst. Paris, 1888. Discovered defensive role of phagocytes; advanced phagocytic (cellular) theory of immunity; many contributions to bacteriology and immunology. Nobel Prize, 1908. Books: *Leçons sur la Pathologie de l'Inflammation*, 1892; *L'immunité dans les Maladies Infectieuses*, 1901; *Etudes sur la Nature Humaine*, 1903 (and others). Biogs: Olga Metchnikoff, *Elie Metchnikoff: Life and Work* (English edition, 1921); *Dict. Sci. Biog.* 9:331, 1974. Obit.: *Bull. Méd. Paris* 30:385, 1916.

MIESCHER, Peter A. [1923–]. Born in Zürich. MD, Lausanne and Zürich, 1948; DSc, Basel, 1950. Head, Immunopathology Res., Basel, 1954; Prof. Medicine (hematology), 1960. Studied detection of antibodies to nucleoprotein; mechanisms of damage in immunologic diseases; role of RE system. Books: *Immunpathologie in Klinik und Forschung* (with Vorlaender), 1957; *1st International Symposium on Immunopathology* (with Grabar), 1958.

MILGROM, Felix [1919–2008]. Born in Rohatyn, Poland. MD, Wroclaw, 1947. Director, Inst. Immunology Exper. Therapy, Polish Acad. Sci., 1954; Silesian Med. School, 1954, Prof. Microbiology, SUNY, Buffalo, 1967. Studied serology of syphilis and rheumatoid arthritis; first proposed (with Dubiski) that rheumatoid factor is autoantibody. Obit.: *Intl Arch, Allergy Immunol.* 146:174, 2008.

MILLER, Jacques Francis Albert Pierre [1931–]. Born in Nice. MB, 1955, Univ. Sydney. Chester Beatty Research Inst., London. PhD, 1960, DSc, 1965, Univ. London. Investigating pathogenesis of lymphocytic leukemia he discovered the important role of the thymus in the development and function of the immune system. Returned to Australia in 1966. Prof., Walter and Eliza Hall Inst. and University of Melbourne; continued to contribute importantly to the functions of the thymus and of T cells. Copley Medal, Royal Soc., 2001.

MITCHISON, Nicholas Avrion [1928–]. PhD, Oxford, 1952. Lecturer Zoology, Edinburgh; Natl Inst. Med. Res., London; University College, London. Demonstrated passive cell transfer of transplant immunity; studied antibody response to defined haptens; proposed two-cell cooperation for antibody formation; discovered low-zone tolerance; tumor immunology.

MONTAGU, Lady Mary Wortley [1689–1762]. Born in London. Woman of Belles Lettres. Wife of British Ambassador to the Porte (Istanbul), 1718, where she observed Turkish practice of inoculation against smallpox. On return to

England, helped to popularize the practice. Book: *Letters*, 1777. Biog.: Robert Halsband, *The Life of Lady Mary Wortley Montagu*, Oxford, Clarendon, 1956.

MORGENROTH, Julius [1871–1924]. Born in Bamberg. Studied with Weigert in Frankfurt. Worked with Ehrlich in Steglitz and Frankfurt; Chemotherapeutic Dept, Koch's Inst., Berlin. Published (with Ehrlich) classical series of papers on immune hemolysis. Obit.: *Deutsch. med. Wochenschr.* 51:159, 1925.

MORO, Ernst [1874–1951]. Born in Laibach. Studied at Graz; worked at the Kinderklinik with Escherich. Pediatrics at Munich 1906; Heidelberg, 1911; Director, Universitäts Kinderklinik. Developed percutaneous tuberculin test (Moro reaction). Biog.: *Pediatrics* 29:643, 1962. Obit.: *Z. Kinderheilk.* 70:323, 1952.

MOURANT, Arthur Ernest [1904–1994]. Born in Jersey. MB, St Bart's, London. Cambridge Univ., 1945; Blood Group Ref. Lab., 1946; Director, Serological Population Genetics Lab., 1965; St Bart's, 1965. Developed "Coombs Test" (with Coombs and Race); many contributions to blood group genetics and anthropology. Obit.: *Biog. Mem. Fell. R. Soc.* 45:329, 1999.

MUIR, Robert [1864–1959]. Born in Balfron, Scotland. MD, Edinburgh, 1890. Head, Pathology Dept, Glasgow. Studied antibody formation; toxin–antitoxin reactions. Books: *Manual of Bacteriology* (with Ritchie), 1897; *Studies in Immunity*, 1909. Festschrift on seventieth birthday, *J. Pathol. Bacteriol.* 39:1, 1934. Obits: *Biog. Mem. Fell. R. Soc.* 5:149, 1959; *Br. Med. J.* 1:976, 1959.

NEISSER, Albert Ludwig Sigesmund [1855–1916]. Born in Schweidnitz, Germany. MD, Breslau. Director, Dermatologic Inst., 1882. Studied leprosy; discovered gonococcus organism; devised (with Wasserman and Bruck) complement fixation test for syphilis; leading syphilologist. Biog.: *Dict. Sci. Biog.* 10:17, 1974. Obit.: *Br. Med. J.* 2:410, 1916.

NISSONOFF, Alfred [1923–2001]. Born in New York. PhD, Johns Hopkins, 1951. Head, Dept Biochemistry, Univ. Illinois Coll. Med., 1969; Brandeis Univ., 1975. Studied enzymatic cleavage of antibody; biosynthesis and genetic control of Ig; idiotypy; properties of antibodies active in allergy. Book: *The Antibody Molecule* (with Hopper and Spring), 1975. Obit.: *Nature Immunol.* 2:469, 2001.

NOSSAL, Gustav Joseph Victor [1931–]. Born in Bad Ischl, Austria. MB, Sydney, 1955; PhD, Melbourne, 1960; studied with Burnet and Lederberg. Director, Walter and Eliza Hall Inst., Melbourne. Major contributions to antibody formation; one cell–one antibody; isotype switching; immunological tolerance; cell–cell interactions. Books: *Antibodies and Immunity*, 1969; *Antigens, Lymphoid Cells, and the Immune Response* (with Ada), 1971.

NOWELL, Peter C. [1928–]. Born in Philadelphia. MD, Univ. Penn. Head, Pathology Dept, Penn., 1967. Discovered role of phytohemagglutinin in stimulating division of lymphocytes; worked on bone marrow transplantation; discovered Philadelphia chromosome (with Hungerford).

NUTTALL, George Henry Falkiner [1862–1937]. Born in San Francisco. Educated in USA, France, Germany, and Switzerland. Lecturer in Bacteriology, Cambridge Univ., 1900; Quick Prof. Biology, 1906; founded Molteno Inst. of Biology and Parasitology, 1921. First to describe natural bactericidal action of blood; major studies in application of serology to forensics and evolution; studied arthropod-borne diseases; founded *J. Hygiene*, 1901, *J. Parasitol.*, 1908. Book: *Blood Immunity and Blood Relationships*, 1904. Obit.: *J. Pathol. Bacteriol.* **46**:389, 1938.

OBERMAYER, Friedrich [1861–1925]. Born in Vienna. MD, Univ. Vienna. Worked at Inst. Medicinal Chemistry. First demonstrated the nature of antibody response to chemically treated proteins (with Pick), leading to future studies of hapten–antihapten interactions and valuable specificity data. Obit.: *Wien klin. Wochenschr.* **38**:542, 1925.

OSTROMUISLENSKY, Ivan Ivanovich [1880–1939]. Born in Moscow. PhD, Zürich, 1902; MD 1906. Assistant Prof Chemistry, Moscow; Prof., Nizhny Novgorod; private res. lab., 1911; Scientific Inst., Moscow, 1916; US Rubber Co., 1922; Ardol Rossium Labs, 1932. Proposed instruction theory of antibody formation, 1915; claimed *in vitro* synthesis of antibody.

OUCHTERLONY, Örjan Thomas Gunnersson [1914–2004]. Studied at Karolinska Inst. Prof. Bacteriology, Univ. Götheburg. Developed technique of two-dimensional double diffusion analysis of antigens and antibodies in gels. Book: *Handbook of Immunodiffusion and Immunoelectrophoresis*, 1968.

OUDIN, Jacques [1908–1986]. Director, Analytical Immunology Dept, Pasteur Inst., Paris. Developed agar single diffusion technique for Ag–Ab reactions 1946; independently discovered idiotypy (with Michel), 1963. Obit.: *J. Immunol.* **136**:2334, 1986.

OWEN, Ray David [1915–]. Born in Genesee, Wisconsin. PhD, Wisconsin, 1941. Univ. Wisconsin, 1940; California Inst. Technol., 1947; Chair, Div. Biology, 1961; Vice President, 1975. Worked in immunogenetics, tissue transplantation, serology; discovered erythrocyte mosaicism of cattle chimeras, leading to Burnet's concept of tolerance.

OVARY, Zoltan [1907–2005]. Born in Kolozsvar, Hungary. MD, Paris, 1935. Studied allergy with Bernard Halpern. Worked at Pasteur Institute; University of Rome. Fellow, Johns Hopkins, 1954; Faculty, NY Univ., 1959; Professor of Pathology, 1964. Discovered passive cutaneous anaphylaxis; many other contributions to immediate hypersensitivity. Passionate student of art history. Obits: *Lancet* **366**:364, 2005; *NY Times* June 17, 2005.

PAPPENHEIMER, Alwin Max Jr [1908–1995]. Born in Cedarhurst, New York. PhD, Harvard, 1932; studied with Dale. Faculty, Univ. Penn., 1939; Prof., NY Univ., 1941; Harvard, 1958. Studied diphtheria toxin–antitoxin reactions; antibody formation; devised (with Uhr and Salvin) method for delayed hypersensitivity to simple proteins. Book: *Nature and Significance of the Antibody Response*, 1953. Biog.: *Cell. Immunol.* **66**:1, 1982.

PASTEUR, Louis [1822–1895]. Born in Dôle, France. Educated at Ecole Normale, Paris. Lycée Prof., Dijon, 1848; Prof. Chemistry, Strasbourg, 1852; Dean of Faculty, Lille, 1854; Sorbonne, 1867; founded Inst. Pasteur, 1888. Crystallized D- and L-tartaric acid; studied mechanism of fermentation and diseases of wine and beer; diseases of silkworms; experimentally disproved spontaneous generation; discovered attenuation of pathogens and their use in preventive immunization (fowl cholera, anthrax, rabies, swine erysipelas, etc.). FRS, Copley Medal, Rumford Medal, and many other awards. Books: *Les Maladies des Vers à Soie*, 1865; *Etudes sur le Vin*, 1866; *Etudes sur la Bière*, 1876; *Oeuvres*, 1922–39. Biogs: Valléry Radot, *La Vie de Pasteur*, 1900; René Dubos, *Louis Pasteur, Freelance of Science*, Boston, Little Brown, 1950; *Dict. Sci. Biog.* **10**:350, 1974; and many others. Obit.: *Nature Lond.* **52**:576, 1895.

PAULING, Linus [1901–1994]. Born in Portland, Oregon. PhD, Cal. Inst. Technol., 1925. Cal. Tech., 1927; Univ. California San Diego, 1967; Stanford Univ., 1969. Studied the chemical bond; structure of crystals and molecules; proposed instructive theory of antibody formation; structure of proteins and DNA; discovered (with Itano) molecular defect in hemoglobin S (sickling); studied (with Pressman) thermodynamics of antibody–hapten interactions; orthomolecular medicine. Nobel Prize for Chemistry, 1954; Nobel Peace Prize, 1963. Books: *The Nature of the Chemical Bond*, 1939; *No more War*, 1958. Biog.: T. Goertzel and B. Goertzel, *Linus Pauling: A Life in Science and Politics*, New York, Basic Books, 1995. Obit.: *NY Times*, August 21, 1994.

PFEIFFER, Richard Friedrich Johannes [1858–1928]. Born in Zduny, Posen. Educated at Schweidnitz, Berlin. Military doctor at Koch's Inst, 1887; Prof. Hygiene, Königsburg, 1899; Breslau, 1909. Discovered influenza bacillus; specific lysis of typhoid and cholera organisms (Pfeiffer phenomenon); developed immunization against typhoid fever; many other contributions to bacteriology and serology. Obits: *Centralbl. Bakt.* **1**:106, 1928; *Münch. med. Wochenschr.* **75**:524, 1928.

PICK, Ernst Peter [1872–1960]. Born in Jaromer, Bohemia. MD, Prague, 1896; worked with Obermayer. Prof. Vienna, 1924; Prof. Pharmacology, Columbia Univ., 1938. Studied experimental pathology; biochemistry of antigens; published extensive review of the chemistry of antigens (1912) that stimulated the study of hapten-coupled antigens. Biog.: see "Festnummer gewidmet Ernst Peter Pick…," *Wien klin. Wochenschr.* **64**(No. 35/36), 1952. Obit.: *Wien klin. Wochenschr.* **72**:109, 1960.

PIRQUET VON CESENATICO, Clemens Peter Freiherr von [1874–1929]. Born in Vienna. MD, Vienna. Assistant to Escherich, Clinic for Children's Diseases; Prof. Pediatrics, Johns Hopkins, 1908; Breslau, 1910; Director, Universitäts Kinderklinik, Vienna, 1911. Defined pathogenesis of serum sickness; introduced term "allergy"; introduced skin test for TB (Pirquet reaction); studied nutrition in children. Books: *Die Serumkrankheit* (with Schick), 1905; *Klinische Studien über Vakzination und Vakzinale Allergie*, 1907; *Allergy*, 1911. Biog.: R. Wagner, *Clemens von Pirquet: His Life and Work*, 1968; (see also *Ann. Allergy* 31:467, 1973). Obit.: *Am. J. Dis. Children* 17:838, 1929.

PORTER, Rodney Robert [1917–1985]. Studied at Cambridge. Natl Inst. Med. Res., Mill Hill, 1949; St Mary's Hospital, 1960; Oxford, 1967. Split Ig with enzymes into Fab and Fc; proposed four-chain structure of Ab; studied order of complement genes in MHC, C4 polymorphism. Nobel Prize, 1972; Royal Soc. Medal, 1973; many other awards. Book: *Defense and Recognition*, 1973. Obit.: *Nature* 317:383, 1985.

PORTIER, Paul Jules [1866–1962]. Born in Bar-sur-Seine. MD, Sorbonne, 1897. Physiologist; assistant to Richet. Assistant in Physiology, Sorbonne, 1901; Prof., Inst. Océanographique, Paris, 1906. Discovered anaphylaxis (with Richet), 1903. Biogs: B. Masse, *Paul Jules Portier: sa Vie et son Oeuvre*, 1969; *Dict. Sci. Biog.* 11:101, 1975; see also *J. Allergy Clin. Immunol.* 75:485, 1985. Obit.: *Bull. Acad. Natl Méd.* 146:246, 1962.

PRAUSNITZ–GILES, Carl [1876–1963]. Born in Hamburg. MD, Leipzig, Kiel, and Breslau, 1903; assistant to Dunbar at State Inst. Hygiene, 1903; Royal Inst. Public Health, London, 1905; State Inst. Hygiene, Breslau, 1910; Dept Bacteriology, Breslau, 1917; fled Germany, 1933; practiced medicine, Isle of Wight (as Giles). Worked on cholera; allergies; discovered (with Küstner) passive transfer of allergy with serum, 1921. Obits: *Intl Arch. Allergy Appl. Immunol.* 23:281, 1963; *Lancet* 1:1058, 1963.

PRESSMAN, David P. [1916–1980]. Born in Detroit. PhD, Cal. Inst. Technol.; studied with Pauling. Chief, Immunochemistry, Sloan-Kettering Inst. Cancer Res., New York, 1947; Roswell Park Inst., Buffalo, 1954; Assoc. Director, 1967. Studied antibody specificity; chemical nature of Ab combining site; first used radioactive antibodies for tumor localization to develop immunotoxins. Book: *The Structural Basis of Antibody Specificity* (with Grossberg), 1968. Obit.: (issue dedicated to Pressman) *Transplant. Proc.* 12:366, 1980.

PUTNAM, Frank W. [1917–]. Born in New Britain, Connecticut. PhD, Univ. Minnesota, 1942. Duke Univ., 1942; Univ. Chicago, 1947; Univ. Florida, 1955; Indiana Univ., 1965. Studied structure of plasma proteins and enzymes; denaturation of proteins; protein synthesis; contributed importantly to structure of Ig. Book: *The Plasma Proteins*, two vols, San Diego, Academic Press, 1960.

RAFFEL, Sidney [1911–]. Born in Baltimore. DSc, 1933; MD, Stanford, 1943. Prof. Medical Microbiology, Stanford. Studied immunology of tuberculosis; delayed hypersensitivity. Book: *Immunity, Hypersensitivity, Serology*, New York, Appleton-Century-Crofts, 1953. Autobiog.: *Ann. Rev. Microbiol.* 36:1, 1982.

RAMON, Gaston [1886–1963]. Born in Bellechaume, France. DVM, Alfort, 1911. Inst. Pasteur (Garches), 1911; Assistant Director, Inst. Pasteur, 1933; Head, Bureau Epizootic Diseases, Paris, 1949. Discovered flocculation assay for diphtheria toxin; produced anatoxins with formol and heat; studied adjuvant action of alum, etc.; many contributions to diphtheria toxin–antitoxin interactions. Autobiog.: *Quarante Années de Recherches et de Travaux*, 1957. Biog.: *Dict. Sci. Biog.* 11:271, 1975. Obit.: *Bull. Acad. Méd. Paris* 147:610, 1963.

RHAZES (Abu Bakr Muhammad ibn Zakariya) [*ca.* 865–932?]. Born in Rayy, Persia. Educated in medicine. Chief physician, Rayy, Baghdad. Held to be greatest physician of Islamic world; celebrated alchemist and philosopher; differentiated smallpox and measles; advanced theory of acquired immunity. Book: *Treatise on the Smallpox and Measles*, London, Sydenham Soc., 1848. Biog.: *Dict. Sci. Biog.* 11:323, 1975.

RICH, Arnold Rice [1893–1968]. Born in Birmingham, Alabama. MD, Johns Hopkins, 1919. Dept Pathology, Johns Hopkins, 1919; Chairman, 1947. Leading authority on pathogenesis of tuberculosis, its immunity and hypersensitivity; demonstrated (with Lewis) migration inhibition by antigen, 1932. Gairdner Award, 1959; many other honors. Book: *The Pathogenesis of Tuberculosis*, 1944. Obit.: *Arch. Pathol.* 86:453, 1968.

RICHET, Charles Robert [1850–1935]. Born in Paris. MD, 1869; DSc, 1878. Prof. Physiology, Paris, 1887. Studied physiology of toxins; thermoregulation in animals; discovered anaphylaxis (with Portier). Nobel Prize, 1913. Book: *L'anaphylaxie*, 1911. Autobiog.: *Souvenirs d'un Physiologiste*, 1933. Obit.: *Bull. Acad. Méd. Paris* 115:51, 1936.

ROITT, Ivan Maurice [1927–]. DSc, Oxford. Prof. and Head, Immunology and Rheumatology Research, Middlesex Hospital, London, 1953. Discovered (with Doniach) autoimmune nature of Hashimoto's disease; many contributions to thyroid and other autoimmune diseases. Books: *Essential Immunology*, 1971; *Immunology* (with Brostoff and Male), 1985.

RÖMER, Paul [1873–1937]. Born in Neundorf, Anhalt. Educated at Jena, Greifswald, and Halle. Worked at Ophthalmology Dept, Würzburg, 1902; Greifswald, 1907; Bonn, 1921. Many contributions to ocular anaphylaxis; advanced theory of autoimmune pathogenesis of sympathetic ophthalmia. Book: *Lehrbuch der Augenheilkunde*, Berlin, 1910. Biog.: *Biographisches Lexikon* 7:1311, 1929.

ROSE, Noel Richard [1927–]. Born in Stamford, Connecticut. PhD, Penn., 1951; MD, Buffalo, 1964. First produced (with Witebsky) experimental autoimmune thyroiditis; many contributions to pathogenesis of autoimmune dieases. Books: *Methods in Immunodiagnosis*, 1973; *Principles of Immunology*, 1973; *Manual of Clinical Immunology*, 1976; *The Autoimmune Diseases* (with Mackay), 1985.

ROSENAU, Milton Joseph [1869–1946]. Born in Philadelphia. MD, Penn., 1889; studied at Hygienic Inst., Berlin; Inst. Pasteur, Paris. US Marine Hospital Service, 1890; Prof. Preventive Med., Harvard, 1909; Director Div. Public Health, Univ. N. Carolina, 1936; Dean, 1939. Research in bacteriology and hygiene; leading early contributor to mechanism of anaphylaxis. Obit.: *J. Am. Med. Assoc.* **130**:1185, 1946.

ROUX, Pierre Paul Emile [1853–1933]. Born in Confolens, France. MD, Clermont-Ferrand and Paris. Préparateur to Pasteur, 1878; Chef de Service, Inst. Pasteur, 1885; Sub-director, 1893; Director, 1904. Worked with Pasteur on anthrax and rabies vaccinations; isolated diphtheria toxin (with Yersin); many contributions to diphtheria and tetanus immunity. Copley Medal, 1913. Biogs: Mary Cressac, *Le Dr Roux, mon Oncle*, 1950; *Dict. Sci. Biog.* **11**:569, 1975. Obit.: *J. Pathol. Bacteriol.* **38**:99, 1934.

ROWLEY, Donald Adams [1923–]. Born in Owatona, Minnesota. MD, Univ. Chicago, 1950. NIH, 1951; Univ. Chicago, 1954; Director, La Rabida Children's Hospital, 1977. Studied regulation of immune response; enhancement and suppression by antigen and antibody.

SABIN, Albert Bruce [1906–1993]. Born in Bialystok, Russia. MD, NY Univ. Res. Assoc., NY Univ., 1926; Rockefeller Inst., 1935; Prof. Pediatrics, 1939; President, Weizmann Inst., 1970; Univ. S. Carolina, 1974. Research on neurotropic viruses, oncogenic viruses; perfected oral polio vaccine; Lasker Award, 1965; many other honors. Obits: *Eur. J. Epidemiol.* **9**:240, 1993; *NY Times*, March 4, 1993.

SACHS, Hans [1877–1945]. Born in Kattowitz, Silesia. Educated at Freiburg, Breslau, and Berlin. Worked with Ehrlich at Frankfurt; Director, Inst. für experimentelle Krebsforschung, Heidelberg, 1920. Numerous contributions to immune hemolysis, complement research, and antitissue antibodies; long-time defender of Ehrlich's theories. Book: *Methoden in Hämolyseforschung* (with Klopstock), 1928. Biog.: *Biographisches Lexikon* 7:1349, 1929. Obit.: *Nature Lond.* **155**:600, 1945.

SALK, Jonas [1914–1995]. Born in New York. MD, NY Univ., 1939. Univ. Michigan, 1947; Univ. Pittsburgh, 1947; founding director, Salk Inst., La Jolla, 1963. Developed killed virus vaccine for polio. Lasker Award, 1956; Koch Medal, 1963. Obits: *The Scientist* **9**:19, 1995; *NY Times*, June 24, 1995.

SALVIN, Samuel Bernard [1915–2006]. Born in Boston. PhD, Harvard, 1941. Instructor, Harvard, 1941; Rocky Mountain Lab, NIAID, 1946; Ciba Corp., 1965; Univ. Pittsburgh, 1967. Studied antibody formation; delayed hypersensitivity; lymphokines; developed (with Uhr and Pappenheimer) method to induce delayed hypersensitivity to simple proteins.

SAMTER, Max [1908–1999]. Born in Berlin. MD, Berlin, 1933. Univ. Illinois, 1946; senior consultant, Max Samter Inst. Allergy Clin. Immunol. Studied role of eosinophiles in allergy; drug reactions; bronchial asthma. Book: *Immunological Diseases*, 1965. Obit.: *NY Times*, Feb. 15, 1999.

SCHICK, Bela [1877–1967]. Born in Boglar, Hungary. MD, Graz. Univ. Vienna Children's Dept, 1902; Director Pediatrics, Mt Sinai Hospital, New York, 1923. Defined serum sickness (with von Pirquet), 1905; developed Schick test for diphtheria, 1913. Book: *Die Serumkrankheit* (with Pirquet), 1905. Festschrift on eightieth birthday, in *Current Problems in Allergy and Immunology*, 1959. Biog.: *Ann. Allergy* 38:1, 1977. Obit.: *Ann. Allergy* 26:625, 1968.

SELA, Michael [1924–]. Born in Tomaszow, Poland. PhD, Hebrew Univ. and Univ. Geneva. Head, Dept Chem. Immunology, Weizmann Inst., 1963; Dean, 1970; President, 1975. Introduced use of synthetic amino acid polymers for study of antibody formation and specificity; many contributions to immunochemistry; synthetic vaccines. Otto Warburg Medal. Books: *Topics in Basic Immunology* (with Prywes), 1969; *The Antigens*, 1973.

SEVAG, Menasseh Giragos [1897–1967]. Born in Sis, Armenia. PhD, Columbia Univ., 1929. Inst. Robert Koch, 1932; research biochemist, Schering-Kahlbaum, 1935; Univ. Penn., 1936. Research in chemotherapy, bacterial physiology; advanced instruction theory of antibody formation involving immunocatalysis. Book: *Immunocatalysis*, 1945.

SHERMAN, William Bowen [1907–1971]. Born in Providence, Rhode Island. MD, Columbia Univ., 1931. Columbia Univ., 1935; Director, Inst. Allergy, Roosevelt Hospital, New York, 1960. Many contributions to mechanisms, diagnosis, and treatment of allergic diseases.

SHULMAN, Sidney [1923–]. Born in Baltimore. PhD, Univ. Wisconsin, 1949. Prof. Immunochemistry and Biophysics, SUNY, Buffalo, 1958; Chairman Microbiology, NY Med. Coll., 1968. Studied tissue proteins; autoantibodies; cryobiology; immunology of reproduction.

SHWARTZMAN, Gregory [1896–1965]. Born in Odessa. MD, Brussels, 1920. Prof. Microbiology, NY Med. Coll., 1923; Head, Dept Microbiology, Mt Sinai Hospital, NY, 1926; Prof., Columbia, 1940. Research on bacterial growth; immunologic reactions in tissue culture; local skin reactions to bacterial filtrates (Shwartzman reaction). Book: *Phenomenon of Local Tissue Reactivity and its Immunological and Clinical Significance*, New York, Hoeber, 1937. Obit.: *NY Times*, July 24, 1965.

SIMONSEN, Morten [1921–2002]. MD, Copenhagen. Copenhagen Univ., 1967; Head, Exper. Immunology Inst. Discovered graft-vs-host reaction on chorioallantoic membrane of chick embryo (Simonsen phenomenon); many contributions to transplantation biology, immunogenetics, and tolerance. Obit.: *Scand. J. Immunol.* 56:543, 2002.

SMITH, Richard Thomas [1924–]. Born in Oklahoma City. MD, Tulane, 1956. Univ. Texas, 1957; Univ. Florida, 1958. Many contributions to tolerance; immunopathology; delayed hypersensitivity; organized many symposia.

SMITH, Theobald [1859–1934]. Born in Albany, New York. MD, Albany Med. Coll. Bureau of Animal Industry, US Dept Agriculture, 1884; Prof. Comparative Pathology, 1896; Rockefeller Inst., 1915. Isolated agent of Texas cattle fever and other animal diseases; first used dead bacteria to immunize; differentiated human and bovine tubercle bacilli; first reported anaphylaxis in guinea pigs

(Theobald Smith phenomenon). Copley Medal, 1933. Obit.: *J. Pathol. Bacteriol.* **40**:621, 1935.

SNELL, George Davis [1903–1996]. Born in Bradford, Massachusetts. DSc, Harvard, 1930. Washington Univ., 1933; Jackson Lab., 1936. Definitive work in mouse genetics; invented concept of congenic mice; defined (with Gorer) the H2 locus; many other contributions to transplantation genetics. Gairdner Award, 1976; Nobel Prize, 1980. Book: *Histocompatibility* (with Dausset and Nathenson), New York, Academic Press, 1976. Obit.: *NY Times*, June 8, 1996.

STAVITZKY, Abram Benjamin [1919–]. Born in Newark, New Jersey. PhD, Univ. Minnesota, 1943; DVM, Univ. Penn., 1946. Prof. Microbiology, Univ. Penn., 1947; Case-Western Reserve, 1983. Studied mechanism of antibody formation; immunity in schistosomiasis.

ŠTERZL, Jaroslav [1925–]. Born in Pilsen. MD, Charles Univ., Prague. Head, Immunology Dept, Czech Academy of Science Inst. of Microbiology. Many contributions to ontogeny of immune response; cellular reactions in antibody formation; germ-free animal research. Books: *Molecular and Cellular Basis of Antibody Formation*, 1965; *Developmental Aspects of Antibody Formation* (with Riha), 1969.

SULZBERGER, Marion Baldur [1895–1983]. Born in New York. MD, Harvard; studied dermatology at Zürich and Breslau. Columbia Univ.; Head, Dermatology, Bellevue-NY Univ. Leading investigator of contact allergies; arsphenamine sensitization; desensitization. Obit.: *J. Allergy Clin. Immunol.* **74**:855, 1983.

SZILARD, Leo [1898–1964]. Born in Budapest. PhD, Berlin, 1922; studied physics in Berlin and England. Worked with Fermi in Chicago, 1937; Prof. Biophysics, Univ. Chicago, 1946. Discovered nuclear chain reaction; wrote letter with Einstein to Roosevelt leading to American atom bomb project; proposed molecular-genetic theory of antibody formation, 1960. Atoms for Peace Award, 1959. Biog.: *Dict. Sci. Biog.* **13**:226, 1976. Obit.: *Biog. Mem. Natl Acad. Sci.* **40**:337, 1969.

TALIAFERRO, William Hay [1895–1973]. Born in Portsmouth, Virginia. PhD, Johns Hopkins. Univ. Chicago, 1924. Long-time editor *J. Infect. Dis.* Studied effect of X-rays on immune response; role of spleen; early worker on immunology of parasitic diseases. Obit.: *J. Infect. Dis.* **130**:312, 1974.

TALMAGE, David Wilson [1919–]. Born in Kwangju, Korea. MD, Washington Univ., 1944; worked with Taliaferro in Chicago. Univ. Pittsburgh, 1951; Prof., Univ. Chicago, 1952; Chairman Microbiology, 1963; Dean, 1968; Director, Webb-Waring Inst., Denver, 1973. Studied antibody formation; transplantation; tolerance to cultured allografts; proposed early selection theory of antibody formation, 1957; helped develop Burnet's clonal selection theory, 1959. Book: *The Chemistry of Immunity in Health and Disease* (with Cann), 1961.

THEILER, Max [1899–1972]. Born in Pretoria. Studied at Capetown, Univ. London. Harvard, 1922; Rockefeller Inst., 1930; Yale, 1964. Many contributions to virology; developed vaccine for yellow fever. Lasker Award, 1949; Nobel Prize 1951. Biog.: *Nobel Lectures Physiol. Med. 1942–62*, Amsterdam, Elsevier, 1964, p. 360. Obit.: *NY Times*, Aug. 12, 1972.

THORBECKE, G. Jeanette [1929–2001]. Born in the Netherlands. MD, Groningen, 1950. Worked at Groningen, 1948; Leiden, 1956; NY Univ., 1957. Many contributions to antibody formation; role and reactions of lymphoid tissues; immunological tolerance. Obit.: *Immunity* 16:329, 2002.

TISELIUS, Arne Wilhelm Kaurin [1902–1971]. Born in Stockholm. Educated at Univ. Uppsala. Worked at Univ. Uppsala, 1935; Inst. Advanced Studies, Princeton, 1934; Swedish Natl Res. Council, 1946; President Nobel Foundation, 1960. Developed electrophoresis; identified (with Kabat) antibodies as gamma globulins; developed synthetic blood plasmas. Nobel Prize in Chemistry, 1948. Biogs: *Biog. Mem. Fellows R. Soc.* 20:401, 1974; *Dict. Sci. Biog.* 13:418, 1976; see also *Festschrift: Perspectives in the Biochemistry of Large Molecules* dedicated to Arne Tiselius on his sixtieth birthday, 1962. Obit.: *J. Chromatog.* 65:345, 1972.

TOPLEY, William Whiteman Carleton [1887–1944]. MD, Cambridge, 1918. Taught at Manchester; pathology at St. Thomas' Hospital; Charing Cross Hospital; Prof. Bacteriology and Immunology, London School Tropical Med. Hygiene. Studied immunology of infectious diseases; argued (against Rich) for importance of hypersensitivity for immunity in tuberculosis. Books: *An Outline of Immunity*, 1933; *Principles of Bacteriology and Immunity* (with Wilson), 1936. Obit.: *Br. Med. J.* 1:201, 1944.

TURK, John Leslie [1930–2006]. Born in Farnborough, Hampshire. MB, 1953; DSc, London, 1967. Worked at London School Hygiene and Tropical Med.; Medical Research Council; Prof. Pathology, Univ. London. Many contributions to delayed hypersensitivity. Books: *Delayed Hypersensitivity*, 1967; *Immunology in Clinical Medicine*, 1969. Obit.: *Clin. Exp. Immunol.* 145:571, 2006.

UHLENHUTH, Paul Theodore [1870–1957]. Born in Hanover. Director, Bacteriology Dept, Reichsgesundheitsamt, 1906; Prof. Strassburg, 1911; Marburg, 1921. Worked on differentiation of proteins; discovered organ specificity of lens antigens; syphilis and other infectious diseases. Book: *Handbuch für Mikrobiologischen Technik* (with Kraus), 1923. Obit.: *Münch. med. Wochenschr.* 107:1204, 1965.

UHR, Jonathan William [1927–]. Born in New York. MD, NY Univ.; worked with Pappenheimer. Prof. Medicine and Microbiology, NY Univ.; Chairman Microbiology, Univ. Texas, Dallas, 1969. Developed (with Salvin and Pappenheimer) method for inducing delayed hypersensitivity to simple proteins; role of antibody in immunoregulation; mechanism of Ig synthesis and secretion; immunotoxins.

VAN ROOD, Jon J. [1926–]. Born near Leiden, Holland. Educated at Univ. Leiden. Practiced internal medicine; worked at Dept Immunohematology and Blood Bank, Univ. Hospital. Many contributions to transplantation immunology and immunogenetics.

VAUGHAN, Victor Clarence [1851–1929]. Born in Mt Airy, Missouri. MD, Harvard, 1943. Director, Hygienic Lab., Prof. Physiological Chemistry, Univ. Michigan. Worked on immunity and allergy; role of cellular toxins. Books: *Ptomaines and Leukomaines* (with Novy), 1888; *Infection and Immunity*, 1915. Obit.: *J. Lab. Clin. Med.* **15**:307, 1929.

VOISIN, Guy André [1920–]. Born in Paris. MD, Univ. Paris, 1945; DSc, Sorbonne, 1958. Hôpital Saint-Antoine, 1955; Director of Research, Claude Bernard Assoc., 1964; Sci. Director, Immunopathology and Exper. Immunology, INSERM, 1970. First described autoimmune aspermatogenic orchitis, 1952; coined term "immunopathology," 1953; many contributions to autoimmunity, immunopathology, tolerance, immunology of reproduction.

WAKSMAN, Byron Halstead [1919–]. Born in New York. MD, Univ. Penn., 1943. Dept Neuropathology, Harvard, 1949; Bacteriology and Immunology, 1952; Chairman Microbiology, Yale Univ., 1963; Prof. Pathology, 1974; Vice President Res., Natl Multiple Sclerosis Soc., 1980. Important contributions to histopathology of delayed hypersensitivity and autoimmunity; role of thymus; antileukocyte serum; immunologic tolerance; neuroimmunology. Book: *Atlas of Experimental Immunobiology and Immunopathology*, New Haven, Yale Univ. Press, 1970.

WALDENSTROM, Jan Gosta [1906–1996]. Born in Stockholm. Educated at Uppsala, Cambridge, Munich. Prof. Theoretical Medicine, Uppsala, 1947; Malmö, 1950; Lund, 1952. Discovered Waldenstrom's macroglobulinemia; many other contributions. Gairdner Award, 1966; Ehrlich Medal, 1972. Obit.: *Blood* **89**:4245, 1997.

von WASSERMANN, August [1866–1925]. Born in Bamberg, Germany. Studied at Strassburg, Vienna, and Berlin. Worked under Koch at Inst. Infectious Diseases, Berlin; Director Serum Dept.; Director, Inst. Exper. Therapy, Dahlem, 1913. Developed (with Bruck and Neisser) serologic test for syphilis (Wassermann reaction). Book: *Handbuch der Pathogenen Mikroorganismen* (with Kolle), 1903. Ennobled, 1910. Biog.: *Dict. Sci. Biog.* **14**:183, 1976. Obit.: *Intl J. Dermatol.* **16**:526, 1977.

WEIGLE, William Oliver [1927–]. Born in Monaca, Pennsylvania. PhD, Univ. Pittsburgh, 1956; worked with Dixon. Univ. Pittsburgh Dept Pathology, 1955; member, Scripps Clinic Res. Foundation, La Jolla, 1961; Head, Div. Cellular Immunology, 1984. Many contributions to mechanisms of tolerance; autoimmune diseases; immunity and immunopathology. Books: *Natural and Acquired Immunological Unresponsiveness*, Cleveland, World Publishing, 1967; *Advances in Immunopathology*, New York, Elsevier, 1981.

WELLS, Harry Gideon [1875–1943]. Born in New Haven. MD, Rush, Chicago, 1898; PhD, Univ. Chicago, 1903. Taught at Rush and Chicago. Studied chemical and general pathology. Books: *Chemical Pathology*, 5th edn, 1925; *The Chemical Aspects of Immunity*, New York, Chemical Catalog Co., 1929. Obit.: *Arch. Pathol.* **36**:331, 1943.

WHITE, Robert George [1917–1978]. Born in Barwell, England. MD, Oxford, 1953. Univ. London; Harvard; Prof. and Chairman, Bacteriology and Immunology Dept, Glasgow. Studied cytology of antibody formation; role of germinal centers; described dendritic cells of germinal centers. Book: *Immunology for Students of Medicine* (with Humphrey), 1963.

WIDAL, Georges Fernand Isidore [1862–1929]. Born in Dellys, Algeria. MD, Paris. Prof., Paris. Developed Widal serodiagnostic test for typhoid. Obits: *Br. Med. J.* **1**:134, 1929; *Med. Hist.* **7**:56, 1963.

WIENER, Alexander [1916–1976]. Born in New York. MD, State Univ. NY, 1930. Head, Blood Transfer Div., Jewish Hospital, 1932; Director, Wiener Labs., 1935; NY Univ., 1961. Discovered Rh factor (with Landsteiner), 1938; described blood groups of chimpanzees; disputed genetics and nomenclature of Rh system with Mollison. Lasker Award, 1946. Biog.: *Hematologia* **6**:7, 1972 (Festschrift on sixty-fifth birthday). Obit.: *Arch. Allergy Appl. Immunol.* **54**:191, 1977.

WITEBSKY, Ernest [1901–1969]. Born in Frankfurt. Educated at Frankfurt and Heidelberg; studied under Hans Sachs. Inst. f. Krebsforschung, Frankfurt, 1929; Mount Sinai Hospital, New York, 1934; Univ. Buffalo, 1936. Worked in blood group serology; transfusion problems; classical demonstration of experimental autoimmune thyroiditis (with Rose). Biog.: *1st International Convocation on Immunology*, Basel, Karger, 1968. Obit.: *Blood* **35**:869, 1970.

WOLFF-EISNER, Alfred [1877–1948]. Born in Berlin. MD, Tübingen, 1901. Worked in Königsberg; Univ. Polyklinik, Berlin. Long-time student of human allergies; first to interpret hayfever as an immunologic process; developed ophthalmoreaction for TB. Books: *Das Heufieber*, 1906; *Handbuch der Serumtherapie und experimentelle Therapie*, 1910. Biog.: *Ciba Symp.* **11**:1398, 1951.

WOODRUFF, Michael Francis Addison [1911–2001]. Born in London. MD, Melbourne, 1941; DSc, 1962. Prof. Surgery, Univ. Otago, Dunedin; Univ. Edinburgh, 1957. Many contributions to transplantation. Lister Medal; Royal Soc. Medal. Book: *The Transplantation of Tissues and Organs*, 1960. Obit.: *The Independent*, March 31, 2001.

WRIGHT, Almroth Edward [1861–1947]. Born in Middleton in Teesdale, Yorkshire. MD, Trinity Coll., Dublin, 1889. Prof. Pathology, Army Med. School, Netley, 1892; Inst. Pathology, St Mary's, London, 1902. Theory of opsonins (with Douglas); developed system of antityphoid inoculations. Books: *Pathology and Treatment of War Wounds*, 1942; *Researches in Clinical Physiology*, 1943; *Studies in Immunology* (two vols), 1944. Biogs: *Br. Med. J.* **2**:516,

1961; *Dict. Sci. Biog.* **15**:511, 1976. Obits: *Br. Med. J.* **1**:699, 1947; *Nature* **159**:731, 1947.

YALOW, Rosalyn Sussman [1921–]. Born in New York. PhD in physics, Univ. Illinois. Hunter Coll., NY, 1946; Chief, Radioisotope Service, Veterans Hospital New York, 1950; Chief, Nuclear Med., 1970; Chairman, Dept Clinical Sci., Montefiore Hospital, 1980. Discovered (with Berson) antibody cause of insulin-resistant diabetes; developed radioimmunoassay for peptide hormones. Nobel Prize, 1977.

YERSIN, Alexandre Emile John [1863–1943]. Born in Rougement, Switzerland. Educated at Lausanne, Marburg, and Paris. Pasteur Inst., Paris; Surgeon, French Colonial Army; Director, Pasteur Inst. Nhatrang, Annam. Discovered diphtheria toxin (with Roux), 1888; discovered plague bacillus, Hong Kong, 1894 (independently of Kitasato). Biog.: *Dict. Sci. Biog.* **15**:551, 1976; see also *Hist. Sci. Med.* **7**:353, 357, 1973. Obit.: *Bacteriol. Rev.* **40**:633, 1976.

ZINSSER, Hans [1878–1940]. Born in New York. MD, Columbia Univ., 1903. Prof. Bacteriology, Columbia Univ.; Stanford Univ.; Harvard Univ., 1923. Leading student of plague immunology, hypersensitivity; advanced unitarian theory of antibodies; distinguished tuberculin from anaphylactic hypersensitivity. Books: *Microbiology* (with Hiss), 1911; *Infection and Immunity*, 1914; *Immunity*, 1939. Autobiog.: *As I Remember Him; The Biography of R.S.*, 1940. Biog.: *Dict. Sci. Biog.* **15**:622, 1976. Obit.: *Science* **92**:276, 1940.

(This bibliographic dictionary was assembled with the generous assistance of Mrs Dorothy Whitcomb and Terrence Fischer of the Health Sciences Library, University of Wisconsin, Madison.)

Author Index

A

Abel, R., 10, 32, 41
Abruzzo, J. L., 212
Abt, G., 416
Ackerknecht, E. H., 22, 40
Ackroyd, J. F., 164, 475
Ada, G. L., 167, 175, 265, 284, 490
Adams, D. D., 176, 495
Adbou, N. I., 228
Adkinson, J., 251, 256
Adler, F. L., 150
Adrian, E. D., 282, 289
Alepa, F. P., 175, 415
Alexander, J., 44, 51, 66, 102, 115, 127, 174, 250, 259, 500
Alexander, N. J., 174
Algire, G. H., 34, 41, 243, 255, 334
Allen, G., 207, 249, 256, 324
Althoff, F., 291, 301
Amos, B., 333, 373, 380, 386, 390, 393, 463, 478
Amyand, C., 297, 298
Amzel, L. M., 137, 149
Anderson, D., 255
Anderson, D. J., 174
Anderson, J. F., 176, 179, 183, 203, 204, 205
Anderson, N. A., 256, 478
Andrejew, P., 159, 174, 415
Anfinsen, C. B., 128
Angelico, F., 233
Apt, L., 3, 66, 208, 305, 317
Archer, O., 207
Aristotle, 419, 442, 456, 488
Arnon, R., 148
Arrhenius, S., 38, 41, 101, 110, 125, 355, 363, 434, 438, 459, 461, 475, 487
Arthus, M., 37, 41, 111, 126, 144, 156, 173, 177, 179, 180, 182, 190, 195, 197, 203, 204, 271, 286, 326, 327, 329, 352, 362, 396, 399, 409, 458, 461
Arthus, N. M., 475
Ascher, M. S., 151
Aschoff, L., 124, 160, 469

Askonas, B. A., 140, 475, 488
Atwell, J. L., 150, 335
Auer, J., 180, 204
Austen, K. F., 205, 286, 373, 419
Avey, H. P., 137, 149
Avicenna (Abu Ali al-Husein ibn Sina), 9
Avrameas, S., 95, 277, 286, 287, 288

B

Badash, L., 64
Bail, O., 50, 51, 52, 65, 66, 207, 382, 475
Bain, B., 207
Baldwin, E. R., 190, 191, 205
Bale, W. F., 266, 284
Baltimore, D., 23, 40, 41, 80, 84, 123, 126, 173, 174, 203, 206, 208, 255, 283, 285, 368, 386, 391, 392, 393, 416, 417, 418, 438, 461, 462, 476, 493, 496
Bar, R. S., 176, 238, 254, 389, 493
Barker, C. F., 245, 255, 400, 417
Barnard, J. H., 205
Baron, J., 22, 484, 485
Bashford, E. F., 254, 264, 283
Bates, D.G., 123
Bauer, K. H., 254
Baumgarten, P., 20, 23, 30, 41
Behring, E., 3, 4, 18, 20, 22, 23, 35, 36, 41, 64, 104, 119, 124, 177, 179, 203, 216, 228, 263, 283, 291, 292, 308, 321, 333, 338, 345, 350, 362, 421, 425, 427, 455, 458, 460, 465, 475, 479, 484, 485
Belfanti, S., 422, 428, 458
Bellers, J., 300, 303
Benacerraf, B., 141, 150, 151, 176, 194, 206, 207, 227, 228, 253, 257, 327, 329, 333, 334, 335, 373, 469, 470, 476, 482, 485
Bendtzen, K., 151, 208
Bennet, J. C., 73, 76, 82, 460
Bennett, B., 94, 150, 198, 208, 328, 334, 412, 420, 472

Bennich, H., 188, 205
Berger, E., 47, 65
Berggard, I., 84
Bergman, K. R., 100
Bergsma, D., 150, 208, 334, 391, 418, 464
Bernard, C., 25, 37, 172, 284, 353, 373,
　　　380, 478, 488, 491, 495, 499
Bernard, J., 94, 284
Berrens, L., 205
Berry, G. P., 174, 363
Berson, S. A., 275, 286, 460, 469, 501
Berthollet, C. L., 100
Besredka, A., 40, 155, 181, 188, 199, 205,
　　　208, 212, 217, 220, 221, 229, 310,
　　　321, 323, 461, 462, 476
Beutner, E. H., 176
Bianco, C., 151
Biedl, A., 180, 204
Bigazzi, P., 176, 228
Bigger, S. L., 233, 253
Billingham, P. B., 364
Billingham, R. E., 66, 166, 175, 207, 242,
　　　243, 245, 248, 253, 255, 256, 334,
　　　373, 387, 400, 416, 417, 418, 457,
　　　459, 468, 476, 477, 488
Bilofsky, H., 83, 149
Binz, H., 228
Bitter, S., 30, 102, 113
Blake, J. B., 228, 302
Blanden, R. V., 42
Bloom, B. R., 150, 198, 208, 328, 334,
　　　412, 420, 460
Boerhaave, H., 15, 20, 23
Boissier de Sauvages, F., 99, 123
Böke, W., 395
Boorstin, D. J., 227
Bordet, J., 19, 32, 35, 41, 49, 54, 65, 106,
　　　107, 108, 109, 110, 111, 112, 114,
　　　116, 119, 124, 125, 126, 144, 151,
　　　154, 157, 158, 173, 180, 201, 209,
　　　212, 216, 218, 219, 220, 225, 226,
　　　228, 229, 230, 270, 285, 305, 307,
　　　308, 309, 310, 311, 312, 313, 316,
　　　318, 319, 320, 321, 322, 324, 338,
　　　345, 350, 362, 396, 410, 413, 414,
　　　420, 422, 423, 424, 425, 426, 427,
　　　428, 429, 433, 438, 458, 461, 466,
　　　476, 482

Bordet, P., 419
Borek, F., 335
Boretius, M. E., 302
Boswell, T., 292, 302, 416
Bouchard, C. J., 154, 172
Bovet, D., 467, 476
Boyd, W. C., 47, 65, 127, 131, 132, 148,
　　　162, 340, 345, 357, 363, 462, 463,
　　　476
Boyden, S. V., 340, 345
Boyle, R., 11, 13, 363
Boylston, Z., Dr., 294
Brambell, F. W. R., 418, 462, 477
Bransom, H. R., 451
Braun, H., 150, 158, 174, 391, 392,
　　　415
Braun, W., 207, 228, 334, 362, 373,
　　　379
Breinl, F., 44, 51, 52, 53, 54, 66, 70,
　　　81, 127, 345, 363, 438, 459,
　　　483
Brenner, S., 72, 82, 446
Brent, L. B., 66, 166, 175, 197, 207,
　　　242, 243, 245, 248, 253, 255, 256,
　　　334, 364, 373, 387, 414, 416,
　　　418, 428, 457, 459, 468, 476, 477,
　　　488
Bretscher, P. M., 95
Bridges, R. A., 418
Brieger, G. H., 37, 41, 303
Broussais, F., 154
Brown, J. B., 154, 197, 207, 208, 334, 365,
　　　388, 389, 390, 391, 417, 438, 462,
　　　463, 468, 492
Bruck, C., 157, 158, 172, 173, 230, 285,
　　　362, 426, 429, 458, 461, 477, 490,
　　　499
Bruton, O. C., 66, 174, 200, 208, 328,
　　　334, 364, 404, 405, 418, 457, 459,
　　　477
Buchner, H., 30, 33, 41, 44, 45, 46, 47,
　　　65, 111, 119, 126, 127, 201,
　　　309, 310, 321, 410, 419, 458,
　　　477
Buckley, R., 406
Bullington, S. J., 399, 416
Bulloch, W., 40, 462
Bulman, N., 134, 148

Burnet, F. M., 54, 55, 56, 60, 61, 62, 63, 64, 66, 67, 70, 72, 81, 85, 86, 87, 88, 89, 90, 91, 92, 93, 94, 121, 122, 127, 142, 163, 165, 166, 173, 174, 175, 207, 211, 213, 227, 232, 242, 243, 248, 255, 256, 265, 284, 330, 337, 339, 341, 343, 344, 346, 356, 358, 359, 363, 364, 378, 387, 404, 406, 413, 418, 432, 433, 435, 436, 437, 439, 441, 442, 447, 457, 460, 462, 463, 468, 471, 472, 477, 481, 487, 490, 491, 497

Bush, V., 448, 452

Bussard, A., 66, 279, 288, 364, 431, 437, 477

Byrt, P., 167, 175, 284

C

Calmette, A., 189, 191, 205, 262, 461, 477, 478

Calmette, L. C. A., 189, 191, 205, 262

Calne, R., 245, 255

Cambrosio, A., 262, 283, 287, 384, 422, 428, 429

Campbell, D. H., 65, 84, 134, 148, 149, 478

Camus, L., 219, 229

Cantacuzène, J., 478

Capra, J. D., 69, 71, 73, 81, 82, 94

Carbone, T., 422, 428, 458

Carmichael, A. E., 22, 301

Carrel, A., 233, 234, 239, 253

Castiglioni, A., 22, 40, 123, 450

Caulfield, J. B., 195, 206

Chamberland, C., 22, 23, 40, 41, 419

Chandler, J. W., 373, 416, 417

Chang, T. S., 207, 334, 459

Chaperon, E. A., 150, 175, 207, 334, 364, 460

Charlesworth, M. J., 385

Chase, M. W., 39, 115, 141, 186, 187, 195, 243, 314, 327, 358, 381, 405, 472

Chauveau, A., 19, 23

Chen, B. L., 137, 149

Chernyak, L., 42, 362, 384, 428, 437

Chess, L., 83, 113, 287

Chew, W. B., 247, 255

Chinitz, A., 252, 257

Christian, C. L., 227

Christmas-Dirckinck-Holmfeld J., 30

Chu, A. C., 287

Churchill, F. B., 437

Cinader, B., 175, 379, 385, 391

Citron, J., 157, 173, 426, 429

Claman, H. N., 150, 175, 207, 334, 364, 460

Clapesattle, H., 450

Clark, E. A., 256

Coca, A. F., 187, 205, 209, 251, 256, 240, 278

Coggi, G., 288

Cohen, I. R., 93, 95, 439

Cohen, P. L., 228

Cohen, S. G., 205

Cohn, F. J., 478

Cohn, M., 66, 67, 82-84, 95, 152, 228, 439

Cohn, S., 151

Cohnheim, J., 30, 40, 478

Colle, D. S., 3, 21

Collins, R. C., 164, 175, 415

Compte, I. A. M. F. X., 172

Condie, R. M., 418

Cone, R. E., 142, 150, 335

Connally, J. M., 175

Conry, Y., 437-438

Cooke, R. A., 186–187, 205, 250-251, 256, 478

Coombs, R. R. A., 164-165, 173, 198, 208, 345, 429, 463, 478, 482, 490

Coons, A. H., 128, 175, 287, 346

Coons, A. S., 364

Cooper, M. D., 208, 283, 334, 371, 439

Cooperband, S. R., 267, 284

Corey, R., 451

Coscas, G., 208, 416

Cosenza, H., 227-228

Cosmas, St., 233

Cottier, H., 391, 418, 463

Coyl, E. B., 385

Craig, G. M., 82

Craig, L. C., 418

Creech, H. J., 287

Creighton, C., 301

Crick, F. H. C., 60, 67, 128, 444-447, 452

Crookshank, E. M., 303

Cruse, J. M., 172

Cuello, A. C., 287-288

Cunningham A. J., 67, 81-82, 94, 28, 429
Cunningham, B. A., 84
Currie, J. R., 181, 204
Cyprian, St., 6, 22

D
Da Fano, C., 236, 254
Dale, H. H., 204, 354, 363, 409, 419, 459
Dameshek, W., 164, 315, 409, 419, 459
Damian, St., 233
Dancis, J., 418
Daniel, P., 362
Danysz, J., 106, 108, 125, 458, 479
Darwin, E., 29, 57, 91, 251, 362, 431, 434-435, 437-438
Dausset, J., 253, 255, 373, 391, 393, 417, 462, 469, 470, 479, 497
David, J. R., 208, 334, 420
Dawkins, R. L., 288
de Gaetano, L., 264, 283
De Kruif, P., 32, 41, 476
de Préval, G., Dr., 13
De Rosa, S. C., 287
Delaney, R., 84, 149
Delbrück, M., 445-448, 451-452
Dell'Orto, P., 288
Delthil, S., 416
Denis, J., 139, 149
Detré, L., 157, 173, 225, 230, 426, 429
Dhermy, F., 416
Dhermy, P., 208
Dienes, L., 39, 42, 141, 150, 163, 192, 194, 203, 206, 327, 333-334, 357, 360, 363, 459, 479
DiGeorge, A. M., 201, 208, 334, 364, 405, 418
Dixon, F. J., 175, 185, 204, 287, 373, 479, 499
Doerr, R., 174, 181, 204, 415, 479
Doganoff, A., 205
Doherty, P. C., 92, 94, 143, 151, 175-176, 335, 460, 472, 473
Donath, J., 93, 126, 173, 229, 320, 322, 429, 480
Doniach, D., 480, 494
Dorf, M. E., 151
Douglas, S. R., 37, 41, 139, 150, 363, 458, 480, 500
Doyle, A. C., 402, 418

Drake, J., 11–12, 22
Dreyer, W. J., 73, 76, 82, 94, 472
Du Pasquier, L., 82, 146, 152
Dubiski, S., 212, 227, 489
Dubos, R., 381, 492
Duclaux, E., 480
Dühren, E., 13, 23, 178
Duke-Elder, S., 416
Dumonde, D. C., 393, 412, 420
Dungern, E. F., 107, 256, 288, 429, 460, 480
Dungern, H. E., 125, 249
Dutton, R. W., 364

E
Eason, J., 316-317, 323
Ecker, E. E., 202, 209
Edelman, G. M., 468-, 72, 80, 82, 84, 97, 124, 135-138, 148-149, 387, 439, 444, 447-448, 451, 460, 469, 480
Edelstein, L., 6, 22
Edwards, J. C., 124, 288
Ehrich, W. E., 39, 42, 480
Ehrlich, P., -1174863, 3, 4, 18, 19, 22, 23, 32, 33, 36, 37, 38, 41, 47–49, 52, 57-61, 64-66, 81, 83, 86, 88, 98, 100-108, 110, 120, 122, 124-126, 147, 153-155, 157, 160, 162, 169, 171-174, 177, 179-180, 182, 201, 212, 217-230, 249, 256, 261, 26-265, 276, 282-283, 291, 306-324, 333, 338, 341, 343-344, 349-350, 352, 355, 356, 359, 362, 398, 402-403, 410-411, 413-414, 418, 422-429, 438, 443-444, 447-448, 450-451, 458, 461, 466, 476, 479, 481, 486, 487, 49, 495, 499
Eichmann, K., 83
Eisen, H. N., 132, 141, 148, 150, 152, 285, 330, 335, 387, 481, 485
Eisenberg, R. A., 228
Elberg, S. S., 150, 333
Elias, J. M., 287
Elschnig, A., 161, 174, 397, 415, 481
Embleton, D., 50, 65
Emmerich, E., 317-318, 323
Enders, J. F., 207
Enrlich, P., 419

Epstein, C. J., 128, 207
Erlenmeyer, H., 47, 65

F
Faber, K., 98, 123
Fabre, J., 288
Fagon, G., 300
Fagraeus, A., 167, 175, 346, 459, 462, 481
Fagraeus-Wallbom, A. E., 481
Fahey, J. L., 148, 283, 286
Fahlberg, W. J., 92, 94
Farley, J., 40, 123
Faulk, W. P., 288
Faunce, K., Jr., 150, 333
Fearon, D.T., 83, 209, 420
Feldman, J. F., 207, 373
Feldmann, M., 150, 286, 331, 335
Félix, C.-F., 300, 331
Fellows, R. E., 84, 149, 477, 482, 498
Felton, L. D., 175
Fenner, F. J., 56, 66, 94, 121, 127, 165,
    174, 242, 248, 255-256, 346, 358,
    363, 418, 439, 462, 468, 477, 481
Ferrata, A., 201, 209
Feyerabend, P., 319, 323, 361
Finger, E., 126
Fischer, E., 47, 100, 115, 123, 129
Fisher, J., 102, 227
Fishman, M., 150
Fitch, W. M., 288, 373
Flax, M. H., 195, 206
Fleck, L., 173, 348, 361, 368, 384
Fleming, D., 438
Flexner, S., 117, 178, 203
Flier, J. S., 176
Follis, R. H., Jr., 334
Font, R. L., 175, 415
Forssman, M. J. K. A., 475
Fosdick, R. B., 301
Fracastoro, G., 4, 7, 9–10, 21, 98, 123,
    408, 419, 481
Franek, F., 148
Franken, A., 233
Franklin, B., 291, 298, 302, 416
Freund, J., 141, 164-165, 174, 193, 206,
    327, 333, 397, 415, 459, 482
Fried, M., 82, 148, 303
Friedberger, E., 180, 186, 204, 224-225,
    229-230, 419, 482

Friedemann, U., 186, 204, 409, 419
Frischknecht, H., 228
Fruton, J., 123, 125, 127, 147
Fuchs, E. J., 93
Fukuyama, F., 452
Fuller, T., 14, 23, 65
Fye, W. B., 386

G
Gaffky, G. T. A., 30, 31, 482
Galison, P., 428
Gally, J. A., 72, 80, 82, 84
Garfield, E., 387
Garrison, F. H., 22
Gatti, A., 16, 23
Gautier, L., 303
Gay, F. P., 114, 125, 126, 180, 181, 204,
    230, 461, 482
Gebb, H., 159, 160, 174, 398, 415
Geckeler, W., 84, 95, 439
Geison, G., 437
Gell, P. G. H., 41, 83, 149, 150, 206, 208,
    227, 333, 334, 335, 419
Gengou, O., 157, 173, 225, 230, 285, 362,
    425, 429, 458, 466, 476, 482
George, M., 208, 420
Germain, R. N., 253, 257
Germuth, F. G., 185, 204, 399, 416
Gershon, R. K., 69, 150, 168, 175, 460
Gery, I., 415, 417
Giblett, E. R., 418
Gibson, T., 240, 241, 242, 255
Gitlin, D., 66, 208, 334, 405, 418
Glanzmann, E., 201, 209, 334, 364, 404, 418
Glass, B., 449, 452
Glazer, A. N., 152
Gley, E., 219, 229
Glick, B., 207, 334, 459, 182
Goldberger, R. F., 128
Goldmann, H., 399, 416
Golowin, S., 160, 161, 174, 362, 415, 424,
    428
Goltz, D., 155, 172, 306, 307, 311, 315,
    317, 320, 321, 323
Good, R. A., 207, 208, 334, 418, 419
Gordon, J., 209, 288
Gorer, P. A., 176, 238, 242, 254, 333, 345,
    459, 470, 483, 497
Gottlieb, A. A., 151

Götze, O., 209, 420
Goudie, R., 176
Gould, S. J., 347, 361
Gourdan, Madame, 13
Gowans, J., 473, 483
Grabar, P., 66, 148, 208, 275, 286, 373,
        378, 389-391, 459, 462, 463, 483,
        484, 489
Grawitz, P., 23, 432, 437
Gray, K., 176
Green, M. I., 149
Greene, H. S. N., 416
Greineder, D., 151, 208
Grey, H. M., 89, 149, 288
Grimm, E., 42
Grogan, J. B., 416
Grossberg, A., 127, 148, 335, 363
Gruber, M., 41, 81, 124, 126, 322, 438
Guillemin, R., 469
Guldberg, O., 100

H
Haaland, M., 264, 283
Haeckel, E., 248
Haendel, Z., 159, 174
Haffkine, W., 483
Halsey, J. F., 228
Hamaoka, T., 151
Hamburger, J., 255, 391
Hanan, R., 175, 256
Hanson, L. A., 205, 208, 272
Hanzlik, P. J., 205
Harris, T. N., 42
Harrison, L. C., 176
Hart, D. A., 227, 228
Harvey, B. A. M., 209
Hasek, M., 256
Haurowitz, F., 66, 81, 127, 345, 363, 438,
        483
Hauschka, T. S., 265, 283
Havers, C., Dr., 293
Hawking, S., 452
Haxthausen, H., 195, 206
Heather, C. J., 206
Hebald, S., 205
Heidelberger, M., 38, 47, 65, 49, 120, 127,
        129, 162, 186, 273, 274, 276, 286,
        287, 355, 363, 379, 389, 434, 459,
        484, 485, 488

Heise, E. R., 416
Hektoen, L., 163, 174, 237, 254, 478
Helmsmen, R. J., 417
Helyer, B. J., 176
Henkind, P., 417
Herberman, R. B., 288
Heremans, J. F., 205, 208
Heremans, M. T., 205, 460
Héricourt, J., 484
Hertzfeld, E., 45, 46, 63
Herzenberg, L. A., 276, 287, 371, 373
Heyde, M., 254
Hildemann, W. H., 256, 373
Hill, R. L., 84, 149
Hilschmann, N., 82
Hinde, I. T., 41, 334
Hinde, J. T., 206
Hinshelwood, C., 447, 448, 452
Hirsch, E. F., 46, 65
Hirszfeld, L., 256, 288, 429
Hitzig, W. H., 405, 418
Hobbs, J. R., 200, 208
Hoffmann, E., 173
Hoffmann, M., 151
Holman, E., 253, 254
Holstein, J., 254
Hood, L., 67, 74, 81, 82, 84, 94, 149, 288,
        364, 439, 472
Hooker, S. B., 47, 65, 131, 148
Hopper, J. E., 227, 490
Horgan, J., 449, 450, 451, 453
Howard-Jones, N., 385
Howie, J. B., 176
Hraba, T., 175, 256
Hübener, W., 403, 418
Humphrey, J. H., 127, 167, 175, 265, 284,
        363, 373, 463, 484, 500
Hunkapiller, T., 364, 439
Hunter, J., 5, 34, 41, 484
Hyde, R. R., 202, 209

I
Ingraham, J. S., 279, 288
Inman, J. K., 145, 147, 151
Ishizaka, K., 205
Ishizaka, T., 205
Issaeff, R., 416
Itami, S., 254
Itano, H., 148, 445, 492

J

Jaap, R. G., 207, 334, 459
Jabs, D. A., 415
Jacobs, J. L., Dr., 314, 322
Jagic, N., 115, 127, 322
James, P., 450
Janeway, C. A., 66, 87, 93, 95, 208, 228,
    339, 381, 405, 406, 463
Jenner, E., 262, 299-303, 349, 455, 458,
    484
Jerne, N. K., 450, 457, 459, 460, 462, 470,
    471, 484, 485
Jobe, A., 152
Johanssen, S. G. O., 188, 205
Jones, B. R., 415, 416
Jones, R. N., 287
Jones, T. D., 42
Jordan, P., 66, 451
Jordon, R. E., 176
Joubert, J., 22, 40
Judson, H. F., 97, 122, 128, 451, 452
Julia, J. F., 324, 418
Jurin, J., 301

K

Kabat, E. A., 41, 73, 82, 83, 127, 134, 138,
    148, 149, 151, 164, 175, 335, 357,
    363, 373, 438, 459, 460, 462, 464,
    468, 485, 488, 498
Kahn, C. R., 176
Kahn, R. L., 285
Kallos, P., 485
Kalsow, C. M., 415
Kane, L.W., 207
Kant, I., 101, 124, 311
Kaplan H. S., 417
Kaplan, M. H., 287
Kappler, J. W., 142, 150, 151, 330, 364,
    420
Karsner, H. T., 205, 254
Karush, F., 53, 66, 133, 141, 148, 150,
    152, 330, 335, 365, 481, 485
Katchalski, E., 256
Katsh, S., 174
Katz, D. H., 84, 151, 228
Kay, A. B., 205, 286, 419
Kay, L. E., 365, 446, 452
Keating, P., 262, 283, 287, 345, 384, 421,
    428, 429

Kehoe, J. M., 73, 82
Keighly, G., 286
Keller, H. U., 175, 284
Kelus, A., 83, 149, 213, 227, 460
Kelvin, Lord, 44, 443, 447, 450
Kendall, F. E., 47, 65, 273, 276, 286, 287,
    288, 459
Key, J. D., 418
Khorazo, D., 208
Kindt, T. J., 71, 81, 83, 94, 227, 283
Kirkpatrick, C. H., 151
Kirkpatrick, J., 15, 22
Kitasato, S., 4, 22, 41, 64, 119, 124, 216,
    228, 283, 321, 333, 345, 362, 455,
    458, 465, 485
Klinger, R., 45, 46, 65
Klinman, N. R., 81, 82, 91, 94, 146, 151
Knight, A., 176
Knorr, A., 45, 47, 65, 127
Koch, H. H. R., 485
Koch, R., 3, 17, 25, 26, 31, 35, 102, 126,
    159, 174, 189, 203, 205, 271, 286,
    312, 455, 458, 459, 465, 466, 496
Köhler, G. F., 125, 228, 452
Köhler, H., 228
Kohn, D., 437
Kolle, W., 127, 203, 230, 363, 481
Kondo, K., 175, 460
Koprowski, H., 333, 390
Kossel, H., 219, 229
Kovalevsky, S., 32
Kowarski, A., 288
Kraus, R., 41, 124, 159, 174, 204, 286,
    415, 458, 461, 486
Kreth, H. W., 81, 151
Kronenberg, M., 151, 364
Krusius, F. F., 159, 174, 398, 415, 459
Kuettner, M. G., 83
Kuhn, T. S., 39, 42, 158, 319, 323, 326,
    333, 347, 358, 361, 364
Kunkel, H. G., 75, 82, 83, 149, 205, 208,
    227, 460, 471, 486
Küstner, H., 187, 205, 271, 286, 333, 459,
    461

L

La Condamine, C. M., 15, 23, 301
Lachmann, P. J., 124, 151, 209
Lagrange, H., 416

Laidlaw, P. P., 204, 363, 409, 419
Lakatos, I., 319, 323, 361
Lamm, M. E., 287
Lampkin, G. H., 255
Lampl, H., 116, 127, 363
Landsteiner, K., 19, 38, 39, 49, 52, 53, 54,
    62, 70, 71, 87, 97, 102, 114–119,
    121, 129, 130, 133, 134, 141, 145,
    156, 162, 186, 187, 195, 196, 216,
    221, 243, 249, 250, 262, 281, 306,
    307, 311–318, 319, 327, 337, 338,
    355, 358, 391, 399, 405, 418, 419,
    420, 428, 460–461, 480
Landy, M., 207, 83, 150, 151, 228, 334,
    367, 373, 374, 377, 378, 384, 386,
    392, 393
Langley, J. N., 169, 176
Langman, R. E., 83, 84, 95, 439
Lattes, L., 256, 288, 289, 429
Laudan, L., 323, 361
Lawrence, H. S., 83, 151, 207, 208, 255,
    284, 333, 334, 373, 384, 389, 392,
    459, 463, 472, 486
Le Fanu, W. R., 5, 22
Leake, C. D., 123
Lebowitz, H. E., 84, 149
Leclef, J., 139, 149
Leder, P., 83, 84, 472
Lederberg, J., 81, 89, 90, 94, 127, 175,
    341, 346, 359, 364, 436, 439, 486,
    490
Leduc, E. H., 128, 175, 287, 346, 364, 459
Leeds, A. A., 385
Leibnitz, G. W., 10, 101, 124, 296, 311
Leiner, K., 315, 323
Lemoine, A. N., 163, 174, 415
Lengerová, A., 175, 256, 390, 392, 463
Lenoir, T., 385, 395, 414
Lepow, I. H., 486
Lesky, E., 126
Lespinats, G., 277, 287, 288
Levaditi, C., 126, 173
Levaditi, J.C., 208, 416
Levine, B. B., 150, 206, 257, 335
Levine, P., 126, 256, 320, 418
Lewis, G. N., 123
Lewis, M. R., 42, 150, 208, 334, 420
Lewis, P. A., 204
Lewis, R. E. Jr., 172

Lewis, T., 204
Libby, R. L., 278, 288
Lichtenstein, L. M., 151, 205, 209, 286,
    419, 420
Lieberman, R., 83
Liebig, J., 100
Lindenmann, J., 215, 227, 228, 487
Lindley, D., 449, 450, 453
Lindsley, H. B., 228
Lipton, M. M., 164, 415
Lischke, R. L., 301
Lister, Lord, 25, 30
Lister, M., Dr., 293
Little, C. C., 254
Loc, T. B., 284
Locke, A. L., 65
Loeb, L., 236, 240, 254
LoGrippo, G. A., 206, 208, 381, 389, 462
Loken, M. R., 287
Loveless, M. H., 187, 205
Lowenstein, L., 207
Luntz, M. H., 417
Lurie, M. B., 140, 150
Lyman, S., 176

M
Mach, E., 101
Mackaness, G. B., 39, 42, 150, 333, 487
Mackay, I. R., 175, 286, 364, 415, 463
Madison, C. R., 288
Madsen, T., 125, 355, 487
Mage, R., 84
Magendie, 178, 203
Main, E. R., 65
Maitland, C., 294, 295, 296, 297, 298,
    302
Mak, T., 346
Mäkelä, O., 94, 379, 387, 393
Mallory, T. B., 194, 206
Manley, S. W., 176
Mannik, M., 149, 227, 460
Manoussakis, M. N., 415
Mantoux, C., 189, 205
Manwaring, W. H., 46, 47, 65, 487
Marak, G. E., 415
Marchalonis, J. J., 142, 150, 331, 335
Margoliash, E., 288
Marie, A., 173
Marquardt, M., 65, 124, 450, 481, 487

Marrack, G. E., 175
Marrack, J. R., 66, 127, 147, 148, 287,
    363
Marrack, P., 150, 335, 364, 420
Mason, D. Y., 288
Massey, E., 299, 303
Masugi, M., 488
Mathé, G., 255, 267, 284, 391, 488
Mather, C., 13, 14, 23, 294, 302, 488
Matoltsy, M., 150, 208, 334, 420
Matzinger, P., 87, 93, 95
Maumenee, A. E., 208, 399, 400, 416
Maurer, P. E., 151
Mayer, M. M., 41, 127, 209, 286, 363,
    420, 438, 488
Mayr, E., 64, 78, 83, 102, 124, 319, 324,
    365, 431, 437
Mazumdar, P. M. H., 23, 41, 102, 123,
    124, 125, 126, 148, 311, 321, 322,
    428
McCluskey, J. W., 151, 207
McCluskey, R. T., 151, 176, 207, 227
McDermott, K., 174, 206, 333, 415, 459
McDevitt, H. O., 151, 176, 252, 257, 387,
    393, 470
McKenzie, I. F. C., 175, 208
McMichael, A. J., 288
McNeill, W. H., 21, 22
Mead, R., 302
Medawar, P. B., 453, 66, 174, 175, 197,
    207, 232, 240-244, 247, 253, 255,
    256, 327, 333, 334, 358, 360, 364,
    373, 387, 400, 416, 418, 460, 483,
    488
Mellors, R. C., 287
Meltzer, S. J., 180, 204, 362, 414, 488
Mendel, G., 44
Mercer, D. W., 288
Mercurialis, H., 22, 123
Mertens, E., 42
Merton, R. K., 318, 323, 347, 361
Metalnikoff, S., 155, 173, 306, 311, 320,
    321, 362, 433, 438, 489
Metchnikoff, E., 3, 23, 35, 41, 27, 28, 32,
    41, 65, 81, 104, 111, 124, 138, 149,
    172, 173, 229, 321, 333, 431, 437,
    465, 489
Metchnikoff, I., 326, 333, 356, 361, 419,
    432, 437, 458, 460, 489

Metchnikoff, O., 40, 419
Metzger, H., 152
Mew, D., 267, 285
Meyer, E., 317, 318, 323
Michaelides, M. C., 152
Michaelis, L., 174, 323
Michel, M., 149, 213, 227, 417, 460
Michelson, A. A., 447, 450
Miciotto, R. J., 40
Miescher, P. A., 373, 378, 389, 390, 391,
    392, 419, 462, 489
Milgrom, F., 212, 227, 373, 379, 392, 489
Miller, G., 13, 22, 23, 301, 302
Miller, J. F. A. P., 150, 175, 207, 334, 418,
    460, 489
Miller, P., 256
Milner, R. D. G., 208
Milstein, C., 72, 82, 123, 460, 470
Mishell, R. I., 288, 364
Mitchell, G. F., 150, 175, 207, 460
Mitchison, N. A., 150, 175, 195, 206, 207,
    228, 243, 255, 334, 335, 364, 373,
    392, 489
Mithridates VI, 4
Miyasaka, N., 176
Mizejewski, G. J., 415
Mochizuki, M., 415
Möller, G., 150, 175, 228, 283, 373
Montagu, E. W., Lord, 293
Montagu, M. W., Lady, 294, 295, 302,
    489, 490
Moolten, F. L., 267, 284
Moon, C. R., 288
Morau, H., 254
Morgenroth, J., 108, 172, 203, 229, 256,
    428, 450
Moro, E., 205, 490
Mote, J. R., 42, 206
Moulin, A.-M., 94, 172, 361, 365, 384,
    428, 438
Mourant, A. E., 164, 256, 288, 345, 490
Moutsopoulos, H. M., 19
Mudd, S., 44, 51, 66, 127
Muir, R., 230, 461
Mulkay, M., 14, 17, 385, 414
Müller, R., 115
Müller-Eberhard, H. J., 105, 202
Murphy, J. B., 237, 254
Murray, J. A., 283

N

Nairn, R. C., 287
Najarian, J. S., 207, 373
Najjar, V. A., 212, 227
Nakane, P. K., 277, 287
Nataf, R., 208, 416
Neilson, E. G., 228
Neisser, A. L., 173, 230, 285, 362, 429
Neisser, M., 125, 230
Nelson, R. A., 209, 286
Nencki, M., 23
Newell, J. M., 194, 206
Nezelof, C., 201, 208, 334, 460
Nicolle, C., 204
Nicolle, M., 182, 416
Nisonoff, A., 82, 83, 148, 227, 228, 283, 387
Nissonoff, A., 490
Nordin, A. A., 128, 228, 364, 471
Nossal, G. J. V., 94, 150, 151, 175, 335
Nowell, P. C., 483, 490
Nussenblatt, R. B., 415, 417
Nussenzweig, V., 151
Nuttall, G. H. F., 126, 173, 288, 430, 458, 491

O

O'Connor, M., 42, 208, 389, 417, 463
Obermayer, E., 70, 81
Obermayer, F., 127, 458
Obermeyer, F., 261, 355, 491
Ochs, H. D., 419
Ohno, S., 84, 149, 364, 417, 439
Ojeda, A., 257
Olby, R., 69, 128, 452
Oliner, H., 323
Oppenheim, J. J., 151, 208, 364, 420
Oppenheimer, C., 322, 345
Orr, R. H., 385
Ortega, L. G., 287
Osborn, J. J., 418
Ostromuislensky, I. I., 65, 51, 491
Ostwald, W., 248
Ottinger, B., 175
Otto, R., 203, 204
Ouchterlony, Ö. T. G., 491
Oudin, J., 75, 83, 148, 149, 213, 226, 227, 286, 459, 460, 471, 491
Ousman, A., 345

Ovary, Z., 373, 491
Owen, R. D., 66, 165, 174, 255, 364, 491
Owens, R. D., 91, 94, 165, 166, 285, 468
Oyama, J., 175, 248, 256

P

Pagel, W., 123
Pappenheimer, A. M., Jr., 150, 193, 206, 207, 335, 381, 388, 491
Parish, C. R., 151
Partington, J. R., 123
Pasteur, L., 16, 17, 18, 22, 23, 28, 29
Pauli, W., 101
Pauling, L., 38, 52, 65, 66, 81, 101, 123, 127, 131
Pawlak, L. L., 227, 228
Pearl, E. R., 208
Pernis, B., 83
Peterson, P., 228
Peterson, P. A., 84
Pfeiffer, R., 35, 41, 102, 113, 229, 362, 396, 410, 414, 422, 428
Pflüger, E., 102
Phillips, S. N., 228
Phizackerley, R. P., 137, 149
Pick, E., 70, 116, 208, 151
Pick, E. P., 81, 127, 283, 363, 458, 492
Pierce, G. B., 277, 287
Pilarsky, L. M., 82
Pillemer, L., 202, 209, 410, 420, 486
Plaut, M., 151
Plescia, Otto, 150, 176, 379, 389, 391
Podolsky, S. H., 91, 94
Polak, J. M., 283, 287, 288
Poljak, R. J., 137, 149
Popper, E., 126
Popper, K., 323, 347, 361
Porter, R. R. , 82, 123, 137, 148, 149, 227, 385, 391, 393, 459, 463, 469, 493
Portier, P., 41, 126, 156, 173, 177, 179, 203, 351, 362, 396, 414, 458, 466, 493
Potter, T., 82, 175, 208
Potzl, O., 126
Poulik, M. D., 84, 135, 148, 460
Prahl, J., 81, 288, 333, 459, 461
Prausnitz, C., 187, 205, 271
Prausnitz, K., 187, 286
Prausnitz–Giles, C., 493

Prehn, R. T., 255, 334
Prendergast, R. A., 82, 151, 207, 208, 415, 416
Press, E. M., 123
Pressman, D., 52, 127, 131, 148, 265, 272, 283, 363, 373, 463, 493
Pressman, D. P., 283, 284, 286, 287, 335
Price, J. D. de S., 323, 361, 387, 450
Procopius, St., 4, 21
Provine, W. B., 64, 124, 324, 365, 437
Putnam, F. W., 21, 41, 81, 82, 123, 148, 149, 172, 297, 387, 419, 493
Pylarini, J., Dr., 12, 13, 293, 294

R
Race, R. R., 102, 164, 256, 345
Rackemann, F. M., 17, 39, 42, 163, 357, 363
Raffel, S., 194, 206, 462, 493
Raison, R. L., 256
Rajewsky, K., 228
Raju, S., 416
Ramon, G., 46, 65, 283, 493
Ramseier, H.,3, 150, 227, 228, 33
Rao, N. A., 415
Rapp, H. J., 209
Raubitschek, H., 322
Reich, M., 49, 65, 322, 345
Rescher, N., 449, 453
Reth, M., 420
Rhazes, A., 8, 9, 10, 22, 93, 123, 408, 419, 494
Rheinberger, H.-J., 282, 289
Rhodes, J. M., 150
Rich, A. R., 13, 35, 39, 42, 140, 150, 163, 167, 192, 194, 198, 206
Richet, C., 41, 126, 156, 173, 177, 179, 180, 181, 183, 203, 204, 254, 264, 283, 351, 362, 396, 414, 427
Richter, M., 283, 288
Ridge, J. P., 93
Riha, I., 497
Riniker, P., 201, 209, 328, 334, 364, 404, 418
Ritz, H., 209
Rivers, T. M., 174, 363
Rockey, J. H., 205, 208
Rocklin, R. E., 151, 208
Rodin, A. E., 418

Roitt, I. M., 373, 494
Rokitansky, C., 27
Romano, E. L., 288
Romano, M., 288
Römer, P., 159, 160, 174, 398, 415, 461, 494
Rose, N. R., 95, 164, 172-173, 175-176, 208, 228, 283, 286, 346, 364, 415
Rosen, F. S., 209, 418
Rosenau, M. J., 179, 183, 203, 204-205, 494
Rosenberg, C., 7, 22
Rössle, R., 317, 323, 488
Rostaine, C. R., 229, 316
Roth, J., 176, 288
Rothen, A., 66
Rotky, H., 50, 65
Rous, P., 237, 254
Roux, E., 21, 23, 41, 65, 127, 419
Roux, P. P. E., 495
Rowley, D., 42, 150, 333
Rowley, D. A., 495
Rudner, E. J., 206

S
Sabin, A. B., 495
Sabin, F. R., 276, 287
Sachs, H., 125, 155, 173, 224-226, 229, 230, 317, 321, 323, 428, 495, 500
Salk, J., 447, 495
Salvati, V., 264, 283
Salvin, S. B., 150, 206, 334, 495
Salvin, S. F., 335
Sammons, R. E., 288
Samter, M., 205, 365, 463, 495
Santucci, S., 160, 161, 174, 362, 415, 424, 429
Sattler, C. H., 398, 416
Sauerbeck, E., 19, 23, 461
Sauerbruch, F., 254
Saul, F., 137, 149
Saunders, M. E., 346, 419
Schaffner, K. F., 91, 94
Schally, A., 469
Scharff, M., 206, 335
Schaudinn, F., 173
Scheibel, I. F., 252, 256
Scheuchzer, J. G., 302

Schick, B., 41, 126, 156, 173, 178, 179, 182, 190, 203-205, 352, 362, 396, 414, 458, 461, 493, 495
Schleiden, M., 101, 112
Schlossman, S. F., 287, 485
Schoenheit, E. W., 42, 141, 150, 206, 327, 333, 363, 459
Schöne, G., 234, 235, 240, 241, 254
Schubert, D., 152
Schucht, A., 157, 173, 230, 285, 362, 458
Schulenberg, E. P., 152
Schulhof, K., 174
Schultz, W. H., 185, 204, 409, 459, 467, 475, 479
Schultze, H. W., 205, 460, 484
Schur, P. H., 208
Schwaber, J., 208
Schwartz, R. S., 164, 175, 230
Schweet, R. S., 57, 66
Schwentker, F. F., 174, 363
Seegal, B. C., 208
Seegal, B. H., 416
Seegal, D., 208, 416
Sege, K., 228
Sela, M., 148, 252, 256, 257, 284, 387, 496
Sell, S., 206, 284
Sevag, M. G., 56–57, 66, 462, 496
Shaffer, J. H., 206, 208, 334, 389, 462
Shapin, S., 347, 361
Shaw, B., 37, 353
Shaw, B. G., 25
Sherman, W. B., 381, 496
Shevach, E. M., 364, 420
Shin, H. S., 209, 420
Shulman, S., 496
Shwartzman, G., 496
Sigal, N. H., 81-82, 91, 94, 146, 151
Sigerist, H. E., 5, 22, 40, 172
Silber, H. A., 148
Silverstein, A. M., 22, 65, 82, 93, 95, 124, 175, 206, 208, 209, 229, 254, 287, 301, 335, 346, 363, 376, 384, 387, 415-419, 428, 429, 438, 450
Simms, E. S., 152
Simon, F. A., 39, 42, 117, 122, 164, 178, 193, 206, 357, 363
Simonsen, M., 83, 207, 255, 255, 256, 373, 496

Simpson, J. A., 176
Simpson, L. O., 256
Singer, S. J., 288
Sjögren, H., 415
Slater, R. J., 82, 149, 227
Sloane, H., 296-298, 302, 458
Small, H. G., 387
Smekal, F. G., 126, 321, 322, 486
Smith, J. L., 172
Smith, R. F., 206, 335
Smith, R. T., 83, 175, 256, 391, 393, 418, 419, 463, 496
Smith, T., 496, 497
Smithies, O., 72, 82
Snell, G. D., 176, 238, 239, 254, 387, 459, 469, 470, 483, 497
Söderqvist, T., 376, 384, 385
Sohma, M., 174, 415
Soothill, J. F., 209
Southard, E. E., 180, 181, 204
Spain, W. C., 251, 256
Spallanzani, L., 28
Sparham, L., 23, 303
Sparrow, E., 255
Speiser, P., 126, 321
Staehelin, T., 288
Starzl, T. E., 256
Stavitzky, A. B., 497
Steigherthal, J. G., Dr., 296, 297, 298
Steinberg, G. M., 417
Steinhardt, E., 188, 205
Stent, G., 345, 452
Sternberger, L. A., 287, 288
Sterzl, J., 419
Stettler, A., 21
Stevenson, L. G., 21
Stiehm, E. R., 418, 419
Streilein, J. W., 207, 255, 417, 417
Stull, A., 205
Stürli, A., 256, 429
Sturm, E., 254
Sulzberger, M. B., 207, 497
Suran, A., 417
Sutton, R., Sir, 293
Suzuki, T., 228
Sydenham, T., 7, 11, 22, 98, 99, 99, 123, 478, 494
Szenberg, A., 92, 94, 207
Szilard, L., 67

T

Tagliacozzi, G., 253
Taliaferro, W. H., 461, 497
Talmage, D. W., 66, 67, 81, 82, 93, 94, 127, 148, 346, 364, 420, 439
Tauber, A. I., 42, 66, 94, 384, 428, 437
Taylor, G., 288
Temkin, O., 22, 172
Teresaki, P. I., 176, 386
Theiler, M., 363
Thiele, F. H., 65
Thomas, L., 83
Thorbecke, G. J., 228
Thorndike, L., 22
Thorpe, N., 450
Thucydides, 3, 21, 98
Timbury, G., 176
Timoni, E., Dr., 12, 13, 293, 294, 296, 458
Tiselius, A., 135, 363, 459, 468, 485, 498
Tiselius, A. W. K., 135, 363, 459, 468, 485, 498
Tobler, R., 418
Todd, C. W., 83, 253
Todes, D. P., 362, 437
Tonegawa, S., 76, 83, 364, 471, 472
Topley, W. W. C., 47, 51, 65, 66, 81, 192, 206, 254, 283, 286, 462, 498
Topley, W. W. G., 418
Towbin, H., 288
Trapeznikoff, 30, 41
Traub, E., 165, 174, 207
Trentin, J., 92, 94, 255, 373
Triplett, R. F., 150, 207, 334, 364, 460
Triplett, R. R., 175
Trnka, Z., 374
Tsuda, K., 50, 65
Turk, J. L., 39, 41, 195, 206, 207, 208, 291, 293, 334, 373, 463, 489, 498
Tyzzer, E. E., 235, 236, 237, 238, 241, 254

U

Uhlenhuth, P. T., 159, 174, 397, 415, 458, 486, 498
Uhr, J. W., 142, 150, 151, 193, 206, 283, 335, 373, 393, 451, 491, 495, 498
Unanue, E. R., 150, 175, 287, 476
Underdown, B. J., 152

V

Vaerman, J. P., 205
Vaillard, L., 45, 65, 127
Vallery-Radot, R., 40
van der Scheer, J., 117, 118, 119, 127, 134, 148, 150, 256
van Helmont, J. B., 11, 22, 98
Van Noorden, S., 283, 287, 288
van Rood, J. J., 373, 391, 499
Vander Veer, A., 250, 251, 256
Varco, R. L., 201, 208
Varndell, I. M., 288
Vas, M., 207
Vaughan, J. H., 180, 181, 198, 208, 412, 420
Vaughan, V. C., 204
Verhoeff, F. H., 163, 174, 415
Viale, G., 288
Vincent, A., 176, 284, 476
Virchow, R., 27, 29, 40, 172, 431, 443, 447, 450
Vitetta, E.S., 142, 151, 284
Voisin, G. A., 164, 175, 373, 499
Voltaire, 13, 296, 302
von Behring, E., 3, 4, 18, 20, 22, 23, 35, 36, 41, 216, 308, 427, 455, 458, 460, 465, 475
von Decastello, A., 249, 256, 429
von Dungern, E., 107, 125, 249, 460
von Dungern, F., 256, 288, 429
von Gruber, M., 81, 102, 112, 115, 124, 126, 311, 312, 322, 438, 458, 480
von Mayer, E., 123
von Nägeli, K., 101–102, 112, 312
von Pirquet, C., 41, 112, 126, 156, 173, 177, 179, 182, 183–185, 189, 190, 192, 203-205, 207, 352, 362, 396, 414, 458, 461, 486–487, 493, 495
von Szily, A., 173, 399, 416, 417, 461
von Wassermann, A., 127, 173, 224, 229, 230, 285, 351, 363, 429, 458, 499

W

Waage, P., 100
Wacker, W. B., 415
Waksman, B. H., 39, 42, 150, 174, 196, 203, 206-208, 334, 373, 399, 416, 418, 420, 457, 462, 499
Waldenstrom, J. G., 484, 499

Wall, H., 228, 403
Wang, A., 83, 227-227, 256
Wangensteen, O. H., 303
Wangensteen, S. D., 303
Ward, S. M., 82, 149, 227
Warwick, W. J., 207
Watson, J., 340, 345, 446, 452
Watt, P. J., 208
Weaver, J. M., 255, 334
Webb, G., 442, 447, 450, 457, 497
Wedgwood, R., 405-406
Weigert, C., 30, 65
Weigert, M., 82, 83, 490
Weigle, W. O., 175, 204, 206, 373, 463, 499
Weil, E., 159, 351, 382, 391
Weil, P. E., 158
Weiser, R. S., 325, 333, 416
Welch, W. H., 368, 433-434, 438, 488
Wells, H. G, 127, 186, 363, 461, 500
Wernicke, E., 35
Wessely, K., 399, 416
Whitcomb, D., 172, 501
White, R. G., 38, 127, 189, 373, 463, 484, 500
Whitman, L., 363
Widal, G. F. I., 229, 316, 323, 500
Wiener, A. S., 102, 112-113, 115, 126, 172, 250, 256, 405, 418, 459, 467, 500
Wigzell, H., 151, 228
Williams, A. F., 439
Williams, C. A., Jr., 148, 286
Williams, R. C., 83, 149, 227
Williamson, A. R., 81, 83, 151
Wilson, D., 83, 94
Wilson, G. S., 192, 206, 418, 462

Winkelstein, J. A., 419
Wintringham, C., 12, 22
Wissler, F. C., 82, 148
Witebsky, E., 155, 164, 172-173, 429, 494, 500
Witmer, R. H., 399, 416
Woernley, D. L., 148
Woglom, W. H., 236, 240, 241, 254
Wolff-Eisner, A., 180, 204, 362, 414, 461, 500
Wolstenholme, G. E. W., 42, 208, 388-391, 463
Woodruff, M. F. A., 253-254, 256, 373, 416, 463, 500
Woods, A. C., 174, 415, 417, 462
Woodward, J., 294, 303
Wright, A. E., 37-38, 41, 139, 150, 287, 363, 442, 458
Wu, T. T., 73, 82-83, 115, 149, 460, 482, 485, 494

Y
Yalow, R. S., 275, 286, 460, 469, 501
Yaron, A., 148
Yersin, A. E. J., 4, 458, 465, 485, 495, 501

Z
Zak, S. J., 334, 418
Zanetti, M., 228
Ziegler, E., 30, 41
Ziff, M., 176
Zinkernagel, R. M., 92, 94, 143, 151, 175, 176, 335, 434, 438, 460, 472-473
Zinsser, H., 23, 34, 39, 41, 42, 81, 191, 205, 325, 333, 334, 357, 363, 461, 501
Zirm, E., 253

Subject Index

A

ABO blood group system, 306
Abrin, 267
Acetylcholine, 271
Acquired immunity, 349
  adaptation theories, 20–21
  cellular-humoral mechanisms, 25–40,
    185
  depletion theories, 12–18
  distension theories, 10–12
  expulsion theories, 7–10
  iatrophysics theory, 10–12
  osmotic and alkalinity theories, 19–20
  phenomenon, 3–5
  prerequisites, 5
  retention theories, 19
  theories, 5–21, 104
Acquired immunodeficiency syndrome
    (AIDS), 201
Acquired immunologic tolerance, 166–167
Active prophylactic immunization,
    262–263
Acute anaphylactic shock, 186
Adaptation theory, of acquired immunity,
    20–21
Adaptive enzyme theory, of antibody
    formation, 54–56
Addison's disease, 271
Adrenalitis, 169
Affinity
  concept of, 100, 106–107
  maturation, 340
Agglutination, 36, 43, 97, 104, 110, 113,
    119
Alexine, 308, 310, 312, 423
Alkalinity theory, of acquired immunity,
    19–20
Allergy, 37, 39, 111, 140, 143, 156
  anaphylatoxin, 186
  B cell defects, 200
  clinical discipline, 186–188
  complement deficiencies, 202
  complement system, 201–202

delayed type hypersensitivity, 193–198
  cell–cell intercommunications,
    198
  histopathologic studies, 194–195
  hypersensitivity to bland proteins,
    193–194
  lymphocyte subset functions, 197
  passive transfer, 195–196
  relationship to other phenomena,
    196–197
  sorting out the mechanisms, 197
desensitization, 188–189
genetics of atopic, 250–251
humoralist view, 186
and immunity, 177–203
immunity relationship, 192–193
immunologic deficiency diseases and,
    200–202
immunopathologic processes and,
    198–199
of infection, 189–190
local organ hypersensitivity and, 199
lymphoproliferative diseases and, 199
mechanism, 180–186
  cellular theory, 185–186
  cytotropic antibody, 185–186
  direct toxicity, 180
  misdirected immunity, 180–185
  unitarian approach, 182–185
observations, 178–180
progress in, 186–188
selective immunoglobulin deficiencies
    and, 200
severe combined immunodeficiency, 201
speculations, 190–192
T cell defects, 201
tuberculin type hypersensitivity,
    193–198
Allograft rejection, 39, 56, 58, 141, 197
Allokeratoplasty, 400
Alpha-fetoprotein and carcinoembryonic
    antigen (CEA), 266
Ambozeptor, 308, 309, 311, 315, 423

American Academy of Allergy, Asthma, and Immunology, 406
American Association of Immunologists, 406
Amethopterin, 267
Anal fistula, 291, 300
Anaphylactic keratitis, 401
Anaphylactin, 180
Anaphylatoxin allergy, 186
Anaphylaxis, 37, 111, 140, 156, 159–162, 177–181, 183, 185–188, 190–192, 198, 201–202, 271, 409–410
  and related diseases, 351–352
Andmomordin, 267
Ankylosing spondylitis, 252
Annales de l'Institut Pasteur, 352–353
Anterior chamber-associated immune deviation (ACAID), 401
Anthrax infection, 19, 31, 34, 69, 262
Anthropology, and antibody, 279–280
Anti-antibodies
  during1890-99, 216–219
  during1899-1904, 219–225
  antibody-combining site, 213, 219–220
  autoanti-antibody immunoregulation, 220–222
  demise during1901–1905, 225–226
  images of antigen, 222–224
  immunoregulation, 211–225
  side-chain theory, 217
Anti-arsanilic acid antibody, 47
Antibody
  anthropology, 279–280
  assay and localization, 273–279
    ELISA assays, 275–276
    fluorescence-activated flow cytometry and sorting, 276
    gel diffusion analysis, 274–275
    immunoblots, 278–279
    immunoelectrophoresis, 275
    immunohistochemistry, 276–278
    immunoprecipitation, 273
    plaquing technics, 279
    radioimmunoassay, 275
  combining site size, 134
  forensic pathology, 280–281
  hapten interaction, 52
  heterogeneity and thermodynamics, 132–134

  identification, 273–279
  in immunodiagnosis, 269–273
    delayed-type skin tests, 271–272
    dermal reactions, 271–272
    diagnostic antibody libraries, 272
    radioimmunodiagnosis, 272–273
    serologic tests, 270–271
  in immunotherapy, 262–269
    active prophylactic immunization, 262–263
    immunotoxic agents, 264–269
    nonspecific immunotherapy, 269
    passive serotherapy, 263–264
  molecule approaches to specificity, 135–138
  site shape, 131–132
  taxonomy, 279–280
  uses, 261–282
Antibody–cell paradox, 331
Antibody diversity
  biochemical data, 73–74
  evolutionary paradox, 77–80
  generation of, 69–80
  germline-somatic mutation debate, 69–80
  germline V region genes and, 77–78
  in historical perspective, 69–70
  immunoglobulin mechanism, 78–79
  ontogenetic data, 72–73
  opposing positions, 70–80
  resolution of debate on, 75–76
  serological data, 75
  speciation problem, 79–80
Antibody formation
  antigen-incorporation theories, 44–47
  first selection theory, 47–49
  immunologic specificity and, 119–122
  instruction theories, 49–57
    adaptive enzyme theory, 54–56
    complement theory, 50–54
    direct template theories, 50–54
    immunocatalysis theory, 56–57
    implications, 57
    indirect template theories, 54–57
    template-inducer theory, 57
  lymphocyte role in, 39
  selection theories, 58–63
    clonal selection theory, 60–62
    molecular-genetic theory, 62–63

natural selection theory, 59–60
quantum-mechanical resonance
theory, 58–59
theories of, 43–64, 119–22
Antibody paradigm
DTH and antibody formation, 328–329
and DTH specificity, 329
high-affinity antibodies, 330
Anticholera antibodies, 422
Anti-erythrocyte antibodies, 426, 427
Anti-erythrocyte isoantibodies, 338
Antigen, images of, 222–224
Antigen-antibody
interaction and immunologic specificity,
101, 104, 106–107, 115, 144
precipitin reaction, 54
reactions, 399, 407, 408, 409,
411, 412
Antigen-incorporation theories, of
antibody formation, 44–47
Antigen molecule
approaches to immunologic specificity,
130–134
antibody heterogeneity and
thermodynamics, 132–134
shape of specific antibody site,
131–132
size of antibody combining site, 134
Anti-idiotypes, 212–216
immunoregulation, 211–226
Antimony, 99
Antisepsis, concept of, 29
Antitoxins, 350
Apoptosis (programmed cell death), 87
Arrhenius–Madsen theory, 110
Arthropathy, 184
Arthus phenomenon, 37, 111, 144, 156,
177, 179, 182, 190, 197, 271, 327,
352
Arthus skin reactivity, 195
Assay and localization
of antibody, 273–279
ELISA assays, 275–276
fluorescence-activated flow cytometry
and sorting, 276
gel diffusion analysis, 274–275
immunoblots, 278–279
immunoelectrophoresis, 275
immunohistochemistry, 276–278

immunoprecipitation, 273
plaquing technics, 279
radioimmunoassay, 275
Association for Research in Vision and
Ophthalmology (ARVO), 401
Asthma allergy, 140, 144, 178, 180,
187–188, 251, 271
Atopic allergy, genetics of, 250–251
Atoxyl hapten, 47
Autoallergic diseases, 39, 141
Autoambozeptor, 317
Autoanti-antibody immunoregulation,
220–222
Autoantibodies, 351
Autohemolysin, 221
Autoimmune dacryoadenitis, 398
Autoimmune diseases, 61, 86–87, 89, 115,
155, 157, 159–160, 162–164,
168–171, 178, 196, 200, 212, 215,
222, 231, 243, 248, 270–271
Autoimmune interstitial nephritis, 215
Autoimmunity, 86
classical period of research, 156–161
to lens proteins, 159–160
paroxysmal cold hemoglobinuria,
156–159
sympathetic ophthalmia, 160–161
Wassermann' antibody, 157–159
concept of, 153–171
dark ages of research, 161–164
modern period of research, 164–171
basis of immunoregulation, 168
conceptual progress, 164–168
immunologic tolerance, 165–18
multiple-system autoimmune disease,
170
phenomenological progress, 169–170
single-organ autoimmune disease,
169–170
technical progress, 170–171
Autointoxication theory, 154
Avidin–biotin systems, 277–278

B
Bacterial agglutination, 36, 43
Bacterial allergy, 39, 141
Bacterial anaphylaxis, 191, 201
Bacteriology, 17–18, 21, 38, 99, 102, 120,
153–154, 455–456

B cell
    defects allergy, 200
    Ig receptor, 331–332, 344–345
    receptor, 359
B cell (bursa/bone marrow), 328
Behring's serotherapeutic doctrine, 352
Beta-human gonadotropin, 266
Biological specificity
    chemistry, 99–101
    in immunologic specificity, 98–102
    medicine and pharmacy, 98–99
    philosophy, 101–102
Black bile (melaine chole), 27
Black Death, 7
Blood groups, 249–250, 281
Blood (sanguis), 27
Bone marrow transplantation, 246, 248
Bordet–Wassermann reaction, 157–158
Bovine serum globulin, 70
Brook Lodge series, 370, 375, 378, 379
Brucellosis, 191
Bubonic plague, 3
Buffalo Convocations, 370, 371, 379
Bullous pemphigoid, 169
Burnet's concept of immunologic
    tolerance, 406
Burnet's theory, 343
Bursa of Fabricius, in birds, 328
Buying smallpox practice, 12

C
Cardiolipin, 159, 270
Carrier effect, 141, 194, 197
Celiac disease, 252
Cell-bound antibodies, 141
Cell-cell intercommunications allergy, 198
Cell Interactions and Receptor Antibodies
    in Immune Responses, 379
Cellular immunity, 25–40, 104, 153, 185,
    190, 195. See also Humoral and
    cellular immunity
    delayed-type hypersensitivity, 140–143
    specificity in, 138–143
    transfer factor, 143
Cellular immunology, 349
Cellular (phagocytic) theory, 19
Cellular theory, of allergy, 185–186
Cervical cancer, 262
Charles University, 382

Chemical affinity, 100
The Chemical Aspects of Immunity (Well),
    357
Chemistry, and biological specificity,
    99–101
The Chemistry of Antigens and Antibodies
    (Marrack), 357
Chemotherapy, 106–107, 170, 226, 247,
    264
Chicken cholera, 29, 69
Chickenpox, 297
Chlamydia trachomatis, 199
Cholera, 7, 29, 34–35, 50, 262, 269
Choline, 45
Choriomeningitis infection, 196–197
Ciba Foundation, 370, 378
Cinchona, 292
Cis- and trans-immunology-problems,
    43–44
Clinical Research Society, 406
Clinical shock syndrome, 179
Clonal abortion mechanism, 166
Clonal selection theory (CST), 122, 134,
    142, 166, 211, 213, 243
    of antibody formation, 60–62, 70–72
    challenges, 85–87, 90–93
        to clonal selection, 85–87
        danger signal, 87
        evaluation, 90–91
        idiotypic networks, 85–86
        immunological homunculus, 86–87
    immunological self, 91–93
    implications, 88–90
Cluster analysis, 374
Cluster of differentiation (CD) markers, of
    lymphocytes, 266, 276
Coccidioidin, 196
Cold Spring Harbor series, 370, 375, 377,
    382
Collegium Internationale Allergolicum
    series, 370, 380
Complement deficiencies allergy, 202
Complement fixation test, 43, 225
Complement system allergy, 201–202
Complement theory, of antibody
    formation, 50–54
Congenital adrenal hypoplasia, 252
Congenital agammaglobulinemia, 61
Congestin, 180

Congresses of the Transplant Society, 380
Conjunctiva, 399–400
Conjunctival ophthalmoreaction, 189
Conjunctivitis, 401
Continuity, law of, 101
Coombs test, 164–165
Corneal transplantation, 233, 239, 245, 400
Cowpox vaccination, 291
Cupping practice, 27
Cutaneous reaction, 189–190
Cyclophosphamide, 247
Cyclosporine A, 247
Cytotoxic antibodies, 350–351
Cytotoxines, 310
Cytotropic antibody mechanism, of
    allergy, 185–186

D
Danish Science and Medical Library,
    Copenhagen, 368
Danysz (Bordet–Danysz) phenomenon,
    106–108
Darwinian evolution, 33, 48, 77–78, 102,
    171
Darwinian evolutionary principles, 349,
    356
Darwinism triumphant, 436–437
Darwin's theory, 91, 431–432
Delayed-type hypersensitivity (DTH), 39,
    193–198, 325, 327
  and antibody formation, 328–329
  cell–cell intercommunications, 198
  histopathologic studies, 194–195
  hypersensitivity to bland proteins,
      193–194
  lymphocyte subset functions, 197
  passive transfer, 195–196
  relationship to other phenomena,
      196–197
  sorting out the mechanisms, 197
  specificity of, 329
Delayed-type skin tests, 271–272
Denkkollektiv, 348, 355
Depletion theory, of acquired immunity,
    12–18
Dermal reactions, 271–272
Dermatitis, 141, 195
  herpetiformis, 252
Desensitization allergy, 188–189

Diagnostic antibody libraries, 272
Differentiating cells, population dynamics
    of, 61
Dinitrofluorobenzene, 327
Dinitrophenyl poly-L lysine (DNP-PLL), 252
Diphtheria, 35, 262–264
  immunology of, 46–48
  infection, 308
  toxin, 4, 20, 35, 48, 69, 103, 106–108,
      154, 177, 179, 218, 267
  toxin–antitoxin reactions, 340
Direct template theories, of antibody
    formation, 50–54
Direct toxicity mechanism, of allergy, 180
Disease
  germ theory, 28, 99, 154, 262
  infectious, 4, 7, 14, 17, 20, 27, 29, 35,
      38, 43, 69, 98, 102, 139, 153, 177,
      182, 189–193, 216, 225, 234, 250,
      262–264, 269, 277, 291
  nature, 27–28
  ontological concept, 98
  systematization and classification, 99
  theurgic origin, 5–7
Distension theory, of acquired immunity,
    10–12
The Doctor's Dilemma (Bernard Shaw),
    353
Donath–Landsteiner report, on
    autoantibody, 315–318
Dye labels, in immunohistochemistry, 276

E
Eel toxin, 179, 219
Egg albumin, 70
Ehrlich–Bordet debates, on immunologic
    specificity, 107–111
Ehrlich–Gruber debate, on immunologic
    specificity, 111–114
Ehrlich's theory, of antibody formation,
    338, 342–343, 355, 409
Ehrlich-type receptor, 341
Einstein's relativity theory, 319
Electron-opaque labels, in
    immunohistochemistry, 277–278
ELISA assays, 275–276
Encephalitis, 162
Encephalomyelitis, 163–164, 169
Endophthalmitis, 160

Endophthalmitis phacoanaphylactica, 398
Enzyme labels, in immunohistochemistry, 277
Epochs in immunology, 455
Equilibrium dialysis technique, 133
Erysipelas, 33
Erythematous skin rashes, 170
Erythroblastosis fetalis, 250
Essentialism, concept of, 101
Essential lymphocytophthisis, 201
European Society for Immunodeficiencies, 407
European Society for Paediatric Haematology and Immunology, 407
European Society of Pediatric Allergy and Clinical Immunology, 407
Exanthematous diseases, 12
Expulsion theory, of acquired immunity, 7–10

F
Ferritin, 278
    lactic dehydrogenase, 266
F1 hybrid disease, 243
First selection theory, of antibody formation, 47–49
Fluorescence activated cell sorter (FACS), 276
Fluorescence-activated flow cytometry and sorting, of antibody, 276
Fluorescent antibody immunohistochemistry, 122
Fluorescent labels, in immunohistochemistry, 276–277
5-Fluorouracil, 247
Forensic medicine, 112
Forensic pathology and antibody, 280–281
Fowl cholera, 262
Fowl sarcoma rejection, 237
Fracastoro, 408
Freund's adjuvants, 141, 165, 193
Fundamentals of Immunology (Boyd), 357

G
Gel diffusion analysis, of antibody, 274–275
Genesis and Development of a Scientific Fact (Ludwik Fleck), 348

Geneticists, and transplantation biology, 238–240
Genetics, of atopic allergy, 250–251
Germinal Center series, 370, 371
Germline V region genes, and antibody diversity, 77–78
Germ theory of disease, 28
Glanders, 191, 272
Glomerulonephritis, 184
Golden Age of bacteriology, 38
Graves disease, 170
Guillemin, R., 469

H
Hanson's bacillus, 272
Hapten–antibody interactions, 133
Haptens, 329, 355
Hashimoto's thyroiditis, 199, 252, 271
Haurowitz's instruction theory, 343
Hayfever allergy, 140, 144, 178, 180, 187–188, 251, 271
Hematohumoral theory, 27
Hematoporphyrin, 267
Hemolysin, 314
Hemolysine, 310
Hemolysis, 97, 113, 144, 157, 219–221, 226. *See also* Immune hemolysis
Hemolytic anemias, 169, 215
Hemolytic autoantibodies, 306
Hemolytic disease, 164
Hemolytic plaque assays, 122
Herpes simplex virus, 196
Heuristics, 305
High-affinity antibodies, 330
Histamine, 144, 185–186, 188, 271
Histocompatibility testing
    series, 370
    workshops, 381
Histologic staining, 102
Histopathologic studies, in allergy, 194–195
HLA and disease susceptibility, 252
Homograft rejection, 240–241
Horror autotoxicus, 154–157, 160–161, 167, 171, 221, 306, 332, 396, 424–425
Horse antidiphtheria toxin, 352
Horse-radish peroxidase, 277
Human papilloma virus, 262

Humoral and cellular immunity. *See also*
      Cellular immunity
   conflict on, 26–27
   debate on, 32–40
   disease and, 27–28
   inflammation and, 27–28
   international politics and, 30–32
   Hybridoma techniques, 247
   Hydrogen-bonding, 52
   Hypersensitivity to bland proteins
      allergy, 193–194
   Hypogammaglobulinemia, 200

I
Iatrochemistry, 100
Iatrophysics and acquired immunity,
      10–12
Idiopathic hemochromatosis, 252
Idiosyncrasies, 178
Idiotypes
   anti-idiotypes and, 212–216
      Jerne's network theory, 168, 214–216
   network theory, 85–86
IgA-deficient patient, allergy in, 187,
      200
IgE antibodies, 178, 188
IgE antibody–allergen interaction, 403
Immune hemolysis, 50, 104, 107, 110,
      155, 198, 201, 216–218, 221, 305,
      308
   autoimmunity, 424–425
   background, 421–422
   cytotoxic antibodies, 423–424
   discovery of blood groups, 426–427
   effects on Metchnikoff's phagocytic
      theory, 425
   Ehrlich–Bordet debate on mechanisms,
      422–423
   horror autotoxicus, 424–425
   Wassermann reaction and serodiagnosis,
      425–426
Immune response (Ir) genes, 252–253
Immunity
   and allergy, 177–203
   evolution of concept, 21
   of infection, 141
   phagocytic theory, 28, 31–33
   in transplantation, 234–236, 243
Immunobiological revolution, 357–359

Immunobiology, 58, 121, 129, 244,
      435–436
   renaissance of, 457
Immunoblots, 278–279
Immunocatalysis theory, of antibody
      formation, 56–57
Immunochemistry, 48, 58, 103, 129, 162,
      354–357, 434–435, 456–457
Immunodiagnosis
   antibody in, 269–273
   delayed-type skin tests, 271–272
   dermal reactions, 271–272
   diagnostic antibody libraries, 272
   radioimmunodiagnosis, 272–273
   serologic tests, 270–271
Immunodrugs, 267–268
Immunoelectrophoresis, 275
Immunofluorescent histochemical staining,
      399
Immunogen, antibody-combining site as,
      219–220
Immunogenetics
   blood groups, 249–250
   genetics of atopic allergy, 250–251
   immunological diversity, 253
   major histocompatibility complex,
      251–253
   transplantation biology, 248–253
Immunoglobulin, 73–74, 79, 135, 138,
      170, 178, 187–189, 200–201,
      212–213, 215, 231, 249, 263–265,
      268, 271, 276–278, 280
Immunohematology, 115
Immunohistochemistry, 276–278
   antibody in, 276–278
   dye labels, 276
   electron-opaque labels, 277–278
   enzyme labels, 277
   fluorescent labels, 276–277
   radioactive labels, 278
Immunological community, problems
      with, 325
   immunological deficiency diseases,
      328
   widening dichotomy, 326–328
Immunological deficiency diseases, 328
Immunological diversity and
      immunogenetics, 253
Immunological homunculus, 86–87

Immunological privilege, 400
Immunological self, in CST, 91–93
Immunological suicide experiments, 167
Immunological tolerance, 39, 56, 58,
        60–61, 69, 76, 91, 121–122,
        165–168, 213, 246
    in transplantation biology, 242–243,
        247–248
Immunologic deficiency diseases, 39, 58,
        121, 141, 164, 200, 243
    and allergy, 200–202
    B cell defects, 200
    complement deficiencies, 202
    complement system, 201–202
    selective immunoglobulin deficiencies,
        200
    severe combined immunodeficiency, 201
    T cell defects, 201
Immunologic paralysis, 166
Immunologic specificity
    approaches via antibody molecule,
        135–138
    approaches via antigen molecule,
        130–134
        antibody heterogeneity and
            thermodynamics, 132–134
        shape of specific antibody site,
            131–132
        size of antibody combining site,
            134
    biological specificity, 98–102
    in cellular immunity, 138–143
        delayed-type hypersensitivity,
            140–143
        transfer factor, 143
    concept of, 97–122
    Ehrlich–Bordet debates, 107–111
    Ehrlich–Gruber debate, 111–114
    Karl Landsteiner and, 114–119
    Paul Ehrlich and, 102–114
    and repertoire size, 145–147
    solutions, 129–147
    specific receptors, 102–107
        antigen–antibody interaction,
            106–107
        side-chain theory, 104–106
    specific triggers and nonspecific
        amplifiers, 143–145
    structural basis, 130–138

    via antibody molecule, 135–138
    via antigen molecule, 130–134
    and theories of antibody formation,
        119–122
Immunology
    anaphylaxis and related diseases,
        351–352
    cellular, 349
    critiques on Donath–Landsteiner report,
        315–318
    cytotoxic antibodies, 350–351
    epochs in, 455
    fate of early research programs,
        352–354
    hemolysis mechanism, 306
    immunobiological revolution, 357–359
    immunochemistry, 354–357
    important discoveries, 460–464
    Landsteiner's scientific style, 313–315
    M, N, and P blood groups, 306
    preventive, 348–349
    problems in, 441–443
    seminal discoveries, 458–460
    serodiagnosis, 351
    serotherapy, 350
    terminology for immunologic reactions
        basis of Bordet, 309–311
        basis of Ehrlich, 308–309
        basis of Landsteiner, 311–313
Immunology and Immunopathology of the
        Eye symposia, 401
Immunology meetings, 1951–1972
    chronology, 395–399
    disciplinary leadership, 371–374
    impacts, 381–383
    meeting clusters, 374–377
    participants, 373
    selection criteria, 368–369
    subdisciplinary relationships, 377–380
    taxonomy of, 369–371
    trends of mainstream immunology,
        380–381
Immunopathology, 39, 121, 145, 163, 170
    local organ hypersensitivity, 199
    lymphoproliferative diseases, 199
    processes and allergy, 198–199
Immunophysiology
    allergic reactions, 409–410
    complement and its products, 410

earliest medical views, 408–409
immunity without antibodies, 411–412
institutionalization of discipline,
    412–413
migration-inhibition factor (MIF),
    411–412
Immunoprecipitation, 273
Immunoregulation
    anti-antibodies, 211–225
    autoanti-antibody, 220–222
    autoimmunity basis of, 168
    idiotype–anti-idiotype network, 155
    network theory, 212, 214–216
Immunoregulatory mechanisms system,
    167
Immunosuppressive therapy, in
    transplantation biology, 246–247
Immunotherapy
    active prophylactic immunization,
        262–263
    antibody in, 262–269
    immunodrugs, 267
    immunotoxic agents, 264–269
    immunotoxins, 266–267
    liposome vehicles, 267–268
    nonspecific immunotherapy, 269
    passive serotherapy, 263–264
    photoimmunotherapy, 267
    radiolabeled antibodies, 265–266
    technical problems, 268–269
Immunotoxic agents, 264–269
Immunotoxins (ITs), 266–268
Impotence, 99
Indirect template theories, of antibody
    formation, 54–57
Infectious diseaseDisease
Infectious immunosuppression, 168
Inflammation, nature and cellular-humoral
    immunity, 27–28
Influenza, 7, 61, 262
Inoculation
    immunization procedure, 291
    introduction in England, 292–295
    Royal experiment on smallpox,
        295–301
Instruction theories, 343, 356
Instruction theories, of antibody
    formation, 49–57, 70, 165
    adaptive enzyme theory, 54–56

complement theory, 50–54
direct template theories, 50–54
immunocatalysis theory, 56–57
implications, 57
indirect template theories, 54–57
template-inducer theory, 57
Insulin-dependent diabetes, 252, 271
Interatomic and intermolecular forces
    theory, 52
Internal medicine, side-chain theory in,
    106
International Congress of Immunology,
    Washington, 370, 377
International Congress series, 370
International Society of Developmental
    and Comparative Immunology, 407
International Union of Immunological
    Societies, 377
Interphotoreceptor retinoid-binding
    protein (IRBP), 397
Intracellular digestion process, 48
Intradermal reaction, 189
Ionic bonding, 52
Ipecacuanha, 292
Isoagglutinins, 315
Isohemagglutinins, 306
Isohemolysins, 315

J
Japanese waltzing mouse, 235, 238
Jerne's network theory, 214–216
Jones–Mote hypersensitivity, 194

K
Karl Landsteiner and immunologic
    specificity, 114–119
Koch phenomenon, 140, 177, 190
Koch reaction, 111
Komplement, 315, 417

L
Lamarckian view, of antibodies, 339, 341
Landsteiner's rule, 249
Laudable pus, 29
Lawrence's Cellular and Humoral Aspects
    of Hypersensitivity States, 381
Leeches application practice, 27
Lens-induced immunogenic disease, 398,
    401
Lens proteins, autoimmunity to, 159–160

Lepromin, 140, 272
Leprosy, 98, 140, 158, 272
Liposome vehicles, 267–268
Liquid hybridization kinetics, 76
Liver tumors, 266
Local ocular hypersensitivity, 398
Local organ hypersensitivity and allergy,
    199
Luetin test, 140, 271
Lupus erythematosus, 215
Lupus nephritis, 170
Lymphocyte
    role in antibody formation, 39
    subset functions allergy, 197
Lymphocytic choriomeningitis (LCM)
    virus infection, 165, 196
Lymphoproliferative diseases and allergy,
    199

M
M, N, and P blood groups, 306
Magic and theurgic origin of disease, 5–7
Major histocompatibility complex
    (MHC), 170
    HLA and disease susceptibility, 252
    immune response genes, 252–253
    immunogenetics and, 251–253
Major immunogene complex (MIC), 251
Malaria, 7, 158
Mallein, 271
Massachusetts witchcraft trials, 13
Measles, 98, 196, 262, 272
Medicine, and biological specificity, 98–99
Menstrual blood theory, for smallpox, 9–10
6-Mercaptopurine, 247
Mercury, 99
Metchnikoff's phagocytic theory, 353,
    408, 425
Methotrexate, 247
Migration-inhibition factor (MIF), 412
Misdirected immunity mechanism, of
    allergy, 180–185
Mithridaticum mixture, 4
Molecular-genetic theory, of antibody
    formation, 62–63
Multiple-system autoimmune disease, 170
Mumps, 196, 262, 272
Murine leukemia, 267
Myasthenia gravis, 170–171

Mycobacteria, 70
Myeloid leukemia, 267
Myeloma proteins, 73, 146, 213
Myxomatosis virus infections, 61

N
National Institute of Child Health and
    Human Development, 378
Natural antibodies
    era of Ehrlich's side-chain theory,
        337–338
    immunobiological era, 340–341
    immunochemical era, 338–340
    resolution of conceptual problems,
        341–342
Natural immunity, 349
Natural selection theory, of antibody
    formation, 59–60
Neo-Darwinian, 91
Network theory immunoregulation, 212,
    214–216
New York Academy of Sciences, 369
N-hydroxylsuccinimidyl
    3-(2-pyridyldithio) propionate, 267
Nitrated proteins, 355
Nitric acid, 99
"Non-sense" antibodies, 340
Nonspecific immunotherapy, 269
Nucleic Acids in Immunology, 379

O
Obstetrics and gynecology, side-chain
    theory in, 106
Ocular immunology
    allergic diseases, 398–400
    autoimmune diseases, 396–398
    corneal transplantation, 400–401
    disciplinary institutionalization, 401
Oncofetal antigens, 266
Ophthalmitis, 351
Ophthalmology, 396
Opsonic indexes, 37
Osmotic theory, of acquired immunity,
    19–20
Overcompensation, law of, 48

P
Parabiosis intoxication, 243
Parenchymatous inflammation, 29
Paresis, 158

Paroxysmal cold hemoglobinuria (PKH), 115, 156–159, 171, 221, 306–307, 315–316, 353, 424
Passive serotherapy, 263–264
Passive transfer allergy, 195–196
Passive-transfer serum therapy, 43
Pasteurella pestis, 3
Pasteurian immunization, 353
The Pathogenesis of Tuberculosis (Arnold Rich), 327–328
Pauci-articular juvenile rheumatoid arthritis, 252
Paul Ehrlich and immunologic specificity, 102–114
Pauling's instruction theory, 343
Pediatric immunology
  institutionalization of the discipline, 406–407
  Paul Ehrlich's studies, 402–403
  preventive vaccinations, 404–406
Pemphigoid, 169
Pemphigus, 169, 252
Peptone, 45
Percutaneous reaction, 189
Pernicious anemia, 252
Pfeiffer phenomenon, 34–35
Phacoanaphylactic endophthalmitis, 163
Phacoanaphylaxis, 164, 169
Phagocytic theory, of immunity, 28, 31–33, 35, 37–38, 48, 139, 171
Phagocytosis, 29, 33–34, 37, 76, 139
Pharmacology, 106
Pharmacy, and biological specificity, 98–99
Philosophy, and biological specificity, 101–102
Phlebotomy practice, 27
Phlegm (pituita), 27
Photoimmunotherapy, 267
Picryl chloride, 327
Plague, 98–99, 291
  of Athens of 430 BC, 3
  of Justinian, 4
Plaquing technics, for antibody assay and localization, 279
Plasmacytosis, 199
Platelet disease, 164
Poisons, mithridatic adaptation of, 4, 20
Poliomyelitis, 115, 262, 291

Polyacrylamide gel electrophoresis (PAGE), 278
Porphyria, 291
Prague series, 371, 375, 378, 382
Precipitin reaction, 35, 43, 54, 104, 110–111, 118–119, 181–182, 188, 271, 273, 275, 279
Preventive immunology, 348–349
The Production of Antibodies (Macfarlane Burnet), 358
Prophylaxis, 27–28, 38
Prostate specific antigen (PSA), 266
Prostate tumors, 266
Prostatic acid phosphatase, 266
Pseudomonas exotoxin, 267
Psoriasis vulgaris, 252
Psylli tribe, resistance to snakebite, 3
Purgatives practice, 27

Q
Quantitative Immunochemistry (Kabat and Mayer), 357
Quantum-mechanical resonance theory, of antibody formation, 58–59
Quantum mechanics, 319

R
Rabies, 69, 262
Radioactive labels, in immunohistochemistry, 278
Radioimmunoassay, 275
Radioimmunodiagnosis, 272–273
Radiolabeled antibodies, 265–266
Red blood cells, 337
Regulation of the Antibody Response, 379
Reiter's disease, 252
Retention theory, of acquired immunity, 19
Rheumatoid arthritis, 170
Rheumatoid synovitis, 170
Rh factor, 405
Ricin, 267
Ring test, 273
Royal experiment
  human experimentation, 299–300
  opposition to inoculation, 298–299
  on smallpox inoculation, 291–301
  validity of experimental evidence, 300–301
Rubella, 262

Rumor, 30
Runt disease, 243
Rutgers University, 379

S
Sandwich technique, 275–276
Sanibel Island series, 370, 375, 377, 378
S-antigen, 397
Saporin-S6, 267
Schultz–Dale phenomenon, 185, 409
Scientific disciplines
  immunophysiology, 407–413
  ocular immunology, 396–401
  pediatric immunology, 402–407
Second Collegium Allergolicum, 375
Selection theories, of antibody formation,
    58–63
  clonal selection theory, 60–62
  molecular-genetic theory, 62–63
  natural selection theory, 59–60
  quantum-mechanical resonance theory,
    58–59
Selective immunoglobulin deficiencies and
    allergy, 200
Self-nonself dichotomy, 56, 87, 89–93,
    165–166, 236
Senile cataract formation, pathogenesis of,
    159
Sensibilisatrice, 308
Sensibilisine, 181
Sequestered antigen theory, 160
Serodiagnosis, 115, 129, 157, 351
  Wassermann reaction and, 425–426
Serologic tests, 270–271
Serotaxonomy, 112, 129
Serotherapy, 350
Serotonin, 144
Serum sickness, 37, 111, 156, 177, 179,
    182–184, 187, 190, 263, 352
Sessile, 325
Severe combined immunodeficiency
    disease (SCID), 201, 328, 404
Sex-linked hypogammaglobulinemia, 200
Sheep red cells, 70
Sicca syndrome, 252
Side-chain theory, 36, 48–49, 59–60, 70,
    104–108, 110–113, 116, 120, 122,
    142, 154, 171, 180, 211–212,
    217–222, 224, 226, 408, 411, 423

Single-organ autoimmune disease,
    169–170
Sjögren's syndrome, 170, 271, 398
Skin grafting, 240, 242
Skin-graft rejection, immunology of, 164
Skin test for allergy, 140
Smallpox, 98–99
  cowpox vaccination and, 291
  depletion theories, 12–18
  iatrophysical theory, 11–12
  inoculation, 13–14, 291–295
  menstrual blood theory, 9–10
  phenomenology of, 14–15
  prevention, 291–292
  prophylactic method for prevention of,
    291
  Royal experiment on inoculation,
    295–301
    human experimentation, 299–300
    opposition to, 298–299
    validity of evidence, 300–301
  vaccination, 3
Society for Pediatric Research, 406
Speciation problem, and antibody
    diversity, 79–80
Specificity
  antibody molecule approaches, 135–138
    combining site size, 134
    heterogeneity and thermodynamics,
    132–134
    site shape, 131–132
  cellular immunity in, 138–143
The Specificity of Serological Reactions
    (Landsteiner), 357
Specific receptors
  antigen–antibody interaction, 106–107
  and immunologic specificity, 102–107
  side-chain theory, 104–106
Spermotoxines, 311
Spirochaeta pallida, 157
Spondylitis, 252
The Structure of Scientific Revolutions
    (Thomas Kuhn), 358
Sulfur, 99
Summer Schools, 369
Surgeons, of transplantation biology,
    232–233
Swinepox, 297
Swiss-type agammaglobulinemia, 201

"Swiss-type" diseaseSevere combined
     immunodeficiency (SCID)
Sympathetic ophthalmia, 160–164, 397,
     401
Syphilis, 7, 13, 98, 115, 140, 157–159,
     182, 262, 270–271, 282, 351
Systemic lupus erythematosus (SLE), 170,
     252, 271
Systemic serum sickness, 263

T
Taxonomy and antibody, 279–280
T cell
     defects allergy, 201
     immunoglobulin receptor (IgT),
          330–331
T cell-mediated immunity, 398
T cell receptor (TCR), 142–143, 326,
     330–331, 359
     structural basis of, 332
T cell (thymus) systems, 328
Template-inducer theory, of antibody
     formation, 57
Tetanus, 35, 262–264
     infection, 308
     toxin, 45, 69, 154, 219
TGAL (tyrosine, glutamic acid, -alanine,
     -lysine), 252
Thalassin, 180
Theriac mixture, 4
Thermodynamics and antibody
     heterogeneity
     in immunologic specificity, 132–134
Thermostable antibody, 305
Thrombocytopenias, 169
Thyroid, 155, 162–164, 169–170, 199,
     215, 252, 271
Thyroiditis, 169
Thyroid-simulating hormone (TSH), 170,
     271
Till–McCulloch spleen colony technique,
     92
Tissues
     employed in transplantation biology,
          245–246
     typing and donor–recipient matching in
          transplantation biology, 246
Toxalbumin, 45, 154
Toxin spectra, 106

Trachoma, 262
Transplantation, 232–253
     geneticists, 238–240
     immunity, 234–236, 243
     immunogenetics and, 248–253
     immunologic tolerance, 247–248
     immunosuppressive therapy, 246–247
     laws, 235, 239, 241
     progress in research, 244–248
     renaissance, 240–244
     research in 1930s, 239–240
     surgeons, 232–233
     tissues employed in, 245–246
     tissue typing and donor–recipient
          matching, 246
     tumor researchers, 234–237
Transplantation Society, 369
Treponema pallidum, 158, 270
Treponemes, 70
Tropical parasites, 70
Trypanosomiasis, 262
Tuberculin, 111, 140, 157, 177, 187,
     189–192, 199, 271–272
     type hypersensitivity allergy, 39,
          192–198, 243
Tuberculin-type "delayed" reactions, 399
Tuberculin-type hypersensitivities, 399
Tuberculosis, 33, 39, 111, 129, 140,
     189–192, 194, 262
     pathogenesis of, 237
     serodiagnosis of, 157
Tumor, researchers and transplantation
     biology, 234–237
Two gene–one polypeptide theory, 76
Typhoid fever, 191
Typhus, 33, 70

U
Unitarian approach mechanism, of allergy,
     182–185
Urpeptide, 79
Uveitis, 398–399

V
Vaccinia virus, 196
Van der Waal's interactions, 52
Vengeful deity, concept of, 6
Viral diseases, 196, 262
Viral infections, immunity in, 39

"Virgin" immunocyte, 337
"Virgin" lymphocytes
   clonal selection, genetics, and receptor
       patterns, 343–345
   Ehrlich's view of cell, 342–343
   instruction theories, 343

W
Wassermann antibody, 157–159
Wassermann reaction test, 115, 157–158,
   182, 270, 282

Welch Library of the Johns Hopkins
   Medical School, Baltimore, 368

Y
Yellow bile (chole), 27

Z
Zeitschrift für Immunitätsforschung, 352
Zwischenkörper, 105, 111, 219, 309, 311,
   423

Printed and bound by CPI Group (UK) Ltd, Croydon, CR0 4YY

15/10/2024

01774669-0001